Integrated Biotechnology: Principles and Practices

Integrated Biotechnology: Principles and Practices

Edited by **Wendell Carter**

SYRAWOOD
PUBLISHING HOUSE
New York

Published by Syrawood Publishing House,
750 Third Avenue, 9th Floor,
New York, NY 10017, USA
www.syrawoodpublishinghouse.com

Integrated Biotechnology: Principles and Practices
Edited by Wendell Carter

International Standard Book Number: 978-1-68286-312-1 (Hardback)

Printed in China.

Contents

Preface

Every book is initially just a concept; it takes months of research and hard work to give it the final shape in which the readers receive it. In its early stages, this book also went through rigorous reviewing. The notable contributions made by experts from across the globe were first molded into patterned chapters and then arranged in a sensibly sequential manner to bring out the best results.

Biotechnology has become a prominent multidisciplinary field of study and is used widely across various fields. This book aims to elucidate the significant concepts and techniques of biotechnology used in different industries and sectors. It provides the information needed to efficiently translate new research findings into applications in the fields of cellular and tissue engineering, genetic engineering, recombinant DNA technology, microbiology, bioinformatics, etc. The researches and case studies incorporated in this text attempt to highlight the recent advancements in different applications of biotechnology. It will serve as a valuable source of reference for graduate and post graduate students.

It has been my immense pleasure to be a part of this project and to contribute my years of learning in such a meaningful form. I would like to take this opportunity to thank all the people who have been associated with the completion of this book at any step.

Editor

Experimental design approach in recombinant protein expression: determining medium composition and induction conditions for expression of pneumolysin from *Streptococcus pneumoniae* in *Escherichia coli* and preliminary purification process

Guillermo Marini[1,2*], Mateus Dalcin Luchese[1], Ana Paula Correa Argondizzo[1], Ana Carolina Magalhães Andrade de Góes[1], Ricardo Galler[1], Tito Lívio Moitinho Alves[2], Marco Alberto Medeiros[1] and Ariane Leites Larentis[1,3*]

Abstract

Background: *Streptococcus pneumoniae (S. pneumoniae)* causes several serious diseases including pneumonia, septicemia and meningitis. The World Health Organization estimates that streptococcal pneumonia is the cause of approximately 1.9 million deaths of children under five years of age each year. The large number of serotypes underlying the disease spectrum, which would be reflected in the high production cost of a commercial vaccine effective to protect against all of them and the higher level of amino acid sequence conservation as compared to polysaccharide structure, has prompted us to attempt to use conserved proteins for the development of a simpler vaccine. One of the most prominent proteins is pneumolysin (Ply), present in almost all the serotypes known at the moment, which shows an effective protection against *S. pneumoniae* infections.

Results: We have cloned the pneumolysin gene from *S. pneumoniae* serotype 14 and studied the effects of eight variables related to medium composition and induction conditions on the soluble expression of rPly in *Escherichia coli (E. coli)* and a 2^{8-4} factorial design was applied. Statistical analysis was carried out to compare the conditions used to evaluate the expression of soluble pneumolysin; rPly activity was evaluated by hemolytic activity assay and served as the main response to evaluate the proper protein expression and folding. The optimized conditions, validated by the use of triplicates, include growth until an absorbance of 0.8 (measured at 600 nm) with 0.1 mM IPTG during 4 h at 25°C in a 5 g/L yeast extract, 5 g/L tryptone, 10 g/L NaCl, 1 g/L glucose medium, with addition of 30 µg/mL kanamycin.

(Continued on next page)

* Correspondence: guillermo@peq.coppe.ufrj.br; arianelarentis@fiocruz.br
[1]Bio-Manguinhos (Instituto de Tecnologia em Imunobiológicos) - Fundação Oswaldo Cruz (FIOCRUZ) - VDTEC (Vice-Diretoria de Desenvolvimento Tecnológico), Av. Brasil 4365, Pavilhão Rockfeller Sala 202 - 21040-360, Manguinhos, Rio de Janeiro, RJ, Brazil
[3]ENSP (Escola Nacional de Saúde Pública Sergio Arouca) - Fundação Oswaldo Cruz (FIOCRUZ) - Centro de Estudos da Saúde do Trabalhador e Ecologia Humana (CESTEH), Av. Leopoldo Bulhões 1480, Prédio 1º de Maio - 21041-210, Manguinhos, Rio de Janeiro, RJ, Brazil
Full list of author information is available at the end of the article

(Continued from previous page)

Conclusions: This experimental design methodology allowed the development of an adequate process condition to attain high levels (250 mg/L) of soluble expression of functional rPly in *E. coli*, which should contribute to reduce operational costs. It was possible to recover the protein in its active form with 75% homogeneity.

Keywords: Soluble expression, Experimental design, Design of experiment (DoE), rPly, Recombinant *E. coli*, Hemolytic activity

Background

Streptococcus pneumoniae and pneumococcal diseases

Since it was first isolated in 1881, *Streptococcus pneumoniae* is one of the most extensively studied microorganisms. In spite of the vast number of publications on the bacterium, many questions about its pathogenicity are still unanswered, and this pathogen remains a major causative agent of serious human diseases such as pneumonia, meningitis and bacteraemia [1]. Indeed, pneumonia is the leading cause of child mortality worldwide, with 1.9 million of the estimated 10 million child deaths each year, with meningitis as the second leading cause of child mortality. Additionally, pneumococcus causes noninvasive diseases such as acute otitis, sinusitis, conjunctivitis, and bronquitis. The incidence of pneumococcal diseases is raising and is a serious global problem. The vast majority of its victims come from the poorest countries in the world [2].

Differences observed in the chemical composition and also immunogenicity of the polysaccharide capsule (PS), make it possible to recognize more than 90 different serotypes of *S. pneumoniae* worldwide. Serotype prevalence may change based on patients age, geographical region and type of infection [3]. *S. pneumoniae* expresses a wide variety of surface proteins to interact with host cell components during the colonization or dissemination stages of the bacterium. They are also involved in the pathogenesis of the disease both as mediators of inflammation and as a direct part on host tissue attack [4].

Pneumococcal vaccines and pneumolysin as a potential protein candidate for a new vaccine

Attempts to develop a vaccine to protect against the pneumococcus based on capsular polysaccharides yielded three licensed vaccines against *S. pneumoniae* which are commercially available: Pneumovax (Merck), Prevnar or Prevenar (Wyeth) and Synflorix (GSK). Pneumovax is a 23-valent PS vaccine with strong efficacy in adults but poor efficacy in infants and young children (<2 years) due to the lack of mature B cells. The other two available vaccines are based on PS conjugated to carrier proteins to enhance the immunologic response, with a broader efficacy in infants and young children, despite the poor protection against less common serotypes. The main disadvantages of these commercial vaccines are the high-cost production process, and the incomplete coverage of serotypes whose incidence may vary among countries [5,6].

Several pneumococcal proteins have been investigated in the past two decades, as an alternative to the expensive conjugate vaccines, as an antigen in potential protein-based vaccine candidates or as carriers for polysaccharides, aiming to broaden serotype coverage with less components in the vaccine [5-8]. In this regard, the most interesting proteins described to date are the pneumococcal toxin pneumolysin (Ply); choline-binding surface protein family (CBPs); pneumococal surface proteins, as the pneumococcal surface adhesin A (PsaA) and pneumococcal surface protein A (PspA); and the Caseinolytic protease (ClpP), among others under study. These proteins are present in the majority of pneumococcal serotypes isolated to date. Combinations of diverse pneumococcal proteins with conserved amino acid sequences may provide the best serotype-independent protection against *S. pneumoniae.*

Ply is a 471 amino acids monomeric protein with a molecular mass of 53 kDa, folded in four globular domains [9]. This protein shows a highly conserved 11 residues sequence in one of the domains, rich in tryptophan residues (ECTGLAWEWWR), suggested by earlier studies [10,11] as the responsible sequence for anchoring at host cell membrane and forming a transmembrane pore complex. Recent study by Soltani *et al.* [12] suggested that the cholesterol recognition motif resided within two structural loops, at the same domain where the 11 residues sequence are located. The pores upset the delicate osmotic balance between the cell and its environment, allowing material to leak in and out freely, quickly leading to lysis of the target cell [9,13,14]. Ply also presents other biological activities, depending on the concentration, especially at the initial stage of the pathogenesis of pneumococcal infection. In high concentrations, the enzyme is toxic to ciliated bronchial epithelial cells, reducing the ciliary movement, destroying the integrity of joints and cellular bronchial epithelial monolayer, facilitating the spread of pneumococcal infection. It also interacts with the epithelial cells of the alveoli and pulmonary endothelial cells, causing alveolar edema and hemorrhage during pneumococcal pneumonia, and easing the penetration from the epithelium to the pulmonary interstitium and ultimately into the bloodstream

[14]. At low concentrations, this enzyme is able to inhibit effector functions (respiratory bursts) of neutrophils and monocytes, chemotaxis, bactericidal activity, and production of lymphokines and immunoglobulins [14]. Consequently, the phagocytic activity of the host cell is inhibited [9,15].

Optimization of heterologous protein production

In order to express target proteins in high yields, recombinant DNA technology is the most appropriate tool and *Escherichia coli* is one of the most used hosts, due to its ability to grow rapidly at high cell density, with relative less expensive substrates, in addition to its well-established genetic background, and the existence of many commercial cloning vectors and strains suitable for expression [16-18]. An improvement in the process of expression of recombinant proteins can facilitate the subsequent step of purification. Consequently, achieving high expression levels of recombinant proteins is an essential step in the development of a bioprocess, towards maximum profitability and economic viability. However, there is no single, universal process for the expression of all recombinant proteins in this host, because the levels of expression of a heterologous gene will depend on multiple variables that are specific to each recombinant system. All these variables should be carefully evaluated because they influence the system [8,19].

The traditional method of process evaluation involves varying one variable at a time, while keeping other variables constant. This strategy requires a relatively large number of experiments and frequently fails to anticipate optimal conditions [20]. Changing one factor at a time does not depict the combined effect of all the variables involved. It is also a time consuming process because a large number of experiments is required. The deficiency can be overcome by applying more efficient, statistic-based experimental design. In this respect, factorial design is an important tool to determine optimal process conditions. The advantage of using the factorial design method is that many more variables can be screened simultaneously and much quantitative information can be extracted with a few experimental trials. Statistical methods through factorial experimental designs not only offer the simultaneous study of many variables, but also allow the study of interactive effects of many factors together, facilitating the prediction of the response for the values of variables not tested in the experiment [7,21]. In recent years, factorial designs - which are statistical techniques for designing experiments, defined significant effects, building models, evaluating the effects of variables and searching for optimum conditions - have successfully been used to optimize many bioprocesses [22,23]; however, they are not yet the most common strategies used to evaluate heterologous protein expression. The use of statistically designed experiments allows the rapid and economical determination of optimal culture conditions with fewer experiments and minimal resources [24]. Using a fraction of the complete factorial design is also advantageous in cases when the number of variables is larger than four, which demands a great number of experiments. The fraction of experiments should conserve the statistical condition of orthogonality that allows the estimation of independent parameters [22,25].

In view of these considerations, the aim of this study was to express recombinant Ply (rPly) in high concentrations and in its soluble form, using *E. coli* as a heterologous expression system. To improve the protein production, the many variables that influence the expression of this protein were evaluated, using experimental design. As such, fractional factorial design was used for eight variables related to medium composition and induction conditions. These variables are the concentrations of yeast extract, tryptone, glucose, glycerol, kanamycin, and inducer, the absorbance at induction moment, and the temperature of expression after the induction moment.

Results and discussion

Multivariant analysis and variables evaluated

The statistical experimental design methodology, where the response is evaluated by changing more than one variable at a time, allow the estimation of the variables that are statistically significant, taking into account the interactions between them. This multivariant method permits a thoroughly analysis compared to the traditional univariant method, where the response is evaluated changing one variable at a time while fixing the others. Furthermore, the multivariant method enables to characterize the experimental error, to compare the effects of variables between themselves when variables are normalized, and hence, to gather high-quality information with as few experiments as possible. All these advantages make this approach a powerful tool to optimize culture medium compositions and process conditions for recombinant protein expression [7,19,21,23,24,26-32].

When the recombinant protein is expressed intracellularly in host cells, it is known that to yield high amounts of that protein, it is necessary to achieve high cell growth. Thus, the higher the cell growth, the more recombinant protein is synthesized [33]. As a consequence, different operational strategies are developed to increase cell growth. However, it is also known that protein may be expressed in a soluble or insoluble form, into inclusion bodies. Several works in the literature consider the formation of these inclusion bodies as a drawback of the expression system in *E. coli*, because they require further addition of a complex process of isolation and purification, in which proteins are denatured and refolded *in vitro*. Therefore, the recovery of protein may be very low, which adds to the losses in all subsequent stages of purification, making the global recovery low [34]. Thus, a major challenge in recombinant expression is to design strategies for soluble production. To evaluate whether the protein expressed in a soluble form is directly related to cell growth, the first approach of this study was to determine the variables that have significant effects on cell growth and then compare that effect pattern with the one obtained for the protein expression, which in turn was evaluated through the pneumolysin biological activity assay.

In order to define a process with high yields and reduced time for the expression of the recombinant protein pneumolysin in its soluble form, we assessed the effects of eight variables related to medium composition and induction conditions on three relevant responses (cell growth, biological activity and productivity of rPly) in bench scale. To that effect, a statistical strategy based on a fractional factorial screening design was used, with two levels for each of the eight variables (2^{8-4}) and replicas at the central point. A previous study of this expression system reported that induction times longer than 6 h were associated with lower productivity, meanwhile induction times between 4 h and 6 h presented similar levels of productivity [35]. Thus, for this system, the expression time was defined in 4 h, which means the highest level of productivity of pneumolysin, for the lowest operational time. A sequence of 24 different experimental conditions was run to gather information about these variables that may affect the rPly soluble expression. Results are presented in Table 1. When summiting the *E. coli* BL21 Star (DE3)™ strain with pET plasmid without the interest gene (negative control) to the same expression process as the *E. coli* BL21 Star (DE3)™/pET28a/*ply*, no band enhancement was detected in SDS-PAGE analysis at the same size as rPly in the total protein extract after 4 h induction, and no biological activity was observed in the hemolytic assay.

Effect of variables on cell growth

After 4 h of induction, culture samples were taken to determine cell growth, measured at an absorbance of 600 nm. As shown in Table 1, cell growth was different among the 24 experimental runs, suggesting that certain variables, related to medium composition and induction conditions, have significant effects on that response. To determine which variables those are, a statistical study of the effects was carried out, considering significant effects with p-value lower than 0.1, as shown in Table 2.

According to the statistical analysis (Table 2), six statistically significant variables could be recognized on the cell growth, since they showed p-values lower than 0.1: induction absorbance, inducer concentration, expression temperature, concentrations of yeast extract, tryptone, and glucose. The two other variables, concentrations of glycerol and kanamycin showed no significant effects. When four variables were considered (induction absorbance, expression temperature, and concentrations of yeast extract and tryptone), it was possible to observe positive effects on the response for higher values.

Based on the statistical analysis of cell growth, there were only linear effects, since curvature was not significant; that is, the highest cell growth could be achieved by defining the significant variables at their extreme levels depending on their positive or negative effects.

Cell concentration at induction is relevant, given the fact that the expression of heterologous proteins may inhibit the growth because of the metabolic burden implied in the expression process that restricts the energy supply for other cell processes. During the stationary phase, when cell growth reached its maximum, induction is not desired, since expression rate is proportional to the growth rate at the induction moment. Therefore, the best moment should happen when the metabolic activity is completely active and consequently the growth rate is peaking, which happens during the exponential growth phase [36]. Based on these facts, expression induction is done when culture is at the exponential phase, but depending on the induction system and the recombinant protein, expression could be done at the initial, intermediate or final part of that phase. For a high cell growth, induction is better in the middle of the exponential phase, according to the statistical evaluation.

In the case of inducer concentration, the effect on cell growth was negative, showing a lower response value as concentration is raised from the inferior to the superior IPTG level. This phenomenon could be explained by the toxic characteristic of IPTG, which is widely reported in literature [31,37]. Noteworthy, expression of other recombinant proteins from *S. pneumoniae* studied in our laboratory (PsaA and ClpP protease), using this same host system, showed a negative effect of IPTG on cell growth [7,31].

Another variable with negative effect on cell growth was glucose concentration. This effect may be explained because of the medium acidification by glucose metabolization, which generates acidic by-products such as acetic acid. Hansen *et al.* [38] showed a direct correlation between acetic acid production and glucose consumption when the medium contains glucose. In contrast, when glucose was replaced for glycerol, no acetic acid was detected in the supernatant. However, in this study the only variable which was statistically significant for medium pH change was glucose concentration; meanwhile glycerol did not presented statistical effect.

On the other hand, a richer media in carbon source (yeast extract and tryptone) and the temperature

Table 1 Fractional factorial screening design and responses

Condition	Induction absorbance	[IPTG]	Expression temperature	[Yeast extract]	[Tryptone]	[Glucose]	[Glycerol]	[Kanamycin]	Cell growth	Hemolytic activity	Process time (*)	Productivity
	Abs_{ind}	*mM*	°C	*g/L*	*g/L*	*g/L*	*% v/v*	*μg/mL*	*Abs*	*HU/mL*	*min*	*HU/mL/min*
1	0.8 (−1)	0.1 (−1)	25 (−1)	5.0 (−1)	0 (−1)	1 (−1)	0.0 (−1)	10 (−1)	2.08	612	345	1.77
2	2.0 (+1)	0.1 (−1)	25 (−1)	5.0 (−1)	0 (−1)	10 (+1)	0.4 (+1)	50 (+1)	2.86	1171	450	2.60
3	0.8 (−1)	1.0 (+1)	25 (−1)	5.0 (−1)	10 (+1)	1 (−1)	0.4 (+1)	50 (+1)	2.62	1306	345	3.79
4	2.0 (+1)	1.0 (+1)	25 (−1)	5.0 (−1)	10 (+1)	10 (+1)	0.0 (−1)	10 (−1)	3.46	1306	425	3.07
5	0.8 (−1)	0.1 (−1)	37 (+1)	5.0 (−1)	10 (+1)	10 (+1)	0.4 (+1)	10 (−1)	3.18	678	340	1.99
6	2.0 (+1)	0.1 (−1)	37 (+1)	5.0 (−1)	10 (+1)	1 (−1)	0.0 (−1)	50 (+1)	6.42	1188	385	3.09
7	0.8 (−1)	1.0 (+1)	37 (+1)	5.0 (−1)	0 (−1)	10 (+1)	0.0 (−1)	50 (+1)	2.78	658	350	1.88
8	2.0 (+1)	1.0 (+1)	37 (+1)	5.0 (−1)	0 (−1)	1 (−1)	0.4 (+1)	10 (−1)	3.76	656	470	1.40
9	0.8 (−1)	0.1 (−1)	25 (−1)	23.5 (+1)	10 (+1)	10 (+1)	0.0 (−1)	50 (+1)	3.45	1446	345	4.19
10	2.0 (+1)	0.1 (−1)	25 (−1)	23.5 (+1)	10 (+1)	1 (−1)	0.4 (+1)	10 (−1)	4.73	1344	403	3.34
11	0.8 (−1)	1.0 (+1)	25 (−1)	23.5 (+1)	0 (−1)	10 (+1)	0.4 (+1)	10 (−1)	2.65	915	345	2.65
12	2.0 (+1)	1.0 (+1)	25 (−1)	23.5 (+1)	0 (−1)	1 (−1)	0.0 (−1)	50 (+1)	4.21	1146	420	2.73
13	0.8 (−1)	0.1 (−1)	37 (+1)	23.6 (+1)	0 (−1)	1 (−1)	0.4 (+1)	50 (+1)	4.09	639	335	1.91
14	2.0 (+1)	0.1 (−1)	37 (+1)	23.6 (+1)	0 (−1)	10 (+1)	0.0 (−1)	10 (−1)	5.47	1013	420	2.41
15	0.8 (−1)	1.0 (+1)	37 (+1)	23.6 (+1)	10 (+1)	1 (−1)	0.0 (−1)	10 (−1)	4.02	335	335	1.00
16	2.0 (+1)	1.0 (+1)	37 (+1)	23.6 (+1)	10 (+1)	10 (+1)	0.4 (+1)	50 (+1)	5.40	1353	420	3.22
17	1.4 (0)	0.55 (0)	31 (0)	14.3 (0)	5 (0)	5.5 (0)	0.2 (0)	30 (0)	3.60	1469	370	3.97
18	1.4 (0)	0.55 (0)	31 (0)	14.3 (0)	5 (0)	5.5 (0)	0.2 (0)	30 (0)	3.79	1557	370	4.21
19	1.4 (0)	0.55 (0)	31 (0)	14.3 (0)	5 (0)	5.5 (0)	0.2 (0)	30 (0)	3.73	1610	360	4.47
20	1.4 (0)	0.55 (0)	31 (0)	14.3 (0)	5 (0)	5.5 (0)	0.2 (0)	30 (0)	3.82	1615	350	4.61
21	1.4 (0)	0.55 (0)	31 (0)	14.3 (0)	5 (0)	5.5 (0)	0.2 (0)	30 (0)	3.82	1506	370	4.07
22	1.4 (0)	0.55 (0)	31 (0)	14.3 (0)	5 (0)	5.5 (0)	0.2 (0)	30 (0)	3.68	1507	370	4.07
23	1.4 (0)	0.55 (0)	31 (0)	14.3 (0)	5 (0)	5.5 (0)	0.2 (0)	30 (0)	3.24	1158	370	3.13
24	1.4 (0)	0.55 (0)	31 (0)	14.3 (0)	5 (0)	5.5 (0)	0.2 (0)	30 (0)	3.32	1263	370	3.41

(*)Process time refers to culture time from the inoculum growth at 37°C to the end of 4 h of induction at different conditions defined by the experimental design.

Table 2 Effects of variables on cell growth, rPly activity and process productivity

	Cell growth		rPly activity		Process productivity	
	Effect	*p*-value	Effect	*p*-value	Effect	*p*-value
Mean/interc.	**3.82**	**<0.0001**	**985.4**	**<0.0001**	**2.57**	**<0.0001**
Induction absorbance	**1.43**	**<0.0001**	**323.5**	**0.0016**	0.33	0.2248
IPTG	**−0.42**	**0.0387**	−52.0	0.5422	−0.19	0.4720
Expression temperature	**1.13**	**<0.0001**	**−340.8**	**0.0011**	**−0.91**	**0.0041**
Yeast extract	**0.86**	**0.0004**	77.0	0.3706	0.23	0.3930
Tryptone	**0.67**	**0.0027**	**268.2**	**0.0061**	**0.79**	**0.0095**
Glucose	**−0.33**	**0.0920**	**164.3**	**0.0685**	0.37	0.1797
Glycerol	−0.32	0.1011	44.8	0.5993	0.09	0.7241
Kanamycin	0.31	0.1163	**256.0**	**0.0082**	**0.72**	**0.0160**
Curvature	−0.40	0.2356	**950.5**	**<0.0001**	**2.85**	**<0.0001**

In bold variables with statistical significance with a confidence level of 90% (p-value < 0.1).

enhanced cell growth but not necessarily the protein's recombinant expression. Finally, kanamycin concentration showed no statistical effect on this response, as it happened when it was evaluated in recombinant ClpP protein expression [31].

Effect of the variables on the pneumolysin biological activity and on process productivity

As it was previously discussed, heterologous protein expression in *E. coli* frequently happens in the cytoplasm. Consequently, to obtain high levels of the foreign protein, a high cell growth culture is necessary [33]. However, recombinant protein expression in high cell growth may happen in an insoluble and inactive way, forming inclusion bodies. Thus, to express biologically active recombinant proteins, such as enzymes, it is necessary to avoid inclusion bodies formation, which requires a thorough study of the medium composition and the expression conditions of the system. Processes in which protein was expressed in its active form were presented by Nikerel *et al.* [26,39], and Volontè *et al.* [29]. Those processes were optimized by using experimental design strategy.

The main disadvantage of expressing a recombinant protein in inclusion bodies is the high operational cost for its recovery in a soluble way when it is needed [33,40]. From an operational point of view, the importance of a correct expression of a heterologous protein is to go directly to the purification step, avoiding additional previous steps, such as solubilization, oxidation, and refolding, during which a great amount of material is lost [34].

In this study, the aim was to define a process to express rPly in its soluble form. As previously discussed, it was necessary to evaluate the soluble fraction of pneumolysin, which is the active fraction, expressed in all 24 different operational conditions, to define the most appropriate expression condition and then compare it with the condition for the best cell growth to determine whether the soluble rPly expression was correlated to cell growth.

The soluble fraction of the protein was assessed by measuring its hemolytic activity, since this is the only way to measure the exact and correct structural folded protein [41]. The statistical analysis of the effect of the eight variables on the biological rPly activity is presented in Table 2.

According to results in Table 2, five variables presented statistically significant effects on rPly activity: induction absorbance, expression temperature, and the concentrations of tryptone, glucose and kanamycin, since they presented p-values lower than 0.1. The curvature was also statistically significant, which means some responses do not have linear behavior, consequently, the best response corresponds to an intermediate value of the variable within the range tested, in contrast with linear responses that were found at the extreme levels of the range tested.

The negative effect of expression temperature was the main effect on rPly activity, since it was the highest value, and is in accordance with other studies, which also observed that lower temperatures reduced the rate of protein expression, thus avoiding the formation of inclusion bodies, enhancing soluble expression [28,29]. Volontè *et al.* [29] and Larentis *et al.* [7] showed that there was growth rate reduction when expression temperature was decreased from 37°C to 25°C, with a concomitant enhancement in the soluble protein quantity expressed. They also recognized that a lower reduction of the expression temperature implies an additional cooling step, making it a more expensive process, besides promoting significant reduction of cell growth. A similar result was found by Swalley *et al.* [21], when they studied medium composition and induction conditions to express a viral protein in *E. coli*. However, depending on the protein, it may be adequate to reduce the temperature to enhance the expression, and, consequently, productivity [7].

Concerning the induction absorbance, to achieve high amount of rPly in its soluble form, induction at high cell concentration is preferable. The concentrations of tryptone and glucose also had positive effect on rPly activity as well as kanamycin concentration. This kanamycin effect can be explained due to its contribution to the plasmid stability, maintaining high concentration of plasmid bearing cells, favoring the expression. At the lowest kanamycin concentration, probably, the plasmid segregation was drastically higher than that at higher antibiotics concentration. Based on a previous study carried out in our laboratory, it is known that plasmid segregation occurs in the same pET expression system for ClpP protease production, even at the highest kanamycin concentration evaluated in this study [31]. However, as the curvature is significant, and the highest activity was observed at the central point, the best condition to achieve a high activity could be defined at middle concentrations of these variables (tryptone, glucose and kanamycin concentrations).

The other three variables, concentrations of IPTG, yeast extract and glycerol, have no statistically significant effects ($p > 0.1$). The lack of statistical effect leads these variables to be defined at their lowest values to reduce operational costs. Therefore, inducer concentration may be reduced 10-fold compared with the traditional IPTG concentration used in most expression systems [26,29,34,36]. That is an advantage because of its toxic effect on cell growth, as discussed before. This was also observed for the IPTG-based expression of PsaA and ClpP proteins (within the 0.1 mM to 1 mM range) in the same cell host studied in our laboratory [7,31]. Yeast extract concentration can be fixed at 5 g/L, which is the lowest value normally used in most studies [26,34]. Glycerol concentration was evaluated as a glucose substitute to avoid acetate formation and to favor protein stability [38], but it showed no beneficial effects on rPly expression.

Comparing the patterns of the significant effects on the cell growth and on the biological activity, we could say there is no direct relationship between the solubility of rPly and cell growth. Thus, to define the medium composition and the induction conditions of a process for the soluble expression of rPly, the activity is the main response to be considered. However, the 24 different processes presented a great variability in terms of process times because they were induced at an absorbance of 0.8, 1.4, and 2.0. Therefore, in order to avoid erroneous conclusions, the statistical analysis should be done taking into account the biological activity regarding the total process times, that is, the statistical analysis should be done considering the different process productivities, as it is shown in Table 2.

The results presented in Table 2 show that expression temperature was the main variable affecting productivity, as observed on the biological activity analysis. The negative effect means the highest productivity level was reached at lower temperatures within the range tested. The concentrations of tryptone and kanamycin also significantly affected process productivity, but in contrast with the expression temperature, their effects were positive, which means that, for higher productivity levels, these variables should be defined at higher levels. On the optimization process for the rPsaA expression, tryptone did not show statistical effect and thus could be possibly removed from the medium [32]. Once more, the curvature was also statistically significant, conferring a nonlinear effect to one or more variables. As curvature was positive, productivity peaks within the evaluated range, rather than at the extremes of that range. Therefore, to define the best medium composition and induction conditions, it is important to consider the productivity levels at the central point, where the response presented the highest level. In contrast, the other variables did not have statistical effects on the process productivity. These variables should be defined at their lowest levels within the tested range, to reduce process costs as much as possible without interfering on the productivity level of the process.

Based on the previous discussion and on the factorial design analysis, it is possible to infer the more accurate condition which allows the maximization of system productivity. Therefore, the medium composition was defined as follows: 5 g/L yeast extract, 5 g/L tryptone, 1 g/L glucose, 30 μg/mL kanamycin, without glycerol, and with 10 g/L NaCl (variable not evaluated in this factorial design). The expression condition was defined by inducing the expression at absorbance of 0.8, with 0.1 mM IPTG, and allowing for 4 h induction at 25°C.

The defined process showed that it is not necessary to use a culture medium richer in yeast extract, tryptone, or glucose, as most systems to express recombinant proteins normally use. In this medium, yeast extract content can be reduced about 5-fold compared to TB, a richer medium used in *E. coli* cultures [32]. Tryptone and kanamycin can also be reduced 2-fold in comparison to the highest concentrations tested, associated with TB and LB media content, and referred in Sambrook and Russell [42] molecular cloning laboratory manual, respectively; inducer, which is an expensive component, can be reduced 10-fold compared with the standard concentration used to induce this kind of system [42], and it is not necessary to add glycerol. Regarding expression temperature, its reduction to 25°C leads to energy saving, meanwhile induction at absorbance 0.8 means shorter process time.

The choice of tryptone and kanamycin concentration levels was due to their significant and positive effects in experiments, being that the highest productivity levels were obtained when these variables were set at their central levels. In addition, as the curvature was significant and positive, both concentrations, statistically significant, leading to a higher response when they are defined at the central levels of the ranges tested. Once the medium composition and induction conditions were defined according to the statistical analysis on the productivity of pneumolysin, the next step was to validate that process.

Validation of the process defined by factorial design

Validation of the process defined by the factorial design was carried out in triplicate during 4 h kinetics (k1, k2, k3). Cultures in the medium grew at 37°C until reaching exponential growth phase (Abs_{ind} 0.8), when they were induced with 0.1 mM IPTG and temperature was reduced at 25°C to express the recombinant pneumolysin. After that, cell culture samples were taken every hour until the end of expression process (4 h). These samples were used to evaluate cell growth and productivity through process time. Results are shown in Table 3.

The average of the pneumolysin activity in the validation condition was 1394 ± 73 HU/mL, which is statistically the same as the highest values obtained in the fractional design experiments. The total process time was 310 min (growth phase of 70 min and expression phase of 240 min), which is the minimum process time obtained in the factorial experiments. Thus, it can be observed that in this validated process, in which most of the medium composition and induction variables were defined at their lowest values, the pneumolysin soluble expression was high even though the medium composition was poorer and the total process time was the shortest.

Figure 1 shows cell growth and the hemolytic activity of the soluble fraction of the pneumolysin, through the total process time. It is important to highlight that there was no hemolytic activity detected in the strain *E. coli* BL21 Star (DE3)™/pET28a culture (negative control), in the same culture and expression conditions of the strain *E. coli* BL21 Star (DE3)™/pET28a/*ply*. Thus, the hemolytic activity is related to the recombinant protein pneumolysin.

Table 3 Cell growth and productivity for the validation experiment

Induction time	rPly activity	Cell growth	Yield per cell
h	*HU/mL*	*Abs*	*HU/mL/Abs*
0*	0	0.82 ± 0.01	0
1	401 ± 18	1.60 ± 0.04	251 ± 7
2	656 ± 62	2.42 ± 0.09	272 ± 33
3	867 ± 20	2.65 ± 0.02	328 ± 6
4	1394 ± 73	2.93 ± 0.17	476 ± 34

Results are averages of three experiments. Expression of rPly was performed at Abs_{ind} 0.8 with 0.1 mM IPTG for 4 h induction at 25°C in a medium culture comprised of 5 g/L yeast extract, 5 g/L tryptone, 1 g/L glucose, 30 μg/mL kanamycin and 10 g/L NaCl.
*Induction moment.

The cell growth kinetic show that after a two-hour induction (190 minutes of process), the cells started to reach the stationary growth phase, while the rPly activity was still rising. As discussed above, the pneumolysin expression, measured through its hemolytic activity, is not directly associated with cell growth, which can be confirmed in this kinetics (Figure 1). Based on the literature, this might happen due to the use of systems with strong promoters, such as T7 promoter. This kind of promoter, during induction, makes the cell energy turn to the protein expression to the detriment of cell growth, consequently the growth rate drops because the host cell metabolism is overburdened [31,43]. The specific growth rate obtained in this study was 0.56 h^{-1}. Possible differences with other recombinant *E. coli* may be related to the different gene expressed, the interaction host/vector as well as the properties of the expressed protein [34,40].

The productivity average obtained in the validated process defined from the statistical fractional design was 4.50 HU/mL/min, with a standard deviation of 0.24 and a relative standard deviation of 5.3%. Comparing this value with the ones of the fractional design experiments, it is statistically equivalent to those obtained in experiment 9 (4.19 HU/mL/min), as well as those experiments performed at the central point (3.99 ± 0.50 HU/mL/min). These values are the highest productivity values obtained among the values from design experiments. That is relevant since highest rPly productivity levels were noticed in a relative low-cost and time-consumption.

Preliminary purification process

The characteristics of recombinant protein production, the molecule location, and protein destination determine its purification process. Within the biopharmaceutical production processes, purification accounts for a high part of product cost, because it often results in low recovery of protein.

In this work, the soluble fraction of rPly expressed in *E. coli* at 25°C was used at the preliminary purification tests. The expression of rPly in total extract, soluble and insoluble fractions are presented in Figure 2. The densitometry analysis of the rPly band performed on the SDS-PAGE indicated that this protein is expressed around 250 mg/L, which corresponded to 25-28% of the total *E. coli* protein extract. The soluble fraction of rPly was around 90% from the total of rPly expressed in this condition.

Recombinant protein purification was conducted using two different strategies, ion exchange and affinity chromatography, without fusion tags. These strategies were run separately. The preliminary purification results of

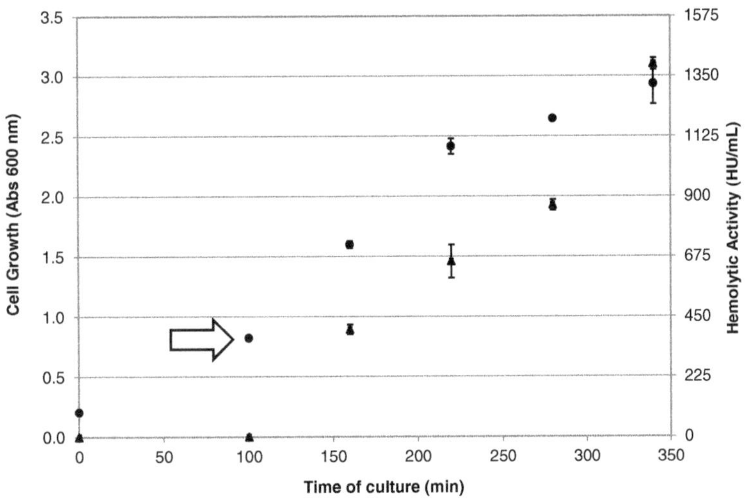

Figure 1 Cell Growth (•) and rPly Activity (▲) through the total process. Arrow is indicating the induction moment. Standard deviation of triplicate runs.

both strategies are also presented in Figure 2, with the recovery of a 75% homogeneity rPly fraction (shown in SDS-PAGE). Considering there are not many works in literature performing purification of rPly, and to our knowledge, neither Q FF column nor Thiopropyl Sepharose column were tested for this protein, these preliminary results could be considered as prominent. However, both methods need to be optimized in order to enhance recovery yields, as well as homogeneity, and to develop a feasible purification method that could be used in a production process of rPly. Hemolytic activity of the protein was confirmed after both processes, which was maintained after 8 days stored at 4°C. These results for one-step purification of rPly without tags were comparable to those homogeneities described in the literature for several purification steps [44], and allowed the production of the recombinant protein with conformation closer to the original fold [7]. One of the major advantages of our cloning strategy without tags for purification is its ability to produce a protein which is ready to

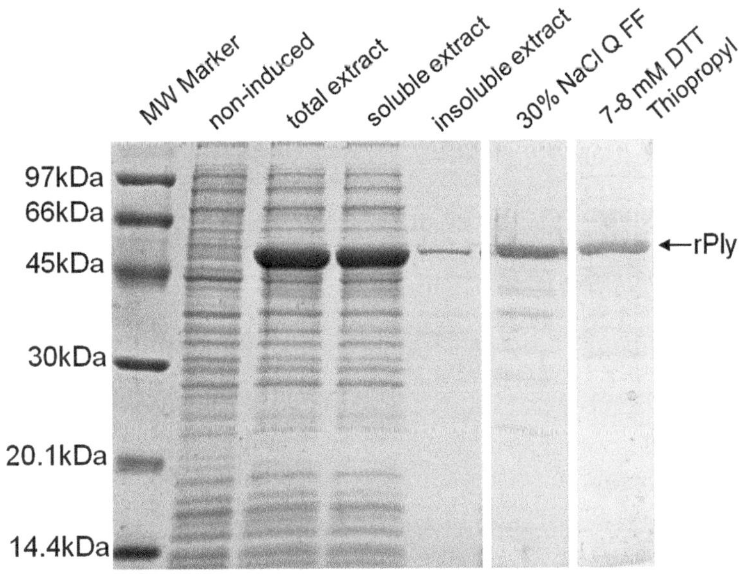

Figure 2 SDS-PAGE of rPly expression and preliminary purification. Non-induced sample and rPly expression at 25°C in TB medium after 4 h induction (total protein extract, soluble and insoluble fractions). Preliminary purification fractions obtained with 30% NaCl elution using Q FF column and 7-8 mM DTT elution using Thiopropyl matrix (MW = molecular weight).

be used in biopharmaceutical applications, eliminating the tag removal step after purification, which is a step that increases process costs.

Conclusions

Different levels of a soluble form of recombinant pneumolysin were obtained in all the experiments of the fractional design used. As there was no direct relationship between cell growth and soluble rPly expressed, and as total process time varied in the different levels of absorbance at induction, biological activity was the main response to be considered to find out the variables with statistically significant effects on productivity. The temperature of expression and the concentrations of tryptone and kanamycin were the three variables which affected productivity. The defined process was successfully validated, which yielded 4.50 HU/mL/min, one of the highest productivity values obtained compared with the productivities of the experimental design and with reduced operational costs. It should, therefore, be possible to reduce the inducer concentration ten-fold, reduce absorbance time of the induction moment and the concentrations of yeast extract, glucose, and glycerol to the minimum levels of the ranges tested, keeping rPly protein expression at high levels (250 mg/L, which corresponded to 25-28% of the total *E. coli* protein extract) and reducing process costs.

Once more, and in accordance with different literature data, the results presented emphasize the importance of experimental design to improve the biotechnological conditions of processes used to express a recombinant protein in *E. coli*. It was also proven it is necessary to have a minimum number of experiments to gather relevant information on the effect of each of the variables under study, as well as the experimental error. In contrast with traditional univariant methods, the assessment of various variables at the same time avoids the lack of considerations about the interactions of the relevant effects. All these advantages make experimental design a powerful tool for defining an optimized recombinant protein production process.

The reduction in the production cost of this protein - as well as other potential target proteins used in a possible protein-based vaccine, through the use of experimental design - demonstrated how essential this statistical tool is to obtain a highly efficient commercial vaccine with low-cost production to be accessible to the world's most underprivileged populations.

Methods

Chemicals

Bacto™ yeast extract and tryptone were purchased from BD (Becton, Dickinson and Company), glucose and NaCl were purchased from Merck, glycerol from Invitrogen, kanamycin from Sigma and IPTG (isopropyl β-D-1-thiogalactopyranoside) was purchased from Promega. Potassium salts (K_2HPO_4 and KH_2PO_4) used in the medium were purchased from Merck.

Cloning procedure

The *Escherichia coli* BL21 Star (DE3)™ strain (Invitrogen) was used as a host for the plasmid pET28a (Novagen) harboring the gene responsible for pneumolysin expression, and kanamycin resistance. Pneumolysin protein-coding gene was isolated from *S. pneumoniae* serotype 14, since it is the most prevalent in Brazil and also one of the most prevalent and virulent serotype in the world. The clone expressing the recombinant protein pneumolysin was developed at LATER/Bio-Manguinhos/FIOCRUZ and cloning techniques were reported in Larentis *et al.* [7] and Einsfeldt *et al.* [31]. The clone cryopreservation was at -70°C in 25% glycerol (Invitrogen). The working vials of the clone *E. coli* BL21 Star (DE3)™/pET28a/*ply*, thawed from the working bank, were evaluated in reference to their cell viability.

Cell viability test

The cell viability of the working vials of recombinant *E. coli* BL21 Star (DE3)™/pET28a/*ply* in LB (5 g/L yeast extract, 10 g/L tryptone, 5 g/L NaCl, pH 7.0) with 25% glycerol, stored at -70°C, was assessed by counting the colony forming units (CFU) for all the experimental design experiments. Serial dilutions were made in PBS pH 7.4 and transferred to Petri plates containing LB Agar and 50 µg/mL kanamycin (concentration of working vials were, on average, 28×10^{10} CFU/mL).

Variables and experimental design

Eight variables of the expression process, related to both the composition of the culture medium and the induction conditions, were chosen to be evaluated by the experimental design: concentrations of yeast extract, tryptone, glucose, glycerol, kanamycin and inducer, absorbance at induction moment, and expression temperature (Table 4).

To evaluate the statistical effects of the variables on the soluble expression of pneumolysin, a fractional factorial design 2^{8-4} was chosen, with replicates on the central point. This fractional factorial design is resolution IV, which means that the main effects can be determined, but not their interactions [22]. The curvature was also evaluated by the factorial design, which allowed us to learn whether effects are linear or not. When a variable effect is linear, the highest response is obtained at the highest level of the variable for a positive effect, or at the lowest level for a negative effect. On the other hand, when a variable effect is not linear, the highest or

Table 4 Variables and levels assayed

	Inferior level (−1)	Central level (0)	Superior level (+1)
Induction absorbance (Abs_{ind})	0.8	1.4	2.0
IPTG (mM)	0.1	0.55	1.0
Expression temperature (°C)	25	31	37
Yeast extract (g/L)	5	14.3	23.6
Tryptone (g/L)	0	5	10
Glucose (g/L)	1	5.5	10
Glycerol (%v/v)	0	0.2	0.4
Kanamycin (μg/mL)	10	30	50

the lowest response values could be obtained at an intermediate level of the variable within the range evaluated.

Due to the large variability inherent to bioprocesses, the parameters were considered significant with *p*-values lower than 10% ($p < 0.1$). Using 90% confidence in the statistical analysis (Software *Statistica* 9.1/Statsoft) allowed the inclusion of possible relevant variables that would be excluded from a strict statistical analysis with a significance level of 5% ($p < 0.05$). The normalization of the levels of each variable allows the comparison of the importance of each variable effect, in reference to the evaluated response. Once the most adequate experimental design was chosen, the next step was to select the range of values to be studied for each of the eight variables, defining the lowest level as −1, the highest level as +1, and 0 as the central point calculated as the middle of these two levels (Table 4). IPTG and kanamycin levels were defined based on information found in literature [35,42]; induction conditions and media composition were chosen according to previous studies with pneumolysin [35] and other *S. pneumoniae* proteins expressed in *E. coli* [7,31,32]; temperature levels were selected evaluating rPly solubility [35]; induction absorbances were defined at the initial (absorbance 0.8 measured at 600 nm) and at the middle (absorbances 1.4 and 2.0) exponential phase from cell growth kinetics in different medium compositions and temperatures [35]; media components levels were chosen based on LB (Luria-Bertani) and TB (Terrific Broth) media compositions [32,35].

Protein expression and purification process

Recombinant *E. coli* BL21 Star (DE3)™/pET28a/*ply* was pre-inoculated (10 μL of working vial) in 10 mL of the TB medium (23.6 g/L yeast extract, 11.8 g/L tryptone, 9.4 g/L K_2HPO_4, 2.2 g/L KH_2PO_4, pH 7.2) enriched with 1% glucose, 0.4% glycerol and 50 μg/mL kanamycin. The pre-inoculum was incubated for 16 hours at 37°C and 200 rpm in 50 mL flasks under agitation; simultaneously, the cell viability test was done. After 16 h, the 24 different expression experiments were done in 6 different sets of experiments and in each group a central point repetition was performed, as discussed below. The inoculum was prepared in 500 mL flasks with 2 mL pre-inoculum and 100 mL of the medium, with the corresponding yeast extract, tryptone, glucose, glycerol, and kanamycin concentrations, according to the experimental design (as described in the previous section). The culture was incubated at 37°C and 200 rpm until it reached the evaluated values of the exponential growth phase. At this point, expression was induced with the corresponding IPTG concentration for four hours under different expression temperature, as the experimental design indicated. In addition to the 16 different expression conditions, 8 replicates of the central point were done to define the experimental error associated to this kind of expression process, as well as the curvature, resulting in the 24 different runs. Samples of 1 mL volume were taken from each experiment before and after the four-hour expression period to assess cell growth and soluble rPly expression (by hemolytic activity).

The same expression process was done using an *E. coli* BL21 Star (DE3)™ strain, with the pET28a plasmid without the interest gene, as a negative control, to confirm the lack of expression of rPly in a SDS-PAGE analysis and also the lack of biological activity in an hemolytic analysis.

The recombinant protein was purified in an ÄKTA Purifier 10 system by strong anion exchange in a 1 mL Q FF column (Hitrap Q Sepharose™ Fast Flow/GE Healthcare) with 50-100 mM NaCl added to the initial sample and eluted with 20 mM Tris HCl and 300 mM NaCl (pH 8.0). Purification was also tested by affinity using a Thiopropyl Sepharose™ 6B matrix (GE Healthcare) that reacts with solutes containing thiol groups, and elutes using the same elution buffer with 5-25 mM DTT.

Samples were analyzed by 12.5% SDS-PAGE stained with Coomassie blue and densitometry was used to evaluate the concentration related to the rPly band in the total protein extract, soluble and insoluble (inclusion bodies) fractions and also to evaluate the homogeneity of purification fractions in a Bio-Rad GS-800 Calibrated Densitometer/QuantiOne 4.4.1 software according to Larentis *et al.* [7].

rPly solubility by hemolytic activity measurements

The cells from the pre-induction (0 h) and after 4 h induction samples were harvested by centrifugation at 20817 × *g* for 10 min at 4°C to separate the culture medium. In order to solubilize the expressed rPly, the cells were resuspended in a lysis buffer (20 mM Tris, 1 mM EDTA, 200 mM NaCl, pH 8.0) restituting the 1 mL sample volume, to obtain the total protein extract. The total extract was disrupted by eight 10-second ultrasound cycles at 60% amplitude in an ultrasonic cell disruptor (Sonics & Materials, Inc.). The soluble and insoluble protein fractions were separated from the cultures by centrifugation (20817 *g* for 10 min at 4°C). The soluble fraction was used to assess the hemolytic activity.

In round-bottomed 96-well microplates, a two-fold serial dilution was done with 100 μL of soluble fraction from each experimental condition using the lysis buffer (20 mM Tris, 1 mM EDTA, 200 mM NaCl, pH 8.0). A volume of 100 μL of triple-washed 2% defibrinated bovine blood cells in the same lysis buffer was added to the dilutions and incubated for 1 h at 37°C and overnight at 4°C, so as to allow unlysed red blood cells to sink to the bottom of the well and form a clear red dot in the transparent solution. On the other hand, lysed cells release their hemoglobin, giving the solution a red color that allows the monitoring of protein lysing activity. A volume of 100 μL supernatant from each well was transferred to a new flat-bottomed 96-well microplate. The hemolytic activity was evaluated spectrophotometrically at 540 nm by assaying the hemoglobin released in the supernatant. Samples with 2% blood cells completely lysed as a 100% lysis reference were serially two-fold diluted in lysis buffer to produce the standard curve; the supernatant of 2% erythrocyte suspension (100 μL) incubated with lysis buffer was used as the blank reference (0% lysis). One Hemolytic Unit (HU) was expressed as the quantity of rPly presented in a sample comparing with the standard curve, which allows the obtainment of complete hemolysis of a 2% erythrocyte suspension, after 1 h incubation at 37°C. Therefore, the first well of the standard curve contained 1 HU. That hemolytic activity was related to 1 mL of sample.

Competing interests

The authors declare that there were no competing interests in this study.

Authors' contributions

GM carried out all the experiments for the evaluation of the statistical influence of medium composition and induction conditions for the soluble expression of the recombinant Ply in *E. coli*, performed the hemolytic activity measurements, the statistical analysis and the preparation of the manuscript. MDL carried out the experiments to define the levels of variables and the duration of induction time based on system productivity and revision of the manuscript. APCA carried out the cloning of the recombinant protein in the host cell and revised the manuscript. ACMAG participated in the purification strategy and experiments. RG made periodic analysis of the project for the development of a protein vaccine against pneumonia and careful revision of the manuscript draft. TLMA participated in the design of the study, discussion and interpretation of the bioprocess and hemolytic activity data. MAM was responsible for bioinformatics analysis and cloning strategy, and also participated in the discussion of the data, coordinated financial support from LATER/Bio-Manguinhos/Fiocruz and revised the manuscript. ALL was responsible for the project and student coordination, as well as for the design of the statistical study, discussion and interpretation of the statistical analysis, for the financial support from PAPES/Fiocruz and helped to draft the manuscript. All authors have approved the manuscript.

Acknowledgements

This work was supported by grant from Bio-Manguinhos, PDTIS (Programa de Desenvolvimento Tecnológico de Insumos para Saúde) and PAPES V (Programa Estratégico de Apoio à Pesquisa em Saúde) from Fundação Oswaldo Cruz (Fiocruz)/Ministério da Saúde/Brasil. We thank the Bio-Manguinhos technicians for their technical support to this study, ENSP/Fiocruz for supporting the manuscript language revision and Vice-Presidência de Ensino, Informação e Comunicação from Fiocruz for supporting the publication in BioMed Central.

Author details

[1]Bio-Manguinhos (Instituto de Tecnologia em Imunobiológicos) - Fundação Oswaldo Cruz (FIOCRUZ) - VDTEC (Vice-Diretoria de Desenvolvimento Tecnológico), Av. Brasil 4365, Pavilhão Rockfeller Sala 202 - 21040-360, Manguinhos, Rio de Janeiro, RJ, Brazil. [2]Programa de Engenharia Química - COPPE - Universidade Federal do Rio de Janeiro (UFRJ), Av. Horácio Macedo 2030, Bloco G, Sala 115 - 21941-972, Ilha do Fundão, Rio de Janeiro, RJ, Brazil. [3]ENSP (Escola Nacional de Saúde Pública Sergio Arouca) - Fundação Oswaldo Cruz (FIOCRUZ) - Centro de Estudos da Saúde do Trabalhador e Ecologia Humana (CESTEH), Av. Leopoldo Bulhões 1480, Prédio 1º de Maio - 21041-210, Manguinhos, Rio de Janeiro, RJ, Brazil.

References

1. Kadioglu A, Weiser JN, Paton JC, Andrew PW: **The role of *Streptococcus pneumoniae* virulence factors in host respiratory colonization and disease.** *Nat Rev Microbiol* 2008, **6:**288–301.
2. World Health Organization: *Treatment and Prevention of Pneumonia, Sixty-Third World Health Assembly*, Report A63/26. ; 2010:1–4.
3. Brandileone MC, Casagrande ST, Guerra ML, Zanella RC, Andrade AL, Di Fabio JL: **Increase in numbers of β-lactam-resistant invasive *Streptococcus pneumoniae* in Brazil and the impact of conjugated vaccine coverage.** *J Med Microbiol* 2006, **55:**567–574.
4. Souza Carmo T: *Otimização da produção de polissacarídeo capsular do Streptococcus pneumoniae sorotipo 6B em biorreator*, PhD thesis. Universidade de São Paulo, Instituto Butantan; 2010.
5. Tai SS: ***Streptococcus pneumoniae* protein vaccine candidates: properties, activities and animal studies.** *Crit Rev Microbiol* 2006, **32:**139–153.
6. Douce G, Ross K, Cowan G, Ma J, Mitchell TJ: **Novel mucosal vaccines generated by genetic conjugation of heterologous proteins to pneumolysin (PLY) from *Streptococcus pneumoniae*.** *Vaccine* 2010, **28:**3231–3237.
7. Larentis AL, Argondizzo AP, Esteves Gdos S, Jessouron E, Galler R, Medeiros MA: **Cloning and optimization of induction conditions for mature PsaA (pneumococcal surface adhesin A) expression in *Escherichia coli* and recombinant protein stability during long-term storage.** *Protein Expr Purif* 2011, **78:**38–47.
8. Barry AM, Paton JC: **Additive attenuation of virulence of *streptococcus pneumoniae* by mutation of the genes encoding pneumolysin and other putative pneumococcal virulence proteins.** *Infect Immun* 2000, **1:**133–140.
9. Marriott HM, Mitchell TJ, Dockrell DH: **Pneumolysin: a double-edged sword during the host-pathogen interaction.** *Curr Mol Med* 2008, **8:**497–509.
10. Jacobs T, Cima-Cabal MD, Darji A, Méndez FJ, Vázquez F, Jacobs AA, Shimada Y, Ohno-Iwashita Y, Weiss S, de los Toyos JR: **The conserved undecapeptide shared by thiol-activated cytolysins is involved in membrane binding.** *FEBS Lett* 1999, **459:**463–466.
11. Sekino-Suzuki N, Nakamura M, Mitsui KI, Ohno-Iwashita Y: **Contribution of individual tryptophan residues to the structure and activity of theta-toxin (perfringolysin O), a cholesterol-binding cytolysin.** *Eur J Biochem* 1996, **241:**941–947.
12. Soltani CE, Hotze EM, Johnson AE, Tweten RK: **Structural elements of the cholesterol-dependent cytolysins that are responsible for their cholesterol-sensitive membrane interactions.** *Proc Natl Acad Sci USA* 2007, **104:**20226–20231.
13. Gilbert RJ, Jiménez JL, Chen S, Tickle IJ, Rossjohn J, Parker M, Andrew PW, Saibil HR: **Two structural transitions in membrane formation by pneumolysin, the pore-forming toxin of *Streptococcus pneumoniae*.** *Cell* 1999, **97:**647–655.
14. Jedrzejas MJ: **Pneumococcal virulence factors: structure and function.** *Microbiol Mol Biol Rev* 2001, **65:**187–207.

15. Rubins JB, Janoff EN: **Pneumolysin: a multifunctional pneumococcal virulence factor.** *J Lab Clin Med* 1998, **131**:21–27.
16. Jana S, Deb JK: **Strategies for efficient production of heterologous proteins in *Escherichia coli*.** *Appl Microbiol Biotechnol* 2005, **67**:289–298.
17. Choi JH, Keum KC, Lee SY: **Production of recombinant proteins by high cell density culture of *Escherichia coli*.** *Chem Eng Sci* 2006, **61**:876–885.
18. Peti W, Page R: **Strategies to maximize heterologous protein expression in *Escherichia coli* with minimal cost.** *Protein Expr Purif* 2007, **51**:1–10.
19. Larentis AL, Sampaio Hde C, Martins OB, Rodrigues MI, Alves TL: **Influence of induction conditions on the expression of carbazole dioxygenase components (CarAa, CarAc, and CarAd) from *Pseudomonas stutzeri* in recombinant *Escherichia coli* using experimental design.** *J Ind Microbiol Biotechnol* 2011, **38**:1045–1054.
20. Montgomery DC: *Design and Analysis of Experiments.* New York: Wiley & Sons; 2004.
21. Swalley SE, Fulghum JR, Chambers SP: **Screening factors effecting a response in soluble protein expression: formalized approach using design of experiments.** *Anal Biochem* 2006, **351**:122–127.
22. Rodrigues MI, Iemma AF: *Experimental Design and Process Optimization.* Campinas: Carita Editora; 2012.
23. Zhao J, Wang Y, Chu J, Zhang S, Zhuang Y, Yuan Z: **Statistical optimization of medium for the production of pyruvate oxidase by the recombinant *Escherichia coli*.** *J Ind Microbiol Biotechnol* 2008, **35**:257–262.
24. Wang YH, Jing CF, Yang B, Mainda G, Dong ML, Xu AL: **Production of a new sea anemone neurotoxin by recombinant *Escherichia coli*: Optimization of culture conditions using response surface methodology.** *Proc Biochem* 2005, **40**:2721–2728.
25. Pinto JC, Schwaab M: *Análise de Dados Experimentais v. II – Planejamento de Experimentos.* Rio de Janeiro: E-Papers Serviços Editoriais Ltda; 2011.
26. Nikerel IE, Toksoy E, Kirdar B, Yildirim R: **Optimizing medium composition for *TaqI* endonuclease production by recombinant *Escherichia coli* cells using response surface methodology.** *Proc Biochem* 2005, **40**:1633–1639.
27. Islam RS, Tisi D, Levy MS, Lye GJ: **Framework for the rapid optimization of soluble protein expression in *Escherichia coli* combining microscale experiments and statistical experimental design.** *Biotechnol Prog* 2007, **23**:785–793.
28. Chen Y, Xing XH, Ye F, Kuang Y, Luo M: **Production of MBP-HepA fusion protein in recombinant *Escherichia coli* by optimization of culture medium.** *Biochem Eng J* 2007, **34**:114–121.
29. Volontè F, Marinelli F, Gastaldo L, Sacchi S, Pilone MS, Pollegioni L, Molla G: **Optimization of glutaryl-7-aminocephalosporanic acid acylase expression in *E. coli*.** *Protein Expr Purif* 2008, **61**:131–137.
30. Pistorino M, Pfeifer BA: **Efficient experimental design and micro-scale medium enhancement of 6-deoxyerythronolide B production through *Escherichia coli*.** *Biotechnol Prog* 2009, **25**:1364–1371.
31. Einsfeldt K, Severo Júnior JB, Argondizzo APC, Medeiros MA, Alves TLM, Almeida RV, Larentis AL: **Cloning and expression of ClpP protease from *Streptococcus pneumoniae* in *Escherichia coli*: study of the influence of kanamycin and IPTG concentration on cell growth, recombinant protein production and plasmid stability.** *Vaccine* 2011, **29**:7136–7143.
32. Larentis AL, Nicolau JFMQ, Argondizzo APC, Galler R, Rodrigues MI, Medeiros MA: **Optimization of medium formulation and seed conditions for expression of mature PsaA (pneumococcal surface adhesin A) in *Escherichia coli* using a sequential experimental design strategy and response surface methodology.** *J Ind Microbiol Biotechnol* 2012, **39**:897–908.
33. Khan MA, Sadaf S, Sajjad M, Waheed Akhtar M: **Production enhancement and refolding of caprine growth hormone expressed in *Escherichia coli*.** *Protein Expr Purif* 2010, **68**:85–89.
34. Shin CS, Hong MS, Bae CS, Lee J: **Enhanced production of human mini-proinsulin in fed-batch cultures at high cell density of *Escherichia coli* BL21(DE3) [pET-3aT2M2].** *Biotechnol Prog* 1997, **13**:249–257.
35. Marini G: *Avaliação por planejamento de experimentos dos efeitos da composição do meio de cultura e das variáveis de indução na expressão de pneumolisina de Streptococcus pneumoniae em Escherichia coli.* MSc. thesis: Universidade Federal de Rio de Janeiro, Programa de Engenharia Química; 2011.
36. Donovan RS, Robinson CW, Glick BR: **Optimizing inducer and culture conditions for expression of foreign proteins under the control of the *lac* promoter.** *J Ind Microbiol* 1996, **16**:145–154.
37. Olaofe OA, Burton SG, Cowan DA, Harrison STL: **Improving the production of a thermostable amidase through optimizing IPTG induction in a highly dense culture of recombinant *Escherichia coli*.** *Biochem Eng J* 2010, **52**:19–24.
38. Hansen R, Eriksen NT: **Activity of recombinant GST in *Escherichia coli* grown on glucose and glycerol.** *Proc Biotechnol* 2007, **42**:1259–1263.
39. Nikerel IE, Öner ET, Kirdar B, Yildirim R: **Optimization of medium composition for biomass production of recombinant *Escherichia coli* cells using response surface methodology.** *Biochem Eng J* 2006, **32**:1–6.
40. Chuan YP, Lua LHL, Middelberg APJ: **High-level expression of soluble viral structural protein in *Escherichia coli*.** *J Biotechnol* 2008, **134**:64–71.
41. Sørensen HP, Mortensen KK: **Advanced genetic strategies for recombinant protein expression in *Escherichia coli*.** *J Biotechnol* 2005, **115**:113–128.
42. Sambrook J, Russell DW: *Molecular Cloning: A Laboratory Manual.* Laboratory Press: Cold Spring Harbor; 2001.
43. Toksoy E, Önsan ZÍ, Kirdar B: **High-level production of *TaqI* restriction endonuclease by three different expression systems in *Escherichia coli* cells using the T7 phage promoter.** *App Microbiol Biotechnol* 2002, **59**:239–245.
44. Kelly SJ, Jedrzejas MJ: **Structure and molecular mechanism of a functional form of pneumolysin: a cholesterol-dependent cytolysin from *Streptococcus pneumoniae*.** *J Struct Biol* 2000, **132**:72–81.

2

Comparison of various microbial inocula for the efficient anaerobic digestion of *Laminaria hyperborea*

Alastair D Sutherland[1,2*] and Joao C Varela[2]

Abstract

Background: The hydrolysis of seaweed polysaccharides is the rate limiting step in anaerobic digestion (AD) of seaweeds. Seven different microbial inocula and a mixture of these (inoculum 8) were therefore compared in triplicate, each grown over four weeks in static culture for the ability to degrade *Laminaria hyperborea* seaweed and produce methane through AD.

Results: All the inocula could degrade *L. hyperborea* and produce methane to some extent. However, an inoculum of slurry from a human sewage anaerobic digester, one of rumen contents from seaweed-eating North Ronaldsay sheep and inoculum 8 used most seaweed volatile solids (VS) (means ranged between 59 and 68% used), suggesting that these each had efficient seaweed polysaccharide digesting bacteria. The human sewage inoculum, an inoculum of anaerobic marine mud mixed with rotting seaweed and inoculum 8 all developed to give higher volumes of methane (means between 41 and 62.5 ml g^{-1} of seaweed VS by week four) ,compared to other inocula (means between 3.5 and 27.5 ml g^{-1} VS). Inoculum 8 also gave the highest acetate production (6.5 mmol g^{-1} VS) in a single-stage fermenter AD system and produced most methane (8.4 mL mmol $acetate^{-1}$) in phase II of a two-stage AD system.

Conclusions: Overall inoculum 8 was found to be the most efficient inoculum for AD of seaweed. The study therefore showed that selection and inclusion of efficient polysaccharide hydrolysing bacteria and methanogenic archaea in an inoculum offer increased methane productivity in AD of *L. hyperborea*. This inoculum will now being tested in larger scale (10L) continuously stirred reactors optimised for feed rate and retention time to determine maximum methane production under single-stage and two-stage AD systems.

Keywords: Seaweed, Microbial inoculum, Anaerobic digestion, *Laminaria hyperborea*, Methane

Background

There is a renewed interest in seaweeds as potential biomass for sustainable biofuel production; particularly as their growth does not compete for agricultural land or require water for irrigation. From the 1970s through to 1990s various groups were involved in assessing the suitability of seaweed (particularly *Macrocystis pyrifera*) as a biomass for anaerobic digestion (AD) to produce methane biofuel (see review [1]). The overall result of these studies indicated that seaweeds were generally a suitable biomass for AD.

In the Northern Atlantic *Laminaria*, *Fucus* and *Ascophyllum* species are particularly common brown algae found growing on the littoral and sub-littoral seashore. Few studies have examined AD of these seaweeds but Moen *et al.* [2,3] successfully demonstrated AD of *Laminaria hyperborea* and *Ascophyllum nodosum*, respectively. The efficient hydrolysis of seaweed polysaccharides, particularly alginates, was however seen as a rate limiting step for efficient AD to proceed [2,4,5]. Animal manure slurry is often utilised as an inoculum for anaerobic digesters, but various reports have suggested that specific bacteria capable of fermenting marine phycocolloids may accelerate and increase biogas production from seaweeds

* Correspondence: a.sutherland@gcal.ac.uk
[1]Department of Biological and Biomedical Sciences, School of Health and Life Sciences, Glasgow Caledonian University, Cowcaddens Road, Glasgow G4 0BA, Scotland, UK
[2]Centre of Marine Sciences, University of Algarve, Campus de Gambelas, 8005-139, Faro, Portugal

[reviewed in 4]. Williams *et al.* [6] studied the ruminal microbiota of the North Ronaldsay breed of sheep (the male adults of which survive almost entirely on a seaweed diet on the remote island of North Ronaldsay in the Orkney Islands, Scotland) and isolated bacteria and crude enzyme extracts that hydrolysed seaweed polysaccharides including laminarin, alginate, carrageenan, cellulose and to a lesser extent fucoidan. Additionally, in serial culture this rumen microbiota hydrolysed a range of seaweed polysaccharides and homogenates of *L. hyperborea*, mixed *Fucus* spp. and *A. nodosum* to produce methane and acetate. This particular rumen microbiota and bacterial isolates were therefore considered to represent potential adjunct organisms or enzymes which may improve hydrolysis of seaweed components and thus improve the efficiency of seaweed anaerobic digestion for methane biofuel production.

With a long term aim of developing and characterising an optimal inoculum with which to degrade seaweeds in AD, we compared the ability of a range of microbial inocula, including North Ronaldsay sheep rumen microbiota [6], to effectively hydrolyse *L. hyperborea* and produce methane biofuel. We examined these inocula in single-stage and two-stage fermenters in order to select the most productive inoculum for future optimisation studies in large scale (10L) fermenter systems. We studied single and two-stage systems because others have recommended [7] that seaweed AD is better conducted in two stages. In the first stage complex polysaccharide hydrolysis and acidogenesis to volatile fatty acids (VFAs) by fast growing and low pH tolerant bacteria is encouraged. The VFAs produced in the leachate of such stage one fermenters occurs at pH 5.5 to 6.0. The leachate is then neutralised and fed into phase II fermenters where acetate is then further converted to methane by acetoclastic methanogens. This two-stage system has been recommended for seaweed AD because seaweeds have a high sugar content and so tend to lower the fermenter pH with rapid growth of acidogenic bacteria, which may inhibit methanogenic archaea [7,8]. We aimed to select the inoculum which was most efficient in both systems.

Results

Comparison of AD inocula in single-stage fermenters

All the microbial inocula (for a description of the inocula see the Methods section) grew to give similar metabolic activity of around 6.0 $\log_{10}$ fluorescence units (Figure 1a) and so they clearly adapted to survive during *in vitro* fermentation of *L. hyperborea* using the mini-fermenters described in the methods section.

The *L. hyperborea* seaweed used in this study was added to fermenters as wet weight per volume. Samples of this feed were examined for the percentage total solids and volatile solids (VS) on three separate occasions giving a mean (±SD) percentage of 23.96 (± 2.32) TS and 20.13 (±0,98) VS, therefore showing the ability to reproducibly supply the same amount of wet weight feed to each fermenter.

All the inocula tested could hydrolyse *L. hyperborea* to some extent, but between 28 and 55% of VS were still undigested in individual fermenters. Figure 1b shows the mean percentage of VS used in triplicate fermenters of the same inoculum. Over the 4 week period Inoculum 6 gave the highest overall utilisation of seaweed VS (mean of 67.74% used) and this was significantly higher ($p < 0.05$) usage than by inocula 2, 3, 4, 5, and 7 (range 46.14% to 56.37%) but not by inocula 1 or 8 with means of 59.50 and 58.69% VS used, respectively.

Triplicate fermenters of each inoculum (1–8) were tested randomly on separate occasions. So it was interesting that all showed decreased biogas production in week 2 (various dates) and then recovered (Figure 1c). Thereafter, all inocula generally produced biogas above initial (week 1) biogas production levels. Inocula 5, 6, 7 and 8 (either from anaerobic digesters or a mixture of all inocula) had developed to give more biogas (mean of 82.0, 81.25, 115.42 and 72.08 mL g^{-1} VS, respectively) than other ruminant derived inocula by week 4. The percentage of methane and carbon dioxide in the biogas was analysed by GC and the volume of methane in biogas calculated from this. All inocula were capable of producing methane from seaweed (Figure 1d) but some gave significantly better yield than others. Inocula 5, 6, 7 and 8 had developed by week 4 to give higher volumes of methane production than other inocula (mean (± SD) of 41.25 (±18.01), 40.76 (±1.84), 58.55 (±16.81) and 63.10 (±20,40) ml g^{-1} VS, respectively). Inoculum 8 was highest but only significantly higher than inocula 1 to 5 [$p < 0.05$], not 6 or 7. There was a drop in methane production by most inocula in week 2, which coincides with the decrease in total biogas produced (Figure 1c).

Acetate production in fermenters over the 4 weeks was significantly greater by inoculum 1 (mean of 0.298 g/L) than by inocula 2, 3, 4, 6 ($p < 0.05$) and 7 ($p < 0.01$) but not inoculum 5 or 8. Inoculum 8 contained microbiota from inoculum 1 and so this is perhaps not surprising. Residual acetate from a single-stage fermenter can be converted to methane by acetoclastic methanogens in a second stage fermenter [7]. The calculated potential methane yield from acetate (mean results of three fermenters per inoculum sampled over four weeks) in the single-stage fermenters is given in Table 1. The theoretical methane production from acetate was calculated as 1 mole of acetate is converted into 1 mole of methane; and 1 mole of any gas at standard laboratory conditions (25°C, 1 atmosphere pressure) equals 22.47 L of gas.

Taking the actual methane produced and combining it with the calculated methane yield from acetate (Table 1). Inoculum 8 gave the highest mean overall calculated

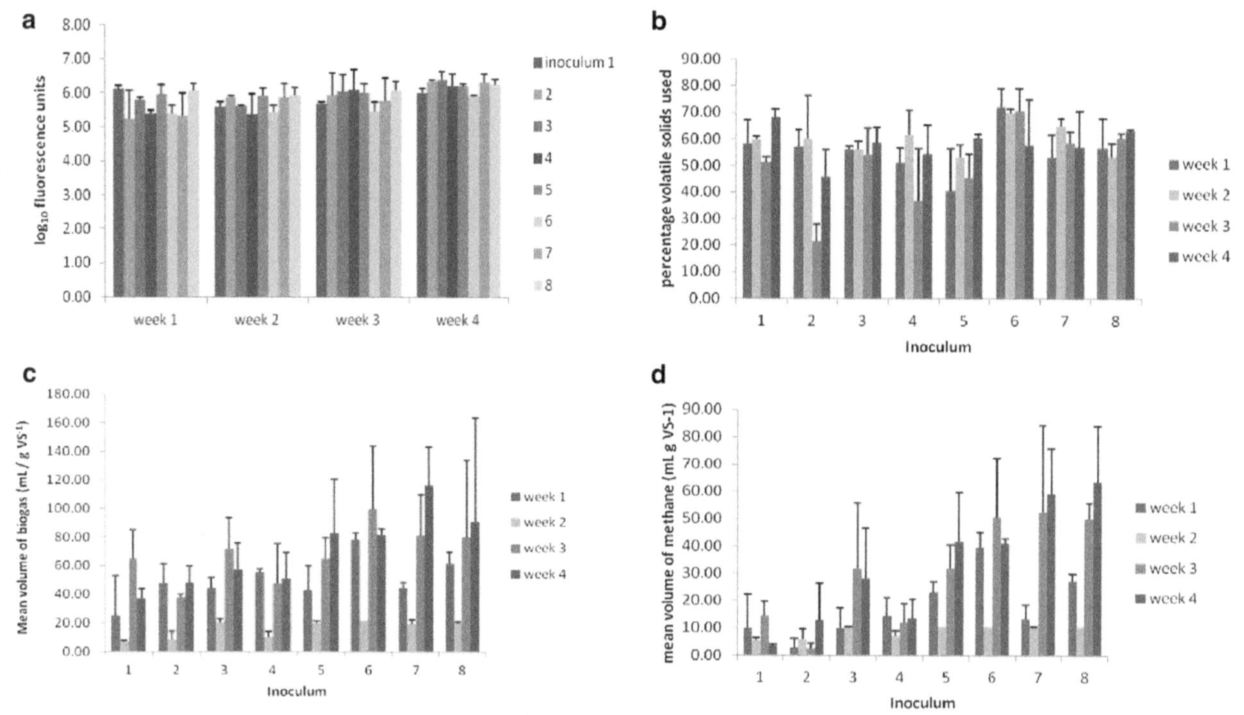

Figure 1 Operational parameters in single-stage fermenters (a) microbial viability, (b) utilisation of volatile solids, (c) biogas produced and (d) methane produced.

methane yield (0.253 L g^{-1} VS), followed by inoculum 5, then inoculum 1 (0.199 L g^{-1} VS, 0.185 L g^{-1} VS, respectively).

Phase I fermenter operation

Even running at a pH adjusted daily with 2M NaOH to be between 5.5 and 6.0 (the range of pH the culture covered during daily fermentation was between 4.96 and 5.88) there was a substantial percentage of methane detected in the headspace biogas most weeks; with 34.80/65.11; 0.54/99.35 and 42.13/57.72 % CH_4/CO_2 in weeks 2,3 and 4, respectively (the week 1 gas sample was lost due to human error). The volume of biogas was not measured in this study. That methane was produced at such low fermenter pH was in agreement with the study of Vergara-Fernández *et al.* [7] who also detected methane gas production in a phase 1 fermenter for seaweed hydrolysis. The growth of inoculum 6 used in the phase I fermenter was stable over the 4 week period with log_{10} fluorescein diacetate (FDA) values of 5.76, 5.72, 5.68 and 5.37 for weeks 1 to 4, respectively; and so showed good survival of the inoculum under the phase I fermenter conditions. The % of total solids (TS) used over the 4 week fermentation was 40.00, 50.78, 60.75 and 62.00%, respectively; which was, at least initially, less than the utilisation by inoculum 6 during 4 weeks of the single-stage fermentations where there was an average of 67% usage (see above). The lower VS usage in the first two weeks of fermentation may indicate that the hydrolytic bacteria had to adapt to the low fermenter pH.

The acetate concentration in the pooled phase I fermenter leachate was 0.518 g/L. This supplied 3.5 mmol of acetate to each phase II fermenter over 4 changes of leachate (400 mL) which theoretically could have given 22.17 mL of methane.

Comparison of inocula in phase II fermenters

The volume of biogas (expressed as mL mmol⁻1 acetate) produced over four weeks from 400 mL of leachate

Table 1 Calculated total methane yield from acetate and actual methane produced in single-stage fermenters

	Inoculum							
	1	2	3	4	5	6	7	8
Acetate* produced (mmol g^{-1} VS)	7.2	4.5	5.3	4.3	5.4	2.9	2.8	6.5
Calculated methane yield from acetate (litres g^{-1} VS)	0.161	0.101	0.120	0.097	0.122	0.066	0.064	0.145
Calculated total methane yield (litres g^{-1} VS)	0.185	0.118	0.177	0.130	0.199	0.167	0.160	0.253

*acetate produced is the mean cumulative amount over four weeks from triplicate single-stage fermenters. The total methane yield is the calculated methane yield from acetate added to the actual methane produced in the same fermenter over four weeks.

(3.5 mM of acetate) generated in a phase I fermenter (see Methods section); the percentage of methane in that biogas and the total methane production in mL $mmol^{-1}$ acetate over four weeks for each inoculum is shown in Table 2. Inoculum 8 produced most methane (8.4 mL $mmol^{-1}$ acetate), followed by inoculum 6 (7.31mL $mmol^{-1}$ acetate). Other inocula gave less than 3.0 mL $mmol^{-1}$ acetate of methane each and inocula 2, 3 4 and 7 had particularly low methane levels.

Discussion

All the inocula tested could hydrolyse *L. hyperborea* to some extent (Figure 1b) but even by week 4, inoculum 6, 1 and 8, the most hydrolytic inocula, still had means of 32.26, 40.50 and 41.31% VS remaining, respectively. It will be interesting to determine in further fermenter studies what the composition of this residue is. It possibly contains a high percentage of alginates since Moen *et al.* [2] reported that between 13 and 50% of alginates were resistant to AD, depending upon levels of *L. hyperborea* polyphenols present and the ratio of calcium and sodium alginate salt forms. The high utilisation of VS by inoculum 6 (human sewage AD leachate) and inoculum 1 (North Ronaldsay sheep rumen microbiota) suggests that they had efficient seaweed polysaccharide hydrolysing bacteria. Moreover, this result is in agreement with preliminary studies (A.D. Sutherland, unreported results) in which it was found that a human sewage AD effluent recovered in Scotland was similarly most effective in AD of *L. hyperborea*; this was thus not a unique occurrence by the Portuguese sourced inoculum 6. Inoculum 1 was previously shown to have bacteria producing polysaccharidases capable of hydrolysing seaweed polysaccharides [6] and Hehemann *et al.* [9] have recently reported the presence of alginate lyase genes in human microbiota which have been derived from marine bacteria. Alginates and other seaweed polysaccharides are used as thickeners in foods, toothpastes and cosmetics and it is conceivable that human gut bacteria with hydrolytic enzymes survive in human sewage digestate (inoculum 6). Therefore, both inoculum 6 and 1 may offer adjunct bacteria or enzymes to improve the efficiency of hydrolysis and are likely to have contributed to the high utilisation of VS by inoculum 8 (a mixture of all the inocula studied).

Since there was no concomitant reduction in the %VS used (Figure 1b) the drop in biogas and methane production seen in all week 2 samples may have been due to an initial reduction in methanogenic archaea. These are very slow growing and very strict anaerobes, the decrease of which may have been overcome by biofilm formation that was observed to have developed on the fermenter walls by week 3.

Since inoculum 8 gave the highest levels of methane production by week 4 it would be advantageous to use this mixed inoculum for AD of *L. hyperborea.* It would be interesting, in further studies, to add selected bacterial isolates rather than the whole rumen microbiota (e.g. alginate lyase producers as isolated by Williams *et al.* [6]) to inoculum 6 to see if it improves VS usage. Also, inoculum 6 should be investigated to determine what endogenous alginate hydrolysers it may possess.

That inoculum 1 had high levels of acetate may be the reason why this inoculum (and to some degree inoculum 3 [normal sheep rumen microbiota]) required most pH adjustment, reaching lows of 5.8 within 24 hours of adjustment. This may also be a reason why methanogenesis was progressively reduced in some replicates of these inocula, since methanogens grow poorly at pH less than 6.0 and would possibly have begun to be washed out of the semi-continuous fermenter. Even without methane production, as long as the microbiota in an inoculum can degrade seaweed solids to produce VFAs, these may be utilisable for biofuel and other valuable chemical feedstock production via the carboxylate platform [10], where lack of methane production is desirable. One example of this would be inoculum 1, which could be a valuable source of hydrolytic bacteria for the latter platform. If the hydraulic retention time in a single-stage fermenter is too short to allow all acetate produced to be converted to methane by acetoclastic methanogens then accumulated acetate can be fed into a second (phase II) fermenter with immobilised acetoclastic methanogens [7].

Inoculum 8 gave the highest methane production when considering actual methane produced plus the calculated methane production from acetate (Table 1). This further showed that inoculum 8 was the most productive inoculum in single-stage fermenters.

Table 2 Biogas and methane production from phase II fermenters fed with phase I fermenter leachate

	Inoculum							
	1	2	3	4	5	6	7	8
Mean biogas* (mL $mmol^{-1}$ acetate)	10.0	12.9	21.4	17.1	14.3	22.9	12.9	22.9
Mean % methane	25.35	0.70	5.40	0.16	18.15	31.98	1.70	36.88
Mean methane volume produced (mL $mmol^{-1}$ acetate)	2.3	0.5	0.6	0.1	0.6	7.3	0.1	8.4

*The volume of biogas (expressed as mL $mmol^{-1}$ acetate) produced over four weeks from 400 mL of leachate (3.5 mM of acetate) generated in a phase I fermenter.

A two-stage fermenter system was operated in which inoculum 6 was used to produce acetate in a phase I fermenter. The leachate of the latter was then fed into individual phase II fermenters containing inocula 1 to 8 in order to determine which inoculum was most productive in acetoclastic methanogenesis.

Inocula 2, 3, 4 and 7 gave poor methane production from acetate in the second phase of a two-stage fermenter system, indicating they do not have sufficient levels of acetoclastic methanogens for operation of a phase II fermenter. Levels of acetoclastic methanogens in normal sheep rumen are for example very low; however, acetoclastic methanogens such as *Methanosarcina barkeri* are present and can grow to significant levels if dietary changes such as high starch content occurs [11]. The results for the phase II fermenters complement the findings for the single stage fermenters where again inocula 8 and 6 were the most efficient inocula and inoculum 8 the best at producing methane from acetate (8.4 mL methane mMol acetate^{-1}). Inoculum 1 was the third best acetoclastic methanogenic inoculum (Table 2) and it will be interesting to determine by metagenomics what archaea are present in inocula 8, 6 and 1 after growth in a phase II fermenter.

In general, laboratory studies have found that methane yield is greater from brown algae than from green algae [4]. There are very few studies conducted on AD of *Laminaria* sp. but Hanssen *et al.* [12] reported only a 36% usage of VS and 280 mL methane gVS^{-1} in an 8L fermentation volume with a retention time of 24 days. More recently Matsui *et al.* [13] using a separated acidic and methanogenic phase commercial scale AD plant of 30M^3 reported 22 mL methane g^{-1} wet weight seaweed. In the present study the maximum % of VS used in any fermenter using inoculum 8 (the most productive inoculum) was 66.54% and a yield of 83 mL methane g^{-1} VS (17 mL methane g^{-1} wet weight seaweed). However, the present study aimed to compare the ability of several inocula to ferment *L.* hyperborea under conditions of low feed rate (1% wet weight *L.* hyperborea day^{-1}) and high turnover of the liquid phase (a complete change in leachate week^{-1}) in order to minimise the effects of acidification and product feedback inhibition previously noted by others [7,13]. The study was not conducted under optimised AD conditions, but was designed to compare several inocula in one study to select the most efficient in the system used. The methane yield by inoculum 8 will likely be increased in a large scale (10 L) continuously stirred fermenter optimised for feed rate, solids and liquids retention time and automatic pH adjustment; and this study is now intended by this group. As hydrolysis of polysaccharides is the rate limiting step in seaweed AD, the need for more effective inocula for AD of brown seaweed is reflected in the conclusions of others [2-4]. Wargacki *et al.* [8] to this end have recently cloned alginate lyase genes into *E. coli* and have produced bioethanol from seaweed at near 80% of the theoretical yield. Like others [14-16] who found marine sediments were efficient inocula in seaweed AD we adopted, instead of gene cloning, to study complex microbiota in inocula from diverse locations in order to select efficient seaweed hydrolysers for AD, and found that a mixture of all the inocula (including North Ronaldsay sheep rumen and human sewage AD slurry microbiota) was most productive.

We are currently employing dissociative gradient gel electrophoresis and next generation sequencing on PCR amplified 16S rRNA genes to characterise the microbial consortia present in the various inocula studied here. DNA was also extracted from each fermenter weekly and metagenomics will be used to examine the temporal and spatial development of microbial consortia. This may indicate, for instance, which microbes from the various inocula developed in inoculum 8 to make it the most efficient in methane production. It may indicate why methane production dropped in week 2 in all fermenters and it may also establish which hydrolytic bacteria and methanogenic archaea were active in single-stage and two-stage (phase I and II) fermenters.

Conclusions

Several microbial inoculum were examined for the ability to hydrolyse polysaccharides and produce methane in AD of *L. hyperborea*. All the inocula could hydrolyse *L. hyperborea* and produce methane to some extent but there were significant differences in the %VS used and the amount of methane produced thus showing that selection of an inoculum can improve AD. Inoculum 8, which corresponded to a mixture of inocula 1–7, proved to be highly hydrolytic and produced the most methane after AD in a single-stage (0.253 Litres methane g^{-1} VS when including calculated methane from acetate)or in phase II of a two-stage digester system (8.4 mL methane mmol^{-1} acetate) (Tables 1 and 2, respectively). This showed that careful selection of a complex microbial inoculum can increase the efficiency of seaweed AD, most likely due to combining effective polysaccharolytic bacteria, in this instance from North Ronaldsay sheep rumen microbiota [6], with appropriate methanogenic archaea which have developed in AD reactors, as in inoculum 6. When examined individually, normal (grass fed) sheep rumen microbiota (inoculum 4) did not perform as well and neither did faecal microbiota from North Ronaldsay sheep or normal sheep (Tables 1 and 2). This group will now use inoculum 8 in optimised single stage and two stage AD systems to determine methane productivity.

Methods

Throughout the study all materials and reagents used were pre-reduced by gassing in a sealed plastic bag flushed with a mixture of 80% nitrogen, 10% carbon dioxide, 10% hydrogen in the presence of a palladium catalyst (Don Whitley, Scientific) to convert oxygen to water. Strict anaerobic conditions were observed by the reduction of a resazurin based indicator strip (Oxoid, UK) from pink to colourless.

Inoculum collection

Samples of the various microbial inocula for AD had been either, collected in Scotland in 100 mL aliquots (completely filled bottles) and frozen and transported by courier at −20°C, which was previously shown to retain microbial viability in these samples [6] or were collected in The Algarve, Portugal and frozen on collection. The inocula were:

1. North Ronaldsay sheep rumen contents, obtained at slaughter in Kirkwell, Scotland.
2. North Ronaldsay sheep faeces, obtained at slaughter in Kirkwell, Scotland.
3. Normal sheep (grass eating) rumen contents, obtained at slaughter in Paisley, Scotland
4. Normal sheep (grass eating) faeces, obtained at slaughter in Paisley, Scotland.
5. A mixture of municipal AD fermenter leachates kindly supplied by Biogen Greenfinch, UK.
6. Human sewage AD leachate kindly supplied by ETAR, Lagos, Portugal.
7. Marine sediments (anaerobic, black mud collected from the Ria de Formosa, Faro, Portugal and mixed with naturally rotting *L. hyperborea* which was anaerobically fermented for 48 hours after collection.
8. A mixture of all the above inocula (in equal proportions).

L. hyperborea preparation

Frozen *L. hyperborea* collected as beach cast seaweed (having no obvious signs of deterioration compared to fresh seaweed) on the Firth of Forth, near Edinburgh, Scotland was stored at −20°C upon courier delivery in frozen condition.

The seaweed was thawed and washed under tap water to remove sand and dirt. The seaweed fronds and stipe (5 Kg) were ground with a kitchen mincer (6 mm diameter holes) and then were thoroughly mixed by hand to obtain a homogeneous mass. Aliquots (100 g) were frozen at −20°C, defrosted as required and again mixed thoroughly by hand. Amounts of 12 g were weighed and reduced to approximately 2 mm diameter particles in 60 mL of phosphate buffered saline (PBS) using a Braun kitchen electric hand mixer with a grinding bowl. The blended seaweed (20% wet weight/volume [w/v]) was autoclaved, gassed under nitrogen and then stored at 4°C and used within 1 week.

Single-stage AD fermenter operation

Each inoculum was tested randomly on three separate occasions (up to 10 fermenters were run at a time, the whole experiment taking three months to complete). Because the inocula had been frozen upon collection, 100-mL aliquots of each were defrosted at 30°C and then immediately placed under anaerobic conditions into sterile pre-reduced 500 mL Duran bottles. Each bottle contained 300 mL of phosphate buffered saline (PBS) supplemented with 0.5 mL each of 20% (w/v) sterile glucose and yeast extract. Incubation was carried out for 48 h at 37°C. Each sample was then split into 4 x100 mL volumes: one volume (avoiding any debris from the original inoculum) was put into a sterile pre-reduced conical-flask fermenter with a water-displacement gas collector (Figure 2) and three were frozen again for further studies. To the fermenters were added 10 mL of 20% (w/v) *L. hyperborea* in PBS and 90 mL of PBS to give a final seaweed concentration of 1.0% (w/v) in a 200 mL volume (this was day 1 week 1 for each fermenter). The samples were adjusted to pH 7.0 with 2M Na OH as required and then incubated at 37°C. These samples were then treated in triplicate semi-continuous culture as in Table 3 for 4 weeks each (4 hydraulic retention times [HRT]).

The fermenter leachates were examined every 24 hours for pH (which was adjusted to pH >6.5 as required with 2M NaOH). Weekly, the microbial metabolic activity (FDA fluorescence), the volume of biogas produced and the biogas composition on a gas chromatograph (% methane and carbon dioxide) was measured. The FDA assay was carried out essentially as published [17]. A 1/50 dilution of FDA stock solution (10 mg/mL in acetone) was made in 100 mM MOPS buffer (pH 7.0). Volumes (100 μL) of this were added to duplicate 100 μL volumes of each inoculum sample (at 10^{-1} and 10^{-2} dilution) in black microtitre plates. Samples were incubated for 1 hour and fluorescence read (emission 494 and excitation 520 nm wavelength) after

Figure 2 Conical-flask fermenter with gas collector on side-arm.

Table 3 Weekly protocol for the semi-continuous fermentation of *L. hyperborea* in single-stage fermenters

Treatment	Day						
	1*	2	3	4	5	6	7
	Mon	Tue	Wed	Thu	Fri	Sat	Sun
Volume of leachate removed (mL)	200 (140 mL returned)	0	60	0	60	0	0
Volume of 20% wet W/V *L. hyperborea* added	10	10	10	10	20	0	0
Volume of PBS added	50	0	50	0	40	0	0
Seaweed removed	yes	no	no	no	no	no	no

*From week two onwards, on day 1 of each week the contents of each fermenter were removed to a sterile plastic pot. A 140 mL volume of the leachate of each fermenter was poured back into the fermenter, 10 mL of *L. hyperborea* seaweed was added and the fermenter volume was made up to 200 mL with PBS. In this manner the fermenter volume was maintained at 200 mL; seaweed was added daily and leachate was removed every second day. The residual seaweed was dried for estimation of residual total solids, then ash and volatile solids (see Methods section). This semi-continuous system was designed to reduce acidity and product feedback inhibition from volatile fatty acids, whilst having a long enough hydraulic retention time to allow maintenance of slow growing and acid sensitive methanogens and also allow digestion of seaweed to occur and was similar to the phase 1 system of a two-stage semi-continuous system adopted by Vergara-Fernández *et al.* [7].

subtraction of the reading of a blank of 100 μL of PBS plus FDA. The fluorescent reading was multiplied by the reciprocal of the dilution factor of the inoculum and then converted to $\log_{10}$. The biogas composition was determined on a gas chromatograph, as previously described [6].

Also, weekly, TS, ash and VS content of residual seaweed in each fermenter were measured by drying the samples at 100°C overnight and weighing (to obtain the TS) and then further drying at 530°C for approximately 6 hours, until there was no further weight loss and weighing again to get ash content (and by subtraction from the TS, the VS content). These figures were compared with the TS, VS and ash content of the same weight of original seaweed to calculate the percentage of each that were removed by AD. A compositional analysis of *Laminaria* sp. has been carried out previously by others [2,4,12,13]. Each fermenter received 60 mL of 20% (w/v) seaweed per week; which equals 12 g of wet weight seaweed (2.4 g VS).

Acetate in the fermenter leachates was measured by an enzymatic kit (BioPharm Ltd.) as previously described [6].

Comparison of AD inocula in a two-stage semi-continuous fermenter system

The same inocula as were compared in the single-stage AD fermenters (see above) were also compared in a two-stage (phase I and II) system to determine which inoculum was optimum for methanogenesis in phase II. To make the stage II system fermenters comparable they were all fed from a large volume of stored leachate as feedstock derived from a phase I fermenter run with inoculum 6, which had the best usage of *L. hyperborea* VS in the single-stage system.

Production of acetate in a semi-continuous phase I AD fermenter

A 100 mL volume of frozen inoculum 6 was defrosted and added to 50 mL of sterile 20% wet weight per volume (w/v) *L. hyperborea* seaweed and 850 mL of sterile distilled water in a pre-reduced one litre Dewar bottle as a fermenter vessel. The culture was adjusted to pH 5.5-6.0 by the addition of 2M NaOH to encourage growth of hydrolytic and acetogenic bacteria. The fermenter therefore had a volume 5-fold higher than that used in single-stage fermentations (see above). The fermenter was then run semi-continuously in the same manner as in Table 3 but with all volumes increased 5-fold. The residual seaweed VS was determined weekly and head-space gases were monitored for methane production. The fermenter was run for 4 weeks with 1 litre of leachate collected weekly, pooled and frozen at −20°C until required. The pooled leachate was defrosted, the pH adjusted to 8.0 with NaOH to convert VFAs to their less volatile salt form and the leachate was then centrifuged at 3,000 × *g* for 30 min to pellet bacteria and the supernatant autoclaved to sterilise it and the pH readjusted to 7.0 and stored at 4°C until used.

Production of methane in semi-continuous phase II AD fermenters

A 100 mL volume of each frozen inoculum (1 to 8) was defrosted and resuscitated with glucose and yeast extract (as for the single-stage fermenters above). A 100 mL volume of each inoculum was then added with 100 mL of phase I leachate as feedstock to sterilised, pre-reduced side-arm flask fermenters with gas collectors (Figure 2). The fermenters additionally had ¼ of a Scotch-Brite plastic scouring pad in them to encourage methanogens in particular to adhere to them and slow wash out. The volume of biogas produced and its composition as % methane/CO_2 was measured and pH was monitored regularly. When biogas production ceased 100 mL of leachate was removed from a fermenter and was replaced with a further 100 mL of phase I fermenter leachate. The process was repeated until 4 cycles of leachate feeding had occurred.

Statistical analysis

All comparisons of data from fermenters were done on SigmaPlot for Windows version 10, using a paired Student's T-test.

Competing interests

The authors declare that they have no competing interests

Authors' contributions

AS and JV were both involved in conception of the study, sample collection, manuscript writing and editing and interpretation of data. AS was responsible for fermenter operations and data analysis. All authors read and approved the final manuscript.

Authors' information

AS is a senior lecturer and researcher in microbiology at Glasgow Caledonian University. He held a Marie Curie senior researcher fellowship and was seconded to Centre of Marine Sciences, University of Algarve for two years during which the work in this manuscript was carried out. JV is a lecturer at the University of Algarve and the head of the MarBiotech research group of the Centre of Marine Sciences of Algarve (CCMAR).

Acknowledgements

The work reported here was carried out at The Centre of Marine Sciences, University of Algarve and was funded by a Marie Curie Senior Researcher Fellowship (SEAWEED AD) held by Prof. Alastair Sutherland.

References

1. Forro J: **Microbial degradation of marine biomass.** In *Seaweed cultivation for renewable resources.* Edited by Bird KT, Benson PH. Amsterdam: Elsevier; 1987:305–325.
2. Moen E, Horn S, Ostgaard K: **Alginate degradation during anaerobic digestion of *Laminaria hyperborea* stipes.** *J Appl Phycol* 1997, **9:**157–166.
3. Moen E, Horn S, Ostgaard K: **Biological degradation of *Ascophllum nodosum*.** *J Appl Phycol* 1997, **9:**347–357.
4. Morand P, Carpentier B, Charlier RH, Maze J, Orlandini M, Plunkett BA, De Vaart J: In *Seaweed resources in Europe: uses and potential.* Edited by Guiry MD, Blunden G. N.Y. and Chichester, UK: John Wiley and Sons Ltd; 1991:95–148.
5. Horn SJ: *Bioenergy from brown seaweeds.* Norway: PhD thesis. Department of Biotechnology Norwegian University of Science and Technology NTNU Trondheim; 2000.
6. Williams AG, Withers S, Sutherland AD: **The potential of bacteria isolated from ruminal contents of seaweed-eating north ronaldsay sheep to hydrolyse seaweed components and produce methane by anaerobic digestion *in vitro*.** *Microbial Biotechnol* 2012, **6**(1):45–52.
7. Vergara-Fernández A, Vargas G, Alarcón N, Velasco A: **Evaluation of marine algae as a source of bio-gas in a two-stage anaerobic reactor system.** *Biomass Bioenergy* 2008, **32:**338–344.
8. Wargacki AJ, Leonard E, Win MN, Regitsky DD, Santos CNS, Kim PB, Cooper SR, Raisner RM, Herman A, Sivitz AV, Lakshmanaswamy A, Kashiyama Y, Baker D, Yoshikuni Y: **An engineered microbial platform for direct biofuel production from brown microalgae.** *Science* 2012, **335:**308–313.
9. Hehemann J-H, Correc G, Barbeyron T, Helbert W, Czjzek M, Michel G: **Transfer of carbohydrate-active enzymes from marine bacteria to Japanese gut microbiota.** *Nature* 2010, **464:**908–912.
10. Agler MT, Wrenn BA, Zinder SH, Angenent LT: **Waste to bioproduct conversion with undefined mixed cultures: the carboxylate platform.** *Trends Biotechnol* 2011, **29**(2):70–78.
11. Stewart CS, Flint HJ, Bryant MP: **The rumen bacteria.** In *The rumen microbial ecosystem.* 2nd edition. Edited by Hobson PN, Stewart CS. London: Blackie Academic & Professional; 1997:10–72.
12. Hanssen JF, Indergaard M, Ostgaard K, Baevre OA, Pederson TA, Jensen A: **Anaerobic digestion of *Laminaria* species and *Ascophyllum nodosum* and application of end products.** *Biomass* 1987, **14:**1–13.
13. Matsui JT, Amano T, Koike Y, Saiganji A, Saito H: *Methane fermentation of seaweed biomass.* San Francisco: American Institute of Chemical Engineers; 2006.
14. Rao PS, Tarwade SJ, Sarma KSR: **Seaweed as source of energy. I: effect of a specific strain on biogas production.** *Bot Mar* 1980, **23:**599–602.
15. Leclercq S, Tine E, Suisse de Saint Claire E: *Fermentations methaniques de macrophytes marins.* Dakar, Senegal: Document scientifique No. 102. Centre de Recherche Oceanographiques de Dakar-Thiaroye/Institut Senegalais de Recherche Agricole; 1985:22.
16. Hanisak MD: **Recycling the residues from anaerobic digesters as a nutrient source for seaweed growth.** *Bot* 1981, **24:**57–61.
17. Peeters E, Nelis HJ, Coenye T: **Comparison of multiple methods for quantification of microbial biofilms grown in microtiter plates.** *J Microbiol Methods* 2008, **72:**157–165.

3

Solubility of recombinant Src homology 2 domains expressed in *E. coli* can be predicted by TANGO

Thorny Cecilie Bie Andersen[1], Kjersti Lindsjø[1†], Cecilie Dahl Hem[1†], Lise Koll[1], Per Eugen Kristiansen[2], Lars Skjeldal[3], Amy H Andreotti[4] and Anne Spurkland[1*]

Abstract

Background: Signalling proteins often contain several well defined and conserved protein domains. Structural analyses of such domains by nuclear magnetic spectroscopy or X-ray crystallography may greatly inform the function of proteins. A limiting step is often the production of sufficient amounts of the recombinant protein. However, there is no particular way to predict whether a protein will be soluble when expressed in *E.coli*. Here we report our experience with expression of a Src homology 2 (SH2) domain.

Results: The SH2 domain of the SH2D2A protein (or T cell specific adapter protein, TSAd) forms insoluble aggregates when expressed as various GST-fusion proteins in *Escherichia coli (E. coli)*. Alteration of the flanking sequences, or growth temperature influenced expression and solubility of TSAd-SH2, however overall yield of soluble protein remained low. The algorithm TANGO, which predicts amyloid fibril formation in eukaryotic cells, identified a hydrophobic sequence within the TSAd-SH2 domain with high propensity for beta-aggregation. Mutation to the corresponding amino acids of the related HSH2- (or ALX) SH2 domain increased the yield of soluble TSAd-SH2 domains. High beta-aggregation values predicted by TANGO correlated with low solubility of recombinant SH2 domains as reported in the literature.

Conclusions: Solubility of recombinant proteins expressed in *E.coli* can be predicted by TANGO, an algorithm developed to determine the aggregation propensity of peptides. Targeted mutations representing corresponding amino acids in similar protein domains may increase solubility of recombinant proteins.

Keywords: Bacterial inclusion bodies, Protein aggregation, Recombinant protein expression, SH2 domain, SH2D2A, Protein solubility

Background

Src homology 2 (SH2) are structurally conserved protein domains of approximately100 amino acids (aa). SH2 domains regulate numerous intracellular signal transduction events through interaction with tyrosine phosphorylated proteins [1]. Within the human genome, 110 distinct genes encode for 120 SH2 domains [2]. SH2 domains consist of an N-terminal alpha helix (αA) flanking a central five-stranded anti-parallel beta sheet (βB-F) followed by a second alpha helix (αB) and a C-terminal beta strand [3] (Figure 1). The five strands of the central beta sheet separate a conserved phosphotyrosine binding pocket from a more variable pocket that typically binds the third amino acid C-terminal to the phosphorylated tyrosine.

Detailed structural analyses have revealed specific and unique features of different SH2 domains suggesting that these domains, while ubiquitous and highly conserved, nevertheless exhibit unique specificity determining features. For example, NMR analysis of the SH2 domain of Inducible Tec Kinase (Itk) showed that its binding specificity is modulated by cis-trans isomerization of a single proline [4]. Structural analysis of the SH2D1A gene product, Slam associated protein (SAP), revealed the structural basis for a phosphotyrosine independent three-pronged

* Correspondence: anne.spurkland@medisin.uio.no
†Equal contributors
[1]Department of Anatomy, Institute of Basal Medical Sciences, University of Oslo, Oslo, Norway
Full list of author information is available at the end of the article

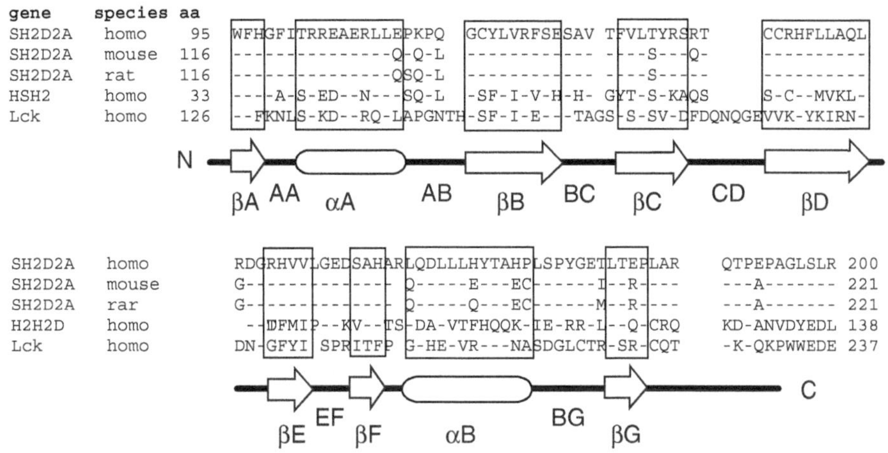

Figure 1 Alignment of SH2 domains. Alpha helices are indicated by open ovals and beta sheets are indicated by open arrows. "-": position is identical to human SH2D2A sequence. " ": gap in the alignment. Nomenclature for SH2 domain segments from [3].

binding mechanism of peptides to its SH2 domain [5]. Similarly, a disulphide bond in the SH2 domain of the C-terminal Src kinase (Csk) was found to be central for the kinase activity [6].

The function and specificity determining elements of many SH2 domain containing proteins, are still not well characterised. One of these is the T cell specific adapter protein (TSAd), encoded by the *SH2D2A* gene and expressed in activated T cells [7] and endothelial cells [8]. TSAd contains one SH2 domain, followed by a proline rich region and possesses ligands for SH2 and SH3 domains [9]. Hitherto only a few ligands for the human TSAd SH2 domain are known, including the phosphorylated vascular endothelial growth factor receptor 2 (VEGFR-2) [8,10], and the phosphorylated valocin containing protein (VCP) [11]. Recruitment of TSAd via SH2 domain binding to phosphorylated Y951 in the VEGFR2 receptor controls migration of [8] as well as permeability [12] of endothelial cells. Similarly, binding of TSAd via its SH2 domain to VCP is required for nuclear translocation of TSAd [11]. Preliminary results from our lab indicate that the TSAd-SH2 domain has additional ligands in activated T cells (Hem, unpublished observations). Given the interesting biology surrounding TSAd, and the role of SH2 domains in mediating important interactions in the context of cell signalling, we wanted to characterize the structure of TSAd SH2. However, when expressed as a GST-fusion protein in *E. coli*, the TSAd SH2 domain was found to be highly insoluble presenting a significant challenge to detailed structural characterization.

An algorithm for prediction of amyloid fibril formation in eukaryote cells, TANGO, has been developed [13], and it has been suggested that TANGO may also predict inclusion body formation in prokaryotic cells [14]. Here we report that TANGO predicts the solubility of recombinant SH2 domains expressed in *E.coli*, and that predictions made by TANGO allowed for targeted mutations that significantly improved the solubility of the recombinant protein.

Results

Low solubility of bacterially expressed human TSAd SH2 domain

To perform biochemical and structural studies on the TSAd SH2 domain, we initially generated a GST-fusion construct (1-TD, Figure 2A) including the flanking amino acids 67–94 and 189–207. The resulting GST-TSAd SH2 domain was expressed mainly as insoluble inclusion bodies (compare lane 1 and 2, Figure 2B). In contrast, the GST-Lck SH2 protein containing only minimal gene-derived flanking sequence was primarily expressed in the soluble fraction (compare lane 4 and 5, Figure 2B). Although small amounts of recombinant GST-TSAd SH2 was present in the lysate, most of this soluble fraction could be captured on glutathione sepharose beads (lane 3, Figure 2B). Anti-GST-immunoblotting of equal amounts of the soluble fractions compared to a defined amount of purified, soluble GST (Figure 2C,i), revealed that the yield of the TSAd SH2 domain) was only 0,15 mg/l bacterial cell culture, while that of Lck-SH2 was estimated to be close to 20 mg/l bacterial culture (Figure 2D).

Flanking sequences, growth temperature and genetic variation influence recombinant expression of TSAd SH2 domains

Having found that the protein yield of the TSAd-SH2 domain construct was only a fraction of that of the Lck-SH2 construct, additional constructs including different lengths of the sequence flanking the human TSAd SH2 domain and three different vector-encoded C-termini (Figure 2A and Table 1) were generated. When expressed at room temperature, the yield of these TSAd SH2 domain constructs varied considerably as judged by

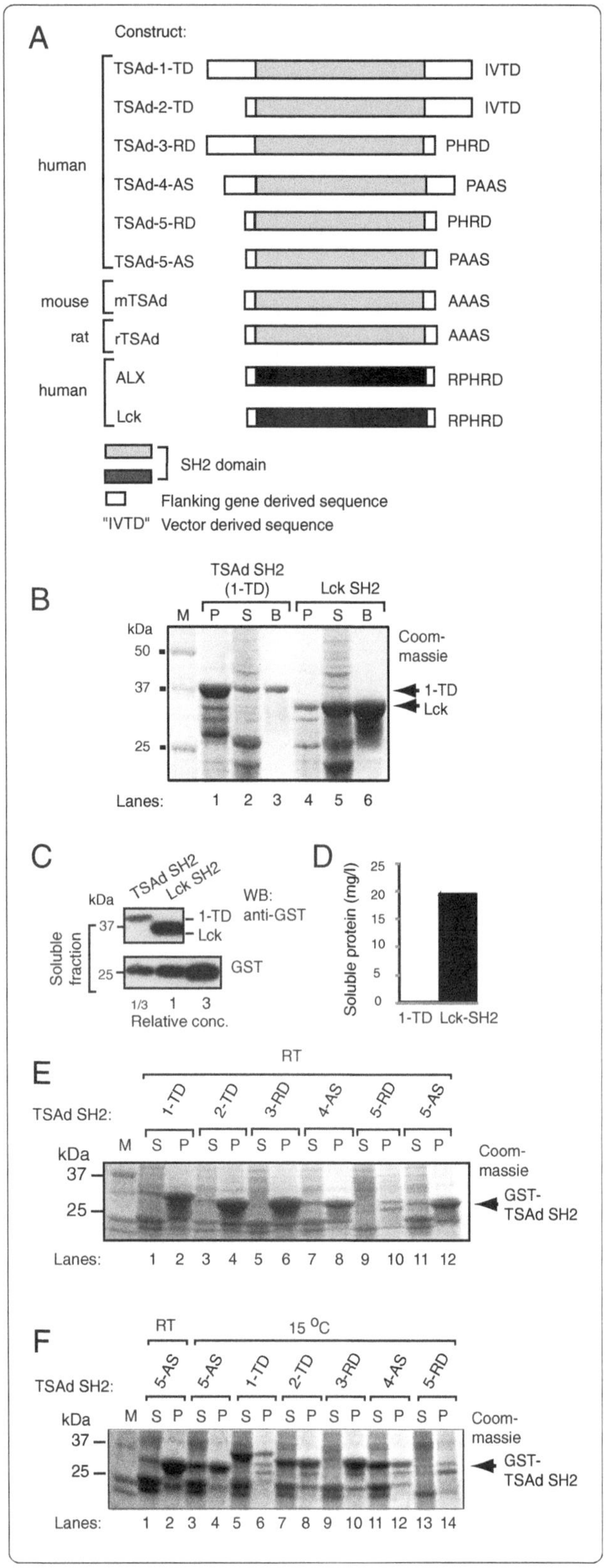

Figure 2 Expression and solubility of GST-TSAd SH2 domains are influenced by flanking sequence and growth temperature. A. Schematic overview of the GST-fusion SH2 constructs (given in Table 1) used in this study. **B**. GST-fusion SH2 constructs of TSAd (1-TD, TSAd-67-207-IVTD) and Lck (Lck-SH2, Lck-124-228-RPHRD) were expressed in *E.coli* at room temperature (RT). Equal volumes of resuspended pellet (P), soluble fraction (S) and glutathione beads **(B)** were separated by 10% SDS-PAGE. Proteins were visualised by Coomassie Brilliant Blue staining. **C**. Quantitation of the amount of soluble GST-SH2 proteins shown in **B**, by immunoblotting with anti-GST and comparison to defined amounts of GST. 5 µl of 2 ml TSAd-SH2 soluble fraction and 0,06 µl of 2 ml Lck-SH2 soluble fraction from 100 ml bacterial cultures were applied on the gel. GST = 1 represents a total amount of 0,11 µg GST applied on the gel. **D**. Quantitation of soluble GST-SH2 proteins using Image J analysis based on **C**. **E** and **F**. Yield of expression of the six GST-TSAd SH2 domain constructs in *E.coli* at RT or 15°C, respectively. Gels processed as in **B**. Constructs are indicated by their short names as listed in Table 1.

Commassie staining of SDS-gels. Compared to the TSAd-67-207-IVTD construct (1-TD), truncation of the sequence N- and/or C-terminal to the TSAd SH2 domain yielded approximately equal levels of expressed protein (Figure 2E, lanes 2, 4, 6 and 12). By contrast the TSAd-90-188-PHRD (5-RD) construct was not expressed at all (Figure 2E, lane 9 and 10). This construct only differed from TSAd-90-188-PAAS (5-AS) in its C-terminal vector-derived sequence, and from TSAd-67-188-PHRD (3-RD) in that the latter includes 37 aa N- terminal to the SH2 domain (Table 1).

Growth temperature affects the solubility of recombinant proteins in *E.coli* [15]. Expression at 15°C resulted in higher yield of soluble TSAd SH2 domain from the TSAd-67-207-IVTD (1-TD), TSAd-90-207-IVTD (2-TD), TSAd-81-193-PAAS (4-AS) and TSAd-90-188-PAAS (5-AS) constructs, whereas solubility and expression of the TSAd-67-188-PHRD (5-RD) and TSAd-90-188-PHRD (3-RD) constructs remained unchanged (Figure 2F). We also tested whether the genetic variation inherent in species differences (Figure 1) affected expression of TSAd SH2 domains. Analysis of solubility and yield of identically cloned TSAd SH2 domains derived from three different species (constructs TSAd-90-188-PAAS (5-AS), mTSAd-111-210-AAAS (mTSAd) and rTSAd-111-210-AAAS (rTSAd), Table 1) revealed slightly improved solubility at 15°C of murine TSAd SH2 compared to human and rat (data not shown). Taken together, both gene- and vector-derived sequence influence the overall expression and solubility of recombinant TSAd-SH2 proteins. However, none of the conditions tested resulted in amounts of soluble recombinant TSAd SH2 domain sufficient for structural analysis.

The TSAd SH2 domain contains a short peptide sequence promoting intermolecular beta-sheet aggregation

In order to examine whether the TSAd SH2 domain has some intrinsic property limiting its solubility in prokaryotic cells [14], we took advantage of an algorithm for prediction

Table 1 Overview of SH2 domain constructs included in this study

Short name	Construct	Gene	Accession number	Species	Start aa	End aa	C-terminal vector derived sequence	Construct total length (aa)
1-TD	TSAd-67-207-IVTD	*SH2D2A*	NP_003966	H sapiens	S67	S207	IVTD	145
2-TD	TSAd-90-207-IVTD	*SH2D2A*	NP_003966	H sapiens	G90	S207	IVTD	122
3-RD	TSAd-67-188-PHRD	*SH2D2A*	NP_003966	H sapiens	S67	R188	PHRD	126
4-AS	TSAd-81-193-PAAS	*SH2D2A*	NP_003966	H sapiens	T81	P193	PAAS	117
5-RD	TSAd-90-188-PHRD	*SH2D2A*	NP_003966	H sapiens	G90	R188	PHRD	103
5-AS	TSAd-90-188-PAAS	*SH2D2A*	NP_003966	H sapiens	G90	R188	PAAS	103
mTSAd	mTSAd-111-210-AAAS	*SH2D2A*	NP_001020742	M musculus	K111	R210	AAAS	104
rTSAd	rTSAd-111-209-AAAS	*SH2D2A*	NP_997488	R norvegicus	R111	R209	AAAS	103
ALX	ALX-31-126-RPHRD	*HSH2*	NP_116244	H sapiens	G30	R126	RPHRD	101
Lck	Lck-124-228-RPHRD	*Lck*	NP_005347	H sapiens	P124	K228	RPHRD	110

of beta-aggregation stretches in proteins [13], TANGO (http://tango.crg.es). TANGO analysis of the human TSAd SH2 domain sequence (S90-R188) revealed that the βC strand of the human TSAd SH2 domain (Figure 1) harbours a nine aa sequence (SAVTFVLTY) with near 100% propensity for intermolecular beta-sheet aggregation (Figure 3A). Similar results were obtained for the murine and rat sequences, whereas no beta-aggregating regions were found in the Lck-SH2 domain, nor in the TSAd homologue ALX SH2 [2,16] (data not shown, and Figure 3A, upper right panel). *In silico* substitution of the TSAd sequence AVTFVLT to the corresponding ALX sequence HVGYTLS and vice versa, reduced the predicted beta-aggregation propensity of the TSAd SH2 domain to near zero, whereas the ALX SH2 domain attained beta-aggregation propensity similar to that of the native TSAd SH2 domain (Figure 3A, bottom panels). Further *in silico* replacements revealed that exchange of the TFV sequence in TSAd to the GYT sequence of ALX, resulted in a 90% reduction in the overall beta-aggregation propensity of TSAd SH2 (Figure 3B). As the structure of the TSAd-SH2 domain is not yet determined, we instead compared the structures of the Lck-SH2 domain and the ALX-SH2 domain to visualize the putative localisation of the TFV sequence in the SH2 domain structure (Figure 3C). The tripeptide region corresponding to the TSAd-SH2 TFV sequence (SFS and GYT respectively) is highlighted. As can be seen from the space filling models the Ser 161 and Ser 163 in Lck, and all three amino acids in the ALX-motif (Gly65, Tyr66 and Thr67), are solvent accessible. Given the structural conservation between SH2 domain folds, it is likely that at least part of the TFV sequence in TSAd-SH2 may be solvent exposed. When the TFV sequence from TSAd-SH2 was *in silico* replaced with the GYT sequence in the ALX-SH2 domain sequence, the TANGO algorithm predicted an increase in beta-aggregation propensity corresponding to the TFV sequence (Table 2). To directly test the predictions made by TANGO, we generated TFV → GYT and FV → YT mutants of the TSAd-90-188-PHRD and TSAd-90-188-PAAS constructs, and compared their expression to that of the ALX SH2 domain. These constructs were chosen, as they contained the shortest flanking sequences compared to the original TSAd 1-TD construct, and were thus better suited for the planned down stream analysis. The TFV → GYT mutation increased both yield and solubility of the TSAd SH2 domains (Additional file 1: Figure S1), most dramatically for the TSAd-90-188-PHRD construct, which was not expressed as soluble protein in its wild type form (Figure 2E and Figure 3D-F). As predicted by TANGO (Figure 3B and Table 2), the FV → YT replacement influenced solubility of the TSAd SH2 domain produced by TSAd-90-188-PAAS to a lesser extent, and had no effect in the TSAd-90-188-PHRD construct (Figure 3D-F). Compared to the original TSAd-SH2 construct (1-TD) which consistently yielded 0,15 mg/l bacterial culture (Figure 2D and 3E), the yield of soluble wild type TSAd-90-188-PAAS protein was doubled, and was further increased by 50% when mutated to GYT (Figure 3D), yielding close to 0,5 mg recombinant protein per liter bacterial cell culture. For the TSAd-90-188-PHRD protein mutated to GYT, the yield of soluble protein was doubled (i.e. 0,3 mg/l bacterial culture) compared to the original 1-TD construct (Figure 3D). When the reverse experiment was performed, replacing GYT sequence in the ALX SH2 domain with TFV, no adverse effect on solubility was observed (Figure 3G). An overview of the TANGO-predicted aggregation scores and yield of soluble protein for the various SH2 domains included in this study are given in Table 2. Thus, taken together TANGO predictions allowed for targeted mutations resulting in increased solubility of the TSAd SH2 domain in prokaryotic cells. However, the insertion of the TFV sequence alone into the ALX-SH2 domain was not enough to render this protein insoluble, indicating that TANGO will not always be able to predict solubility of recombinant proteins in *E.coli*.

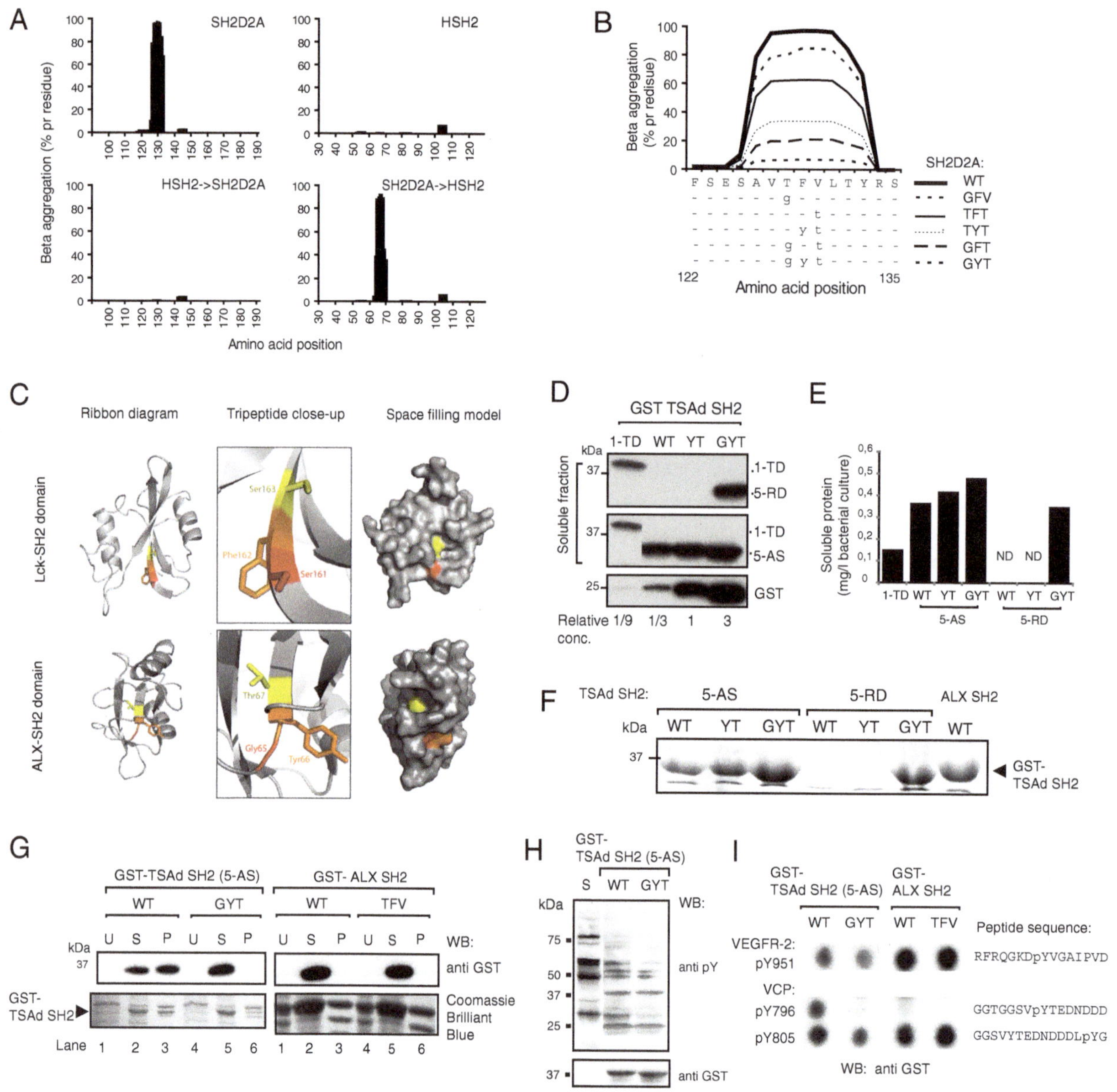

Figure 3 Beta-aggregation prediction of the SH2D2A SH2 domain. A. Amino acid positions of the indicated SH2 domains (X-axis) plotted against the corresponding beta-aggregating value (Y-axis). **B**. Beta-aggregation values of amino acids 122–135 in wild type (WT) and *in silico* mutated SH2D2A encoding TSAd. **C**. Ribbon diagrams and space filling models of the Lck SH2 domain (PDB ID: 1BHH) and the ALX SH2 domain (PDB ID: 2CS0) showing the location of Ser-Phe-Ser (Lck) and Gly-Tyr-Thr (ALX) corresponding to the TSAd TFV sequence. Models show that the side chains of Ser 161 and Ser 163 in Lck SH2 and Gly65, Tyr66 and Thr67 of ALX SH2 are solvent accessible. **D.-F**. WT and mutated SH2 domains expressed in *E.coli* at 15°C. **D**. Equal amount (5 µl) of soluble fraction from a 100 ml bacterial culture was separated by SDS-page and immunoblotted with anti-GST antibodies for quantification of soluble recombinant protein. GST = 1 equals 0,11 µg purified GST protein. One representative experiment out of three is shown. **E**. Quantitation of soluble GST-SH2 proteins by Image J analysis based on D. ND; not detected. **F**. Soluble protein captured on glutathione beads were processed as in Fig. 2B. **G**. Solubility of WT and mutated TSAd SH2 (5-AS, TSAd-90-188-PAAS) (GYT) and ALX SH2 (TFV) expressed in *E.coli*. Uninduced *E.coli* (U), soluble (S) and pellet (P) fractions were processed as in Fig. 2B. Amount of soluble and insoluble SH2 domain visualized by anti-GST immunoblotting (upper panel). Total protein content in samples was visualized by Coomassie Brilliant Blue Staining (lower panel). **H**. CD3 stimulated Jurkat cell lysates (S) and proteins pulled down with the indicated SH2 domains were separated by SDS-page and immunoblotted. **I**. Peptide array with VEGFR-2 and VCP phosphotyrosine peptides probed with the indicated SH2 domains. Bound SH2 domains detected using anti-GST antibody.

Table 2 Overview of predicted TANGO aggregation score, production yield and pull-down activity for SH2 domain constructs included in this study

Short name	Construct	TANGO predicted aggregation score	Production yield (mg/L bacterial culture)	pY-protein binding activity (pull-down)
1-TD	TSAd-67-207-IVTD	793	0,15	ND
2-TD	TSAd-90-207-IVTD	774	ND	ND
3-RD	TSAd-67-188-PHRD	791	ND	ND
4-AS	TSAd-81-193-PAAS	771	ND	ND
5-RD	TSAd-90-188-PHRD	773	—	ND
5-RD YT	TSAd-90-188-PHRD	342	—	ND
5-RD GYT	TSAd-90-188-PHRD	103	0,35	ND
5-AS	TSAd-90-188-PHRD	772	0,37	+++
5-AS YT	TSAd-90-188-PAAS	339	0,42	ND
5-AS GYT	TSAd-90-188-PAAS	102	0,48	++
mTSAd	mTSAd-111-210-AAAS	745	ND	ND
rTSAd	rTSAd-111-209-AAAS	744	ND	+++*
ALX	ALX-31-126-RPHRD	63	ND	ND
ALX TFV	ALX-31-126-RPHRD	488	ND	ND
Lck	Lck-124-228-RPHRD	24	19,46	+++*

ND, not determined; —, below detection limit; +++, strong signal; ++, signal.
*, data not shown.

Mutant GST-TSAd SH2 domain binds to tyrosine phosphorylated proteins

Pull-down assays were performed to test whether the targeted mutations negatively influenced the function of the TSAd SH2 domain. Tyrosine phosphorylation of cellular proteins was induced in Jurkat T cells by anti-CD3 triggering. Lysates from stimulated cells were subjected to pull-down assay, using wild type and GYT-mutated GST-TSAd SH2 (TSAd-90-188-PAAS). Both constructs pulled down the same tyrosine phosphorylated proteins, although smaller amounts were pulled down using the GYT- mutated protein. However, some protein bands seemed to be more affected by the GYT mutation than others (Figure 3H). To further analyze the binding specificity of the mutated TSAd SH2 domain, peptide arrays representing phosphopeptides from VEGFR-2 and VCP, the two known TSAd SH2 domain ligands, were probed with wild type or mutated TSAd SH2 (TSAd-90-188-PAAS) domain. The results show that both TSAd wild type and GYT-mutated protein bind to VEGFR-2 phosphopeptide (pY951) and VCP (pY805), whereas binding to VCP pY796 was lost for the GYT-mutant (Figure 3I). The GYT → TFV replacement in ALX SH2 domain did not alter the specificity towards the selected phosphopeptides (Figure 3I). Taken together, the results show that although the mutant protein may have retained all or most of the assessed functions of the wild type protein, the specificity determining elements of the mutated protein domain may be altered. This should be carefully addressed when mutations are introduced for the purpose of increasing the solubility of the expressed protein.

Low beta-aggregation values predict high solubility of recombinant SH2 domains

Having found that TANGO could predict increased solubility of the TSAd-SH2 domain, we then asked whether beta-aggregation propensity is a common characteristic of SH2 domains. Most of the 120 SH2 domains encoded by the human genome have been expressed as recombinant His- [17] or GST-fusion proteins [18,19]. All available human SH2 domain sequences [2] were analysed by TANGO (Additional file 2: Table S1) and results were correlated to the recently published information on SH2 solubility. SH2 domains displaying low solubility had significantly higher beta-aggregation values than SH2 domains which were soluble [18] (Figure 4A) and [19] (data not shown). Similarly, the percentage of recombinant SH2 domains present as monomers (an indicator of protein solubility) as evaluated by analytical size exclusion chromatography [17], was inversely correlated with high beta-aggregation values (Figure 4B).

Discussion

Expression of recombinant proteins is an important step forward for structural and functional analysis of entire proteins and isolated protein domains. The choice of tag may affect the solubility of the protein [20] and various parameters for successful expression of heterologous

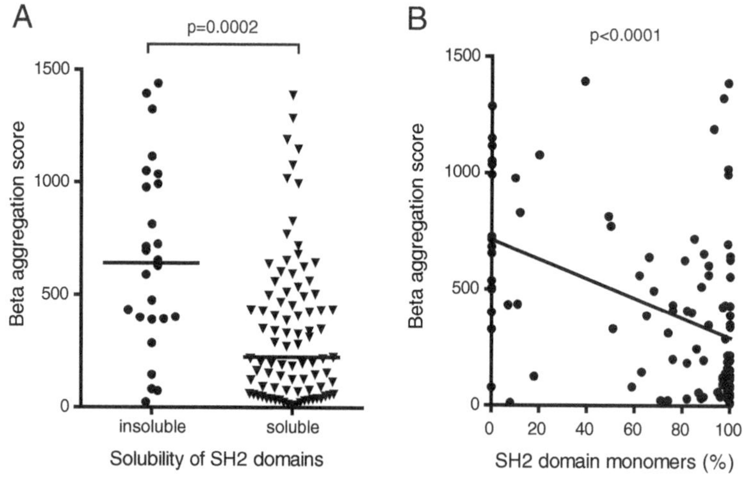

Figure 4 Solubility of SH2 domains can be predicted by TANGO. A. Median beta aggregation values (Y axis) of SH2 domains grouped according to solubility as reported by [19] (Mann-Whithney test $p = 0.0002$) **B**. Median beta aggregation values (+ SD) of SH2 domains plotted against corresponding reported fraction of monomers (%) [17]. Line is fitted with linear regression, using GraphPad Prism ($R^2 = 0.19$, $p < 0.0001$).

recombinant proteins have previously been reported [15,21]. However, there is no particular way to predict whether a protein will be soluble when expressed in *E.coli*. Having an interest in elucidating the function of the SH2 domain containing protein TSAd, we discovered that its SH2 domain was mainly insoluble when expressed in *E.coli*, as was also noted by others [17,18] but not all [19]. Although alteration in flanking sequence and growth temperature influenced expression and solubility of TSAd SH2, overall yield of soluble protein remained low.

Protein aggregation and amyloid fibril formation in degenerative diseases such as Alzheimer's disease resembles the process of inclusion body formation in *E.coli* [14]. To our knowledge targeted mutation to relieve propensity for intermolecular beta-aggregation and increase solubility of recombinant proteins has not previously been reported, although the opposite experiment has been performed: de Groot and Ventura fused green fluorescent protein (GFP) to a mutated Aβ42 Alzheimer peptide to generate an aggregation prone reporter for monitoring protein quality in inclusion bodies by measuring GFP emission spectra [22]. Our results show that identification of aggregation prone proteins *in silico* allow for targeted mutations to increase solubility of the recombinant protein. Provided the overall structure and function of the recombinant protein are retained, this is a potentially efficient strategy to generate soluble recombinant protein for downstream purposes.

Aggregating peptide sequences are common within the hydrophobic core of globular proteins [23]. Accordingly, the beta-aggregating sequence identified by TANGO in the TSAd-SH2 domain (and also in many of the other SH2 domains with high beta-aggregating propensity (not shown)) was found in βC located in the core of the SH2 domain. Histidines, arginines or prolines are preferred immediately prior to, and after, aggregating sequences in the proteome. These gatekeeper amino acids may counteract beta-aggregating propensity, and may also serve as recognition sites for chaperones [24]. Although SH2 domains share overall structure, there is large sequence variability between individual domains of the family [2]. Interestingly, although most SH2 domains did not harbour aggregating βC sequences, the large majority still had histidine, arginine or proline flanking the βC sequence on each side (see alignment in [2]). This suggests that protein domain families may have evolved to resist the potential deleterious effects of beta-aggregating propensity of globular proteins.

Conclusions

We have found that TANGO, an algorithm developed to predict amyloid fibril formation and aggregation of peptides, may also predict the solubility of recombinant SH2 domains expressed in *E.coli*. Predictions made by TANGO may allow for targeted mutations that significantly improve the solubility of the recombinant protein. Mutations introduced for the purpose of increasing the solubility of the expressed protein may alter specificity determining elements of the mutated protein domain.

Methods

Plasmid constructs

TSAd SH2, Lck SH2 and ALX SH2 encoding DNA were isolated from intact cDNA clones [9,16,25] by *in vitro* amplification, and cloned into the *BamHI/NotI* or *EcoR/NotI* site of the bacterial expression vector pGEX-6P-1 (GE Healthcare). The aa sequence alignment of the TSAd SH2 domains encoded by human, mouse and rat *SH2D2A* genes as well as the human Lck SH2 and ALX

SH2 domains are shown in Figure 1. Additionally, Figure 1. includes the predicted secondary structures of these SH2 domains with nomenclature as defined in [3].The different constructs made are shown in Table 1 and are schematically represented in Figure 2A. Mutagenesis was performed using Quickchange (Clontech). All constructs were verified by sequencing.

Protein expression

Glutathion S-transferase (GST) SH2 domain fusion proteins were made by isolating single clones of *E.coli* BL21-CodonPlus-RP (Stratagene,) transformed with appropriate cDNA constructs. Colonies were grown in LB medium at 37°C or in M9 minimal salt medium at 25°C to OD_{600} = 0.6-0.8, prior to addition of isopropyl-beta-D-thiogalactoside (IPTG) (Sigma-Aldrich) to a final concentration of 0.2 mM followed by incubation at room temperature or 15°C for 4–16 hours. Cultures were harvested by centrifugation, lysed in PBS with 0,5 mg/ml lysozyme followed by one freeze-thaw cycle (−80°C over night) and further lysis in GST-lysis buffer (1% Triton X-100, 1 mM DTT, 1 mM PMSF, 0,5 mg/ml lysozyme, 5 μg/ml DNAse, PBS pH 7.4) for ~16 h at 4°C with agitation. Following centrifugation, soluble and insoluble (pellet) fractions were separated. The pellet was resuspended in an equal volume of PBS pH 7.4. Soluble and pellet fractions were analysed by SDS-PAGE followed by Coomassie Brilliant Blue staining or by western blotting using antibodies specific for GST (clone B14) (Santa Cruz Biotechnology, Santa Cruz, CA). Signals were developed using horse peroxidase-conjugated goat anti-mouse IgG (Jackson ImmunoResearch Laboratories, West Grove, PA) followed by Super Signal® west Pico Stable Peroxide Solution (Pierce, Rockford, IL).

Protein purification

Protein was purified in bulk from lysates using Glutathione-Sepharose 4B beads (GE Healthcare) as described by the manufacturer. Beads were stored at 4°C in PBS pH 7.4. Expression level and purity of protein was examined by SDS-PAGE. Protein bands were directly visualised by staining of the gel with Coomassie Brilliant Blue.

Antibodies, cell stimulation, and pulldown experiments

The following mAb were used; anti-human CD3ε (OKT3, American Type Culture collection, Manassas, VA), anti-GST (B-14,Santa Cruz Biotechnology, Santa Cruz, CA), anti-phosphotyrosine (anti-pY) (clone 4G10, Upstate Biotechnology, Lake Placid, NY), and horseradish peroxidase-conjugated goat-anti mouse IgG mAb (Sigma-Aldrich). Pull-down experiments using anti-CD3 stimulated Jurkat E6.1 cells (American Type Culture Collection) and GST-TSAd SH2 fusion proteins on glutathione sepharose beads were performed as previously described [9].

Semi quantitative assessment of GST-fusion proteins

For semi-quantitative assessment of soluble GST-fusion proteins in bacterial lysate, GST was expressed from expression vectors without insert, purified in bulk as described above and then eluted from glutathione beads using glutathione as described by the manufacturer. The eluted protein was quantified using NanoDrop 2000c spectrophotometry (Thermo scientific, Wilmington, DE). A standard curve of 3 fold dilutions of the GST protein was included on the SDS-PAGE and the immunoblot signals from anti-GST (clone B14) (Santa Cruz Biotechnology, Santa Cruz, CA) were visually compared. Additionally, the amount of soluble protein was assessed by signal intensity meassurements using the image processing ImageJ software (http://rsb.info.nih.gov/ij/index.html). At least two independent membranes were examined for each quantitation.

Peptide spot array analysis

Peptide arrays were synthesized on nitrocellulose membranes using a MultiPep automated peptide synthesizer (INTAVIS Bioanalytical Instruments AG, Germany) as described [26,27]. Membranes were probed with GST-tagged SH2 fusion proteins (80 μg/ml), and signals were detected using relevant antibodies followed by chemiluminescent detection.

Prediction of protein aggregation

Prediction of aggregating regions of unfolded polypeptide chains were performed using the algorithm TANGO (http://tango.crg.es), using the default conditions pH 7.0, 25°C and 0.02 mM ionic strength [13,23,28].

Additional files

Additional file 1: Figure S1. Yield and solubility of mutated TSAd-SH2 domains. WT and mutated SH2 domains as indicated were expressed in E.coli at 15°C. Soluble and pellet fractions were separated by 10% SDS-PAGE. Proteins were visualised by Coomassie Brilliant Blue staining.

Additional file 2: Table S1. SH2 domains analysed by TANGO.

Abbreviations

aa: Amino acid; TSAd: T cell specific adapter protein; SH2: Src homology 2; mAb: Monoclonal antibody.

Competing interests

The authors declare no financial or non-financial competing interests.

Authors' contributions

TCBA participated in the design of the study, did much of the laboratory work, analyzed the data and wrote the paper. KL did part of the laboratory work and analyzed the data. CDH did part of the laboratory work and analyzed the data, participated in the writing of the paper. LK did part of the laboratory work and analyzed the data. PEK participated in the design of the study and analyzed the data. LS participated in the design of the study. AHA participated in the design of the study and the analysis of data. ASP conceived of the study, analyzed data and participated in the writing of the paper. All authors have revised the manuscript critically for important intellectual content and have approved of the final version of the manuscript.

Acknowledgements
This study was supported by grants from the University of Oslo EMBIO, the Norwegian Research Council, Novo Nordisk, and the Norwegian Cancer Society. The technical assistance of Bjørg Simonsen and Christine Regnell is highly appreciated.

Author details
[1]Department of Anatomy, Institute of Basal Medical Sciences, University of Oslo, Oslo, Norway. [2]Department of Biosciences, University of Oslo, Oslo, Norway. [3]Department of Chemistry, Biochemistry and Food science, Norwegian University of Life Sciences, Ås, Norway. [4]Department of Biochemistry, Biophysics and Molecular Biology, Iowa State University, Ames, Iowa, USA.

References

1. Songyang Z, Shoelson SE, Chaudhuri M, Gish G, Pawson T, Haser WG, King F, Roberts T, Ratnofsky S, Lechleider RJ: **SH2 domains recognize specific phosphopeptide sequences.** *Cell* 1993, **72**:767–778.
2. Liu BA, Jablonowski K, Raina M, Arce M, Pawson T, Nash PD: **The human and mouse complement of SH2 domain proteins-establishing the boundaries of phosphotyrosine signaling.** *Mol Cell* 2006, **22**:851–868.
3. Kuriyan J, Cowburn D: **Modular peptide recognition domains in eukaryotic signaling.** *Annu Rev Biophys Biomol Struct* 1997, **26**:259–288.
4. Mallis RJ, Brazin KN, Fulton DB, Andreotti AH: **Structural characterization of a proline-driven conformational switch within the Itk SH2 domain.** *Nat Struct Biol* 2002, **9**:900–905.
5. Hwang PM, Li C, Morra M, Lillywhite J, Muhandiram DR, Gertler F, Terhorst C, Kay LE, Pawson T, Forman-Kay JD, Li SC: **A "three-pronged" binding mechanism for the SAP/SH2D1A SH2 domain: structural basis and relevance to the XLP syndrome.** *EMBO J* 2002, **21**:314–323.
6. Mills JE, Whitford PC, Shaffer J, Onuchic JN, Adams JA, Jennings PA: **A novel disulfide bond in the SH2 Domain of the C-terminal Src kinase controls catalytic activity.** *J Mol Biol* 2007, **365**:1460–1468.
7. Sundvold V, Torgersen KM, Post NH, Marti F, King PD, Rottingen JA, Spurkland A, Lea T: **T cell-specific adapter protein inhibits T cell activation by modulating Lck activity.** *J Immunol* 2000, **165**:2927–2931.
8. Matsumoto T, Bohman S, Dixelius J, Berge T, Dimberg A, Magnusson P, Wang L, Wikner C, Qi JH, Wernstedt C, Wu J, Bruheim S, Mugishima H, Mukhopadhyay D, Spurkland A, Claesson-Welsh L: **VEGF receptor-2 Y951 signaling and a role for the adapter molecule TSAd in tumor angiogenesis.** *EMBO J* 2005, **24**:2342–2353.
9. Granum S, Sundvold-Gjerstad V, Dai KZ, Kolltveit KM, Hildebrand K, Huitfeldt HS, Lea T, Spurkland A: **Structure function analysis of *SH2D2A* isoforms expressed in T cells reveals a crucial role for the proline rich region encoded by *SH2D2A* exon 7.** *BMC Immunol* 2006, **7**:15.
10. Wu LW, Mayo LD, Dunbar JD, Kessler KM, Ozes ON, Warren RS, Donner DB: **VRAP Is an Adaptor Protein That Binds KDR, a Receptor for Vascular Endothelial CellΓÇëGrowthΓÇëFactor.** *J Biol Chem* 2000, **275**:6059–6062.
11. Marti F, King PD: **The p95-100 kDa ligand of the T cell-specific adaptor (TSAd) protein Src-homology-2 (SH2) domain implicated in TSAd nuclear import is p97 Valosin-containing protein (VCP).** *Immunol Lett* 2005, **97**:235–243.
12. Sun Y, Li X, Massena S, Kutschera S, Padhan N, Gualandi L, Sundvold-Gjerstad V, Gustavsson K, Choy WW, Zang G, Quach M, Jansson L, Phillipson M, Abid MR, Spurkland A, Claesson-Welsh L: **VEGFR2 – TSAd – c-Src signaling regulates vascular permeability *in vivo*.** *J Ex Med* 2012, **209**:1363–1377.
13. Fernandez-Escamilla AM, Rousseau F, Schymkowitz J, Serrano L: **Prediction of sequence-dependent and mutational effects on the aggregation of peptides and proteins.** *Nat Biotechnol* 2004, **22**:1302–1306.
14. Ventura S: **Sequence determinants of protein aggregation: tools to increase protein solubility.** *Microb Cell Fact* 2005, **4**:11.
15. Sorensen HP, Mortensen KK: **Soluble expression of recombinant proteins in the cytoplasm of Escherichia coli.** *Microb Cell Fact* 2005, **4**:1.
16. Greene TA, Powell P, Nzerem C, Shapiro MJ, Shapiro VS: **Cloning and characterization of ALX, an adaptor downstream of CD28.** *J Biol Chem* 2003, **278**:45128–45134.
17. Jones RB, Gordus A, Krall JA, MacBeath G: **A quantitative protein interaction network for the ErbB receptors using protein microarrays.** *Nature* 2006, **439**:168–174.
18. Machida K, Thompson CM, Dierck K, Jablonowski K, Karkkainen S, Liu B, Zhang H, Nash PD, Newman DK, Nollau P, Pawson T, Renkema GH, Saksela K, Schiller MR, Shin DG, Mayer BJ: **High-throughput phosphotyrosine profiling using SH2 domains.** *Mol Cell* 2007, **26**:899–915.
19. Huang H, Li L, Wu C, Schibli D, Colwill K, Ma S, Li C, Roy P, Ho K, Songyang Z, Pawson T, Gao Y, Li SS: **Defining the specificity space of the human SRC homology 2 domain.** *Mol Cell Proteomics* 2008, **7**:768–784.
20. Makrides SC: **Strategies for achieving high-level expression of genes in Escherichia coli.** *Microbiol Rev* 1996, **60**:512–538.
21. Murby M, Uhlen M, Stahl S: **Upstream strategies to minimize proteolytic degradation upon recombinant production in Escherichia coli.** *Protein Expr Purif* 1996, **7**:129–136.
22. de Groot NS, Ventura S: **Protein activity in bacterial inclusion bodies correlates with predicted aggregation rates.** *J Biotechnol* 2006, **125**:110–113.
23. Linding R, Schymkowitz J, Rousseau F, Diella F, Serrano L: **A comparative study of the relationship between protein structure and beta-aggregation in globular and intrinsically disordered proteins.** *J Mol Biol* 2004, **342**:345–353.
24. Rousseau F, Serrano L, Schymkowitz JW: **How evolutionary pressure against protein aggregation shaped chaperone specificity.** *J Mol Biol* 2006, **355**:1037–1047.
25. Nejad S, Bryceson Y, Dissen E, Sundvold-Gjerstad V, Naper C, Rolstad B, Spurkland A, Vaage JT: **cDNA cloning of a rat orthologue of SH2D2A encoding T-cell-specific adaptor protein (TSAd): expression in T and NK cells.** *Immunogenetics* 2004, **56**:338–342.
26. Kramer A, Schneider-Mergener J: **Synthesis and screening of peptide libraries on continuous cellulose membrane supports.** *Methods Mol Biol* 1998, **87**:25–39.
27. Frank R: **High-density synthetic peptide microarrays: emerging tools for functional genomics and proteomics.** *Comb Chem High Throughput Screen* 2002, **5**:429–440.
28. Rousseau F, Schymkowitz J, Serrano L: **Protein aggregation and amyloidosis: confusion of the kinds?** *Curr Opin Struct Biol* 2006, **16**:118–126.

4

Enhancement of fermentable sugar yield by competitive adsorption of non-enzymatic substances from yeast and cellulase on lignin

Yong Tang[1,2], Fuhou Lei[2], Carrasco Cristhian[3], Zuguang Liu[2], Hailong Yu[1] and Jianxin Jiang[1*]

Abstract

Background: Enhancement of enzymatic digestibility by some supplementations could reduce enzyme loading and cost, which is still too high to realize economical production of lignocellulosic biofuels. A recent study indicates that yeast hydrolysates (YH) have improved the efficiency of cellulases on digestibility of furfural residues (FR). In the current work, the components of YH were separated by centrifugation and size exclusion chromatography and finally characterized in order to better understand this positive effect.

Results: A 60.8% of nitrogen of yeast cells was remained in the slurry (YHS) after hydrothermal treatment. In the supernatant of YH (YHL), substances of high molecular weight were identified as proteins and other UV-absorbing compounds, which showed close molecular weight to components of cellulases. Those substances attributed to a synergetic positive effect on enzymatic hydrolysis of FR. The fraction of YHL ranged from 1.19 to 2.19 mL (elution volume) contained over 50% of proteins in YHL and had the best performance in stimulating the release of glucose. Experiment results proved the adsorption of proteins in YHL on lignin.

Conclusions: Supplementation of cellulases with YH enhances enzymatic digestibility of FR mainly by a competitive adsorption of non-enzymatic substances on lignin. The molecular weight of these substances has a significant impact on their performance. Different strategies can be used for a good utilization of yeast cells in terms of biorefinery concept.

Keywords: Cellulases, Enzymatic hydrolysis, Yeast hydrolysate, Competitive adsorption

Background

Expected shortage of fossil fuels and the concern for climate change has increased the interest for the production of fuels and chemicals from renewable sources. Ethanol is already being produced by the fermentation of sucrose (from sugarcane or sugar beets) or starch (from corn or wheat) carbohydrate-rich substances [1]. However, to meet increasing demands for fuel, and to avoid conflict between food and energy production, it will be necessary to use other kinds of biomass, and in this context, lignocellulosic materials have attracted interest as sources of fermentable sugars [2]. Furfural residue (FR) is an industrial waste in China. The corncobs are heated at high temperature ranged from 170°C to 185°C under acidic conditions to hydrolyze arabinoxylans into xylose, and then the xylose are converted into furfural. The cellulose and lignin in the cobs are relatively stable under these conditions, so the residues left over after the furfural production are enriched in cellulose (about 45%), which can be easily hydrolyzed by cellulases. Currently, FR is only recycled as boiler fuel, but a potential use could be as raw materials for biofuel production [3].

The utilization of lignocellulosic feedstocks for biofuel production requires the deconstruction of their polysaccharides to free sugars, mainly by acid and enzymatic hydrolysis. Enzymatic biodegradation is one of the most promising options for production of lignocellulosic biofuels as mild and environmentally friendly process. However at the same time, the lignin-carbohydrate complexes pose much challenge to decrease enzyme dosage in

* Correspondence: jiangjx@bjfu.edu.cn
[1]Department of Chemistry and Chemical Engineering, Beijing Forestry University, Beijing, China
Full list of author information is available at the end of the article

lignocellulose digestibility, while reducing the cost of ethanol production [4-6]. Other properties, including surface properties and crystallinity, impede enzymatic hydrolysis, which has been investigated in the literature recently [7].

It is well know that pretreatment is required to improve enzymatic digestibility on lignocellulose [2]. Previous studies have been showed that post-pretreatments and addition of some supplements increase the hydrophilicity of lignin [8-10]. Some examples are the hemicellulase supplementations which stimulate glucose release by depolymerisation of hemicelluloses [11,12]. Surfactants enhance release of free glucose by reducing the interaction between enzymes and lignin [13]. Non-protein surfactants contain synthetic surfactants as Tween-20, Tween-80, PEG 4000 and PEG 6000 [13-15], and natural surfactants as *Gleditsia* saponins [16]. Bovine serum albumin (BSA) is a protein surfactant and its enhancement is mainly due to surface activity than catalytic activities [11].

As the main microorganism for bioethanol production, yeast contains about 8% of nitrogen (equivalent to 50% of protein). Being rich in proteins attribute to that co-product of 1G-ethanol industry could be marketed as animal fodder and the feasibility of the use of spent cells as nutrient sources [17]. Occasionally, it was found that yeast hydrolysates (YH), obtained by hydrothermal treatment, can increase enzymatic digestibility of furfural residues (FR) [18]. In the current work, YH was used as a cheap supplementation of commercial cellulases and the components of YH were separated by centrifugation and size exclusion chromatography and then characterized. The aim of this study is to better understand hydrothermal treatment process of yeast, find out the mechanism for YH enhancement of enzymatic digestibility and evaluate the absorption capacity of YHL protein on lignin.

Results

Hydrothermal treatment of yeast

After centrifugation, 159 g of YH was separated into two phases, 132.3 g of supernatant (YHL, 83.8%) and 24.6 g of slurry (YHS, 16.2%) (Figure 1). The total mass loss was about 1.4% after hydrothermal treatment and centrifugation. The dry matter content of YHS and dissolved matter content (DMC) of YHL were 23.5 and 2.07%, respectively. As seen in Figure 1, the nitrogen content of YHS was 7.25% and the protein concentration of YHL was 0.75 g/L, accounting for 2.28% of total yeast nitrogen. One explanation for the relative small nitrogen in YHL could be that there is nitrogen loss during hydrothermal treatment and some nitrogen may not be detected by the Bradford method. Soluble sugars were not detected in the YHL according to the HPLC analysis. The matter recovery after hydrothermal treatment and centrifugation was 94.6%.

The effect of YHS on enzymatic hydrolysis and the relative roles of pH control and YHL addition

For comparison, the addition of YHS and YHL in the FR was tested in hydrolysis experiments. YHL and YHS were separated by centrifugation, followed by being added to hydrolysis. The yield of enzymatic hydrolysis was calculated based on FR cellulose (Figure 2). Based on the datas of enzymatic hydrolysis of YH [18], when the YHS was supplemented for the FR hydrolysis, the glucose yield from FR was calculated to be 23.1%, which was lower than that without YHS and YHL (31.9%) at neutral pH (Figure 2). At neutral pH, the glucose yield with YHL (71.6%) was two times higher than that without YHS and YHL (31.9%) (Figure 2). Thus, these results could support that YHS had almost no effect on the digestibility [18].

The pH and YHL addition are likely to have an effect on the enzymatic hydrolysis of FR. In general, the optimized pH of enzymatic hydrolysis is well recognized to be 4.8. When neither of YHL and YHS was added, controlling pH at 4.8 increased the glucose yield from 31.9 to 67.2%. The addition of YHL shortened the gap between the glucose yields at different pH values (Figure 2). For instance, the yield of hydrolysis at neutral pH with YHL was higher than that at pH 4.8 without YHL. These results indicated that the interference of lignin on enzymes had much higher impact on enzymatic hydrolysis than non-optimum pH. The enzyme activity was not detected in YHL. When YHL was added to control experiments that use Cellulolytic Enzyme Lignin as sole substrate, Filter paper activity (FPU) in the liquid fraction samples at 4, 24 and 72 h were increased by 0.1%, 6.6% and 5.7%, respectively, indicating that YHL is a protein surfactant and its enhancement is mainly due to surface activity than catalytic activities [11,13].

Comparison of molecular weight distribution between YHL and cellulases

As seen in Figure 3, two similar peaks were presented in elution curves (UV at 280 nm) of Celluclast 1.5 L and YHL, indicating that the components of cellulase and YHL showed a close molecular weight. It should be noted that the conductivity curve of Celluclast 1.5 L had a positive peak in retention time of 1.67 min (accumulated volume 5 mL) while an adverse peak was found for YHL. The elution curve of Novozym 188 was similar to that of Celluclast 1.5 L except for a straight line of conductivity curve (data not shown).

Protein and mass distribution between fractions of YHL

Fractions of YHL were collected in 32 tubes separately (Figure 1). The protein concentration of fraction in each tube was detected with the Bradford method, except a relatively small fraction, 0.15 mL, in tube 4 (Figure 1). Fraction in tube 2 contained 96.97 mg/L protein, accounting for

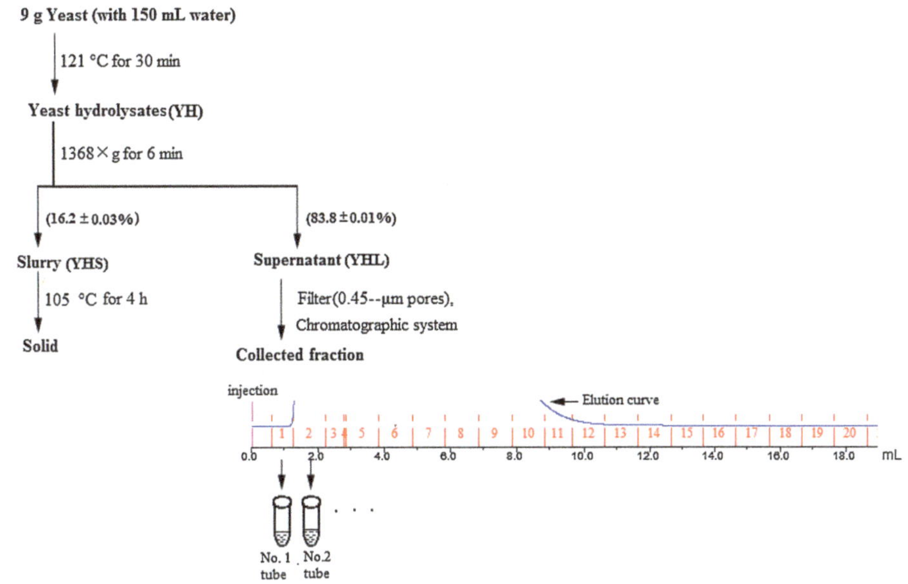

Figure 1 Scheme for separating components of yeast hydrolysate (YH).

about 51.8% of sample proteins. The protein concentration in tube 3 was out of linear range of the Bradford method and proteins were not detected in other tubes. Figure 4 shows the relative purity of protein in YHL at different UV elution curves (254 and 280 nm) [19,20]. The lowest A254/A280, about 1, was found at the top of peak 1 (elution volume of 1.48 mL), meaning that the purity of protein in tube 2 was the highest (Figure 4). These results were consistent with protein concentration by the Bradford method. A254/A280 increased with increasing mobile phase volume from 2.11 to 6.25 mL, indicating that the fractions in tube 5 and 6 contained certain amount of other UV-absorbing compounds (Figures 4 and 5). Those compounds are inferred to attribute to the adverse conductivity peak in Figure 3. However the molecular weight of proteins and other UV-absorbing compounds in tube 2, 5 and 6 showed the close molecular weight to components of Celluclast 1.5 L and Novozym 188.

DMC was detected to investigate the mass distribution between fractions of YHL as UV detector cannot determine substances without ultraviolet absorption (Figure 5). DMC of sodium citrate buffer is theoretically calculated to be 25.54 mg/mL. In practice, it was 21.68 mg/mL, showed as dotted line in Figure 5. One explanation for this could be that some reactions during the oven-dried process (105°C) decrease the value. Figure 5 shows that the DMC of fractions in tube 2, 5, 15, 21 and 30 were higher than 21.68 mg/mL. A close DMC was found between fractions in tubes 9, 13 and 27 and sodium citrate buffer, indicating that the minimum components of YHL were transferred into those tubes. Results demonstrated that substances with different molecular weight were formed during hydrothermal treatment and those of high molecular weight were predominating.

The effect of YHL fractions with different molecular weight on enzymatic hydrolysis

The effect of fractions in tubes 2, 5, 6, 10, 15, 21 and 30 on enzymatic hydrolysis were investigated at a relatively high DMC (Figure 5). As seen in Figure 6, the glucan digestibility was improved in tubes 2, 5 and 6 at 24 and 48 h. Fractions in tube 2 had a better performance than fractions between tubes 5 and 6. The highest concentration of protein is supposed to be the reason for the best performance of fraction in tube 2 [3]. After 72 h, the glucose yield fraction in tube 5 seems not be improved, probably due to that enzyme effectiveness dropped over time by heat-induced denaturation [21,22]. No significant effect of fractions eluting after 7.5 mL (tube 10, 15, 21 and 30) on hydrolysis was observed (Figure 6). The molecular weight of substances in fractions is supposed to have an effect on their performance. Hydrolysis with the whole YHL obtained the highest yield, indicating that there were a synergy between fractions in tube 2, 5 and 6. However when the amount of YHL was decreased from 12.6 g (Figure 2) to 6.29 g (Figure 6), the hydrolysis efficiency decreased slightly, which is consistent with the previous study [18].

Competitive adsorption of non-enzymatic protein from YHL and cellulases on lignin

After hydrolysis of 3% FR, the nitrogen content of the dried residues was 0.84%. For that with YHL, it was 1.27%. There was 3.61 gram of liquid per gram of dried residue in the slurry. When YHL were used in enzymatic hydrolysis, the nitrogen in the residual liquid should be deducted to

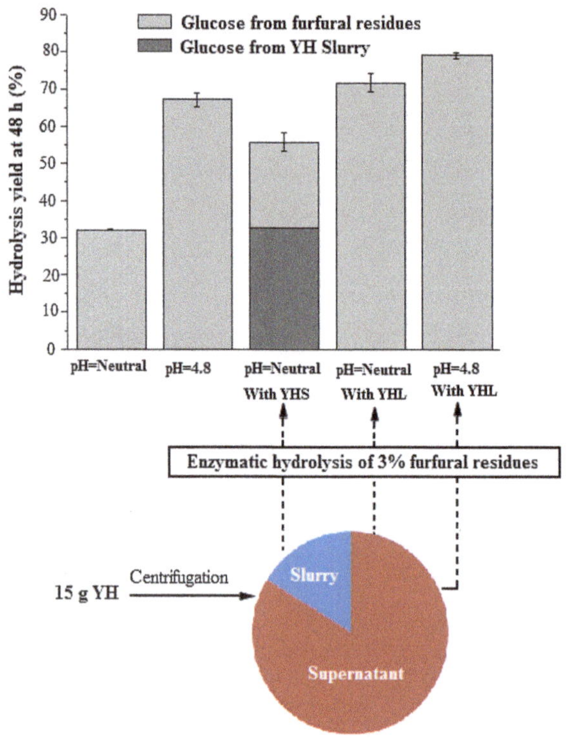

Figure 2 The effect of 2.43 g (YHS) and 12.6 g (YHL) on enzymatic hydrolysis of 3% furfural residues at 45°C.

determine how many proteins were absorbed on lignin. Assuming that all nitrogen from yeast distributes in YHL and YHS and there is no nitrogen loss during hydrothermal treatment, the maximum nitrogen content of YHL is 2.05 mg/mL based on the total nitrogen of yeast and the amount of nitrogen of YHS. When YHL is diluted by 2 times in enzymatic hydrolysis (15 g YH were added to enzymatic hydrolysis with a total working weight of 30 g), the maximum nitrogen content in liquid phase is 1.03 mg/g hydrolysate. A 3.61 gram of hydrolysate contained 3.72 mg of nitrogen. The amount of nitrogen absorbed on lignin was indirectly estimated by subtracting the amount of free nitrogen in the residual hydrolysate. More than 0.63 mg nitrogen from YHL (12.7-8.35-3.72 = 0.63 mg) was absorbed on 1 gram of lignin. Competitive adsorption of non-enzymatic proteins and cellulases on lignin is probably a good reason for the enhancement of glucose yield [4,13].

Discussion

The hemicellulase addition is not suitable for hydrolysis of materials that contain a little hemicellulose content as FR. The supplementations of commercial cellulases with other enzymes as hemicellulases that contribute novel catalytic activities are simple methods to enhance the fermentable sugar yields. Lignin controlled cellulose accessibility by restricting xylan accessibility to some extent, so hemicellulases promote enzymatic hydrolysis of cellulose by depolymerisation of hemicelluloses [11]. However, the pretreatment severity, fermentation mode and substrate composition would affect the efficiency of hemicellulase addition [12]. The xylosidase enhancement was not a general protein effect because BSA at similar concentration did not increase hydrolysis yields from corn stover [11]. The xylanases supplementation was observed to be only positive effect in SHF (separate hydrolysis and fermentation) but not in SSF (Simultaneous saccharification and fermentation) mode [5].

Lignin poses a bigger challenge to low enzyme dosage during the enzymatic hydrolysis process than hemicelluloses. Two strategies, including increasing the hydrophilicity

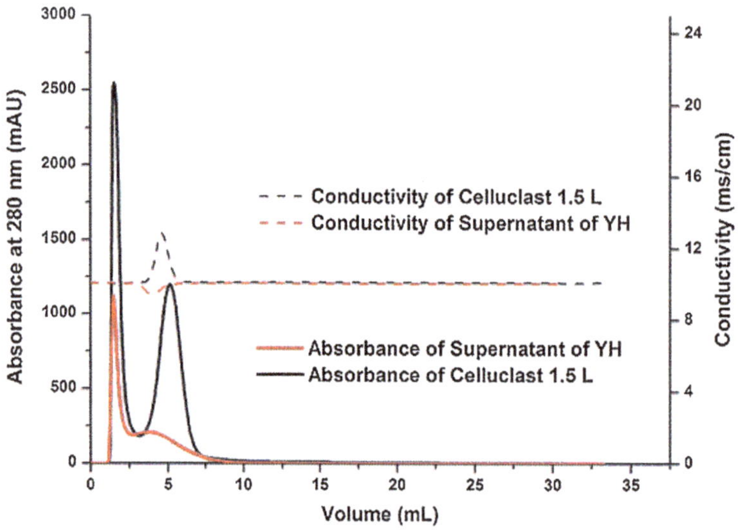

Figure 3 Comparison of the molecular weight distribution of cellulases and YHL.

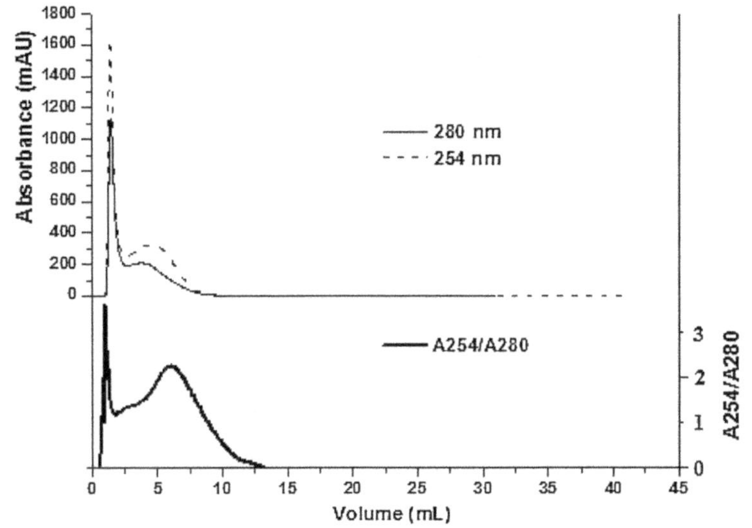

Figure 4 Comparison of elution curves of YHL at 254 and 280 nm.

of lignin and removing lignin by post-treatment, have been employed in reducing the interference of lignin on enzymes [8,15]. When different substrates are used, quantifying the relative roles that lignin might play in either binding the enzyme or limiting the swelling of the cellulose is helpful to decide which methods should be used [4]. Besides the end product type affects the severity of post-treatment and the degree of delignification [23]. The surface active additives are the most common supplementations used during hydrolysis and pretreatment process [15]. According to a previous study, addition of 2.5-5.0 mg/mL (123–245 mg/g of the substrate) of BSA was found to be a highest digestibility in pretreated softwood [4]. Other study showed an optimum BSA concentration of 25 mg/g DM using wheat straw [13]. Both studies suggested that BSA to reduce the adsorption of cellulases as surfactant on enzymatic hydrolysis. For instance, the maximum adsorption capacities of pretreated corn stover were ranged from 38.7 to 67.5 mg/g cellulose (corresponding to 77.8-124.6 mg/g substrates) [21]. In this study, the protein concentration used in this study ranged from 0 to11.77 mg/g FR. Since non-protein substances (fractions in tube 5 and 6) could enhance the sugar yield simultaneously, it is inferred that the potential sites in FR were not occupied by proteins completely, meaning that the optimum protein range of FR would

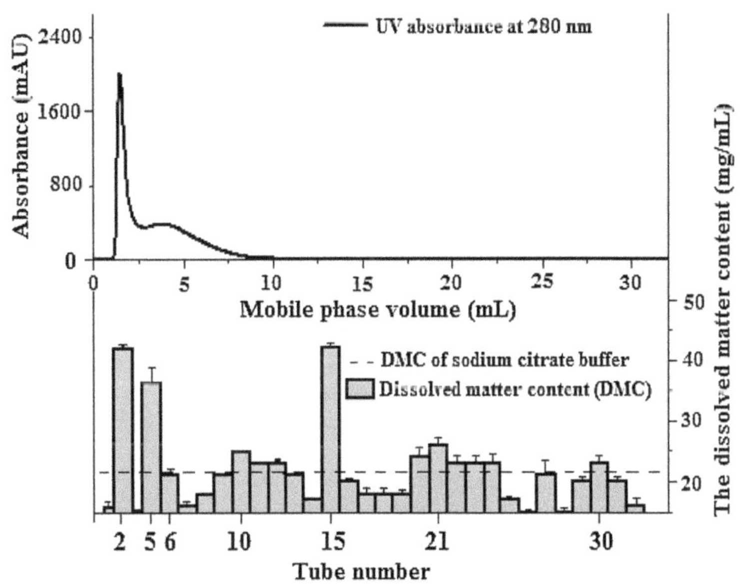

Figure 5 The total dissolved mater content (DMC) of fractions of YHL.

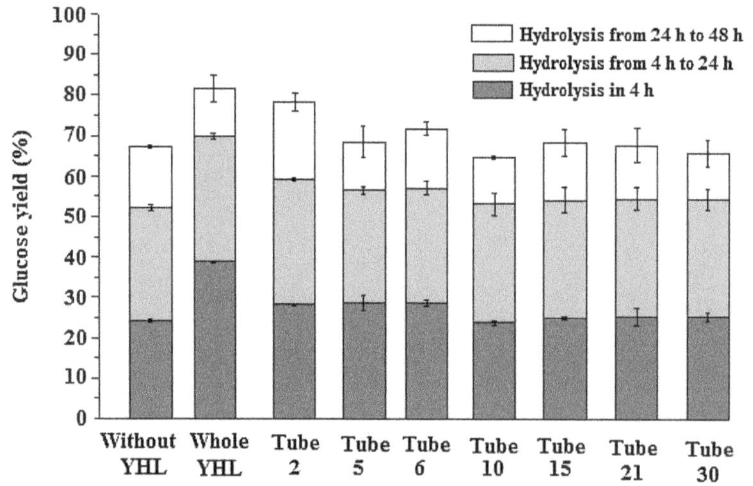

Figure 6 The effect of fractions of 6.29 g YHL on enzymatic hydrolysis of 3% of furfural residues at 45°C.

be higher than 11.77 mg/g FR. However, although these sites were not all associated with proteins, further additions increased hydrolysis yield slightly, indicating that non-protein substances played an important role in enhancement of sugar yield. Interestingly, it has been found that the levelling off concentration of the surface active performance of YHL was about 30 g/L (15 g YH were added to enzymatic hydrolysis with a working weight of 30 g) [18]. Possibly, it is difficult to improve the performance of YHL further only by changing the severity of hydrothermal treatment. The use of diluted acid can improve the performance of YH as nutrients but may decrease the molecular weight of product, thus resulting in losing the function of enhancement of enzymatic hydrolysis [17].

This study showed that fermentation residues from bioethanol process can be reused to produce YHL or proteins that can improve enzymatic hydrolysis. The process had such advantages that there is no need to dry YHL and to deal with storage and transport, thus resulting in less production cost. When byproduct of dry-mill starch to ethanol production is used, the majority of yeast nitrogen will be dissolved in YHS, thus indicating that the feasibility of marketing YHS as animal fodder. Other studies gave methods to utilize carbohydrates in YHS [3-27]. Integration process of 2G bioethanol with 1G has been proved to be competitive with 1G bioethanol [25]. This study could provide a potential way to utilize the fermentation residues, which is difficult to be marketed as animal fodder [3].

The YHL supplementation is supposed to be positive effect on SSF mode. Hydrolysis experiments revealed that there were few inhibitors produced during hydrothermal treatment. The addition of YHL will reduce the decrease of enzyme activity due to non-optimal pH in SSF as the interference of lignin on enzymes has much higher impact on enzymatic hydrolysis than the decrease of enzyme activity due to non-optimum pH. Besides YHL are inferred to be nutrients for fermentation according to a previous study [18]. Further investigations are needed, however, to use SDS-PAGE and mass spectrometry for characterization of the elution fractions to balance the relative contribution of YHL as fermentation nutrients and hydrolysis promoter. Besides more works are needed to investigate the effect of substances with low molecule weight on the impurities of fermentation broth and product purification cost, especially for lactic acid fermentation [28].

Conclusions

The majority of nitrogen and carbohydrate of yeast cells were transferred into YHS during hydrothermal treatment. The increase in enzymatic convertibility of FR is only due to YHL. There were soluble substances of different molecule weight in YHL and those of high molecular weight were predominating, including non-enzymatic proteins and other UV-absorbing compounds, which showed close molecule weight to components of cellulases. Supplementation of cellulases with YH enhanced enzymatic convertibility mainly by competitive adsorption of proteins and cellulase on lignin. Since YH can be used as nutrients for fermentation, different strategies can be used for a good utilization of yeast cells in terms of biorefinery concept.

Methods

Raw material preparation

Raw furfural residue was kindly provided by Chunlei Company (Hebei province, China). The initial pH of the material was between 2 and 3. After raw furfural residue was water-rinsed until neutral pH and then dried at 60°C for 12 h. *Saccharomyces cerevisiae* was used in the form of dry yeast (Angel Yeast Company Ltd, China).

Fractionation of yeast hydrolysate (YH)

Nine grams of yeast cells were added into a 180 mL-sealed pressure flask. Each flask was contained 150 mL of distilled water and then autoclaved at 121°C for 30 min [18]. The YHS and YHL were collected by centrifugation at 1368 × g for 6 min. The determination of moisture content of YHS was made in an oven-dried at 105°C (Figure 1). The pH of YHL was about 6. A 1 mL supernatant was oven-dried (105°C) and the residual solid was quantified to determine DMC of aqueous phase. DMC were calculated as follows:

$$\mathrm{DMC}\ (\mathrm{mg/mL}) = \frac{\text{The residual solid after oven-dried (mg)}}{\text{The liquid volume (mL)}} \times 100\%$$

Component separation of YHL

Before separation, YHL was filtered using 0.45-μm filters. The separation of each component was made on an AKTA purifier (Amersham Pharmacia Biotech; Uppsala, Sweden) equipped with a pump (P-920), UV monitor (UPC-900), a valve (INV-907), a mixer (M-925), and a fraction collector (Frac-920). The separation was performed in a 5 mL HiTrap desalting size exclusion column at room temperature with 3 mL/min of sodium citrate buffer (pH 4.80 and 50 mM) as mobile phase. The injection volume was 0.5 mL and injections were carried out 14 times to separate 6.29 g of YHL (7.5 g YH contains 6.29 g YHL). When there is no peak, Frac-920 collects 1 mL fraction in each tube (Figure 1). When there is a peak, the peak will be collected as one large fraction or in several tubes each collecting at most 1 mL fraction (Figure 1). A 0.33 mL of Celluclast 1.5 L was added to a 10-mL volumetric flask and diluted with sodium citrate buffer to obtain 40 g/L solution for component separation.

Enzymatic hydrolysis

The enzymatic hydrolysis (EH) of furfural residue was performed using Celluclast 1.5 L (with activity of 75 FPU/mL) and supplemented with the β-glucosidase preparation Novozym 188 (43.9 IU/mL). Both enzyme preparations were provided by Novozymes A/S, Bagsvaerd, Denmark. The filter paper units (FPU) were determined according to the IUPAC method [29] and β-glucosidase activity by [30]. The EH was made using 3% FR in 60-mL conical flasks with a working weight of 30 g. The amount of Celluclast 1.5 L and Novozym 188 added corresponded to a 12.5 FPU and 13.5 IU/g cellulose of FR, respectively. The flasks were kept in a shaker bath at a temperature of 45 ± 0.1°C at 150 rpm. YHL was used in hydrolysis experiments at one of two scales. Small and large scale hydrolyses were done using 6.29 and 12.6 g of YHL, respectively. The initial reaction mass was kept 30 g by adding sodium citrate buffer (pH 4.80 and 50 mM). After 144 h, the insoluble phase of hydrolysate (mainly rich in lignin) was collected by centrifugation at 1368 × g for 6 min and then oven-dried at 105°C. Duplicate samples were withdrawn at 0, 4, 6, 24 and 48 h for carbohydrate analysis by HPLC. Liquid fractions were separated by filtration and subsequently stored at −18°C awaiting analysis.

In control experiments, EH were carried out by using 0.425 gram of cellulolytic enzyme lignin as sole substrate with or without 6.29 g of YHL. Duplicate samples were withdrawn at 4, 24 and 72 h for FPU analysis.

Cellulolytic enzyme lignin isolation

Enzyme hydrolysis of FR was carried out as above without the addition of YHL. The hydrolysate at 96 h was filtered and the residues were collected for a second hydrolysis. After the second hydrolysis using identical amounts of cellulase and β-glucosidase, the residues were recovered by filtration and washed with 400 mL of distilled water (about 50°C), followed by aqueous dioxane extraction of ball-milled wood meal [31]. The final lignin was put into vacuum oven and dried at room temperature. The lignin samples were ground and screened through 40 mesh.

Compositional analysis

The chemical composition of water-rinsed furfural residue (FR) and yeast were analysed and prepared according to the standard procedure by NREL [32]. FR composition on an original dry weight (DW) residue basis was glucan, 48.2%; Klason lignin, 43.3%; and ash, 6.42%. Yeast has a composition of glucan, 18.4%; protein, 48.1%; and mannan, 2.77%. The total nitrogen content of YHS and the lignin residue was determined by the conventional Kjeldahl method [33]. The protein content was obtained by multiplying the elemental N content by the universal factor of 6.25. The protein content of YHL was measured with the Bradford method using BSA as a standard [34].

Sugar analysis

Previous analysis, the hydrolysates were centrifuged at 1368 × g for 6 min and filtered using 0.20-μm filters. Monosaccharides were quantitatively determined on an Aminex HPX-87P column at 85°C with 0.6 mL/min water as eluent. The HPLC system used as a Waters 2695e (Milford, MA, USA), comprising a Waters performance-PLUS vacuum degasser, a Waters four channels solvent delivery system, a Waters high temperature column oven, a Waters 2414 refractive index (RI) detector and a Waters plus auto-injector. All experiments were performed in duplicates, and the analysis was carried out at least three times for each sample.

Abbreviations

YH: Yeast hydrolysates; FR: Water-rinsed furfural residue; YHS: The solid phase of yeast hydrolysates obtained by centrifugation; BSA: Bovine serum albumin; YHL: The liquid phase of yeast hydrolysates obtained by centrifugation; DMC: Dissolved matter content; SSF: Simultaneous saccharification and fermentation; SHF: Separate hydrolysis and fermentation; EH: Enzymatic hydrolysis; FPU: Filter paper activity.

Competing interests

The authors declare that they have no competing interests.

Authors' contributions

YT, ZL and FL designed and coordinated the study. YT carried out all the experiments. YT and HY analysed the results. YT wrote the paper, and JJ, FL and CC were reviewed the manuscript. All authors read and approved the final manuscript.

Acknowledgements

The China Ministry of Science and Technology (2014DFG32550), the Guangxi Key Laboratory of Chemistry and Engineering of Forest Products (GXFC12-07) and the National Science Foundation of China (31070510) are gratefully acknowledged for their financial support of this research project.

Author details

[1]Department of Chemistry and Chemical Engineering, Beijing Forestry University, Beijing, China. [2]GuangXi Key Laboratory of Chemistry and Engineering of Forest Products, Nanning 530006, China. [3]Department of Chemical Engineering, Lund University, Lund, Sweden.

References

1. Ghosh T, Prelas M: **Ethanol.** In *Energy Resources and Systems.* Netherlands: Springer; 2011:419–493.
2. Galbe M, Zacchi G, Pretreatment: **The key to efficient utilization of lignocellulosic materials.** *Biomass Bioenergy* 2012, **46:**70–78.
3. Tang Y, Zhao DQ, Cristhian C, Jiang JX: **Simultaneous saccharification and cofermentation of lignocellulosic residues from commercial furfural production and corn kernels using different nutrient media.** *Biotechnol Biofuels* 2011, **4:**22.
4. Kumar L, Arantes V, Chandra R, Saddler JN: **The lignin present in steam pretreated softwood binds enzymes and limits cellulose accessibility.** *Bioresour Technol* 2012, **103:**201–208.
5. Shen F, Kumar L, Hu JG, Saddler JN: **Evaluation of hemicellulose removal by xylanase and delignification on SHF and SSF for bioethanol production with steam-pretreated substrates.** *Bioresour Technol* 2011, **102:**8945–8951.
6. John T, Ishizawa CI, Davis MF, Himmel ME, Adney WS, Johnson DK: **Cellulase digestibility of pretreated biomass is limited by cellulose accessibility.** *Biotechnol Prog* 2007, **98**(1):112–122.
7. Kristensen JB, Thygesen LG, Felby C, Jørgensen H, Elder T: **Cell-wall structural changes in wheat straw pretreated for bioethanol production.** *Biotechnol Biofuels* 2008, **1**(5):1–9.
8. Kumar L, Tooyserkani Z, Sokhansanj S, Saddler JN: **Does densification influence the steam pretreatment and enzymatic hydrolysis of softwoods to sugars?** *Bioresour Technol* 2012, **121:**190–198.
9. Bu LX, Xing Y, Yu HL, Gao YX, Jiang JX: **Comparative study of sulfite pretreatment for robust enzymatic saccharification of corn cob residue.** *Biotechnol Biofuels* 2012, **5:**87.
10. Cullis FI, Mansfield SD: **Optimized Delignification of Wood-Derived Lignocellulosics for Improved Enzymatic Hydrolysis.** *Biotechnol Bioeng* 2010, **106**(6):884–893.
11. Jabbour D, Borrusch MS, Banerjee G, Walton JD: **Enhancement of fermentable sugar yields by alpha-xylosidase supplementation of commercial cellulases.** *Biotechnology Biofuels* 2013, **6:**58.
12. Pryor SW, Karki B, Nahar N: **Effect of hemicellulase addition during enzymatic hydrolysis of switchgrass pretreated by soaking in aqueous ammonia.** *Bioresour Technol* 2012, **123:**620–626.
13. Kristensen JB, Borjesson J, Bruun MH, Tjerneld F, Jorgensen H: **Use of surface active additives in enzymatic hydrolysis of wheat straw lignocelluloses.** *Enzyme Microb Technol* 2006, **40:**888–895.
14. Alkasrawi M, Eriksson T, Borjesson J, Wingren A, Galbe M, Tjerneld F, Zacchi G: **The effect of Tween-20 on simultaneous saccharification and fermentation of softwood to ethanol.** *Enzyme Microb Technol* 2003, **33:**71–78.
15. Cas S, Aita GM: **Enzymatic hydrolysis and ethanol yield of combined surfactant and dilute ammonia treated sugarcane bagasse.** *Bioresour Technol* 2013, **131:**357–364.
16. Tang Y, Su ZQ, Zhao DQ, Qi X, Jiang JX: **The effect of *Gleditsia* Saponin on simultaneous saccharification and fermentation of furfural residue for ethanol production.** *Adv Mater Res* 2011, **236–238:**108–111.
17. Gao MT, Hirata M, Toorisaka E, Hano T: **Study on acid-hydrolysis of spent cells for lactic acid fermentation.** *Biochem Eng J* 2006, **28:**87–91.
18. Tang Y, Bu LX, He J, Jiang JX: **L(+)-Lactic acid production from furfural residues and corn kernels with treated yeast as nutrients.** *Eur Food Res Technol* 2013, **236**(2):365–371.
19. Simpson RJ: *Purifying Protein s for Proteomics:A Laboratory Manual (Ru BG, Trans.).* Beijing: Chemical Industry Press; 2009:622. Original work published 2004.
20. Li YH: *A Laboratory Manual for Protein analysis and experiments.* Beijing: Higher Education Press; 2011.
21. Kumar R, Wyman CE: **Cellulase adsorption and relationship to features of corn stover solids produced by leading pretreatments.** *Biotechnol Bioeng* 2009, **103**(1):252–267.
22. Rahikainen J, Mikander S, Marjamaa K, Tamminen T, Lappas A, Viikari L, Kruus K: **Inhibition of enzymatic hydrolysis by residual lignins from softwood-study of enzyme binding and inactivation on lignin-rich surface.** *Biotechnol Bioeng* 2011, **108**(12):2823–2834.
23. Tang Y, Bu LX, Deng LH, Zhu LW, Jiang JX: **The effect of delignification process with alkaline peroxide on lactic acid production from furfural residues.** *Bioresoures* 2012, **7**(4):5211–5221.
24. Kumar D, Murthy GS: **Impact of pretreatment and downstream processing technologies on ecomomics and energy in cellulosic ethanol production.** *Biotechnol Biofuels* 2011, **4:**27.
25. Macrelli S, Mogensen J, Zacchi G: **Techno-economic evaluation of 2nd generation bioethanol production from sugar cane bagasse and leaves integrated with the sugar-based ethanol process.** *Biotechnol Biofuels* 2012, **5:**22.
26. Linde M, Galbe M, Zacchi G: **Bioethanol production from non-starch carbohydrate residues in process streams from a dry-mill ethanol plant.** *Bioresour Technol* 2008, **99:**6505–6511.
27. Kim TH, Nghiem NP, Taylor F, Hicks KB: **Conversion of hulled barley into fermentable sugars using chemical, thermal, and enzymatic (CTE) treatment.** *App Biochem Biotechnol* 2011, **164:**534–545.
28. Gao MT, Hirata M, Toorisaka E, Hano T: **Lactic acid production with the supplementation of spent cells and fish wastes for the purpose of reducing impurities in fermentation broth.** *Biochem Eng J* 2007, **36:**276–280.
29. Ghose TK: **Measurement of cellulase activities.** *Pure Appl Chem* 1987, **2:**257–268.
30. Berghem LER, Petterson LG: **Mechanism of enzymatic cellulose degradation-isolation and some properties of a beta-glucosidase from *Trichoderma viride*.** *Eur J Biochem* 1974, **2:**295–305.
31. Chang HM, Cowling EB, Brown W: **Comparative studies on cellulolytic enzyme lignin and milled wood lignin of sweetgum and spruce.** *Holzforschung* 1975, **29**(5):153–159.
32. Sluiter A: *Determination of structural carbohydrates and lignin in biomass. NREL Laboratory Analytical Procedures.* Golden, CO, USA: National Renewable Energy Laboratory; 2004.
33. Kjeldahl J: **A new method for the determination of nitrogen in organic matter.** *Z Anal Chem* 1883, **22:**366–382.
34. Branford MM: **A rapid and sensitive method for the quantitation of microgram quantities of protein utilizing the principle of protein-dye binding.** *Anal Biochem* 1976, **72:**248–254.

5

High-throughput strategies for penicillin G acylase production in *rE. coli* fed-batch cultivations

Ana Maria Vélez[1†], Adilson José da Silva[1†], Antonio Carlos Luperni Horta[2†], Cintia Regina Sargo[1†], Gilson Campani[1†], Gabriel Gonçalves Silva[1†], Raquel de Lima Camargo Giordano[1†] and Teresa Cristina Zangirolami[1*]

Abstract

Background: Penicillin G acylase (PGA) is used industrially to catalyze the hydrolysis of penicillin G to obtain 6-aminopenicillanic acid. In *Escherichia coli*, the most-studied microorganism for PGA production, this enzyme accumulates in the periplasmic cell space, and temperature plays an important role in the correct synthesis of its subunits.

Results: This work investigates the influence of medium composition, cultivation strategy, and temperature on PGA production by recombinant *E. coli* cells. Shake flask cultures carried out using induction temperatures ranging from 18 to 28°C revealed that the specific enzyme activity achieved at 20°C (3000 IU $gDCW^{-1}$) was 6-fold higher than the value obtained at 28°C. Auto-induction and high cell density fed-batch bioreactor cultures were performed using the selected induction temperature, with both defined and complex media, and IPTG and lactose as inducers. Final biomass concentrations of 100 and 120 gDCW L^{-1}, and maximum enzyme productivities of 7800 and 5556 IU L^{-1} h^{-1}, were achieved for high cell density cultures using complex and defined media, respectively.

Conclusions: To the best of our knowledge, the volumetric enzyme activity and productivity values achieved using the complex medium are the highest ever reported for PGA production using *E. coli*. Overall PGA recovery yields of 64 and 72% after purification were achieved for crude extracts obtained from cells cultivated in defined and complex media, respectively. The complex medium was the most cost-effective for PGA production, and could be used in both high cell density and straightforward auto-induction protocols.

Keywords: Penicillin G acylase, Recombinant *E. coli*, Auto-induction medium, Fed-batch culture

Background

Penicillin G acylase (PGA) is used industrially to catalyze the hydrolysis of penicillin G to obtain 6-aminopenicillanic acid (6-APA), which is a key intermediate for the synthesis of β-lactam antibiotics [1]. These drugs account for the largest fraction of global sales of antibiotics, and comprised 60% of the 5×10^7 kg/year produced worldwide in 2003 [1]. The production of semi-synthetic β-lactam antibiotics requires ever-increasing quantities of PGA, and the annual consumption of this enzyme is estimated to be in the range of 10–30 million tons [2]. The unmet demand for PGA at an acceptable cost could be supplied by improving the processes used for its production.

PGA can be produced by a variety of microorganisms, including bacteria, fungi and yeasts. The microorganism that has been most widely studied for this purpose is *Escherichia coli*, which accumulates the enzyme in the periplasmic cell space. The enzyme precursor is synthesized as an inactive 96 kDa pre-pro-PA, which contains a signal peptide at its N terminus [3] that mediates the translocation of pro-PA into the periplasm. This intermediate is further processed into α (23 kDa) and β (63 kDa) PGA chains during autoproteolytic reactions [4].

In the past, PGA has mainly been manufactured by fed-batch as well as by batch processes, using recombinant *E. coli*, *Bacillus megaterium*, or *Arthrobacter viscosus*, and sucrose or glucose as carbohydrate substrates [5].

Optimization of the performance of cultures for recombinant enzyme production has two main objectives:

* Correspondence: teresacz@ufscar.br
†Equal contributors
[1]Chemical Engineering Department, Federal University of São Carlos (UFSCar), Rodovia Washington Luís, km 235, C.P. 676, CEP 13565-905 São Carlos, SP, Brazil
Full list of author information is available at the end of the article

high-level gene expression, and intense formation of biomass containing a high activity level of soluble, correctly-folded intracellular enzyme [6].

Strategies for high-level gene expression involve increasing the efficiency of one or more gene expression steps [3]. In the case of the *pac* gene, encoding penicillin G acylase of *E. coli, B. megaterium* and other organisms [7,8], many different procedures have been employed to this end. The efficiency of transcription of the *pac* gene can be genetically modulated by mutation or removal of the regulatory region, mutations in the transcription initiation region, or by replacement of the *pac* promoter [9-11]. The post-translational processes, which are crucial for obtaining soluble active PGA, have been investigated to identify optimum host/vector combinations that can efficiently produce the mature protein [3,6], with co-expression of helper proteins that assist the formation of correctly-folded active PGA [6,12] . In addition, cloning the *pac* gene from different microorganisms in *E. coli* is another strategy that has been used to improve enzyme production [7,13,14].

Temperature plays a very important role in modulating the expression of key genes involved in recombinant protein production [15-17]. The rate of protein synthesis is reduced at low temperatures, which increases the formation of correctly-folded biomolecules [18]. Selection of a suitable temperature is also crucial to ensure that a balanced flux is maintained throughout the stages of protein synthesis (transcription) and maturation (translocation and periplasmic processing), in order to avoid accumulation of immature precursors and optimize expression of the *pac* gene [10]. Otherwise, the formation of insoluble protein aggregates (inclusion bodies) can occur [19]. Keilmann et al. [18] also reported that the initiation step of translation of the *pac* mRNA is blocked at high temperatures.

The production of PGA in bioreactor cultivations has been widely studied using both batch [20-22] and high cell density fed-batch cultures [6,23,24]. Fed-batch cultivation is considered to be the most efficient process for the production of inducible heterologous proteins, including PGA, because the nutrient supply can be modulated to firstly achieve high biomass formation and then proceed with the induction [25]. However, performing a high cell density culture (HCDC) is a challenge, mainly because control of both the dissolved oxygen concentration and the flow rate of the feed medium is problematic at high biomass concentrations [24,26].

The auto-induction approach [27,28] offers an alternative to fed-batch cultures. Biomass concentrations of up to 40 gDCW L^{-1} [29,30] can be obtained directly from batch experiments employing a mixture of glucose, glycerol, and lactose, with minimal requirements for handling of the expression culture [31]. However, this strategy has not yet been reported for PGA production in shake flask or bioreactor cultures of recombinant *E. coli*.

The composition of the culture medium is another important consideration in design of a suitable cultivation strategy, because it not only affects cell growth, but also influences gene expression and, consequently, the pool of proteins accumulated in the cells. Both defined and complex media have been employed for PGA production, but no reported studies have addressed the influence of medium composition on the performance of processes used for PGA recovery from cell lysates. In addition, the composition of the medium, as well as the induction strategy (especially the inducer selected), play important roles in the post-induction cell response. So far, the majority of published works have focused on the use of isopropyl-β-D-thiogalactopyranoside (IPTG) as the inducer [6,21,22,24]. Nonetheless, the fact that IPTG is an expensive and potentially toxic chemical restricts its application as an inducer in the industrial production of recombinant proteins. Lactose, on the other hand, is an inexpensive, natural, environmentally friendly inducer that can be used in expression systems based on the *lac* promoter. Despite these advantages, only one study has so far considered PGA production induced by lactose [23].

With the goal of developing an optimized protocol for PGA production, the main objectives of the present study were therefore to: i) assess the influence of induction phase temperature on PGA production; ii) compare auto-induction and conventional fed-batch cultivation strategies for PGA production in a bioreactor; and iii) analyze the influence of the culture medium composition and inducers on protein synthesis and purification.

Results

Influence of temperature on PGA production

As already stated above, temperature plays an important role in PGA synthesis by *E. coli* cells, and values between 22 and 37°C have been reported for the induction phase [7,32]. Here, induction temperatures ranging from 18 to 28°C were tested in the shake flask cultures. Figure 1 shows the specific and volumetric enzyme activities after 24 h of induction at different temperatures. The specific enzyme activity was ~3000 IU $gDCW^{-1}$ at 20°C, ~2500 IU $gDCW^{-1}$ at 22 and 24°C, and decreased to less than 1000 IU $gDCW^{-1}$ at 28°C. As expected, the final biomass concentration increased at higher temperatures, ranging from 2.3 (at 18°C) to 3.3 gDCW L^{-1} (at 28°C).

Despite the reduced biomass formation observed at lower temperatures, the enzyme concentration per volume of culture medium was significantly improved at 20°C. The enzyme production achieved at this temperature was 2 to 3-fold superior to values reported in the literature [6,23,33], probably due to the temperature employed during the induction phase in the earlier studies, which ranged from

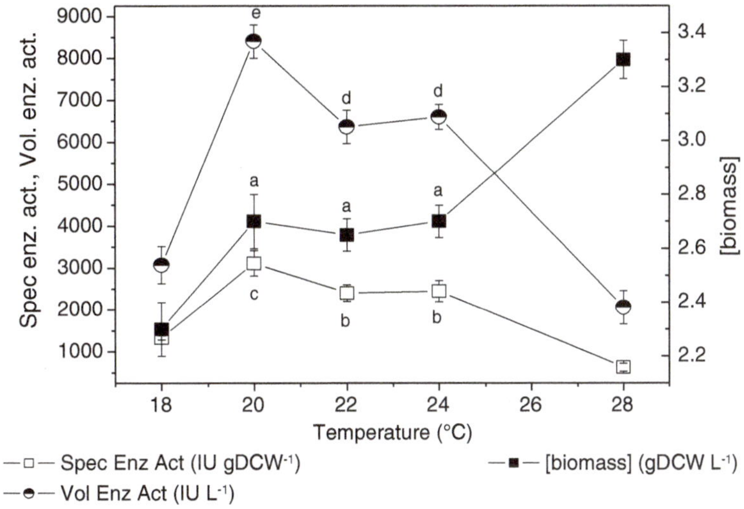

Figure 1 Effect of induction temperature upon specific activity (Eq. 4) and volumetric enzyme activity (Eq. 5) of PGA produced by recombinant *E. coli* grown in shake flasks containing LB medium after 24 h of induction. All experiments induced with 0.25 mM IPTG, T_{growth} = 37°C, $OD_{induction}$ = 1.5. Error bars for specific enzymatic activity, volumetric enzyme activity and biomass concentration are standard deviations from triplicates. Means with different letters (a–e) differ significantly according to the Tukey's test ($p < 0.05$).

28 to 30°C. Plasmid stability was also favored at 20°C, with values of 70-80% obtained after 24 h of induction. When a temperature of 24°C was used, plasmid retention decreased to 53% at the end of the induction period.

These results pointed out 20°C as the best temperature for PGA production by r*E. coli*. This value is lower than the optimum temperature (28°C and 22°C) identified by Cheng et al. and Dai et al. [7,32], respectively, for maximizing PGA production with IPTG as inducer. In these studies, constructs were based on plasmids pET24a and pMLB1023, whereas in the present work, plasmid pT101/D-TOPO was used. So, the results suggest that the optimum temperature for PGA synthesis may also depend on the construct (plasmid and gene). Further evidence of the strong influence of temperature on PGA synthesis was obtained from the SDS-PAGE analyses of disrupted cells after 24 h of induction at 28 and 20°C (Figure 2). An intense band corresponding to pre-pro-enzyme (~90 kDa) was found for the cells induced at 28°C, for both soluble and insoluble fractions (lanes 2 and 3). The formation of large amounts of pre-secretory product could be due to overloading of the transportation machinery of the cell. As a result, untranslocated pre-pro-enzyme may have accumulated and formed aggregates of an inactive form of the enzyme [21]. On the other hand, when the induction took place at 20°C, subunits α and β were observed (lanes 4 and 5), but the pre-pro-enzyme was not present.

Growth and protein production in bioreactor culture using the auto-induction strategy

Before investigating an induction strategy for high cell density cultures (HCDC) based on a low induction temperature, the alternative auto-induction protocol was used to perform bioreactor experiment B1 and validate the shake flask results. Despite being frequently used for protein production using T7-based *E. coli* expression systems in shake flask cultures [27,34], only a few investigations have considered bioreactor cultures based on auto-induction complex medium formulations [29,30,35].

Figure 3 shows the main results obtained for the auto-induction culture. Intense biomass formation was observed in the first 6 h of culture (μ_{max} = 0.73 h^{-1}), assisted by the high growth temperature (37°C) and fueled by the consumption of glucose and proteins/amino acids from the yeast extract and tryptone (components of the complex medium). Glucose prevents the uptake of lactose, and its exhaustion (at ~6.5 h) signaled the beginning

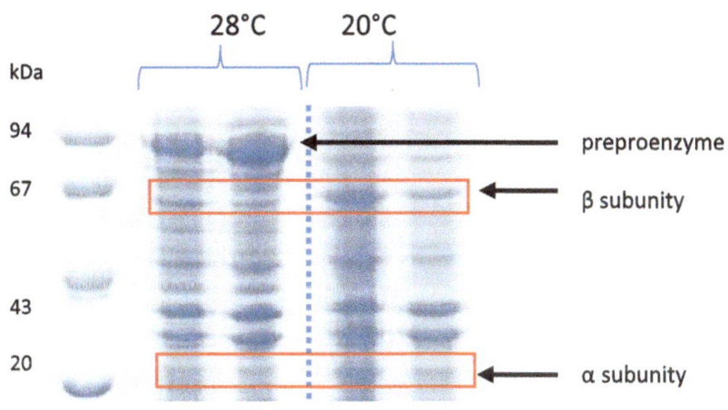

Figure 2 SDS-PAGE of cell lysates for recombinant *E. coli* grown in shake flasks containing LB medium after 24 h of induction. Experiments induced with 0.25 mM IPTG. (1) protein standards; (2) insoluble proteins (28°C); (3) soluble proteins (28°C) (4) insoluble proteins (20°C); (5) soluble proteins (20°C).

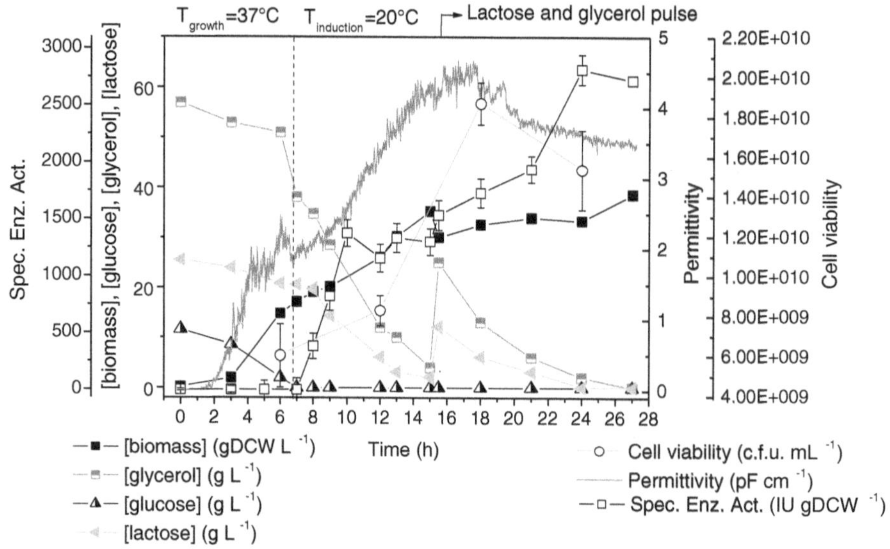

Figure 3 Growth, carbon sources consumption and PGA production by recombinant *E. coli* in intermittent fed-batch culture B1 with auto-induction medium. T_{growth} = 37°C and $T_{induction}$ =20°C. Pulse of glycerol and lactose added after 15.5 h of culture. Error bars for cell viability and specific enzymatic activity are standard deviations from triplicates.

($t_{induction}$ = 0 h) of heterologous protein synthesis induced by the lactose present in the medium to induce expression [27]. Shortly after glucose depletion, the temperature was reduced to 20°C to reproduce the induction strategy selected from the shake flask results. The PGA specific activity data (Figure 3) were indicative of fast enzyme accumulation in the cells, reaching 2800 IU $gDCW^{-1}$ after 17.5 h of induction. This value was similar to the maximum specific activity observed in the flask experiments (Figure 1) carried out at the same temperature (20°C).

The biomass concentration increased continuously up to 36 gDCW L^{-1}, with a moderate growth rate ($\mu = \sim 0.09\ h^{-1}$) after the temperature decrease (between 7 and 15 h), and a slow growth rate ($\mu = \sim 0.02\ h^{-1}$) during the last 12 h of induction due to intensification of the metabolic burden related to protein expression [36-38]. The permittivity signal (Figure 3) closely followed the biomass formation profile up to ~16 h of culture. The capacitance sensor measures the dielectric permittivity produced by a polarized cell membrane when it is subjected to low radio frequencies. Since this only occurs for viable cells [39], the shift in the permittivity curve reflected a decrease in the viable cell concentration due to the stress associated with protein production, as also shown by the c.f.u. counts [36]. The measured loss of viability was also influenced by the decrease in the proportion of cells retaining the plasmid, which changed from 90% (at 5.5-11.5 h of induction) to 85% (at 17.5 h of induction).

The scale-up auto-induction medium composition proved to be effective. The high initial glucose concentration prevented early protein expression, but did not cause formation of high levels of inhibitory metabolites: the concentrations of acetic, formic, and lactic acids remained below 2 g L^{-1} throughout the cultivation [40,41]. The low formation of organic acids was assisted by the choice of glycerol as the main carbon source. Together with glucose released by lactose hydrolysis due to β-galactosidase, glycerol was steadily consumed and both carbon sources were used by the cells to support growth and protein production during the entire induction phase. Galactose, which is not assimilated by *E. coli* BL21(DE3) cells, accumulated in the culture medium (data not shown). At 15.5 h of cultivation, an additional pulse of a solution containing lactose and glycerol was added in order to avoid limitation of protein synthesis due to a lack of either carbon source or inducer. Use of this simple cultivation strategy yielded a bulk PGA volumetric enzyme activity exceeding 93775 IU L^{-1} within 17.5 h of induction, corresponding to a productivity of 3907 IU L^{-1} h^{-1} (or 5358 IU L^{-1} h^{-1}, considering the elapsed induction period).

Growth and protein production for scaled-up high cell density cultures

The lactose-based induction strategy was further scaled up as a high cell density culture (HCDC) using complex medium (culture B2). For comparison, a HCDC using defined medium and IPTG as inducer (culture B3) was also carried out. The feeding media for both fed-batch cultures contained exclusively glycerol as carbon source (Table 1), and were supplied following an exponential profile (Equation 7) based on the actual growth rates, as described below (Methods section). Induction was triggered by pulses of lactose (B2) or IPTG (B3), shortly after the temperature had been reduced to 20°C.

Table 1 Composition of media used in bioreactor cultivations

	Auto-induction		Complex media		Defined media	
Nutrient (g L^{-1})	Batch	Pulse	Batch	Fed-batch	Batch	Fed-batch
Glucose	10.0	—	10.0	—	—	—
Glycerol	60.0	258.3	40.0	800	40.0	800.0
Lactose	20	130.0	—	—	—	—
Tryptone	10.0	10.0	10.0	10.0	—	—
Yeast extract	5.0	5.0	5.0	5.0	—	—
$MgSO_4.7H_2O$	0.5	0.5	0.5	40.0	1.6	20.0
KH_2PO_4	3.4	3.4	3.4	3.4	17.7	21.28
$(NH_4)_2HPO_4$	—	—	—	—	5.3	6.4
$Na_2HPO_4.12H_2O$	9.0	9.0	9.0	9.0	—	—
Citric acid	—	—	—	—	2.27	2.27
NH_4Cl	2.7	2.7	2.7	2.7	—	—
Na_2SO_4	0.7	0.7	0.7	0.7	—	—
Thiamine	—	—	—	—	45.0	45.0

Auto-induction medium: cultivation B1; complex medium: cultivation B2; defined medium: cultivation B3.

The results obtained for the complex medium fed-batch culture (B2) are shown in Figure 4. Due to the low solubility of lactose and the increasing biomass concentration, the inducer was supplied in the form of multiple 400 mL pulses of solutions containing 200 g L^{-1} lactose, which were added after every 20 gDCW L^{-1} (~40 OD_{600nm} units) increase in the cell concentration. This strategy was based on the results obtained in the previous auto-induction culture (B1), relating biomass formation to lactose uptake (Figure 3). Lactose concentration measures during cultivation B2 showed that, despite continuous hydrolysis of lactose, it was possible to maintain a residual inducer concentration capable of promoting *pac* gene expression throughout the induction phase.

The cultivation strategy adopted was effective for both biomass formation and enzyme production. The initial growth phase was carried out at 37°C, lasted 7.3 h, and led to a biomass formation of 33 gDCW L^{-1} (μ_{max} = 0.68 h^{-1}). The continuous fed-batch phase was then started, and supported continuous cell accumulation up to 105 gDCW L^{-1} during the next 14.8 h, with high viability, as confirmed by the permittivity data and the c.f.u counts. Enzyme production started after the first pulse addition (at 8 h of culture), at a biomass concentration of 54 gDCW L^{-1}, and the specific enzyme activity rapidly increased to 1800 IU $gDCW^{-1}$ within the first 12 h of induction, with complete hydrolysis of the lactose supplied. During the last 10 h of induction, the enzyme synthesis rate slowed, and the specific enzyme activity reached a final value of 2061 IU $gDCW^{-1}$. This shift in the enzyme production profile was mainly due to intensified stress associated with protein synthesis [36], which caused a reduction in plasmid retention from 97% (at 4-8 h of induction) to 89% (at 12 h of induction). Once again, the formation of inhibitory metabolites was negligible (2.1 g L^{-1} acetate),

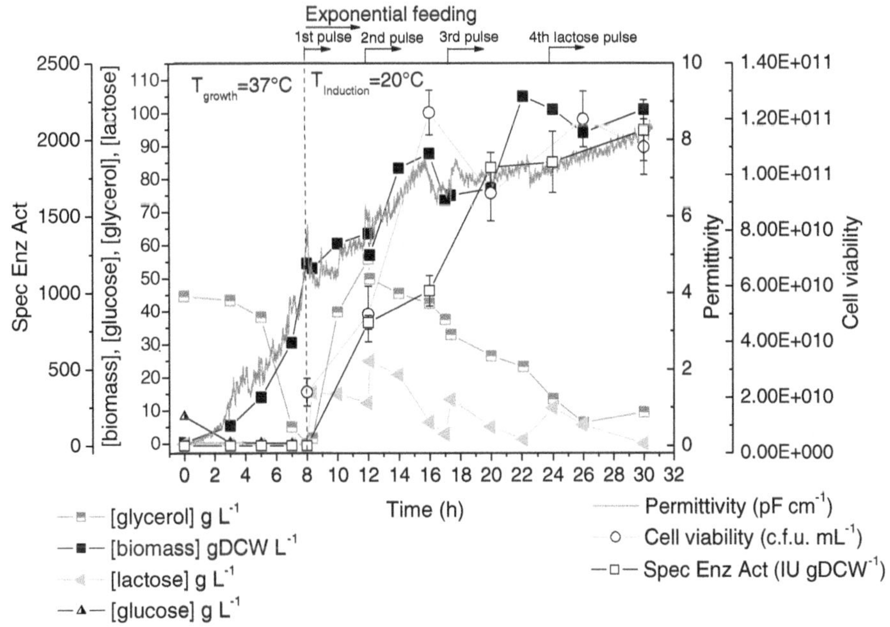

Figure 4 Growth, carbon sources consumption and PGA production by recombinant *E. coli* using fed-batch culture B2 with complex medium. T_{growth} = 37°C and $T_{induction}$ = 20°C. Lactose pulses added at 8, 12, 17 and 24 h. Feeding started at 7.3 h. Error bars for cell viability and specific enzymatic activity are standard deviations from triplicates.

despite glycerol accumulation during the first 4 h of feeding [40,41]. It is also important to clarify that the fluctuations in cell concentration and viability were caused by the withdrawal of large volumes of culture medium (~400 mL) prior to the addition of the lactose pulses, in order to ensure that the reactor volume remained within the operational range.

In summary, the lactose-based complex medium fed-batch strategy yielded a PGA volumetric enzyme activity equivalent to 208222 IU L^{-1} within 24 h of induction, which corresponded to a productivity of 6941 IU L^{-1} h^{-1} (or 9465 IU L^{-1} h^{-1}, considering the elapsed induction period). These values are about twice those observed for the B1 auto-induction culture, because the biomass formation phase was extended in the continuous fed-batch operation mode and occurred simultaneously with intense protein production during the first 15 h of induction, due to the low induction temperature employed.

The influence of inducer and medium composition on PGA production and purification was assessed using continuous fed-batch cultivation B3 with defined medium and IPTG as inducer. The main results are presented in Figure 5. Glycerol was the only carbon source during the initial batch and fed-batch phases. The feed supply was initiated 10 h after the beginning of the culture, as soon as the glycerol from the batch culture was exhausted, yielding a biomass concentration of 18 gDCW L^{-1} and a maximum specific growth rate of 0.45 h^{-1}. The temperature was then reduced from 37 to 20°C, and a pulse of IPTG was applied at 13 h (at a biomass concentration of 54 gDCW L^{-1}) to initiate the induction phase. Despite the temperature reduction, the cells continued to grow at a moderate rate (~ 0.1 h^{-1}), reaching 82 gDCW L^{-1} within 4 h of induction.

The B2 and B3 continuous fed-batch cultures showed equivalent performance (Figures 4 and 5) up to the moment of induction. Obviously, due to the inherently lower biomass yield for the defined medium ($Y_{X/S}$ values of 0.44 and 0.55 g g^{-1} were obtained for the defined and complex media, respectively), in the case of culture B3 a longer growth phase (13 h) was required before the desired biomass concentration was attained (~54 g L^{-1}) and the induction phase could be initiated. However, throughout the induction phase, the B3 culture showed fluctuations on the permittivity data and accumulation of glycerol. The instability can be partially attributed to the sudden temperature decrease from 37°C to 20°C prior to IPTG addition. Due to the reduced growth rate imposed at 20°C, glycerol started to accumulate in the broth. Glycerol accumulation was probably even intensified by the salt precipitation observed after the temperature decrease, due to the low aqueous solubility of several of the defined medium components, especially KH_2PO_4 (Table 1). In such a situation, growth became limited by a nutrient other than the carbon source. The accumulation of glycerol at concentrations of up to approximately 80 g L^{-1} was tolerated by the cells [29], but inhibition occurred when this value was exceeded, causing stagnation of growth at between 16 and 25 h of cultivation. After feeding was interrupted, the cells started to consume the excess glycerol (and salts) accumulated in the medium, and cell growth was restored at the end of induction phase, leading to a final biomass concentration of 120 gDCW L^{-1}. It is also noteworthy that, once again, despite the high concentration of glycerol accumulated in the medium, the concentration of the organic acids that are commonly present did not exceed 2.5 g L^{-1}. This is the

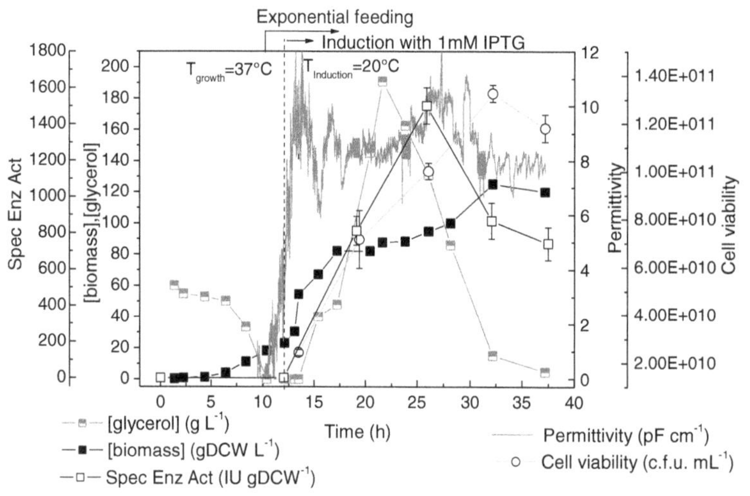

Figure 5 Growth, glycerol consumption and PGA production by recombinant *E. coli* in fed-batch culture B3 with defined medium. T_{growth} = 37°C and $T_{induction}$ =20°C. Induction by IPTG pulse at t = 13 h. Feeding started at 10.1 h. Error bars for cell viability and specific enzymatic activity are standard deviations from triplicates.

main advantage of using glycerol as carbon source for HCDC. In contrast to glucose, its uptake by the cells does not trigger overflow metabolism and, as long as adequate aeration conditions are provided, there is no risk of inhibition of growth or protein synthesis due to metabolite formation [42].

The facts described above also affected protein production. The specific enzyme activity increased to a maximum of 1500 IU gDCW^{-1} after 12 h of induction, and subsequently decreased continuously. This activity level was significantly lower than the maximum values observed for experiments B1 and B2 (Figures 3 and 4), as well as for the shake flask cultures performed using the same induction temperature (Figure 1). The gradual decline in the enzyme yield also showed a relationship with plasmid stability, which diminished from 87% (after 5 h of induction) to 75% (after 13 h of induction) and 65% (after 19 h of induction, at the end of the cultivation). Plasmid retention during this experiment was inferior compared to the B1 and B2 cultures carried out with complex medium, which showed retentions of >80% after 18 h of induction. It is known that plasmid stability is generally favored in amino acid-enriched media [43]. Nevertheless, mainly due to the high final biomass concentration achieved, fed-batch cultivation B3 yielded a PGA volumetric enzyme activity equivalent to 89000 IU L^{-1} within 24 h of induction, corresponding to a productivity of 2393 IU L^{-1} h^{-1} (or 3680 IU L^{-1} h^{-1}, considering the elapsed induction period). On the other hand, care must be taken before choosing the conventional defined high cell density medium (HDF medium) [42] for heterologous protein production when the synthesis is favored by low temperature.

The electrophoresis results for the three bioreactor cultures (Figure 6) showed that the fraction corresponding to contaminant proteins present in the enzyme extracts is

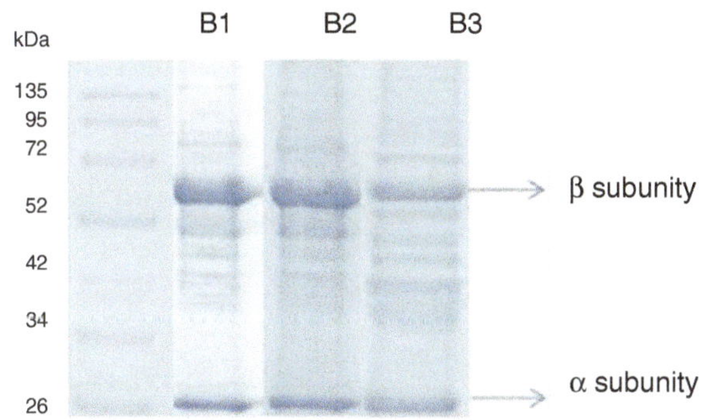

Figure 6 SDS-PAGE of clarified fraction obtained after lysis of recombinant *E. coli* cells cultivated in different media after 24 h of induction. B1: auto-induction; B2: Fed-batch with complex medium and B3: Fed-batch with defined medium. All samples contained 1 mg/mL of protein.

similar, regardless of either the cultivation strategy or the inducer used. This was due to the low temperature employed during the induction phase (20°C), which helped to reduce the production of endogenous proteins by *E. coli.*

Partial purification of recombinant PGA from complex and defined media cultivations

Besides identification of the most suitable cultivation strategy for recombinant enzyme production, it is also important to assess the impact of the chosen cultivation protocol on the performance of the purification process. One of the drawbacks often associated with the use of complex medium is possible stimulation of the synthesis of diverse contaminant proteins, which could impair purification of the target protein.

The results obtained for the partial purification of PGA present in crude extracts obtained from cells grown in both types of medium (complex and defined) are shown in Table 2. The clarified extract generated from the cells cultivated in complex medium had a higher protein content, and consequently a lower specific enzyme activity. This extract retained ~30% of the PGA activity initially present, due to saturation of the resin. The recovery of PGA could be improved by reducing the loading of enzyme per gram of resin. Consequently, the global yields for the purification process were 72 and 64% for the extracts derived from the complex and defined media, respectively.

The eluted activity yields all exceeded 97%, indicative of similar interactions between the resin and the clarified fractions, irrespective of their source. Furthermore, the salt concentration in the eluate buffer was satisfactory, and almost all the recombinant PGA was detached, leading to the high recovery observed in this step. Overall, 2.3-fold and 2.8-fold purity increases were obtained for the extracts from the complex and defined media cultivations, respectively.

These results are similar to those obtained by Pinotti et al. [44], using the same protocol and resin for PGA purification. Although higher purification factors have been reported for penicillin acylase [45], the main objective of this study was not to optimize the methodology, but rather to compare the performance of the purification process when handling cell lysates derived from cells cultivated and induced under different conditions. The results presented here demonstrate that cultivation in complex medium is a better strategy for PGA production, and does not jeopardize the performance of the purification process.

Discussion

Comparison of enzyme production using different cultivation strategies

A comparison of all the cultivation strategies, in terms of enzyme production, is provided in Figure 7. As already

Table 2 Results obtained for the batch adsorption-desorption process of PGA from recombinant *E. coli* using the Streamline SP XL resin at pH 4.8

Samples	Cells harvested from complex medium			Cells harvested from defined medium		
	Enz. Act. (IU L^{-1})	Prot. Conc. (mg_{prot} L^{-1})	Spec. Act. (IU mg_{prot}^{-1})	Enz. Act. (IU L^{-1})	Prot. Conc. (mg_{prot} L^{-1})	Spec. Act. (IU mg_{prot}^{-1})
Clarified	500 ± 45	480.8 ± 32.1	1.04 ± 0.70	420 ± 51	305.1 ± 15.6	1.38 ± 0.42
Extract	130 ± 21	213.1 ± 43.4	0.61 ± 0.07	150 ± 32	235.1 ± 43.1	0.64 ± 0.21
Washing	0.2 ± 0.1	3 ± 1	0.3 ± 0.2	0.2 ± 0.1	0.4 ± 0.1	0.1 ± 0.3
Eluate	1200 ± 82	510.6 ± 14.2	2.35 ± 0.03	893 ± 76	231.1 ± 16.2	3.86 ± 0.34
η_{act}^{el} (%)		97 ± 2			99 ± 2	
η_{act}^{ads} (%)		74 ± 6			64 ± 4	
$\eta^{process}$ (%)		72 ± 5			64 ± 6	
P.F.		2.3 ± 0.3			2.8 ± 0.4	

mentioned, the continuous fed-batch strategy with complex medium (experiment B2) provided the best volumetric enzyme activity and productivity (Figures 7a and 7b). The auto-induction culture (experiment B1) exhibited a stationary protein productivity profile after 6.5 h of induction (Figure 7b), because the biomass concentration remained almost constant, while the specific enzyme activity increased linearly with induction time (Figure 3). Conversely, for the continuous fed-batch cultivation (B2), both biomass concentration and specific PGA activity increased during the first 16 h of induction (Figure 4), resulting in a steeper increase in productivity (Figure 7b). Despite the low induction temperature used, Figure 7b shows that 16 - 18 h of lactose induction was sufficient to achieve high enzyme productivity for both auto-induction (B1) and HCDC (B2) strategies. Figure 7a shows that the initial rate of enzyme accumulation was similar for HCDC cultures B2 and B3 during the first 13 h of induction, due to the increasing biomass concentration and specific enzyme activity (Figures 4 and 5, respectively). The shift in the enzyme production pattern observed in culture B3 after ~13 h of induction (Figure 7a) was caused by the decay in specific enzyme activity, as shown in Figure 5, associated to plasmid instability and viability loss. A similar trend can be seen in the productivity curve obtained for culture B3 (Figure 7b).

Process economics is an important factor to be considered when selecting a cultivation strategy and a simple estimate of the costs associated only to the different media formulation, using the same approach described at [46], was performed. The main results, summarized at Table 3, point out the intermittent fed-batch auto-induction protocol (B1) as the most cost-effective cultivation strategy, thanks to its low media cost. Although the continuous fed-batch strategy based on complex medium (B2) led to the highest volumetric enzyme activity, superior production costs related to the large amount of reagents required to achieve and maintain high cell densities restricted its economic performance. Concerning the fed-batch strategy based on defined medium (B3), the 3.4 fold increase of enzyme cost is mainly due to the lower volumetric enzyme activity achieved at this operation mode, since the cost of the culture medium is similar to B2.

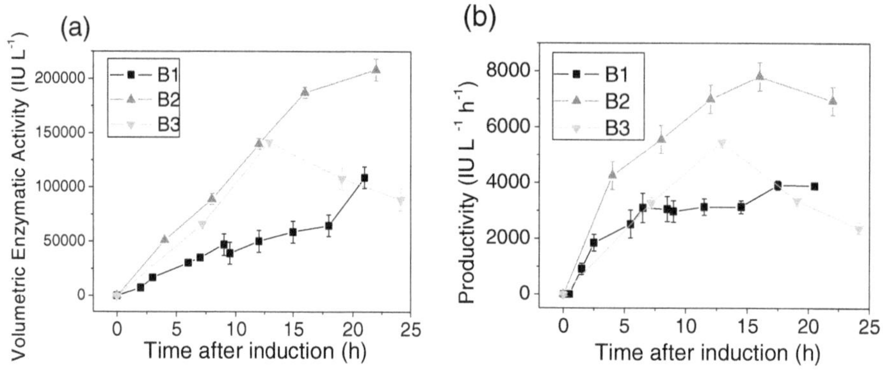

Figure 7 Profiles of (a) volumetric enzyme activity estimated from Eq. 5 (IU/L) and (b) productivity (Eq. 6) of recombinant PGA in *E. coli* during induction phase. Error bars are standard deviations from triplicates.

Table 3 Total cost associated to media used at B1, B2 and B3 cultures and the corresponding estimate of the cost per unit of PGA produced

Culture ID	Batch ($)	Pulse ($)	Feed ($)	Total ($)	Total volume (L)	Medium cost (per L)	Vol Enz Act (U L^{-1})	Enzyme cost ($ U^{-1})
B1	69.30	21.50	—	90.81	4.2	21.62	94000	0.00023
B2	55.23	16.11	321.06	392.40	5.6	70.10	208222	0.00034
B3	52.58	71.10	254.36	378.04	5.4	70.00	89000	0.00079

Comparison of PGA expression for different clones of recombinant *E. coli*

The most recent and relevant results reported for PGA expression using different clones of *E. coli* and different induction strategies are presented in Table 4. Due to the high cell concentration reached, the volumetric enzyme activity and productivity obtained in this work were superior to all the values reported using *rE. coli,* in the literature to date. The specific PGA activity (HCDC and auto-induction) was higher than the best published values [6,33]. These results demonstrate the importance of identifying cultivation conditions that can contribute to the twin goals of increased biomass formation and enhanced gene expression. In the present study, the use of a low induction temperature allied to a fed-batch cultivation strategy using complex medium was crucial for achievement of a superior enzyme titer and maximization of productivity.

Conclusions

High enzyme concentration (208222 IU L^{-1}) and productivity (6941 IU L^{-1} h^{-1}) were achieved for the continuous fed-batch culture with complex medium. Using the auto-induction strategy, both enzyme productivity and concentration can be limited by the accumulation of biomass that may occur in this operational mode. Nevertheless, it remains a very attractive approach, because an enzyme extract with high PGA activity (130 IU per mL of centrifuged cell lysate) can be obtained in just 15 h of simple batch culture, even without an additional pulse of lactose. The results obtained indicate that the innovative high cell density strategy used in this work, combining exponential glycerol feeding, an intermittent lactose supply, and a low induction temperature (20°C), is a promising technique for on-demand penicillin G acylase production by recombinant *E. coli* cells at lower production cost.

Methods

Microorganism

Escherichia coli BL21 (DE3) was donated by the Laboratory of Biochemistry, Department of Physiological Sciences, UFSCar, São Carlos, Brazil. The pT101/D-TOPO plasmid with the *pac* gene from *E. coli* ATCC11105 responsible for PGA production [47,48] was kindly provided by the Laboratory of Biocatalysis, ICP-CSIC, Madrid, Spain. The transformation of *E. coli* was performed by heat shock in the presence of $CaCl_2$.

Culture media

Shake flask cultivations were carried out using LB medium containing tryptone (10 g L^{-1}), yeast extract (5 g L^{-1}),

Table 4 PGA production in various *E. coli* expression systems for bioreactor cultures carried out under different induction strategies

E. coli strain/construction	Ind.	Op. mode	T_{ind} (°C)	[Biomass] (gDCW L^{-1})	EA_{sp} (IU DCW^{-1})	Vol. Enz.Act. (IU mL^{-1})	Product. (IU mL^{-1} h^{-1})	Ref.
RE3/ pKA18	Phenyl acetic acid	Batch	28	5	1000	4.5	0.28*	[33]
MDDP7/ pTrcKnPAC2902	IPTG	Fed-batch	28	33	1020*	37.5	0.48*	[6]
W3110/pPA102	Cheese whey	Batch	29	1.5	781	0.901	0.053*	[23]
9633/pGL-5	IPTG	Fed-batch	30	96.9*	157*	15	0.63*	[24]
BL21(DE3)/pET28b	IPTG	Batch	28	12	800	10	0.42*	[21]
DH5α/ Psmlfpga	IPTG	Batch	30	21.7*	153 mg $gDCW^{-1}$	79.88	0.67*	[22]
DH5α/pET30b	Constitutive	Batch	28.5	4.5	NA	43	2.58*	[20]
BL21(DE3)/pT101/D-TOPO	Lactose	Continuous fed-batch	20	100	2000	190	7.8	This work
BL21(DE3)/pT101/D-TOPO	Lactose	Intermittent fed-batch	20	38	2800	94	3.9	This work

*Estimated from reported data.
NR Not reported.

sodium chloride (10 g L^{-1}), and ampicillin (100 µg mL^{-1}), at pH 7. For bioreactor cultures, intermittent fed-batch B1 was performed using auto-induction medium (adapted from Studier and Silva et al. [27,29]) with pulse supplementation. Continuous fed-batch cultures B2 and B3 were carried out using complex medium [29] and defined medium [26], respectively. The compositions of all the media used are provided in Table 1. Ampicillin (150 µg mL^{-1}), 30% propylene glycol antifoaming agent (1 mL L^{-1}), and a metal solution [29] were added to the media used for the bioreactor cultivations.

Analytical methods

Cell concentration

During the bioreactor cultivations, the cell concentration was monitored using measurements of the optical density (OD) of the culture broth (λ = 600 nm), dry cell weight (gDCW L^{-1}) determinations, and viable cell counts (c.f.u. mL^{-1}), as well as by on-line recording of broth permittivity (pF cm^{-1}) using a biomass sensor [26]. In the shake flask experiments, the cell concentration (C_X) was related to the optical density using the expression:

$$C_X(\text{gDCW L}^{-1}) = (0.49 \pm 0.01) * (OD_{600\text{ nm}}) \quad (1)$$

Plasmid stability and cell viability

Diluted samples of culture broth were spread aseptically onto plates of LB agar, in some cases supplemented with ampicillin, and incubated for 16 h at 37°C. Cell viability and plasmid stability were evaluated by counting the final number of colonies formed [30].

Enzymatic activity and related variables

Samples were centrifuged at 10,000 × g for 5 min at 4°C, re-suspended in 50 mM phosphate buffer (pH 7.5), and disrupted by sonication using 6 pulses of 30 s (8-10 watts) at 20 mHz, with intervals of 20 s. After further centrifugation at 14,000 × g for 20 min at 4°C, the supernatant was collected and used for analyses of enzymatic activity and total soluble protein concentration (Bradford method [49]). In addition, SDS-PAGE [50] was employed to identify the presence of inactive pre-pro-PA and the α and β PGA subunits in the supernatant and pellet derived from the cell lyses.

PGA activity was determined by hydrolysis of a chromogenic substrate, 1 mM 6-nitro-3-(phenylacetamido)benzoic acid (NIPAB), in 0.1 M phosphate buffer (pH 7.5) at 37°C, according to the method described by Kutzbach and Rauenbus [51]. A calibration curve (Equation (2)) was prepared to further convert PGA activity (in U_{NIPAB} mL^{-1}) to the equivalent penicillin G activity (in IU mL^{-1}). The enzymatic reaction product (6-APA) was quantified using the colorimetric method described by Balasingham et al. [52].

$$EA_{penG} = (2.2 \pm 0.1) * EA_{NIPAB} \quad (2)$$

The specific enzyme activities were expressed in terms of the total soluble protein content (EA_{sp_Prot}) and the cell mass (EA_{sp_Biom}) using Equations (3) and (4), respectively. The corresponding enzyme concentrations (C_{enz}) and productivities (Pr_{enz}) per volume of cultivation medium were calculated using Equations (5) and (6), respectively. In these expressions, EA (IU mL^{-1}) is the enzyme activity obtained from Equation (2), C_{X_sonic} (gDCW L^{-1}) is the biomass concentration in the disrupted cell suspension, C_X (gDCW L^{-1}) is the biomass concentration in the cultivation broth, and Δt is the elapsed cultivation time.

$$EA_{sp\_Prot}\left(U \cdot mg_{prot}{}^{-1}\right) = \frac{EA}{C_{sol\_prot}} \quad (3)$$

$$EA_{sp\_Biom}\left(U \cdot gDCW^{-1}\right) = \frac{EA \cdot 1000}{C_{X\_sonic}} \quad (4)$$

$$C_{enz}\left(U \cdot L^{-1}\right) = EA_{sp\_Biom} \cdot C_X \quad (5)$$

$$Pr_{enz}\left(U \cdot L^{-1} \cdot h^{-1}\right) = \frac{C_{enz}}{\Delta t} \quad (6)$$

Concentrations of organic acids

The concentrations of organic acids were determined by HPLC, with UV detection at 210 nm. The stationary phase was an Aminex HPX-87H column (maintained at 50°C) and the mobile phase was 5 mM sulfuric acid, at a flow rate of 0.6 mL min^{-1}.

Partial purification of recombinant PGA

The purification procedure followed the protocol developed by Pinotti et al. [44]. In brief, the biomass harvested at the end of the cultivations in defined or complex media was disrupted by sonication in 20 mM citrate buffer at pH 4.8 (used as the adsorption buffer) and centrifuged at 7000 × g for 10 min to obtain the crude soluble enzyme extract. PGA adsorption was carried out using Streamline SP cation exchange resin. The desorption buffer employed for enzyme elution was 150 mM NaCl. The parameters $\eta^{process}$ (overall yield) and PF (purification factor) were used to compare the performance of the purification process for the different enzyme extracts. Full details of the calculation procedures can be found in Pinotti et al. [44].

Experimental procedure

Preliminary shake flask cultures were conducted to assess the influence of induction phase temperature on enzyme production. Five conditions were investigated (in triplicate), with temperatures ranging from 18 to 28°C. Intermittent fed-batch (B1) and continuous fed-batch (B2 and B3) bioreactor cultures were carried out using the selected induction temperature and different induction strategies.

Shake flask cultures

The shake flask cultivations were carried out for ~30 h using 1 L flasks containing 100 mL of LB medium, incubated at 37°C with agitation at 250 rpm during the initial growth phase. Once the OD_{600} of the culture reached 1.5, the flasks were transferred to another incubator kept at the desired temperature (18, 20, 22, 24 or 28°C), and induction was performed by adding a pulse of 0.25 mM IPTG.

Bioreactor cultures

A 300 mL aliquot of exponentially-growing cell suspension (OD_{600} ~2.0) was transferred to a 5 L laboratory bioreactor containing 3.5 L of the complex or defined auto-induction media described in Table 1. All cultivations were monitored using SuperSys_HCDC^R [26,53]. The temperature was maintained at 37°C during the growth phase, and reduced to 20°C prior to induction. The pH was maintained at 6.9 by addition of 28% (w/v) NH_4OH solution. The dissolved oxygen concentration was controlled at 30% of saturation by automatically adjusting both the agitation speed (in the range 200-600 rpm) and the composition of the gas stream supplied to the bioreactor (by mixing pure oxygen with air). The total inlet gas flow rate was maintained at 5 L min^{-1} using two mass flow controllers. The broth permittivity was monitored using a biomass sensor (Fogale Nanotech), as described by Horta et al. [26,53].

For the continuous fed-batch cultures (B2 and B3), the exponential feed flow was initiated once the carbon sources in the batch medium had been exhausted. The feed flow rates of the complex (culture B2) or defined (culture B3) media were calculated using Equation (7), which assumes that the growth rate is only dependent on the limiting substrate [54], and were automatically controlled by the supervisory system [26].

$$F = \left(\frac{\mu}{Y_{x/s}} + m\right) \cdot \frac{C_{xo} - V_o}{C_{SF} - C_{SR}} \cdot e^{(\mu_{set} - t)} \qquad (7)$$

In Equation (7), F (L h^{-1}) is the feed flow rate, μ (h^{-1}) is the actual specific growth rate, μ_{SET} (h^{-1}) is the desired specific growth rate, $Y_{X/S}$ (gDCW $g_{substrate}^{-1}$) is the biomass yield coefficient, m (h^{-1}) is the maintenance coefficient, C_{XO} (gDCW L^{-1}) and V_0 (L) correspond to the cell concentration and volume, respectively, at the beginning of the fed-batch phase, C_{SF} ($g_{substrate}$ L^{-1}) is the glycerol concentration in the supplementary medium (Table 1), C_{SR} ($g_{substrate}$ L^{-1}) is the expected residual concentration of glycerol in the reactor, and t (h) is the elapsed feeding time. The parameters $Y_{X/S}$ and m were estimated from the previous batch data, and are included in Table 5.

Table 5 Parameter values used for setting up the feeding flow rate (Eq. 7) for fed-batch bioreactor cultures B2 (complex medium) and B3 (defined medium)

Parameter	B2	B3
m (h^{-1})	0.001	0.05
$Y_{X/S}$ (gDCW $g_{substrate}^{-1}$)	0.717	0.485

Analogously, C_{SR} was set to 5 g L^{-1}, based on information obtained in earlier experiments [29].

The exponential feeding profile was regulated using SuperSys_HCDC^R, which automatically re-tuned the parameter μ_{set} every 10 min, using the values of μ inferred from on-line permittivity measurements provided by the capacitance biomass sensor [26].

In intermittent fed-batch culture B1, additional glycerol and lactose were supplied as a pulse (Table 1) when the OD_{600} value reached ~70, in order to provide enough inducer and carbon source for the prolonged induction phase, which lasted 24 h. In continuous fed-batch cultivations B2 and B3, the induction phase was started when the OD_{600} reached ~100, and also lasted 24 h. For experiment B3 (defined medium), the induction was performed by adding an IPTG pulse (1 mM final concentration). For experiment B2 (complex medium), induction was carried out by adding multiple pulses of 400 mL of lactose solution (200 g L^{-1}). In all bioreactor cultures, ampicillin pulses were applied every 6 h after induction to maintain selective pressure on the cells.

Competing interests

The authors declare that they have no competing interests.

Authors' contributions

AMV participated in the performance of all the experiments and drafted the manuscript. AJS was involved in revising the manuscript critically for important intellectual content. ACLH implemented the control and instrumentation tool used to supervise the bioreactor experiments. CRS helped to perform the continuous fed-batch experiments. GC and GGS helped to perform the auto-induction experiment. RLCG helped in discussion of the purification results. TCZ conceived the study and helped to draft the manuscript. All authors read and approved the final manuscript.

Acknowledgements

The authors would like to thank the São Paulo State Research Foundation (FAPESP) (process 2009/00177-2), the National Council for Scientific and Technological Development (CNPq/Brazil) and the Brazilian Federal Agency for Support and Evaluation of Graduate Education (CAPES) for funding this work. We also acknowledge Prof. Dr. Laura Pinotti and Amadeus Gomes de Azevedo for technical support as well as Prof. Dr. Ilton José Baraldi for helping us with statistical analysis.

Author details

[1]Chemical Engineering Department, Federal University of São Carlos (UFSCar), Rodovia Washington Luís, km 235, C.P. 676, CEP 13565-905 São Carlos, SP, Brazil. [2]Institute of Chemistry, Federal University of Goiás, Campus Samambaia, C.P. 131, CEP 74001-970 Goiânia, GO, Brazil.

References

1. Elander RP: **Industrial production of beta-lactam antibiotics.** *Appl Microbiol Biotechnol* 2003, **61**:385–392.
2. de Souza VR, Silva ACG, Pinotti LM, Sobreiro H, Araújo S, de Lima R, Giordano C: **Characterization of the Penicillin G Acylase from *Bacillus megaterium* ATCC 14945.** *Braz Arch Biol Tech* 2005, **48**(June):105–111.
3. Ignatova Z, Mahsunah A, Georgieva M, Kasche V: **Improvement of posttranslational bottlenecks in the production of penicillin amidase in recombinant *Escherichia coli* strains.** *Appl Environ Microbiol* 2003, **69**:1237–1245.
4. Hewitt L, Kasche V, Lummer K, Lewis RJ, Murshudov GN, Verma CS, Dodson GG, Wilson KS: **Structure of a slow processing precursor penicillin acylase from *Escherichia coli* reveals the linker peptide blocking the active-site cleft.** *J Mol Biol* 2000, **302**:887–898.
5. Rajendhran J, Gunasekaran P: **Recent biotechnological interventions for developing improved penicillin G acylases.** *J Biosci Bioeng* 2004, **97**:1–13.
6. Lin Y-H, Hsiao H-C, Chou CP: **Strain improvement to enhance the production of recombinant penicillin acylase in high-cell-density *Escherichia coli* cultures.** *Biotechnol Prog* 2002, **18**:1458–1461.
7. Cheng T, Chen M, Zheng H, Wang J, Yang S, Jiang W: **Expression and purification of penicillin G acylase enzymes from four different micro-organisms, and a comparative evaluation of their synthesis/ hydrolysis ratios for cephalexin.** *Protein Expr Purif* 2006, **46**:107–113.
8. Grulich M, Štěpánek V, Kyslík P: **Perspectives and industrial potential of PGA selectivity and promiscuity.** *Biotechnol Adv* 2013, **31**:1458–1472.
9. Chou CP, Tseng JH, Lin MI, Lin HK, Yu CC: **Manipulation of carbon assimilation with respect to expression of the *pac* gene for improving production of penicillin acylase in *Escherichia coli*.** *J Biotechnol* 1999, **69**:27–38.
10. Chou CP, Yu CC, Tseng JH, Lin MI, Lin HK: **Genetic manipulation to identify limiting steps and develop strategies for high-level expression of penicillin acylase in *Escherichia coli*.** *Biotechnol Bioeng* 1999, **63**:263–272.
11. Akkaya Ö, Öztürk SI, Bolhuis A, Gümüşel F: **Mutations in the translation initiation region of the *pac* gene resulting in increased levels of activity of penicillin G acylase.** *World J Microbiol Biotechnol* 2012, **28**:2159–2164.
12. Wang T, Zhu H, Ma X, Fei Z, Ma Y, Wei D: **Enhancing enzymatic activity of penicillin G acylase by coexpressing *pcm* gene.** *Appl Microbiol Biotechnol* 2006, **72**:953–958.
13. Martín L, Prieto M, Cortés E, García JL: **Cloning and sequencing of the *pac* gene encoding the penicillin G acylase of *Bacillus megaterium* ATCC 14945.** *FEMS Microbiol Lett* 1995, **125**:287–292.
14. Cai G, Zhu S, Yang S, Zhao G, Jiang W: **Cloning, overexpression and characterization of a novel thermostable Penicillin G Acylase from *Achromobacter xylosoxidans*: probing the molecular basis for its high thermostability.** *Appl Environ Microbiol* 2004, **70**:2764–2770.
15. Kim E-K, Moon JC, Lee JM, Jeong MS, Oh C, Ahn S-M, Yoo YJ, Jang HH: **Large-scale production of soluble recombinant amyloid-β peptide 1-42 using cold-inducible expression system.** *Protein Expr Purif* 2012, **86**:53–57.
16. Weng Y-P, Hsu F-C, Yang W-S, Chen H-P: **Optimization of the overexpression of glutamate mutase S component under the control of T7 system by using lactose and IPTG as the inducers.** *Enzyme Microb Technol* 2006, **38**:465–469.
17. Rodríguez-Carmona E, Cano-Garrido O, Dragosits M, Maurer M, Mader A, Kunert R, Mattanovich D, Villaverde A, Vázquez F: **Recombinant Fab expression and secretion in *Escherichia coli* continuous culture at medium cell densities: Influence of temperature.** *Process Biochem* 2012, **47**:446–452.
18. Keilmann C, Wanner G, Böck A: **Molecular basis of the exclusive low-temperature synthesis of an enzyme in *E. coli*: penicillin acylase.** *Biol Chem Hoppe Seyler* 1993, **374**:983–992.
19. Sørensen HP, Mortensen KK: **Soluble expression of recombinant proteins in the cytoplasm of *Escherichia coli*.** *Microb Cell Fact* 2005, **4**:1.
20. Karthikeyan R, Surianarayanan M, Sudharshan S, Gunasekaran P, Asit Baran M: **Biocalorimetric and respirometric studies on production of Penicillin G acylase from *Bacillus badius pac* in *E. coli* DH5α.** *Biochem Eng J* 2011, **55**:223–229.
21. Wen Y, Feng M, Yuan Z, Zhou P: **Expression and overproduction of recombinant penicillin G acylase from *Kluyvera citrophila* in *Escherichia coli*.** *Enzyme Microb Technol* 2005, **37**:233–237.
22. Cheng S, Song Q, Wei D, Gao B: **High-level production penicillin G acylase from *Alcaligenes faecalis* in recombinant *Escherichia coli* with optimization of carbon sources.** *Enzyme Microb Technol* 2007, **41**:326–330.
23. de León-Rodríguez A, Rivera-Pastrana D, Medina-Rivero E, Flores-Flores JL, Estrada-Baltazar A, Ordóñez-Acevedo LG, de la Rosa APB: **Production of penicillin acylase by a recombinant *Escherichia coli* using cheese whey as substrate and inducer.** *Biomol Eng* 2006, **23**:299–305.
24. Liu Y-C, Chang W-M, Lee C-Y: **Effect of oxygen enrichment aeration on penicillin G acylase production in high cell density culture of recombinant *E. coli*.** *Bioprocess Engineering* 1999, **21**:227–230.
25. Chandel AK, Rao LV, Narasu ML, Singh OV: **The realm of penicillin G acylase in β-lactam antibiotics.** *Enzyme Microb Technol* 2008, **42**:199–207.
26. Horta ACL, Sargo CR, da Silva AJ, de Carvalho Gonzaga M, Dos Santos MP, Gonçalves VM, Zangirolami TC, de Campos Giordano R: **Intensification of high cell-density cultivations of *rE. coli* for production of *S. pneumoniae* antigenic surface protein, PspA3, using model-based adaptive control.** *Bioprocess Biosyst Eng* 2012, **35**:1269–1280.
27. Studier FW: **Protein production by auto-induction in high-density shaking cultures.** *Protein Expr Purif* 2005, **41**:207–234.
28. Lee SK, Keasling JD: **Heterologous protein production in *Escherichia coli* using the propionate-inducible pPro system by conventional and auto-induction methods.** *Protein Expr Purif* 2008, **61**:197–203.
29. da Silva AJ, Horta ACL, Velez AM, Iemma MRC, Sargo CR, Giordano RL, Novo MTM, Giordano RC, Zangirolami TC: **Non-conventional induction strategies for production of subunit swine erysipelas vaccine antigen in *rE. coli* fed-batch cultures.** *Springer Plus* 2013, **2**:322.
30. Jia L, Cheng H, Wang H, Luo H, Yan H: **From shake flasks to bioreactors: survival of *E. coli* cells harboring pGST-hPTH through auto-induction by controlling initial content of yeast extract.** *Appl Microbiol Biotechnol* 2011, **90**:1419–1428.
31. Tyler RC, Sreenath HK, Singh S, Aceti DJ, Bingman CA, Markley JL, Fox BG: **Auto-induction medium for the production of [U-15N]- and [U-13C, U-15N]-labeled proteins for NMR screening and structure determination.** *Protein Expr Purif* 2005, **40**:268–278.
32. Dai M, Zhu Y, Yang Y, Wang E, Xie Y, Zhao G, Jiang W: **Expression of penicillin G acylase from the cloned *pac* gene of *Escherichia coli* ATCC11105. Effects of *pac*R and temperature.** *Eur J Biochem* 2001, **268**:1298–1303.
33. Sobotková L, Stepánek K, Plhácková K, Kyslík P: **Development of a high-expression system for penicillin G acylase based on the recombinant *Escherichia coli* strain RE3 (pKAlS).** *Enzyme Microb Technol* 1996, **19**:389–397.
34. da Silva AJ, da Costa Iemma MR, Luperni Horta AC, Sargo CR, de Lima Camargo Giordano R, de Campos Giordano R, Zangirolami TC, Marques Novo MT: **Cloning, auto-induction expression, and purification of rSpaA swine erysipelas antigen.** *Curr Microbiol* 2012, **65**:369–374.
35. Xu J, Banerjee A, Pan S-H, Li ZJ: **Galactose can be an inducer for production of therapeutic proteins by auto-induction using *E. coli* BL21 strains.** *Protein Expr Purif* 2012, **83**:30–36.
36. Summers DK: **The kinetics of plasmid loss.** *Trends Biotechnol* 1991, **9**:273–278.
37. Carneiro S, Ferreira EC, Rocha I: **Metabolic responses to recombinant bioprocesses in *Escherichia coli*.** *J Biotechnol* 2013, **164**:396–408.
38. Lecina M, Sarró E, Casablancas A, Gòdia F, Cairó JJ: **IPTG limitation avoids metabolic burden and acetic acid accumulation in induced fed-batch cultures of *Escherichia coli* M15 under glucose limiting conditions.** *Biochem Eng J* 2013, **70**:78–83.
39. Matanguihan RM, Konstantinov KB, Yoshida T: **Dielectric measurement to monitor the growth and the physiological states of biological cells.** *Bioprocess Engineering* 1994, **11**:213–222.
40. Luli GW, Strohl WR: **Comparison of growth, acetate production and acetate inhibition of *Escherichia coli* strains in batch and fed-batch fermentations.** *Appl Environ Microbiol* 1990, **56**:1004–1011.
41. Li Z, Kessler W, van den Heuvel J, Rinas U: **Simple defined autoinduction medium for high-level recombinant protein production using T7-based *Escherichia coli* expression systems.** *Appl Microbiol Biotechnol* 2011, **91**:1203–1213.
42. Korz DJ, Rinas U, Hellmuth K, Sanders E a, Deckwer WD: **Simple fed-batch technique for high cell density cultivation of *Escherichia coli*.** *J Biotechnol* 1995, **39**:59–65.
43. Ramirez DM, Bentley WE: **Enhancement of recombinant protein-synthesis and stability via coordinated amino-acid addition.** *Biotechnol Bioeng* 1993, **41**:557–565.

44. Pinotti LM, Fonseca LP, Prazeres DMF, Rodrigues DS, Nucci ER, Giordano RLC: **Recovery and partial purification of penicillin G acylase from *E. coli* homogenate and *B. megaterium* culture medium using an expanded bed adsorption column.** *Biochem Eng J* 2009, **44:**111–118.
45. Marcos J, Fonseca L, Ramalho M, Cabral JM: **Application of surface response analysis to the optimization of penicillin acylase purification in aqueous two-phase systems.** *Enzyme Microb Technol* 2002, **31:**1006–1014.
46. Fong BA, Wood DW: **Expression and purification of ELP-intein-tagged target proteins in high cell density *E. coli* fermentation.** *Microb Cell Fact* 2010, **9:**77.
47. Abian O, Grazu V, Hermoso J, García L, Ferna R, Guisán M: **Stabilization of Penicillin G Acylase from *Escherichia coli* : Site-Directed Mutagenesis of the Protein Surface To Increase Multipoint Covalent Attachment.** *Appl Environ Microbiol* 2004, **70:**1249–1251.
48. Montes T, Grazú V, López-Gallego F, Hermoso JA, García JL, Manso I, Galán B, González R, Fernández-Lafuente R, Guisán JM: **Genetic modification of the penicillin G acylase surface to improve its reversible immobilization on ionic exchangers.** *Appl Environ Microbiol* 2007, **73:**312–319.
49. Bradford MM: **A rapid and sensitive method for the quantitation of microgram quantities of protein utilizing the principle of protein-dye binding.** *Anal Biochem* 1976, **72:**248–254.
50. Laemmli UK: **Cleavage of structural proteins during the assembly of the head of Bacteriophage T4.** *Nature* 1970, **227:**680–685.
51. Kutzbach C, Rauenbus E: **Preparation and general properties of crystalline penicillin acylase from *Escherichia coli* ATCC-11 105.** *Hoppe-seylers Zeitschrift Für Physiologische Chemie* 1974, **355:**45–53.
52. Balasingham K, Warburto D, Dunnil P, Lilly MD: **Isolation and kinetics of penicillin amidase from *Escherichia coli.*** *Biochim Biophys Acta* 1972, **276:**250–256.
53. Horta ACL, da Silva AJ, Sargo CR, Gonçalves VM, Zangirolami TC, Giordano RDC: **Robust artificial intelligence tool for automatic start-up of the supplementary medium feeding in recombinant *E. coli* cultivations.** *Bioprocess Biosyst Eng* 2011, **34:**891–901.
54. Nielsen J, Villadsen J, Lidén G: *Principles Bioreaction Engineering Principles Second Edition.* New York: Kluwer academic/Plenum publishers; 2003.

6

Biosynthesis of methylated resveratrol analogs through the construction of an artificial biosynthetic pathway in *E. coli*

Sun-Young Kang[1,2†], Jae Kyoung Lee[1,2†], Oksik Choi[1], Cha Young Kim[3], Jae-Hyuk Jang[1], Bang Yeon Hwang[2] and Young-Soo Hong[1*]

Abstract

Background: Methylated resveratrol analogs show similar biological activities that are comparable with those of the resveratrol. However, the methylated resveratrol analogs exhibit better bioavailability as they are more easily transported into the cell and more resistant to degradation. Although these compounds are widely used in human health care and in industrial materials, at present they are mainly obtained by extraction from raw plant sources. Accordingly their production can suffer from a variety of economic problems, including low levels of productivity and/or heterogeneous quality. On this backdrop, large-scale production of plant metabolites via microbial approaches is a promising alternative to chemical synthesis and extraction from plant sources.

Results: An *Escherichia coli* system containing an artificial biosynthetic pathway that produces methylated resveratrol analogues, such as pinostilbene (3,4′-dihydroxy-5-methoxystilbene), 3,5-dihydroxy-4′-methoxystilbene, 3,4′-dimethoxy-5-hydroxystilbene, and 3,5,4′-trimethoxystilbene, from simple carbon sources is developed. These artificial biosynthetic pathways contain a series of codon-optimized *O*-methyltransferase genes from sorghum in addition to the resveratrol biosynthetic genes. The *E. coli* cells that harbor pET-opTLO1S or pET-opTLO3S produce the one-methyl resveratrol analogues of 3,5-dihydroxy-4′-methoxystilbene and pinostilbene, respectively. Furthermore, the *E. coli* cells that harbor pET-opTLO13S produce 3,5-dihydroxy-4′-methoxystilbene, *bis*-methyl resveratrol (3,4′-dimethoxy-5-hydroxystilbene), and *tri*-methyl resveratrol (3,5,4′-trimethoxystilbene).

Conclusions: Our strategy demonstrates the first harness microorganisms for *de novo* synthesis of methylated resveratrol analogs used a single vector system joined with resveratrol biosynthetic genes and sorghum two resveratrol *O*-methyltransferase genes. Thus, this is also the first report on the production of the methylated resveratrol compounds *bis*-methyl and *tri*-methyl resveratrol (3,4′-dimethoxy-5-hydroxystilbene and 3,5,4′-trimethoxystilbene) in the *E. coli* culture. Thus, the production of the methylated resveratrol compounds was performed on the simple *E. coli* medium without precursor feeding in the culture.

Background

Resveratrol (3,5,4′-trihydroxystilbene), which is found in red wine and grapes as well as in other plants, has been the subject of intensive studies that focus on its possible role in preventing cardiovascular heart diseases and cancer [1]. Resveratrol and its analogs have important functions as antimicrobial and antioxidant compounds in plant defense responses to environmental stresses, such as UV irradiation and fungal infection. However, they also exhibit diverse beneficial properties in humans, including anti-inflammatory effects, anti-tumor activities, and anti-aging effects [2]. In particular, over the past decade, resveratrol has been used as a dietary supplement that has been demonstrated to possess a broad spectrum of pharmacological properties [3,4]. However, Walle *et al.* reported the high absorption but rapid metabolism of resveratrol when it was administered orally to humans [5]. Several reports have demonstrated that the methylation of

* Correspondence: hongsoo@kribb.re.kr
†Equal contributors
[1]Chemical Biology Research Center, Korea Research Institute of Bioscience and Biotechnology(KRIBB), 30 Yeongudanji-ro, Ochang-eup, Chungbuk 363-883, Republic of Korea
Full list of author information is available at the end of the article

resveratrol results in the enhancement of its bioavailability and bioactivity [6,7]. For example, pterostilbene is a 3,5 *bis*-methylated resveratrol that is present in blueberries and it has been investigated extensively [8]. The substitution of the hydroxy with the methoxy groups increases the lipophilicity of pterostilbene over the resveratrol, which results in high bioavailability. These differences in the pharmacokinetics might explain the higher biological activity of pterostilbene over its parental compound resveratrol [9]. Furthermore, 3,5, 4'-trimethoxystilbene has emerged as the most potent proapoptotic analog of resveratrol [10]. In addition, 3,5,4'-trimethoxystilbene and pinostilbene (3,4'-dihydroxy-5-methoxystilbene) were reported to be up to 100-fold more cytotoxic than resveratrol in cancer cell lines [11]. Consequently, methylated resveratrol analogs, which can possess greater oral bioavailability due to their decreased metabolism and increased absorption, have garnered increasing attention as alternative chemopreventive agents. Therefore, methylated resveratrols have become an attractive target for bioengineering, but few attempts to characterize the methylation enzyme of resveratrol have been reported thus far [12-17].

Methylated resveratrol is biosynthesized from the general phenylpropanoid pathway beginning with phenylalanine or tyrosine, or both. The methylated resveratrol biosynthesis pathway designed for use in this study is depicted in Figure 1. In this pathway, tyrosine is converted into 4-coumaric acid using tyrosine ammonia lyase (TAL). The 4-coumaric acid is then activated to 4-coumaroyl-CoA using the 4-coumarate-CoA ligase (4CL). This 4-coumaroyl-CoA is condensed with three molecules of malonyl-CoA via stilbene synthase (STS), which is the key enzyme in resveratrol synthesis. Resveratrol is converted into its methylated derivatives through the resveratrol *O*-methyltransferase. Meanwhile, several attempts to produce resveratrol using recombinant microorganisms, such as *Escherichia coli* [18-23] and *Saccharomyces cerevisiae* [24,25], have been reported. However, little has been reported about the characterization of the resveratrol *O*-methyltransferase function in plants [13-15,17]. Recently, the *sbOMT1* and *sbOMT3* genes from

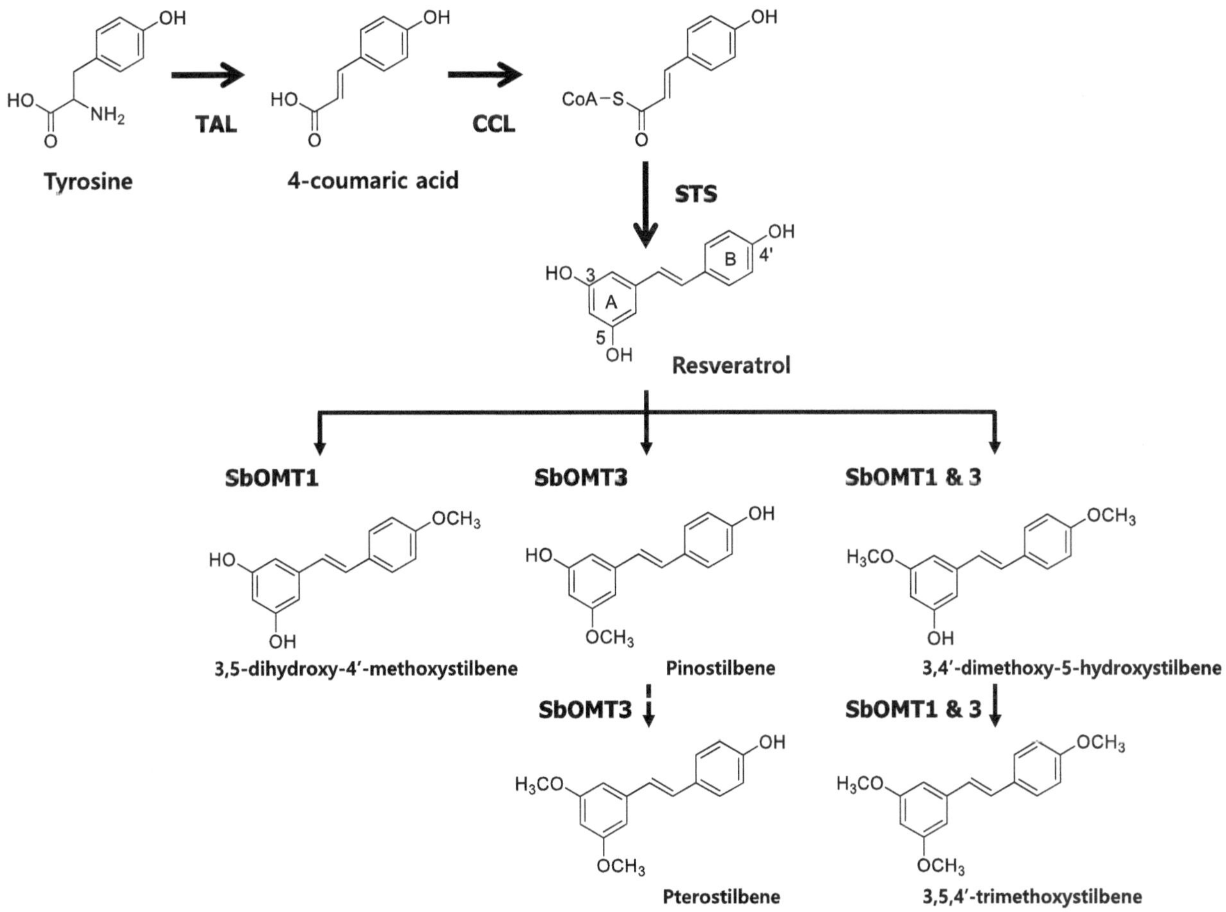

Figure 1 Engineered biosynthetic pathways for the methylated resveratrol analogs starting from tyrosine in *E. coli*. TAL (tyrosine ammonia-lyase) from *S. espanaensis*, CCL (4-coumarate:CoA ligase) from *S. coelicolor*, STS (stilbene synthases) from *A. hypogaea*, and sbOMT1 & sbOMT3 (resveratrol *O*-methyltransferases) from *S. bicolor*.

Sorghum bicolor were reported and their functions were characterized as different resveratrol *O*-methyltransferases (ROMTs), respectively [12,13,16]. Furthermore, Katsuyama *et al.* reported the production of the methylated resveratrols pinostilbene and pterostilbene in recombinant *E. coli* using the pinosylvin methyltransferase (OsPMT) gene from *Oryza sativa* and the tyrosine feeding method in a culture medium [15].

In this study, we describe the production of the methylated resveratrol analogs of pinostilbene, 3,5-dihydroxy-4'-methoxystilbene, 3,4'-dimethoxy-5-hydroxystil bene, and 3,5,4'-trimethoxystilbene using the recom binant *E. coli* that harbors an artificial biosynthetic pathway. These artificial biosynthetic pathways contain a series of codon-optimized *sbOMT1* or *sbOMT3 O*-methyltransferase genes, and both *O*-methyltransferase genes from sorghum in addition to the resveratrol biosynthetic gene cluster. The recombinant *E. coli* produces methylated resveratrol analogs beginning with the simple sugar fermentation, but not in the bioconversion of the intermediates.

Results and discussion

Characterization of the resveratrol *O*-methyltransferase via a bioconversion experiment in recombinant *E. coli*

Previous reports have noted that the two resveratrol *O*-methyltransferase genes (*sbOMT1* and *sbOMT3*) from *Sorghum bicolor* are capable of using resveratrol as a substrate that yields methylated analogs of resveratrol [12,13]. It was claimed that the *sbOMT3 O*-methyltransferase catalyzes the A-ring specific 3,5-*bis*-*O*-methylation of resveratrol, which in turn yields pterostilbene (3,5-dimethoxy-4'-hydroxystilbene) in the co-expression system of *sbOMT3* with a stilbene synthase from peanuts (AhSTS3) [13]. In addition, *sbOMT1*, which had previously been identified as a potential eugenol *O*-methyltransferase, predominantly catalyzes the resveratrol B-ring (4'-*O*-methylation), which yields 3,5-dihydroxy-4'-methoxystilbene [12,13].

In this study, the functions of the codon-optimized resveratrol *O*-methyltransferase genes (*sbOMT1* and *sbOMT3*) were re-evaluated using a bioconversion experiment with resveratrol in recombinant *E. coli*. The codon-optimized synthetic *sbOMT1* and *sbOMT3* genes were cloned in the expression vector pET-22b(+) on the *NdeI*/*Hind*III sites (pET22-sbCOM1 and pET22-sbCOM3, respectively; Figure 2). Each construct was transformed in the *E. coli* C41 (DE3) cells, and the recombinant *E. coli* was selected based on the expression analysis using SDS-PAGE (Additional file 1: Figure S1). In order to investigate whether the recombinant *O*-methyltransferases can catalyze the production of methylated resveratrol derivatives, resveratrol was added to the cultured recombinant *E. coli* harboring pET22-sbCOM1 and pET22-sbCOM3. The culture broth and bacterial cells were collected after 36 hours and were then subjected to HPLC and LC/MS analyses (Figure 3). Under the bioconversion conditions employed in this study most of the feeding resveratrol disappeared and each methylated form was detected as a main peak in the HPLC (Figure 3A). However, when the bioconversion rate was calculated based on a quantitative comparison with feeding substrates and the products, the recombinant *E. coli* harboring pET22-sbCOM1 and pET22-sbCOM3 showed roughly 42% and 12% conversion ratios, respectively (data not shown). These bioconversion ratios

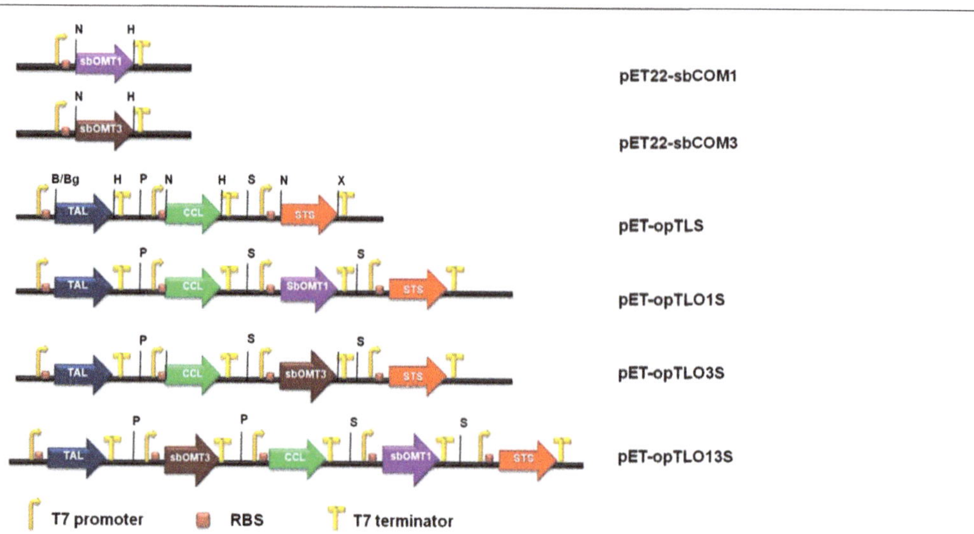

Figure 2 Expression vectors of *O*-methyltransferase and organization of the artificial gene clusters used for the production of each methylated analogs of resveratrol in *E. coli*. All constructs contained the T7 promoter, RBS in front of each gene, and T7 terminator at the rear of each gene. B, *Bam*HI; Bg, *Bgl*II; H, *Hind*III; P, *Pac*I; S, *Spe*I; N, *Nde*I; X, *Xho*I.

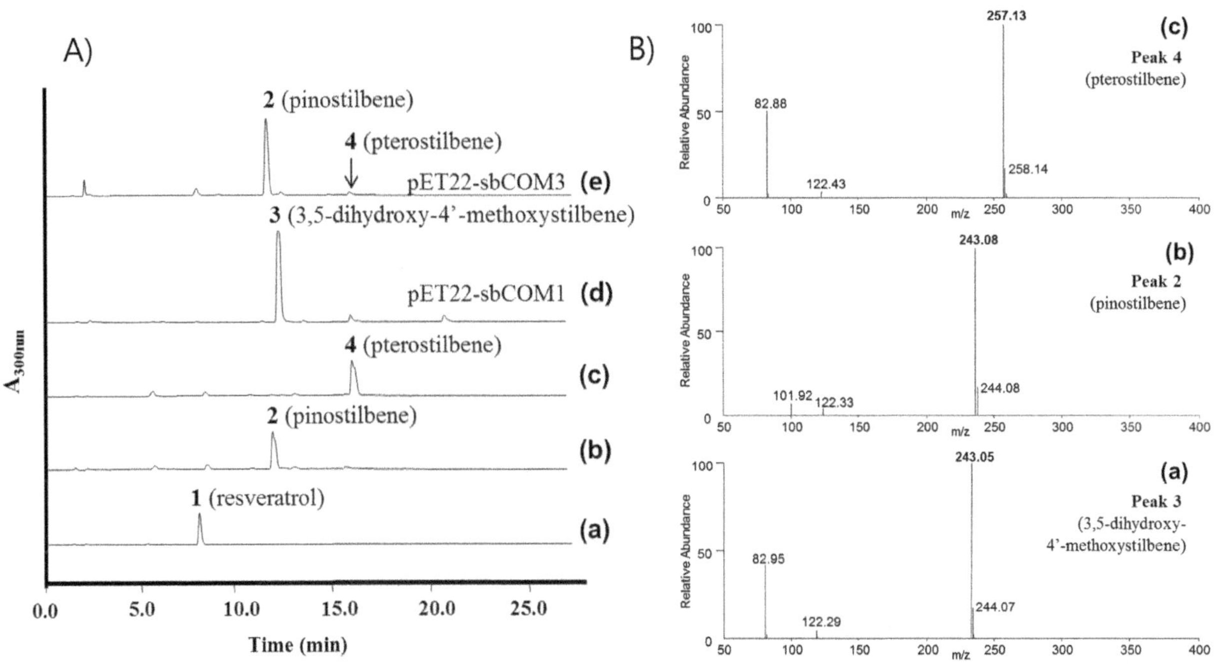

Figure 3 HPLC profile (A) and selected mass ion chromatogram (B) of bioconversion experiments. A; HPLC profile of the **(a)** standard resveratrol, **(b)** pinostilbene, and **(c)** pterostilbene; **(d)** resveratrol supplemented *E. coli* harboring pET22-sbCOM1; **(e)** resveratrol supplemented *E. coli* harboring pET22-sbCOM3. The absorbance was monitored at 300 nm. Peak 1, resveratrol; peak 2, pinostilbene; peak 3, 3,5-dihydroxy-4'-methoxystilbene; peak 4, pterostilbene. **B**; Selected mass ion chromatogram of **(a)** 3,5-dihydroxy-4'-methoxystilbene (*m/z* 243.05 $[M+H]^+$) produced by *E. coli* harboring pET22-sbCOM1; **(b)** pinostilbene (*m/z* 243.08 $[M+H]^+$) and **(c)** pterostilbene (*m/z* 257.13 $[M+H]^+$) produced by *E. coli* harboring pET22-sbCOM3. The minor peak of 15.9 min in the HPLC profile **(d)** was not seen in the mass ion peaks of pterostilbene.

are consistent with results recently reported by Jeong *et al.* [16]. It is presumed that a significant amount of additive resveratrol was decomposed in the *E. coli* culture medium.

The recombinant *E. coli* cells that harbored the pET22-sbCOM1 plasmid produced a 12.2 min retention time peak, which is a slightly later retention time than the pinostilbene (3,4'-dihydroxy-5-methoxystilbene) as seen in Figure 3A(d). The major peak in Figure 3A(d) exhibited parent mass ion peaks at *m/z* 243.05 $[M+H]^+$, which corresponded to one methylation of resveratrol (an addition of 14 Da; Figure 3B(a)). It was expected that this methylated compound could have a methoxy group located in the 4' position in the B-ring (Figure 1). In the pET22-sbCOM3 clone, a major peak was present at the same retention time (11.6 min) as that in the HPLC analysis and at the same mass ion peaks at *m/z* 243.08 $[M+H]^+$ with authentic pinostilbene, which is one methylation of resveratrol (Figure 3B(b)). However, this recombinant produced a very low level minor peak of 15.9 min in the HPLC, which was the same retention time as the pterostilbene (3,5-dimethoxy-4'-hydroxystilbene; Figure 3A(e)). This minor peak was also accepted based on the mass spectra as *m/z* 257.13 $[M+H]^+$, which corresponded to two methylations of resveratrol (an addition of 28 Da; Figure 3B(c)). However, the peak intensity was not sufficient to allow a detailed structural characterization. These results exhibit a controversial conclusion that contrasts with the results reported previously by Rimando *et al.*, who demonstrated that the sbOMT3 produced pterostilbene as a major product [13]. However, our results agreed with Jeong *et al.*'s reported results with the *in vitro* and bioconversion activity of the sbOMT3 [16].

Construction of artificial biosynthetic pathways for the production of methylated resveratrol analogs in *E. coli*

We have previously produced resveratrol in *E. coli* harboring an artificial biosynthetic gene cluster in which the TAL from *Saccharothrix espanaensis*, CCL from *Streptomyces coelicolor*, and STS from the peanut plant *Arachis hypogaea* were contained [18]. In order to produce the methylated resveratrol analogs in *E. coli* using a simple sugar medium, a series of plasmids containing the artificial biosynthetic pathway were constructed (Figure 2). The artificial resveratrol and methylated resveratrol biosynthetic plasmids were constructed using the previously described cloning methods [18], and each plasmid contained genes with their own T7 promoter, ribosome-binding site (RBS), and terminator sequence as in the parental vectors. In this study, a new resveratrol-producing construct of pET-opTLS, which contained codon-optimized *tal* and *sts* genes, and cloned *ccl* gene from *S. coelicolor*, was constructed using previously reported cloning method of the parental vector pET-TLkS [18]. The *E. coli* cells with the pET-opTLS clone exhibited a higher production yield of resveratrol (5.2 mg/L) in the culture system with a modified M9 medium compared with the original pET-TLkS clone (1.4 mg/L) [18]. The cause of this improvement remains unknown, but it is possible that these protein expression ratios may have better optimized the resveratrol production than the original combination.

In order to construct an expression vector that contains the additional *O*-methyltransferase gene(s) that are under the control of the T7 promoter, the DNA fragment containing the promoter, *O*-methyltransferase coding region, and terminator using pET22-sbCOM1 and pET22-sbCOM3 plasmids as templates were amplified. Then, the amplified fragments were inserted into pET-opTLS, which resulted in pET-opTLO1S and pET-opTLO3S, respectively. Similarly, in order to construct the plasmid containing the two *O*-methyltransferases biosynthetic pathways, the *O*-methyltransferase fragment containing the *sbOMT3* coding region was inserted into the pET-opTLO1S, which resulted in pET-opTLO13S. This is useful assembly method for the reconstruction of multi-enzyme biosynthetic pathways such as stilbenes and flavones biosynthesis.

The recombination cells that harbor the artificial biosynthetic gene cluster were cultured in a modified M9 medium. A comparison of the fermentation products of the *E. coli* cells that harbored pET-opTLS, pET-opTLO1S, pET-opTLO3S, and pET-opTLO13S revealed that new peaks were detected in the engineered strain (Figure 4). The retention times for peaks 1 and 2, which were produced in the recombinant pET-opTLO3S, were identical to those of the original resveratrol and pinostilbene, respectively (Figure 4(e)). These peaks were further analyzed using LC/MS/MS in the positive ion mode. Using the LC/MS analysis, peak 1 (resveratrol) and peak 2 (pinostilbene) were identified as *m/z* 229.02 $[M + H]^+$ and *m/z* 243.08 $[M + H]^+$, respectively, when comparing the obtained fragmentation pattern with that of the original standards. Furthermore, peak 3, which was produced in the recombinant pET-opTLO1S and pET-opTLO13S, was identified as *m/z* 243.05 $[M + H]^+$ using the LC/MS. However, the MS/MS analysis of peak 3 exhibited different fragments in the ion pattern with the pinostilbene, and the peaks of *m/z* 134.88 $[M + H]^+$ that represent the two hydroxyl group located in the A-ring was also identified (Figure 5(a)). The fragment ion at *m/z* 134.88 $[M + H]^+$, which contained the phenolic A-ring of a compound with a triple bond, has already been reported as a distinctive mass fragment of resveratrol [26]. The A-ring cleavage of the resveratrol produced a mass peak of a triple bond (A-ring) at *m/z* 134.86 $[M + H]^+$ due to the

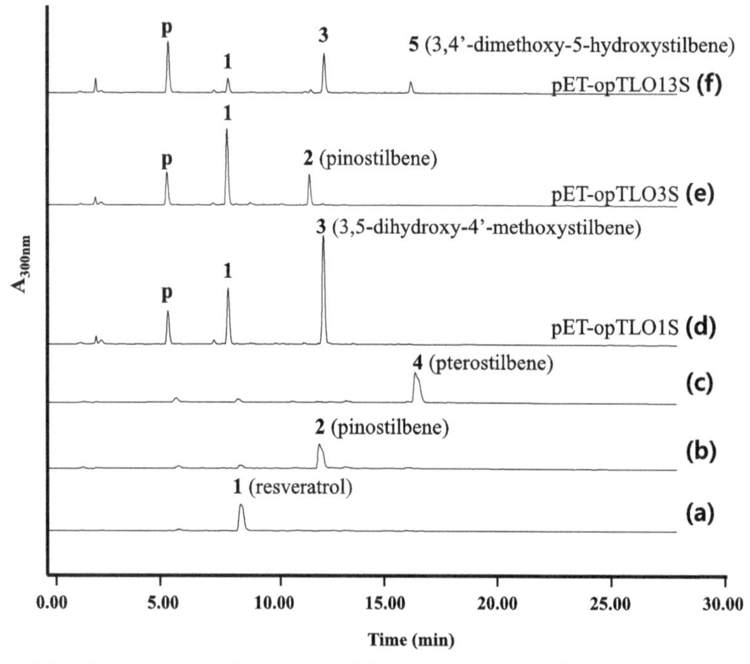

Figure 4 HPLC profile of methylated resveratrol analogs in recombinant *E. coli*. HPLC profile of **(a)** the standard resveratrol, **(b)** pinostilbene, and **(c)** pterostilbene; **(d)** culture extract of *E. coli* harboring pET-opTLO1S; **(e)** culture extract of *E. coli* harboring pET-opTLO3S; **(f)** culture extract of *E. coli* harboring pET-opTLO13S in M9 medium. The absorbance was monitored at 300 nm. p, 4-coumaric acid; peak 1, resveratrol; peak 2, pinostilbene; peak 3, 3,5-dihydroxy-4'-methoxystilbene; peak 4, pterostilbene; peak 5, 3,4'-dimethoxy-5-hydroxystilbene.

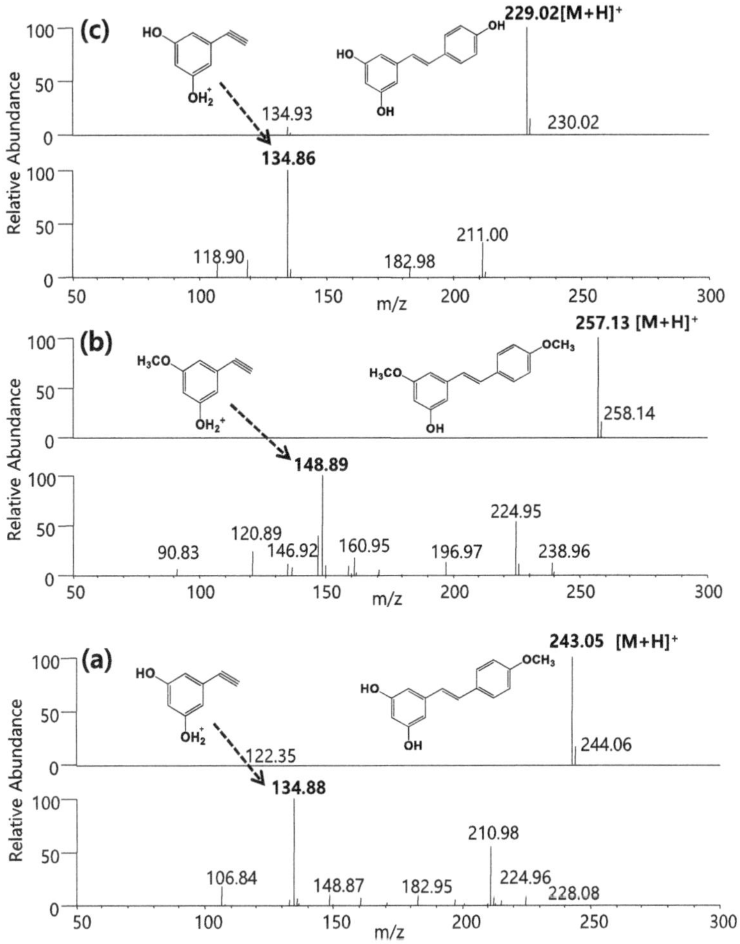

Figure 5 Mass fragmentation analysis of methylated resveratrol analogs in recombinant *E. coli*. Structures and MS^n spectra of the predicted analogs produced using the artificial biosynthetic pathways. **(a)** 3,5-dihydroxy-4'-methoxystilbene; **(b)** 3,4'-dimethoxy-5-hydroxystilbene; and **(c)** standard resveratrol.

hydroxy group at the C-3 and C-5 positions (Figure 5(c)). However, due to the presence of a methoxy group in the A-ring in pinostilbene, the triple bond mass peak appeared at *m/z* 148.89 $[M + H]^+$. In addition, peak 5 exhibited a very similar retention time and the same mass peak (*m/z* 257.13 $[M + H]^+$) as pterostilbene (Additional file 1: Figure S2); however, the major daughter ion peaks of *m/z* 148.89 $[M + H]^+$ represented one methoxy group located in the A-ring (Figure 5(b)). Interestingly, the mass fragmentation pattern of pterostilbene, which was found in a detectable amount in the bioconversion experiment with the recombinant pET22-sbCOM3 as mentioned earlier, was not detected in the recombinant pET-opTLO3S and pET-opTLO13S. These results support that peak 5 was not pterostilbene (3,5-dimethoxy-4'-hydroxystilbene), but rather that it was 3,4'-dimethoxy-5-hydroxystilbene, which has methoxy groups located in both the A-ring and B-ring.

Production and purification of methylated resveratrol analogs in *E. coli* via artificial biosynthetic pathways

The 3,4'-dimethoxy-5-hydroxystilbene synthesis levels in the *E. coli* cells that harbored pET-opTLO13S were not sufficient for structure elucidation. For a higher production of the *bis*-methylated resveratrol, the recombinant *E. coli* cells that harbored pET-opTLO13S were investigated using metabolite pattern analyses according to the culture times for 85 hours (Figure 6). This recombinant strain produced a relatively large amount of 3,5-dihydroxy-4'-methoxystilbene from the early stages of the culture. However, at the late stage of the culture, the resveratrol was consumed and the production of methylated resveratrol analogs slightly increased. The production level of 3,4'-dimethoxy-5-hydroxystilbene (peak 5) increased following the culturing and a new peak (peak 6) was also detected that exhibited a more delayed retention time (Figure 6A). The new peak was identified via LC/MS (*m/z*

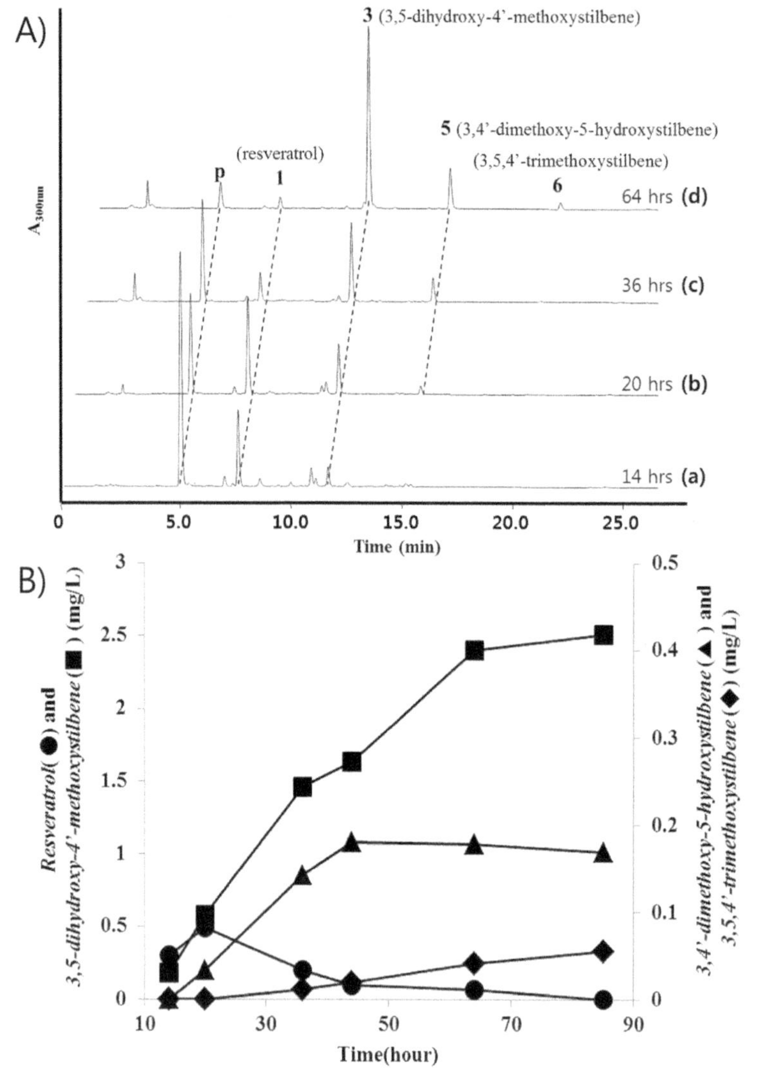

Figure 6 Accumulation of 3,4′-dimethoxy-5-hydroxystilbene and 3,5,4′-trimethoxystilbene in the culture of *E. coli* harboring pET-opTLO13S. A. HPLC profile of the time-course cultivation (14, 20, 36, and 64 hours) of *E. coli* harboring pET-opTLO13S. The cell was cultured in modified M9 medium at 26°C for **(a)**, 14 hours; **(b)**, 20 hours; **(c)**, 36 hours; **(d)**, 64 hours. p, 4-coumaric acid; peak 1, resveratrol; peak 3, 3,5-dihydroxy-4′-methoxystilbene; peak 5, 3,4′-dimethoxy-5-hydroxystilbene; peak 6, 3,5,4′-trimethoxystilbene. **B**. Production pattern of one, *bis*, and *tri*-methylated resveratrol in the each culture time. •, resveratrol; ■, 3,5-dihydroxy-4′-methoxystilbene; ▲, 3,4′-dimethoxy-5-hydroxystilbene; ♦. 3,5,4′-trimethoxystilbene. The concentrations were determined through the use of the corresponding chemical standards. All experiments were performed in triplicates.

271.14 $[M + H]^+$) and it corresponded to the three methylations of resveratrol (an addition of 42 Da; Additional file 1: Figure S3). This result represents the methoxy moiety located at the C-3 and C-5 positions in the A-ring and the C-4′ position in the B-ring (Figure 1).

In order to obtain NMR-accessible amounts from the present culture conditions, it was scaled up to 4 liters of *E. coli* fermentation that harbored pET-opTLO13S. Three methylated resveratrol analogs were purified from the large culture broth. The mass spectra of the purified methylated resveratrol analogs (3,5-dihydroxy-4′-methoxystilbene, 3,4′-dimethoxy-5-hydroxystilbene, and 3,5,4′-trimethoxystilbene) are characterized with mass peaks at *m/z* 243, 257, and 271 $[M + H]^+$, respectively. The ^1H NMR spectra of the purified compounds were similar to those of the one, *bi*, and *tri*-methyl analogs that were isolated, respectively (Additional file 2: Table S1). The structures of the purified resveratrol analogs were identified through spectral data interpretation and comparison with the values reported in the literature [27]. The coexistence of sbOMT1 and sbOMT3 produced a relatively large amount of 3,5-dihydroxy-4′-methoxystilbene from the early stage of the culture; however, the pinostilbene was not significantly detected

(Figure 6). These results indicate that sbOMT1 has superior methylation activity against resveratrol compared with that of sbOMT3. However, at the late stage of the culture, the production yield of 3,4'-dimethoxy-5-hydroxystilbene increased; it finally produced *tri*-methyl resveratrol (3,5,4'-trimethoxystilbene) (Figure 6B). Therefore, 3,4'-dimethoxy-5-hydroxystilbene is a major *bis*-methylated intermediate in the biosynthesis of 3,5,4'-trimethoxystilbene in this *E. coli* system. These results indicate that sbOMT3 has methylation activity in the A ring hydroxyl group against resveratrol as well as 3,5-dihydroxy-4'-methoxystilbene. Furthermore, the presence of 3,5,4'-trimethoxystilbene without the detection of pterostilbene indicates that the C-5 methylation is expected to be the final modification step in the coexistence of the sbOMT1 and sbOMT3 enzymes.

Although the production yield of these compounds is too low for commercial purposes, the results presented here were obtained using the wild type *E. coil*. This is the first report of an artificial biosynthetic pathway to obtain methylated resveratrol compounds used a single vector system joined with resveratrol biosynthetic genes and two resveratrol *O*-methyltransferase genes. The recombinant *E. coli* produces methylated resveratrol analogs beginning with simple fermentation rather than with bioconversion of intermediates. Recently, large-scale production of plant metabolites via metabolic engineered microbes has provided a promising alternative to chemical synthesis and extraction from raw plant sources [28,29]. Further application with a metabolically optimized host strain, i.e. a tyrosine and/or malonyl CoA high producer [23,24], may be very useful for obtaining industrial scale production yields.

Conclusions

We developed new artificial biosynthetic pathway for methylated resveratrol compounds using a single vector system joined with resveratrol biosynthetic genes and sorghum two resveratrol *O*-methyltransferase genes. An *E. coli* system containing an artificial biosynthetic pathway produced methylated resveratrol analogs of pinostilbene, 3,5-dihydroxy-4'-methoxystilbene, 3,4'-dimethoxy-5-hydro xystilbene, and 3,5,4'-trimethoxystilbene from a simple sugar medium without any precursor feeding. The *E. coli* cells that harbored the pET-opTLO1S plasmid (combined with *O*-methyltransferase (sbOMT1) and a resveratrol biosynthetic pathway) produced the one-methyl resveratrol analog, 3,5-dihydroxy-4'-methoxystilbene. The *E. coli* cells that harbored the pET-opTLO3S plasmid (combined with *O*-methyltransferase (sbOMT3) and a resveratrol biosynthetic pathway) produced the other one-methyl resveratrol analog, pinostilbene. Furthermore, the *E. coli* cells that harbored the pET-opTLO13S (combined with two sbOMT1 and sbOMT3 *O*-methyltransferases and a resveratrol biosynthetic pathway) produced 3,5-dihydroxy-4'-methoxystilbene, a *bis*-methyl resveratrol (3,4'-dimethoxy-5-hydroxystilbene), and a *tri*-methyl resveratrol (3,5,4'-trimethoxystilbene) (Figure 1). The coexistence of sbOMT1 and sbOMT3 produced 3,5-dihydroxy-4'-methoxystilbene as a major product in the early stage of the culture; however, the pinostilbene was not significantly detected (Figure 6). Katsuyama *et al.* reported the production of the pinostilbene and pterostilbene from *E. coli* culture using the fungal *O*-methyltransferase (OsPMT) gene and the tyrosine feeding medium [15]. Our described strategy exhibits the first harnessing of microorganisms for the *de novo* synthesis of methylated resveratrol analogs using a one-vector system from a simple sugar without precursor feeding in the culture medium. Thus, this is also the first report on the production of the methylated resveratrol compounds *bis*-methyl and *tri*-methyl resveratrol (3,4'-dimethoxy-5-hydroxystilbene and 3,5,4'-trimethoxystilbene) in the *E. coli* culture. These results provide an opportunity to create an economic incentive to develop strains capable of converting cheaper feedstock into high value compounds.

Methods

Bacterial strains, plasmids, and chemicals

E. coli DH5α and *E. coli* C41 (DE3) [30] were used in the general DNA manipulation and expression of biosynthetic genes, respectively. T-blunt vector (Solgent, Korea) was used in the polymerase chain reaction (PCR) cloning. pET-22b(+) and pET-28a(+) were purchased from Novagen (USA). 4-coumaric acid, resveratrol, pterostilbene, and 3,5,4'-trimethoxystilbene were purchased from Sigma-Aldrich (USA). Pinostilbene was purchased from Tokyo Chemical Industry, Co (Japan). IPTG was purchased from A.G. Scientific, Inc (USA). NMR solvent was purchased from Cambridge Isotope Laboratories, Inc. (USA). Hypergrade solvent for LC-MS was purchased from Merck KGaA (Germany).

DNA manipulation

The restriction enzymes (NEB, Takara), Ex taq polymerase (Takara) pfu taq polymerase (Solgent), and an AccuPower Ligation kit (Bioneer, Korea) were used according to the instructions provided by the manufacturers. The codon optimization and synthesis of *tal* gene from *Saccharothrix espanaensis* (KCTC9392) and the *sbOMT1 & sbOMT3* genes [GenBank:ABP01563.1, ABP01564.1] from *Sorghum bicolor* were performed using the GeneGPS™ program (DNA2.0). The *sts* gene from *Arachis hypogaea* [GenBank: AB027606] was codon-optimized and synthesized using Codon Devices (USA). The *ccl* gene from *Streptomyces coelicolor* A3(2) [GenBank: NP628552] was cloned and characterized in the laboratory. After the DNA manipulation, the absence of undesired alterations during the PCR was verified using nucleotide sequencing on an automated nucleotide sequencer.

Construction of O-methyltransferase expression vectors and assembly of the artificial biosynthetic pathways

The list of plasmids and strains used in this study can be found in Table 1. The plasmids were assembled using a serial stepwise cloning process as previously reported [18,31]. Briefly, the five genes (*TAL, CCL, STS, sbOMT1,* and *sbOMT3*) were independently cloned into pET-22b(+) or pET-28a(+) vectors. In order to construct an expression vector containing the *O*-methyltransferase genes that were under the control of independent T7 promoters, the 1.13-kb and 1.12-kb DNA fragments, which contain the sbOMT1 and sbOMT3 coding regions, were synthesized and then cloned into the *Nde*I and *Hind*III sites on pET-22b(+), which resulted in pET22-sbCOM1 and pET22-sbCOM3, respectively. In order to assemble the vector containing the methylated resveratrol artificial biosynthetic pathways, the sbOMT1 and sbOMT3 coding regions were amplified using pET22-sbCOM1 and pET22-sbCOM3, respectively, as templates with the primer NSpe (the sequence is located upstream of the T7 promoter region of the pET vector and contains the designed *Spe*I site: ACTAGTAGGTTGAGGCCGTTGAGCACCG CC) and CSpe (the sequence is located downstream of the T7 terminator region of the pET vector and contains the designed *Spe*I site: ACTAGTTCCTCCTTTCAGC AAAAAACCCCTC). Each of the amplified fragments were digested with *Spe*I and cloned between the *Spe*I digested pET-opTLS via ligation, which resulted in pET-opTLO1S and pET-opTLO3S. In order to construct the pET-opTLO13S vectors, the 2.1-kb DNA fragment containing the sbOMT3 coding region was PCR-amplified with the NPac (the sequence is located upstream of the T7 promoter region of the pET vector and contains the designed *Pac*I site: TTAATTAATCGCCGCGACAAT TTGCGACGG) and CPac (the sequence is located downstream of the T7 terminator region of the pET vector and contains the designed *Pac*I site: TTAAT-TAATGCGCCGCTACAGGGCGCGTCC) primers, respectively, using pET22-sbCOM3 as a template. The amplified fragment was digested with *Pac*I and cloned between the *Pac*I digested pET-opTLO1S, which resulted in pET-opTLO13S. The gene sequences and orientations were verified via sequencing after each round of cloning.

Production of methylated resveratrol analogs

The recombinant *E. coli* C41 (DE3) strains that harbored plasmids were precultured overnight at 37°C in a *Luria-Bertani* (LB) medium containing 50 μg/mL of appropriate antibiotics (ampicillin for to maintain the pET-22b(+)-derived plasmids and kanamycin for the pET-28a(+)-derived plasmids). The overnight culture was inoculated (1.5%) into a fresh LB medium supplemented with the same concentration of appropriate antibiotics. The culture was grown at 37°C to an optical density of 600 nm (OD_{600}) with 0.6. IPTG being added to the final concentration of 1 mM, and the culture was incubated for 6 hours. The cells were harvested via centrifugation, and then suspended and incubated at 26°C for 36 hours in a modified M9 medium (M9 medium supplemented with 15 g/L glucose, 25 g/L $CaCO_3$, 1 mM IPTG and 50 μg/mL appropriate antibiotics). For the feeding

Table 1 Plasmids and strains used in this study

Plasmid or strain	Relevant characteristics	Source
Plasmids		
pET-22b(+)	f1 ori, T7 promoter, Amp^R	Novagen
pET-28a(+)	f1 ori, T7 promoter, Kan^R	Novagen
pET-opTAL	pET-28a(+) carrying codon-optimized *Saccharothrix espanaensis* TAL	Kang *et al.* [31]
pET-CCL	pET-28a(+) carrying CCL from *Streptomyces coelicolor*	Choi *et al.* [18]
pET-KSTS	pET-28a(+) carrying codon-optimized *Arachis hypogaea* STS	Choi *et al.* [18]
pET22-sbCOM1	pET-22b(+) carrying codon-optimized *Sorghum bicolor* sbOMT1	This study
pET22-sbCOM3	pET-22b(+) carrying codon-optimized *Sorghum bicolor* sbOMT3	This study
pET-opTLS	pET-28a(+) carrying opTAL, CCL, and KSTS	This study
pET-opTLO1S	pET-28a(+) carrying opTAL, CCL, KSTS, and sbCOM1	This study
pET-opTLO3S	pET-28a(+) carrying opTAL, CCL, KSTS, and sbCOM3	This study
pET-opTLO13S	pET-28a(+) carrying opTAL, CCL, KSTS, sbCOM1, and sbCOM3	This study
Stranis		
E. coli DH5a	cloning host	Invitrogen
E. coli C41(DE3)	derivative strain of *E. coli* BL21(DE3)	Miroux B & Walker JE [30]

experiments, the cultures were supplemented with resveratrol (final concentration: 70 μM) and allowed to grow for an additional 36 hours.

Twenty milliliters of culture was extracted with an equal volume of ethyl acetate. The ethyl acetate was dried and resuspended in 200 μL of methanol. Twenty microliters of the extract was applied to a J'sphere ODS-H80 column (4.6 × 150 mm i.d., 5 μm; YMC, Japan) using a high-performance liquid chromatography (HPLC) system [CH_3CN-H_2O (0.05% trifluoroacetic acid(TFA)) 20% to 100% acetonitrile (CH_3CN) for 25 min, 100% CH_3CN for 5 min, at flow rate of 1 mL/min; Dionex, USA] equipped with a photodiode array detector. A liquid chromato graphy-mass spectrometry (LC-MS) was performed using an LTQ XL linear ion trap (Thermo Scientific, USA) equipped with an electrospray ionization (ESI) source that was coupled to a rapid separation LC (RSLC; ultimate 3000, Thermo Scientific) system (ESI-LC–MS) using a Cosmosil 2.5 Cholester column (Nacalai Tesque, Japan) (2.0 × 100 mm; 2.5 μm particle size) with a linear gradient of the binary solvent system under the same HPLC conditions as described above. The ESI (positive ion) parameters for the methylated resveratrol compounds were the source voltage (+5KV), entrance capillary voltage (+18 V) entrance capillary temperature (275°C), and tube lens voltage (+120 V). The scan range was fixed from m/z 50 to 1000. The data-dependent mass spectrometry experiments were controlled using the menu driven software provide with the Xcalibur system (version 2.2 SP1.48; Thermo Scientific). The compounds were identified through comparisons with the standard compounds using the observed retention time, ultraviolet spectra, and mass chromatogram.

Purification and structural elucidation of the methylated resveratrol analogs

The recombinant *E. coli* strains that harbored the pET-opTLO13S plasmid were cultured via the same method as described earlier, but the culture volume was increased to 4 liters. The isolation processes were undertaken as described above. The EtOAc-soluble material (0.9 g) was further purified by reverse-phase HPLC (Waters Co., USA) using the YMC J'sphere ODS-H80 (10 × 250 mm, 3 mL/min) with a linear gradient from 20% to 100% CH_3CN containing 0.05% TFA in order to yield 3,5-dihydroxy-4'-methoxystilbene (4.2 mg), 3,4'-dimethoxy-5-hydroxystilbene (2.1 mg), and 3,5,4'-trimethoxystilbene (0.9 mg). The structural elucidation of the purified compounds was undertaken using ^{1}H NMR spectroscopy. The NMR experiments were performed on a Bruker AVANCE 800 spectrometer (800 MHz; Bruker Inc., USA) and Barianunity^{Ionva}-400 instrument. The structures of 3,5-dihydroxy-4'-methoxystilbene, 3,4'-dimethoxy-5-hy droxystilbene and 3,5,4'-trimethoxystilbene were determined based on the NMR data.

Additional files

Additional file 1: Figure S1. SDS-PAGE analysis of the expression of sbOMT1 and sbOMT3 enzymes in *E. coli*. **Figure S2.** Selected mass ion chromatograms of the pterostilbene (*m/z* 257.13). **Figure S3.** Selected mass ion chromatograms of the 3,5,4'-trimethoxystilbene (*m/z* 271.14) produced by *E. coli* harboring pET-opTLO13S.

Additional file 2: Table S1. Structure elucidation of methylated resveratrol analogs. 1H NMR spectra of the purified 3,5-dihydroxy-4'-methoxystilbene, 3,4'-dimethoxy-5-hydroxystilbene, and 3,5,4'-trimethoxystilbene and data from the literature for resveratrol.

Competing interests

The authors declare that they have no competing interests.

Authors' contributions

SK and JL performed the experiments and wrote the manuscript. SK, OC, and CY coperformed the experiments on the metabolite analysis, molecular cloning, and bioconversion. JL and CY performed the experiments on the metabolite LC/MS analysis, purification, and structure elucidation. BH and JJ contributed general advice, particularly on the metabolite analysis and resource support. YH designed all the experiments and wrote the manuscript. All authors read and approved the final manuscript.

Acknowledgements

This work was supported in part by Basic Science Research program (2012-0001421) and Global R&D Center program funded by the NRF and by the Next-Generation BioGreen 21 Program (SSAC, PJ009549022014) funded by the RDA, Republic of Korea. We thank the Korea Basic Science Institute for the NMR measurements.

Author details

[1]Chemical Biology Research Center, Korea Research Institute of Bioscience and Biotechnology(KRIBB), 30 Yeongudanji-ro, Ochang-eup, Chungbuk 363-883, Republic of Korea. [2]Department of Pharmacy Graduate School, Chungbuk National University, Cheongju 361-763, Republic of Korea. [3]Eco-friendly Bio-Material Research Center, KRIBB, Jeongeup 580 1853, Republic of Korea.

References

1. Jang M, Cai L, Udeani GO, Slowing KV, Thomas CF, Beecher CW, Fong HH, Farnsworth NR, Kinghorn AD, Mehta RG, Moon RC, Pezzuto JM: **Cancer chemopreventive activity of resveratrol, a natural product derived from grapes.** *Science* 1997, **275**(5297):218–220.
2. Baur JA, Sinclair DA: **Therapeutic potential of resveratrol: the *in vivo* evidence.** *Nat Rev Drug Discov* 2006, **5**(6):493–506.
3. Baur JA, Pearson KJ, Price NL, Jamieson HA, Lerin C, Kalra A, Prabhu VV, Allard JS, Lopez-Lluch G, Lewis K, Pistell PJ, Poosala S, Becker KG, Boss O, Gwinn D, Wang M, Ramaswamy S, Fishbein KW, Spencer RG, Lakatta EG, Le Couteur D, Shaw RJ, Navas P, Puigserver P, Ingram DK, de Cabo R, Sinclair DA: **Resveratrol improves health and survival of mice on a high-calorie diet.** *Nature* 2006, **444**(7117):337–342.
4. Legg K: **Metabolic disease: identifying novel targets of resveratrol.** *Nat Rev Drug Discov* 2012, **11**(4):273.
5. Walle T, Hsieh F, DeLegge MH, Oatis JE Jr, Walle UK: **High absorption but very low bioavailability of oral resveratrol in humans.** *Drug Metab Dispos* 2004, **32**(12):1377–1382.
6. Tolomeo M, Grimaudo S, Di Cristina A, Roberti M, Pizzirani D, Meli M, Dusonchet L, Gebbia N, Abbadessa V, Crosta L, Barucchello R, Grisolia G, Invidiata F, Simoni D: **Pterostilbene and 3'-hydroxypterostilbene are effective apoptosis-inducing agents in MDR and BCR-ABL-expressing leukemia cells.** *Int J Biochem Cell Biol* 2005, **37**(8):1709–1726.
7. Mikstacka R, Przybylska D, Rimando AM, Baer-Dubowska W: **Inhibition of human recombinant cytochromes P450 CYP1A1 and CYP1B1 by trans-resveratrol methyl ethers.** *Mol Nutr Food Res* 2007, **51**(5):517–524.
8. Rimando AM, Suh N: **Biological/chemopreventive activity of stilbenes and their effect on colon cancer.** *Planta Med* 2008, **74**(13):1635–1643.

9. Kapetanovic IM, Muzzio M, Huang Z, Thompson TN, McCormick DL: **Pharmacokinetics, oral bioavailability, and metabolic profile of resveratrol and its dimethylether analog, pterostilbene, in rats.** *Cancer Chemother Pharmacol* 2011, **68**(3):593–601.
10. Simoni D, Roberti M, Invidiata FP, Aiello E, Aiello S, Marchetti P, Baruchello R, Eleopra M, Di Cristina A, Grimaudo S, Gebbia N, Crosta L, Dieli F, Tolomeo M: **Stilbene-based anticancer agents: resveratrol analogues active toward HL60 leukemic cells with a non-specific phase mechanism.** *Bioorg Med Chem Lett* 2006, **16**(12):3245–3248.
11. Schneider Y, Chabert P, Stutzmann J, Coelho D, Fougerousse A, Gosse F, Launay JF, Brouillard R, Raul F: **Resveratrol analog (*Z*)-3,5,4'-trimethoxystilbene is a potent anti-mitotic drug inhibiting tubulin polymerization.** *Int J Cancer* 2003, **107**(2):189–196.
12. Baerson SR, Dayan FE, Rimando AM, Nanayakkara NP, Liu CJ, Schroder J, Fishbein M, Pan Z, Kagan IA, Pratt LH, Cordonnier-Pratt MM, Duke SO: **A functional genomics investigation of allelochemical biosynthesis in *Sorghum bicolor* root hairs.** *J Biol Chem* 2008, **283**(6):3231–3247.
13. Rimando AM, Pan Z, Polashock JJ, Dayan FE, Mizuno CS, Snook ME, Liu CJ, Baerson SR: **In planta production of the highly potent resveratrol analogue pterostilbene via stilbene synthase and O-methyltransferase co-expression.** *Plant Biotechnol J* 2012, **10**(3):269–283.
14. Chiron H, Drouet A, Claudot AC, Eckerskorn C, Trost M, Heller W, Ernst D, Sandermann H Jr: **Molecular cloning and functional expression of a stress-induced multifunctional O-methyltransferase with pinosylvin methyltransferase activity from Scots pine (*Pinus sylvestris* L.).** *Plant Mol Biol* 2000, **44**(6):733–745.
15. Katsuyama Y, Funa N, Horinouchi S: **Precursor-directed biosynthesis of stilbene methyl ethers in *Escherichia coli*.** *Biotechnol J* 2007, **2**(10):1286–1293.
16. Jeong YJ, An CH, Woo SG, Jeong HJ, Kim YM, Park SJ, Yoon BD, Kim CY: **Production of pinostilbene compounds by the expression of resveratrol O-methyltransferase genes in *Escherichia coli*.** *Enzyme Microb Technol* 2014, **54**:8–14.
17. Schmidlin L, Poutaraud A, Claudel P, Mestre P, Prado E, Santos-Rosa M, Wiedemann-Merdinoglu S, Karst F, Merdinoglu D, Hugueney P: **A stress-inducible resveratrol O-methyltransferase involved in the biosynthesis of pterostilbene in grapevine.** *Plant Physiol* 2008, **148**(3):1630–1639.
18. Choi O, Wu CZ, Kang SY, Ahn JS, Uhm TB, Hong YS: **Biosynthesis of plant-specific phenylpropanoids by construction of an artificial biosynthetic pathway in *Escherichia coli*.** *J Ind Microbiol Biotechnol* 2011, **38**(10):1657–1665.
19. Horinouchi S: **Combinatorial biosynthesis of plant medicinal polyketides by microorganisms.** *Curr Opin Chem Biol* 2009, **13**(2):197–204.
20. Katsuyama Y, Funa N, Miyahisa I, Horinouchi S: **Synthesis of unnatural flavonoids and stilbenes by exploiting the plant biosynthetic pathway in *Escherichia coli*.** *Chem Biol* 2007, **14**(6):613–621.
21. Watts KT, Lee PC, Schmidt-Dannert C: **Biosynthesis of plant-specific stilbene polyketides in metabolically engineered *Escherichia coli*.** *BMC Biotechnol* 2006, **6**:22.
22. Wu J, Liu P, Fan Y, Bao H, Du G, Zhou J, Chen J: **Multivariate modular metabolic engineering of *Escherichia coli* to produce resveratrol from L-tyrosine.** *J Biotechnol* 2013, **167**(4):404–411.
23. Lim CG, Fowler ZL, Hueller T, Schaffer S, Koffas MA: **High-yield resveratrol production in engineered *Escherichia coli*.** *Appl Environ Microbiol* 2011, **77**(10):3451–3460.
24. Trantas E, Panopoulos N, Ververidis F: **Metabolic engineering of the complete pathway leading to heterologous biosynthesis of various flavonoids and stilbenoids in *Saccharomyces cerevisiae*.** *Metab Eng* 2009, **11**(6):355–366.
25. Zhang Y, Li SZ, Li J, Pan X, Cahoon RE, Jaworski JG, Wang X, Jez JM, Chen F, Yu O: **Using unnatural protein fusions to engineer resveratrol biosynthesis in yeast and Mammalian cells.** *J Am Chem Soc* 2006, **128**(40):13030–13031.
26. Gergely M, Pour NMS, Zoltan S, Katalin B, Tamas L, Robert O, Laszlo M: **Determination of products derived from trans-resveratrol UV photoisomerisation by means of HPLC–APCI-MS.** *J Photochem Photobiol Chem* 2008, **196**:44–50.
27. Sivakumar B, Murugan R, Baskaran A, Khadangale BP, Murugan S, Senthilkumar UP: **Identification and characterization of process-related impurities of trans-resveratrol.** *Sci Pharm* 2013, **81**(3):683–695.
28. Fowler ZL, Koffas MA: **Biosynthesis and biotechnological production of flavanones: current state and perspectives.** *Appl Microbiol Biotechnol* 2009, **83**(5):799–808.
29. Kaneko M, Hwang EI, Ohnishi Y, Horinouchi S: **Heterologous production of flavanones in *Escherichia coli*: potential for combinatorial biosynthesis of flavonoids in bacteria.** *J Ind Microbiol Biotechnol* 2003, **30**(8):456–461.
30. Miroux B, Walker JE: **Over-production of proteins in *Escherichia coli*: mutant hosts that allow synthesis of some membrane proteins and globular proteins at high levels.** *J Mol Biol* 1996, **260**(3):289–298.
31. Kang SY, Choi O, Lee JK, Hwang BY, Uhm TB, Hong YS: **Artificial biosynthesis of phenylpropanoic acids in a tyrosine overproducing *Escherichia coli* strain.** *Microb Cell Fact* 2012, **11**:153.

7

De-novo synthesis of 2-phenylethanol by *Enterobacter* sp. CGMCC 5087

Haibo Zhang[1], Mingle Cao[2], Xinglin Jiang[1], Huibin Zou[1], Cong Wang[1], Xin Xu[1] and Mo Xian[1*]

Abstract

Background: 2-phenylethanl (2-PE) and its derivatives are important chemicals, which are widely used in food materials and fine chemical industries and polymers and it's also a potentially valuable alcohol for next-generation biofuel. However, the biosynthesis of 2-PE are mainly biotransformed from phenylalanine, the price of which barred the production. Therefore, it is necessary to seek more sustainable technologies for 2-PE production.

Results: A new strain which produces 2-PE through the phenylpyruvate pathway was isolated and identified as *Enterobacter* sp. CGMCC 5087. The strain is able to use renewable monosaccharide as the carbon source and NH_4Cl as the nitrogen source to produce 2-PE. Two genes of rate-limiting enzymes, chorismate mutase *p*-prephenate dehydratase (PheA) and 3-deoxy-*d*-arabino-heptulosonic acid 7-phosphate synthase (DAHP), were cloned from *Escherichia coli* and overexpressed in *E.* sp. CGMCC 5087. The engineered *E.* sp. CGMCC 5087 produces 334.9 mg L^{-1} 2-PE in 12 h, which is 3.26 times as high as the wild strain.

Conclusions: The phenylpyruvate pathway and the substrate specificity of 2-keto-acid decarboxylase towards phenylpyruvate were found in *E.* sp. CGMCC 5087. Combined with the low-cost monosaccharide as the substrate, the finding provides a novel and potential way for 2-PE production.

Keywords: 2-Phenylethanol, Biosynthesis, Ehrlich pathway, *Enterobacter* sp

Background

2-Phenylethanol (2-PE), an aromatic alcohol with rose-like odour, is widely used in the food, drink and cosmetic industry, and it's also a potentially valuable alcohol for next-generation biofuel [1]. Additionally, 2-PE is an important raw material for its derivatives, among which phenylethyl acetate is a valuable fragrance compound, and *p*-hydroxyphenylethanol is widely used in pharmaceutical and fine chemical industries [2,3].

2-PE can be synthesized by chemical and biological methods. Up to now, the vast majority of 2-PE is produced chemically, while purification of the product is still a key problem in these chemical processes. The quality of the chemical synthesized 2-PE is reduced greatly because of its harsh conditions and the toxic reagents. 2-PE naturally exists in rose flower, and a number of fungi are known as 2-PE producers including *Phellinus ignarius, Ischnoderma benzoinum, Geotrichum penicillatum, Kluyveromyces marxianus, Saccharomyces cerevisiae, Aspergillus niger* etc. [4–7]. *S. cerevisiae* could produce 4.5 g L^{-1} 2-PE in fed-batch fermentation [4]. When oleic acid was used as the organic phase for two-phase fed-batch cultures, *S. cerevisiae* could produce 12.6 g L^{-1} 2-PE [8]. However, all these fermentations use phenylalanine as feedstock, the prices of which barred the industrial scale production of 2-PE. Very few bacteria were found to synthesize 2-PE. When using aromatic amino acids as nitrogen source, *Erwinia carotovora* was proved to produce 2-PE [9]. Jollivet *et al.* found 2-PE in the culture of *Microbacterium* sp. and *Brevibacterium linens*, while no further progress was reported [10].

2-PE can be biosynthesized from phenylalanine through the Phenylethylamine pathway and the Ehrlich pathway in eukaryotes (Figure 1, http://www.kegg.jp/kegg-bin/highlight_pathway?scale=1.0&map=map00360&keyword=phenethyl alcohol), and the latter one is thought to be the most significant pathway [2]. The shikimate pathway and the Ehrlich pathway can form a phenylpyruvate pathway, which can *de-novo* synthesize 2-PE from glucose. Hwang

* Correspondence: xianmo@qibebt.ac.cn
[1]CAS Key Laboratory of Biobased Materials, Qingdao Institute of Bioenergy and Bioprocess Technology, Chinese Academy of Sciences, No.189 Songling Road, Laoshan District, Qingdao, PR 266101, China
Full list of author information is available at the end of the article

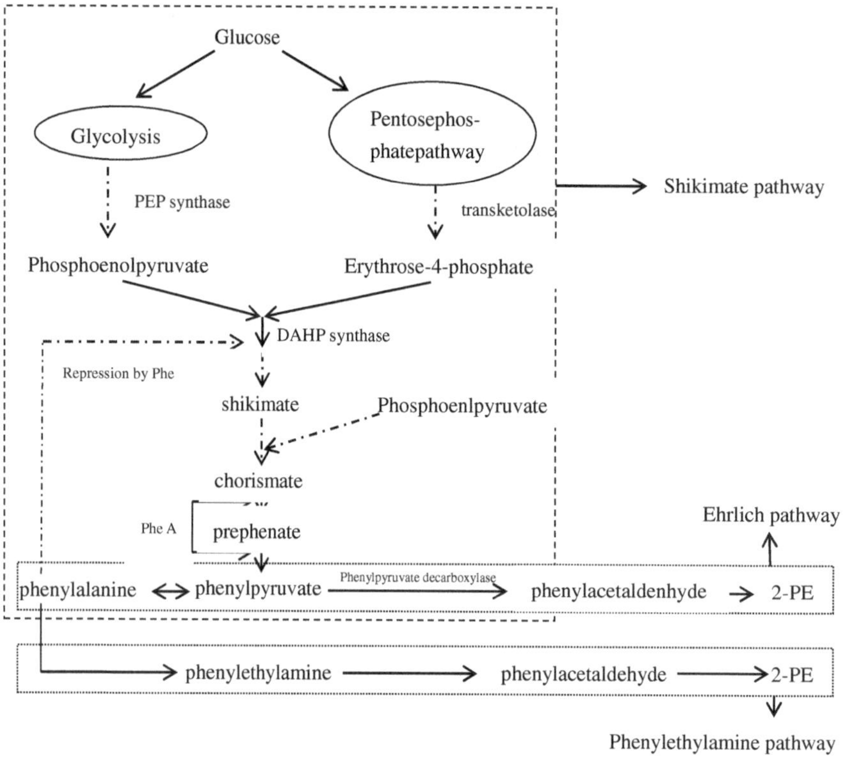

Figure 1 Phenylpyruvate pathway for 2-PE synthesis. 2-PE can be biosynthesized from phenylalanine through the Phenylethylamine pathway and the Ehrlich pathway. The shikimate pathway and the Ehrlich pathway form a phenylpyruvate pathway, which can *de-novo* synthesize 2-PE from glucose.

J-Y *et al.* constructed the yeast Ehrlich pathway into *E. coli,* but the failure of overexpressing all the foreign proteins in soluble and active form blocked the high production of 2-PE [11], and they also used phenylalanine as feedstock. James C. Liao *et al.* engineered *E. coli* to produce varied alcohols by overexpressing different heterologous 2-keto-acid decarboxylases (KDCs) and alcohol dehydrogenases, and 2-PE was detected among these alcohols, but the substrate specificity of KDCs was broad and no further research was performed for 2-PE [12]. Whether increasing carbon flux to the 2-keto acids could improve the productivity of the alcohol or not is still needed to be verified.

In our study, a new strain *Enterobacter* sp. CGMCC 5087 was isolated and identified. The strain was verified to produce 2-PE using *de-novo* synthetic pathway with monosaccharide as a carbon source and NH_4Cl as a nitrogen source. Two rate-limiting enzymes, PheA and DAHP synthase, were overexpressed to increase the flux to 2-keto phenylpyruvate to further increase production of 2-PE.

Results and discussion

Identification of the strain

The bacterium was accidentally obtained when we screened strains against 5 g L^{-1} resorcinol, and 2-PE was detected when it was cultured in the selective medium. The screened strain is rod-shaped, with width from 0.3 to 0.5 μm and length from 1 to 2 μm. And it is an aerobic bacterium. 16S rDNA gene sequence (1412 bp) was blasted at NCBI against Genome Survey Sequence (GSS). Based on these results, this strain was identified to be *E.* sp. and preserved in China General Microbiological Culture Collection Center (CGMCC) as *E.* sp. CGMCC 5087. 2-PE has been used as a bactericide in pharmaceutical preparations for many years [13]. 2-PE was thought to induce the breakdown of the cell membrane limitedly and inhibit the synthesis of deoxyribonucleic acid [14]. Secretion of 2-PE by this bacteria may be a bactericidal substance for interspecies competition.

Many microorganisms, especially yeasts, have been reported to produce 2-PE by normal metabolism, however, the final concentration of 2-PE in the culture broth of selected strains remains very low. That was why all the fermentation for 2-PE used phenylalanine as nitrogen source. Besides, very few wild prokaryotic organisms were reported to secrete 2-PE. Therefore, validation of 2-PE *de-novo* biosynthesis pathway is not only useful metabolic engineering but also valuable for theoretical research.

Validating of 2-PE *de-novo* biosynthesis in *E.* sp. CGMCC 5087

As the selective medium contains beef and yeast extract, M9 medium was used to validate whether *E.* sp. CGMCC 5087 can biosynthesize 2-PE using glucose as the sole carbon source. After *E.* sp. CGMCC 5087 was grown in M9 medium at 37°C for 24 h, 2-PE was detected. Characteristic ion (m/z = 122) spectra was shown in Figure 2b. The production of 2-PE in M9 medium was much lower (Figure. 2b) than in LB medium (70 mg L^{-1}, Figure 2a).

The results indicated that *E.* sp. CGMCC 5087 can synthesize 2-PE using *de-novo* pathway with monosaccharide as carbon source and NH_4Cl as nitrogen source. To the best of our knowledge, this is the first wild bacterium validated to produce 2-PE using glucose as sole carbon source. Although phenylalanine is mostly used as fermentation stock for 2-PE biosynthesis, the high price is still a huge disadvantage. The cheaper glucose makes this strain more competitive for industrial production.

Most microorganisms can synthesize phenylalanine via shikimate pathway, and subsequently phenylalanine can be converted to 2-PE by many microorganisms through Phenylethylamine pathway or Ehrlich pathway [2]. The shikimate pathway and the Ehrlich pathway are connected by the phenylpyruvate to form a phenylpyruvate pathway, which can *de-novo* synthesize 2-PE from glucose. Validating the synthetic pathway of 2-PE would be helpful for the further research.

Validating phenylpyruvate pathway in *E.* sp. CGMCC 5087

We examined the activity of crude enzymes to identify 2-PE biosynthesis pathway involved in *E.* sp. CGMCC 5087. The results indicated that crude enzymes can catalyze phenylpyruvate to 2-PE (Figure 3). As phenylpyruvate is not the intermediate product of Phenylethylamine pathway, the occurrence of 2-PE suggested that it is biotransformed from phenylpyruvate through Ehrlich pathway, which is consistent with the pathway that James C. liao and Hwang J-Y constructed [11,12].

Meanwhile, the results indicate that 2-keto-acid decarboxylase (KDC) is produced by *E.* sp. CGMCC 5087. KDCs have a broad substrate range and widely exist in plants, yeasts and fungi, but less so in bacteria [15]. No other alcohols such as 1-butanol were detected (data not shown) indicates that the KDC has a good substrate specificity [12]. In the Ehrlich pathway, phenylpyruvate was converted to phenylacetaldehyde by KDC firstly, and then converted to 2-PE by alcohol dehydrogenase (ADH) with nicotinamide adenine dinucleotide phosphate (NADPH) as reducing agent. As importing and expressing non-native pathways may lead to the metabolic imbalance, and also heterologous metabolites may exhibit cytotoxicity [3,16,17], co-existence of phenylpyruvate pathway and the two enzymes KDC and ADH makes *E.* sp. CGMCC 5087 a good candidate to produce 2-PE.

Biosynthesis of 2-PE with engineered *E.* sp. CGMCC 5087

The existing shikimate pathway was then genetically modified to increase 2-PE production by enhancing the formation of phenylpyruvate. Both chorismate mutase *p*-prephenate dehydratase (PheA) and 3-deoxy-*d*-arabino-heptulosonic acid 7-phosphate synthase (DAHP synthase) are proved to be rate-limiting enzymes in the L-phenylalanine synthesis pathway. Therefore, enhancing the expression level of these two enzymes will increase the flux to 2-keto-phenylbutyric acid. As is shown in Figure 1, Carbon flow through the shikimate pathway is controlled by modulation of DAHP. In *E. coli*, three unlinked genes *aro*F, *aro*G and *aro*H encode three isoenzymes of DAHP synthase, respectively. Among these genes, *aro*F was proved to play a positive effect for biosynthesis of phenylalanine [18,19]. Chorismate mutase *p*-prephenate dehydratase was also identified as a rate-limiting enzyme for the synthesis of phenylalanine [14,20]. pMW13, an expression vector identified to be able to replicate in *E.* sp. CGMCC 5087 was used to express pheA and aroF. Expression of the proteins was evaluated by SDS-PAGE analysis. Two specific banDs,

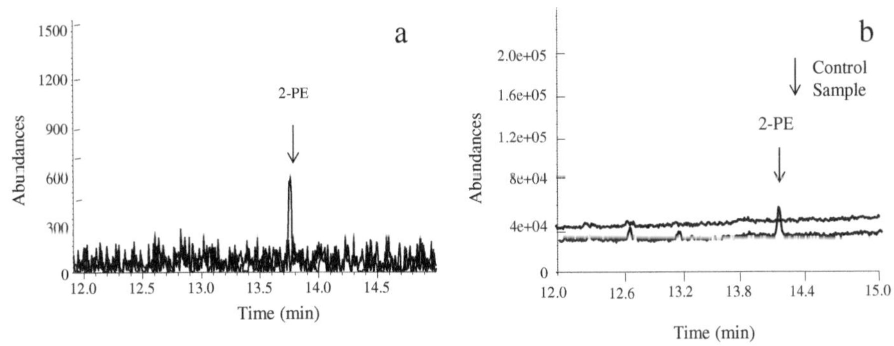

Figure 2 Validating of biosynthesis of 2-PE by *E.* sp. CGMCC 5087. a. *E.* sp. CGMCC 5087 was cultured in M9 medium and detected with GC-MS; **b**. *E.* sp. CGMCC 5087 was cultured in LB medium and detected with GC.

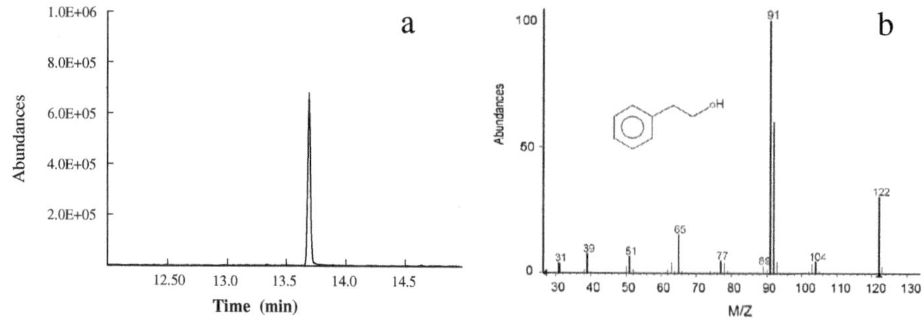

Figure 3 Validating Ehrlich pathway in *E.* sp. CGMCC 5087. a. GC chromatography of 2-PE biocatalyzed with crude enzymes with phenylpyruvate as substrate; **b.** MS spectrum of 2-PE biocatalyzed with crude enzymes with phenylpyruvate as substrate.

with the right molecular masses, were able to be observed apparently at the recombinant strain, compared to the negative control (data not shown) [21].

The wild type *E.* sp. CGMCC 5087 can produce 102.6 mg L^{-1} 2-PE when cultured in production medium for about 12 h (Figure 4). The plasmid pMW13_*pheA*_*aro*F was transformed to enhance phenylpyruvate biosynthesis to combine with the alcohol producing pathway to achieve 2-PE production. The 2-PE production was about 334.9 mg L^{-1}after incubation in shake flask for 12 h, and it is 3.26 times as high as the wild strain. When phenylalanine was added to the production medium at the concentration of 5 g L^{-1}, 2-PE production was 530 mg L^{-1}. It's much higher than the 469 μM (about 57 mg L^{-1}) that James C. Liao *et al.* reported [12].

According to phenylpyruvate pathway, two mol phosphoenolpyruvate, one mol erythrose-4-phosphate and one mol adenosine triphosphate can form one mol phenylpyruvate which can be then converted to one mol 2-PE with two mol NADPH then. According to the maximum theoretical yield, when 1 g 2-PE was produced, 2.53 g glucose or 1.34 g phenylalanine was needed. However, the price of phenylalanine (55, 000 ¥/ton) was about 17 folds that of glucose (3, 200 ¥/ton) in China. The cheaper sugars make the strain more competitive for industrial production. Though the titer of 334.9 mg L^{-1} was still lower than the previous report by *S. cerevisiae,* further optimization, including fed-batch fermentation, addition of extractive solvent, could be done to elevate the yield of 2-PE [4,8].

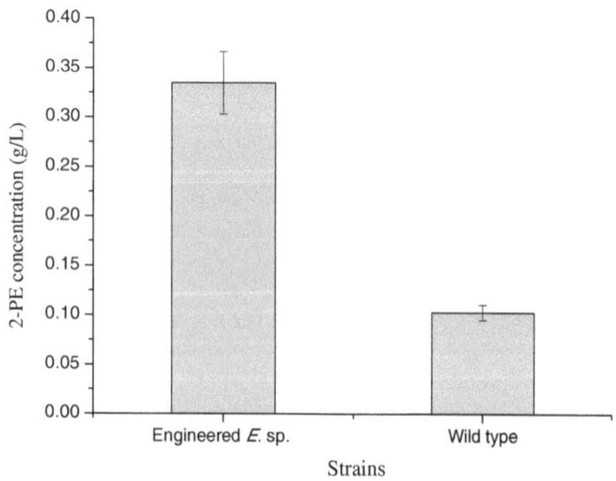

Figure 4 The production of 2-PE by the wild and engineered *E.* sp. CGMCC 5087. Control: wild type *E.* sp. CGMCC 5087; Engineered *E.* sp. CGMCC 5087: *E.* sp. CGMCC 5087 harboring pMW13. The wild type *E.* sp. CGMCC 5087 and the engineered strain can produce 102.6 and 334.9 mg L^{-1} 2-PE, respectively. All the experiments were performed in triplicate, and error bars indicate standard deviations.

Conclusions

A new strain *E.* sp. CGMCC 5087, which is able to produce 2-PE using *de-novo* pathway with monosaccharide as carbon source and NH_4Cl as nitrogen source was isolated and identified. The strain can synthesize 2-PE through phenylpyruvate pathway. Two rate-limiting enzymes, PheA and DAHP synthase from *E. coli* were then overexpressed in *E.* sp. CGMCC 5087 to construct engineered *E.* sp for 2-PE production. The engineered strain can produce 334.9 mg L^{-1} 2-PE for about 12 h. The substrate specificity of 2-keto-acid decarboxylase and the existence of phenylpyruvate pathway make *E.* sp. CGMCC 5087 a good candidate for 2-PE production.

Methods

Chemicals

Phenylpyruvate, phenylethanol and 2-ethylbutyric acid were purchased from Sigma-Aldrich (China). NADPH was obtained from Roche Products (China). The Bacterial DNA Kit and the Plasmid Mega Kit were bought from OMEGA. The Competent Cell Preparation Kit was from TaKaRa (Dalian, China). The other reagents were all analytical grades.

Strains

E. sp. CGMCC 5087 was isolated from the sewage sludge of dyes wastewater with a selective medium. The sewage sludge was diluted with sterile water and incubated on the selective medium with 1.2% agar. Grown clones were then incubated into 50 ml of LB medium, cultured for 24 h, and extracted with ethyl acetate. Extractions were identified with gas chromatography–mass spectrometry as mentioned below. The genomic DNA of the isolated strain was extracted using Bacterial DNA Kit. 16S rDNA was amplified using the consensus primers fD1 (5′-AGAGTT TGATCCTGGCTCAG-3′) and rP2 (5′-CGGCTACCTT GTTACGACTT-3′), and PCR products were cloned into pMD™ 18-T vector. Clones derived from the PCR products were confirmed by restriction analyses and sequenced. *Escherichia coli* MWPEC13-60(ATCC 67459) with plasmid pMW13_*phe*A_*aro*F was purchased from the American Type Culture Collection [22]. The strains and plasmids are shown in Table 1.

Medium

The composition of the selective medium contains: resorcinol, 5 g L^{-1}; glucose, 5 g L^{-1}; beef extract, 2 g L^{-1}; tryptone, 7 g L^{-1}; NaC1, 4 g L^{-1}; pH 7.0 ~ 7.4. With glucose as carbon source and NH_4Cl as nitrogen source, M9 medium (NH_4Cl, 1 g L^{-1}; $Na_2HPO_4 \cdot 7H_2O$, 11 g L^{-1}; KH_2PO_4, 3 g L^{-1}; NaC1, 0.5 g L^{-1}; glucose, 20 g L^{-1}; $MgSO_4$, 120 mg L^{-1}; $CaC1_2$, 10 mg L^{-1}; pH 7.0 ~ 7.4) was used to verify *de-novo* biosynthesis of 2-PE by *E.* sp. CGMCC 5087. Luria-Bertani (LB) medium (tryptone 1%, yeast extract 0.5%, sodium chloride 1%, pH 7.0 ~ 7.4) was used to culture *E.* sp. CGMCC 5087 and *E. coli*. production medium for 2-PE contains the following components: sucrose, 20 g L^{-1}; yeast extraction, 5 g L^{-1}; KH_2PO_4, 3 g L^{-1}; K_2HPO_4; 3.82 g L^{-1}; $MgSO_4 \cdot 7H_2O$, 0.5 g L^{-1}; Na_2MoO_4, 0.01 g L^{-1}; $CaCl_2$, 0.5 g L^{-1}; NaCl, 1.0 g L^{-1}, pH 7.2 ~ 7.5. LB mediums with ampicillin (50 μg ml^{-1}) and kanamycin (50 μg ml^{-1}) were used to culture the strains with pMD™ 18-T vector and pMW13_*phe*A_*aro*F, respectively.

Analytical methods

2-PE was quantified by gas chromatography (GC), which has a FID and HP Innowax capillary column (30 m × 0.25 mm, 0.25 μm film thickness). The chromatographic conditions were: column temperature 50°C (1 min), 50–240°C (15°C/min), 240°C (15 min), injector temperature 250°C, detector temperature 280°C, carrier gas nitrogen (0.2 kPa), flow rate 1.0 ml min^{-1}, volume injected 0.4 μl. 2-Ethylbutyric acid was used as interior standard. 2-PE was identified with gas chromatography–mass spectrometry (GC-MS; Agilent 6890 N GC and 5975 inert MSD) using a pulsed pressure injection of 1 μl onto an Innowax-101025 column (30 m × 0.25 mm, 0.25 μm film thickness), ion source (EI) ionization, and a dwell time of 40 msec were applied for each characteristic ion. The GC-MS chromatographic conditions were the same as GC conditions mentioned above.

Table 1 Bacterial strains and plasmids used in this study

Plasmids or strains	Relevant genotype or description	Reference
Plasmids		
pMDTM18-T	pUC origin; Amp^R	Takara
pMW13	pMW origin; $Kana^R$; P_{PL-PR}::*pheA, aroF*	[20]
Strains		
DH5α	deoR, recA1, endA1, hsdR17(rk⁻, mk⁺), phoA, supE44, λ⁻, thi⁻1, gyrA96, relA1	Takara
E. sp.	*Enterobacter* sp. CGMCC 5087 Amp^R	This study
E.coli MWPEC13-60	*E. coli* harboring pMW13	ATCC® 67459™
Engineered *Ec* sp.	*E.* sp. harboring pMW13	This study

Verifying *de-novo* biosynthesis of 2-PE by *E.* sp. CGMCC 5087

E. sp. CGMCC 5087 was inoculated to M9 medium (50 ml) and incubated at 37°C on a shaker (150 rpm) for 24 h. The culture was then centrifuged at 12, 000 rpm for 1 min, and the supernatant with a volume of 45 ml was extracted with 5 ml of n-butyl alcohol. Extraction was filtrated with 0.22 μm polyether sulphone membrane and qualitatively analyzed the 2-PE content with GC-MS. M9 medium without any inoculation and dealt with the same conditions was used as control.

Verifying phenylpyruvate pathway in *E.* sp. CGMCC 5087

As the key intermediate of the phenylpyruvate pathway, phenylpyruvate was added to verify the pathway in *E.* sp. CGMCC 5087. The strain was inoculated to LB medium and incubated at 37°C on a shaker (150 rpm) for 12 h. The culture was centrifuged at 12, 000 rpm for 2 min. Deposition cell was washed and suspended with 50 mM phosphate buffers (pH 7.2). The cells were lysed by ultrasonication in the ice bath. The lysed solution was centrifuged at 12,000 rpm for 1 min and supernatant was used as crude enzyme to catalyze the conversion of phenylpyruvate to 2-PE with1 μM NADPH. The reaction was incubated at 37°C for 2 h and ended by heating up at 75°C for 30 min. The control was made by heating the crude enzymes and NADPH mixture before the substrate was added. The mixtures were centrifuged at 12, 000 rpm for 20 min, and the supernatant were detected with GC-MS for the 2-PE contents.

Plasmid and gene expression

Plasmid pMW13_*pheA_aroF* was constructed by Lee *et al.* [22], which contains the *pheA* (GI: 254547524), and *aroF* (GI: 145363), under a temperature sensitive promoter (PL-PR promoter of the lambda phage). *AroF* and *pheA* encode DAHP synthase and PheA, respectively. *E.* sp. CGMCC 5087 competent cell was prepare by the Competent Cell Preparation Kit (TaKaRa, Dalian, China). The plasmid pMW13_*pheA_aroF* was transformed into the competent cells, and the positive transformants were screened against kanamycin (50 μM). The engineered *E.* sp. CGMCC 5087 was incubated at 32°C to overexpress proteins.

Biosynthesis of 2-PE with engineered *E.* sp. CGMCC 5087

The engineered *E.* sp. CGMCC 5087 was inoculated into production medium and incubated at 32°C on a shaker (150 rpm) for 12 h. The culture was centrifuged at 12, 000 rpm for 2 min, and the filtered supernatant was quantified its 2-PE content with GC. Wild type *E.* sp. CGMCC 5087 was cultured and measured its 2PE content as the control. Wild type *E.* sp. CGMCC 5087 was cultured in the production medium with 5 g L^{-1} phenylalanine as another control to verify the effects of phenylalanine.

Calculations

The maximum theoretical yield is calculated by assuming that the branching pathways are blocked and that the carbon flow is directed by the most efficient pathways with minimum loss to carbon dioxide and other metabolites [23,24]. At the theoretical level, 1 mol of 2-PE and 4 mol of NADPH can be produced from 2 mol of glucose and 2 mol of ATP.

The following equation was used to calculate the maximum theoretical yield.

$$Q_p = \frac{N_p M_p}{N_g M_g} \times 100\%$$

where Q_p = the maximum theoretical yield; N = the number of moles (mol); M = molecular weight; p = 2-PE; g = glucose.

Abbreviations

ADH: Alcohol dehydrogenase; CGMCC: China general microbiological culture collection center; DAHP: 3-deoxy-*d*-arabino-heptulosonic acid 7-phosphate synthase; FID: Flame ionization detector; GC: Gas chromatography; GC-MS: Gas chromatography-mass spectrum; GSS: Genome Survey Sequence; 2-PE: 2-phenylethanl; KDC: 2-keto-acid decarboxylase; KDCs: 2-keto-acid decarboxylases; NADPH: Nicotinamide adenine dinucleotide phosphate; PCR: Polymerase chain reaction; PheA: Chorismate mutase *p*-prephenate dehydratase.

Competing interests

The authors declare that they have no competing interests.

Authors' contributions

MX developed the idea for the study. HZh designed the research, did the literature review and prepared the manuscript. MX helped to revise the manuscript. MC, XJ, HZo, CW and XX did the lab work, plasmid construction, strain cultivation and product detection. All authors read and approved the final manuscript.

Acknowledgements

We greatly appreciate the financially support from the Basic Research Project of Qingdao Municipal Science and Technology Commission (No. 12-1-4-9-(3)-jch), National Key Technology R&D Program (No. 2012BAD32B06), National High Technology Research and Development Program of China (863 Program, No. SS2013AA050703-2), Knowledge Innovation Program of the Chinese Academy of Sciences (Y112131105).

Author details

[1]CAS Key Laboratory of Biobased Materials, Qingdao Institute of Bioenergy and Bioprocess Technology, Chinese Academy of Sciences, No.189 Songling Road, Laoshan District, Qingdao, PR 266101, China. [2]University of Marburg, Marburg, Germany.

References

1. Keasling JD, Chou H: **Metabolic engineering delivers next-generation biofuels.** *Nat Biotechnol* 2008, **26**(3):298–299.
2. Etschmann MM, Bluemke W, Sell D, Schrader J: **Sell, schrader: biotechnological production of 2-phenylethanol.** *Appl Microbiol Biotechnol* 2002, **59**(1):1–8.
3. Barbirato F, Grivet J, Soucaille P, Bories A: **3-Hydroxypropionaldehyde, an inhibitory metabolite of glycerol fermentation to 1,3-propanediol by enterobacterial species.** *Appl Environ Microbiol* 1996, **62**(4):1448–1451.
4. Eshkol N, Sendovski M, Bahalul M, Katz-Ezov T, Kashi Y, Fishman A: **Production of 2-phenylethanol from L-phenylalanine by a stress tolerant saccharomyces cerevisiae strain.** *J Appl Microbiol* 2009, **106**(2):534–542.
5. Etschmann MMW, Sell D, Schrader J: **Screening of yeasts for the production of the aroma compound 2-phenylethanol in a molasses-based medium.** *Biotechnol Lett* 2003, **25**(7):531–536.
6. Wani MA, Sanjana K, Kumar DM, Lal DK: **GC-MS analysis reveals production of 2-phenylethanol from aspergillus niger endophytic in rose.** *J Basic Microbiol* 2010, **50**(1):110–114.
7. Albertazzi E, Cardillo R, Servi S, Zucchi G: **Biogeneration of 2-phenylethanol and 2-phenylethylacetate important aroma components.** *Biotechnol Lett* 1994, **16**(5):491–496.
8. Stark D, Münch T, Sonnleitner B, Marison IW, Stockar UV: **Extractive bioconversion of 2-phenylethanol from l-phenylalanine by saccharomyces cerevisiae.** *Biotechnol Prog* 2002, **18**(3):514–523.
9. Spinnler H, Djian A: **Bioconversion of amino acids into flavouring alcohols and esters by erwinia carotovora subsp. Atroseptica.** *Appl Microbiol Biotechnol* 1991, **35**(2):264–269.
10. Jollivet N, Bézenger M-C, Vayssier Y, Belin J-M: **Production of volatile compounds in liquid cultures by six strains of coryneform bacteria.** *Appl Microbiol Biotechnol* 1992, **36**(6):790–794.
11. Hwang J-Y, Park J, Seo J-H, Cha M, Cho B-K, Kim J, Kim B-G: **Simultaneous synthesis of 2-phenylethanol and L-homophenylalanine using aromatic transaminase with yeast ehrlich pathway.** *Biotechnol Bioeng* 2009, **102**(5):1323–1329.
12. Atsumi S, Hanai T, Liao JC: **Non-fermentative pathways for synthesis of branched-chain higher alcohols as biofuels.** *Nature* 2008, **451**(7174):86–89.
13. Wilson JRJ, Lyall J, McBride RJ, Murray JB, Smith G: **Partition coefficients of some aromatic alcohols in an n-heptane/water system and their relationship to minimum inhibitory concentration against pseudomonas aeruginosa and staphylococcus aureus.** *J Clin Pharm Ther* 1981, **6**(1):63–66.
14. Silver S, Wendt L: **Mechanism of action of phenethyl alcohol: breakdown of the cellular permeability barrier.** *J Bacteriol* 1967, **93**(2):560–566.
15. Gowrishankar J, Pittard J: **Regulation of phenylalanine biosynthesis in Escherichia coli K-12: control of transcription of the pheA operon.** *J Bacteriol* 1982, **150**(3):1130–1137.
16. Zhu MM, Lawman PD, Cameron DC: **Improving 1,3-propanediol production from glycerol in a metabolically engineered Escherichia coli by reducing accumulation of sn-glycerol-3-phosphate.** *Biotechnol Prog* 2002, **18**(4):694–699.
17. Pitera DJ, Paddon CJ, Newman JD, Keasling JD: **Balancing a heterologous mevalonate pathway for improved isoprenoid production in escherichia coli.** *Metab Eng* 2007, **9**(2):193–207.
18. Ikeda M, Ozaki A, Katsumata R: **Phenylalanine production by metabolically engineered corynebacterium glutamicum with the pheA gene of escherichia coli.** *Appl Microbiol Biotechnol* 1993, **39**(3):318–323.
19. Jossek R, Bongaerts J, Sprenger GA: **Characterization of a new feedback-resistant 3-deoxy-d-arabino-heptulosonate 7-phosphate synthase AroF of escherichia coli.** *FEMS Microbiol Lett* 2001, **202**(1):145–148.
20. Stephan K: **Subunit structure, function and organisation of pyruvate decarboxylases from various organisms.** *Biochim Biophys Acta* 1998, **1385**(2):271–286.

21. Zhou H, Liao X, Wang T, Du G, Chen J: **Enhanced l-phenylalanine biosynthesis by co-expression of pheA^{fbr} and aroF^{wt}.** *Bioresour Technol* 2010, **101**(11):4151–4156.
22. Sae Bae Lee CHW, Park C, Lim BS: **Method for production of L-phenylalanine by recombinant *E. coli*.** *patent* 1991, **5008190**. Patent Number:5,304,475.
23. Patnaik R, Liao JC: **Engineering of escherichia coli central metabolism for aromatic metabolite production with near theoretical yield.** *Appl Environ Microbiol* 1994, **60**(11):3903–3908.
24. Förberg C, Eliaeson T, Häggström L: **Correlation of theoretical and experimental yields of phenylalanine from non-growing cells of a rec escherichia coli strain.** *J Biotechnol* 1988, **7**(4):319–331.

8

Cultivation to improve *in vivo* solubility of overexpressed arginine deiminases in *Escherichia coli* and the enzyme characteristics

Ying Wang and Yue-Zhong Li*

Abstract

Background: Overexpression of foreign genes in *Escherichia coli* cells is an efficient means to obtain recombinant proteins. The technique is, however, often hampered by misfolding, degradation, aggregation and formation in inclusion bodies of products.

Results: In this study, we reported that *in vivo* solubility of overexpressed arginine deiminases (ADI) improved by changing the cultivation conditions. ADI is enzymes that convert L-arginine to L-citrulline. After codon optimization, we synthesized the ADI gene of *Pseudomonas putida* and constructed it for overexpression in *E. coli* cells. The rADI products were mainly in inclusion body forms. We performed a series of optimization to enhance solubility of the protein. Co-expression with the GroES-GroEL chaperone team increased approximately 5-fold of the rADI activity. In addition the combination of L-arginine and D-glucose in the Luria-Bertani (LB) growth medium further increased the total activity to about 15 times. Separate L-arginine and D-glucose or the addition of other saccharides or amino acids had no such effects. The solubilization effects of the combination of L-arginine and D-glucose were further confirmed in the overexpression of another ADI from *Listeria welshimeri*. The enzymatic and conversion characteristics of the rADI products were further determined.

Conclusions: Combined addition of L-arginine and D-glucose in the LB medium significantly improved *in vivo* solubility of rADI proteins. The present study suggested a new strategy to increase the solubilization of overexpressed recombinant proteins in *E. coli* cells.

Keywords: Arginine deiminase, Heterologous expression, *Escherichia coli*, Solubilization, Cultivation, Conversion, L-arginine, L-citrulline

Background

Escherichia coli cells are an efficient factory to produce foreign proteins because of high-cell density, well-known genetics background and the simplest genetic manipulation [1,2]. However, high expressions of recombinant proteins often result in misfolding, aggregation, degradation, low activity and formation in inclusion-bodies of products, which greatly hampers associated researches and industrial and medical applications [2-6]. In order to improve the solubility of recombinant proteins, many strategies have been developed, such as co-expression with chaperones, secretion expression of recombinant proteins and protein fusion techniques [7,8]. In this study, we reported that combined addition of D-glucose and L-arginine in the Luria-Bertani (LB) medium to cultivate *E. coli* cells were efficient to enhance the solubility of recombinant arginine deiminases.

Arginine deiminase (EC3.5.3.6; ADI) hydrolyzes L-arginine to L-citrulline, releasing ammonia. Arginine deiminase, combining with ornithine transcarbamylase and carbamate kinase forms the arginine deiminase pathway [9], which is widely distributed in the microorganisms that are able to utilize arginine as an energy source [10]. By depleting extracellular supply of L-arginine, ADI is a potential anti-tumor agent for its efficient inhibition effects on the growth of arginine-auxotrophic tumor cells, especially hepatocellular carcinomas and melanomas [10-12]. ADI is also considered a useful producer

* Correspondence: lilab@sdu.edu.cn
State Key Laboratory of Microbial Technology, School of Life Science, Shandong University, Jinan 250100, P. R. China

of L-citrulline at industrial scale [13]. So, many ADI genes from different sources, including *Mycoplasma arginini* [11,12,14-16], *Pseudomonas plecoglossicida* [17], *Lactococcus lactis* [18], *Streptococcus sanguis* [19] and *Giardia intestinalis* [20,21] have been cloned and expressed in *E. coli* cells. However, the recombinant ADI proteins were mainly existed in inclusion body forms, which is a bottleneck in its research and development. We overexpressed arginine deiminases from different resources in the *E. coli* cells for industrial conversion L-arginine to L-citrulline. Our recombined ADI proteins named rADI, were similarly mostly in inclusion bodies. In this study, we made comprehensive optimizations to improve solubility of the rADI proteins and found that the combined addition of L-arginine and D-glucose in Luria-Bertani (LB) medium distinctly increased the *in vivo* solubility of rADI in *E. coli* cells. We studied the enzymatic and conversion characteristics of the rADI.

Results

Construction and expression of ADI gene in *E. coli*

The arginine deiminase of *P. putida* had been studied due to its characteristics by purification from its original host [22]. Because of its high specific activity (58.8 U/mg) and *in-vitro* solubility, we selected the enzyme to investigate its potential in industrial conversion from L-arginine to L-citrulline. We optimized the gene sequence according to the codon frequency table of *E. coli* BL21(DE3) using the "codon randomization" strategy [23] and synthesized it. The artificial ADI gene (Figure 1) was constructed into different expression plasmids, including pET-30a(+), pBAD/gIIIA, pColdTF and pGEX-6P-1, for overexpression in *E. coli* BL21(DE3). After a normal optimization process according to the previous report [24], 0.1 mM IPTG and 0.04% L-arabinose incubated at 16°C were used to induce the rADI expression in *E. coli* cells. These expression plasmids, except pET-30a(+), exhibited nearly no solubility

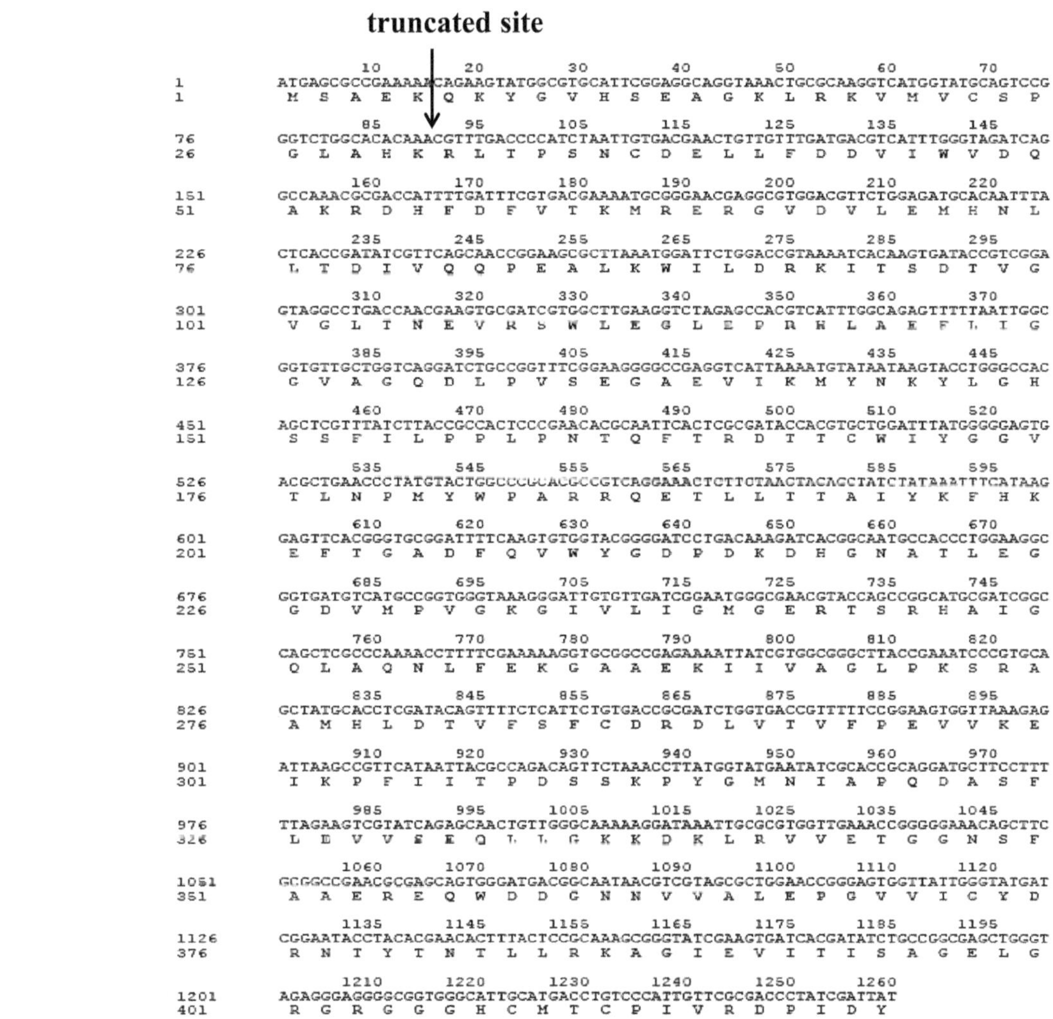

Figure 1 Nucleotide and corresponding amino acid sequences of the artificial ADI gene. The black arrow indicates the truncation site of T-rADI.

and enzymatic activity to hydrolyze L-arginine to L-citrulline. The pET30a-ADI produced overexpressed rADI proteins. However, there was only a weak corresponding band (approximately 53 kDa) in the soluble fractions (Figure 2A, lane 3), which gave the hydrolytic activity of 0.64 ± 0.04 U/ml. Most of the expressed proteins were existed in the form of inactive inclusion bodies (Figure 2A, lane 4).

Wen *et al.* [25] reported that truncation of a non-catalytic region in a 1,3-1,4-β-D-glucanase from *Fibrobacter succinogenes* improved enzymatic activity and thermotolerance. The crystal structures of ADI from *Mycoplasma arginini* and *Pseudomonas aeruginosa* showed that the enzymes shared the same catalytic triad (Cys-His-Glu) and the Cys active site [26,27]. The catalytic region mainly located in the C-terminal region of ADI. To explore the effects of deleting the N-terminal non-catalytic fragment on solubility and activity of rADI, our construction based on pET30a-ADI, a plasmid that expressed a truncated ADI protein lacking the N-terminal 30 amino acids (T-rADI, see Figure 1). Unfortunately, instead of improving solubility of the product, the truncation resulted in complete inactivation of the enzyme (Figure 2B, lanes 2 & 4).

Co-expression with chaperone teams

It is well known that chaperone proteins, such as DnaK-DnaJ-GrpE and GroES-GroEL are efficient to assist protein folding. Thus, improve the production of active recombinant proteins [1,6,28,29]. We co-expressed the pET30a-ADI with four plasmids encoding different chaperone teams respectively. While pGro7 (expressing GroES-GroEL) and pG-KJE8 (expressing DnaK-DnaJ-GrpE and GroES-GroEL) had a similar effect to improve solubility of the rADI proteins, pKJE7 (expressing DnaK-DnaJ-GrpE) and pTf16 (expressing Tig) didn't play any significant role to improve the solubility (Table 1). It suggested that only the GroES-GroEL chaperone team contributed to the improvement of soluble rADI proteins. For example, when the pET30a-ADI was co-expressed with pGro7, more rADI proteins formed as compared to the single pET30a-ADI plasmid that were present in the soluble part (Figure 2A, lane 5). The total activity was accordingly improved to 3.22 ± 0.06 U/ml, approximately 5-fold higher than that of the rADI expressed alone. However, the rADI proteins were still mostly in inclusion body forms (Figure 2A, lane 6). The further optimizations were performed based on the co-expression with pGro7.

Cultivation-improved solubilization of rADI

Kozai *et al.* [30] reported that the addition of L-arginine in the refolding solution was able to clearly increase the enzymatic activity of denatured ADI proteins. However, our experiments showed that the presence of L-arginine had no effect on the *in-vitro* solubilization of rADI proteins in the inclusion body form. We further tested whether L-arginine is helpful to improve the *in vivo* solubilization of ADI proteins in *E. coli* cells before the formation of inclusion bodies. *E. coli* cells are routinely cultivated in the LB medium, which contains peptones

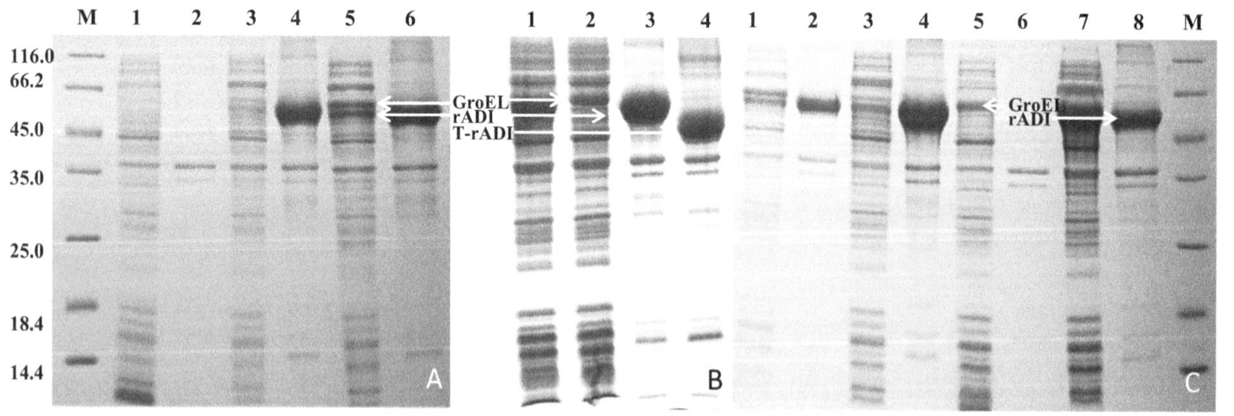

Figure 2 SDS-PAGE analysis of the production of rADI. (A) Co-expression with GroES-GroEL chaperones **(B)** Truncation of the N-terminal 30 amino acids. **(C)** The combined addition of D-glucose and L-arginine in LB medium. All the protein samples were normalized at OD_{600}. In **(A)**, M, molecular-mass marker protein; lanes 1–2: supernatant and cell debris of the pET-30a(+)-harboring *E. coli* BL21(DE3) cells; lanes 3–4: supernatant and cell debris of pET30a-ADI-harboring *E. coli* BL21(DE3) alone; lanes 5–6: supernatant and cell debris of pET30a-ADI-harboring *E. coli* BL21(DE3) cells co-expressed with GroES-GroEL. In **(B)**, lanes 1–2: supernatants of cells harboring rADI and truncated rADI; lanes 3–4: cell debris of cells harboring rADI and truncated rADI. In **(C)**, lanes 1–2: supernatant and cell debris of cells harboring pET30a-ADI and pGro7 from cultures containing 0.5% L-arginine; lanes 3–4: supernatant and cell debris of cells harboring pET30a-ADI and pGro7 from cultures containing 0.5% D-glucose; lanes 5–6: supernatant and cell debris of cells harboring pET-30a(+) and pGro7 from cultures containing 0.5% L-arginine and 0.5% D-glucose; lanes 7–8: supernatant and cell debris of cells harboring pET30a-ADI and pGro7 from cultures containing 0.5% L-arginine and 0.5% D-glucose; lane M: molecular-mass marker protein.

Table 1 The total activity of rADI co-expressed with different molecular chaperone

Plasmid	pET30a-ADI	+ pG-KJE8	+ pGro7	+ pKJE7	+ pTf16
Total activity (U/ml)	0.64 ± 0.04	3.14 ± 0.15	3.22 ± 0.06	0.60 ± 0.08	0.74 ± 0.01

and yeast extracts as the carbon and nitrogen sources. We added different concentrations of L-arginine in the LB medium to cultivate the pET30a-ADI plasmid-harboring *E. coli* BL21(DE3) cells. Low concentrations of L-arginine (less than 1%) had no effect on the solubilization of rADI proteins, while higher concentrations (1%-5%) caused alkalization of the medium (pH10.0), in which *E. coli* cells developed poorly or no growth at all. If we adjusted the media containing high L-arginine concentrations to pH 7.0 or added L-arginine hydrochloride to replace L-arginine, which did not change the pH of the medium significantly, *E. coli* cells were able to grow normally but the solubilization of rADI proteins did not improve.

The bacteria, such as *Streptococcus lactis* and *Oenococcus oeni*, that are able to utilize L-arginine as energy sources were unable to take up L-arginine in the absence of fermentable sugars [31-34]. We suggested that less amount of fermentable sugars in the LB medium might cause no effect of the addition of L-arginine on the solubilization of rADI proteins. Thus, we supplemented D-glucose together with L-arginine in the LB medium. Interestingly, the expression of soluble ADI proteins varied significantly with different concentrations of the supplements. The combination of 0.5% L-arginine and 0.5% D-glucose gave the highest solubility of the protein (Figure 2C), the enzymatic activity of the crude extracts reached 9.31 ± 0.01 to 9.43 ± 0.02 U/ml, which were much higher than that in the other combinations (Table 2). Under this condition, the yield of total and soluble rADI proteins were measured with the Bio-Rad' Image lab software, increased approximately 1.47 and 7.34-fold higher than that of the alone-expressed rADI and 1.28 and 2.87-fold higher than that of the rADI co-expressed with pGro7 respectively. The total activity increased approximately 15-fold higher than the rADI expressed alone (0.64 ± 0.04 U/ml) and 3-fold higher than that of the co-expression with pGro7 (3.16 ± 0.02 U/ml). In the absence of pGro7, the combination of 0.5% L-arginine and 0.5% D-glucose also enhanced the total activity to 2.37 ± 0.03 U/ml.

It was noted that higher concentrations of D-glucose (1% or more) supported better growth of cells, but the activities of crude extracts were prominently decreased. Higher concentrations of supplemented L-arginine (1% or more) inhibited the growth of cells because of alkalization of the medium, even with the presence of D-glucose, which caused no effect in the improvement of enzymatic activities (Table 2). For example, at the concentration of 1% L-arginine (combined with 0.5% D-glucose), the OD_{600} of culture broth was less than 1.0 after 24 hrs of incubation and at the 2% concentration, there were no growth. Surprisingly, when we added different concentrations of L-arginine hydrochloride, instead of L-arginine in the LB medium with D-glucose, no improvement was observed for the solubilization of rADI proteins (Table 2). When L-arginine was combined with other sugars, such as D-galactose, D-sucrose and D-maltose, the solubilization of rADI proteins was not improved. When L-arginine was replaced with non-substrate amino acids of ADI, such as L-tryptophan, L-citrulline and L-glycine no improvement was observed for the solubilization of rADI proteins (Table 2).

We also assayed the effects of the combined addition of 0.5% L-arginine and 0.5% D-glucose on the expression of another ADI, which was from *Listeria welshimeri* serovar 6b str. SLCC5334 (NC_008555). The total activity of the enzyme was 0.1187 ± 0.04 U/ml in the presence of 0.5% L-arginine and 0.5% D-glucose, which was 7.2-fold higher than it expressed alone (0.0165 ± 0.01 U/ml) and 2.9-fold higher than it co-expressed with pGro7 (0.0412 ± 0.02 U/ml). The increase rate was similar to that for the rADI from *Pseudomonas putida*. These results suggested that the combined addition of L-arginine and D-glucose was applicable in solubilization of different ADI proteins.

Enzymatic and conversion characteristics of rADI

After purification using Ni-NTA affinity chromatography, the recombinant proteins showed as a single band with a size of 53 kDa on SDS-PAGE gel (Figure 3A). We assayed enzymatic characteristics of the purified rADI proteins. The rADI, which contained some additional terminal fragments such as a His-tag and some restriction sites, showed similar characteristics as the original ADI obtained from *Pseudomonas putida* [22]. For example, the stability of proteins was observed between pH 7.2 and pH 9.0 after 12 hrs incubation (Figure 3B). The optimal pH was 6.0 for enzymatic activity, but the activity was sharply decreased with the variation of pH value (above 7.2 or below 4.5) (Figure 3C). The enzymatic activity of rADI was mostly retained after incubation at 30°C or 40°C for 90 min, clearly decreased after incubation at 50°C for 90 min (Figure 3D). Cu^{2+} (10 mM), SDS (1 mM, 10 mM) and DTT (1 mM, 10 mM) strongly inhibited the activity of rADI, Cr^{3+} (10 mM), Co^{2+} (10 mM) had a slight inhibition for rADI, while other metal ions had no remarkable effect observed on rADI (Table 3). Interestingly, the suitable temperature for the enzymatic activity of rADI was between 30°C and 50°C and the optimal temperature was 40°C (Figure 3E). The

Table 2 Effects of combined additions of different sugars and amino acids on the enzymatic activities of rADI condition optimization

Combined addition[a]				Enzymatic activity[b] (U/ml)
Amino acid	Percentage	Sugar	Percentage	
			0%	2.08 ± 0.05
			0.5%	9.43 ± 0.02
L-arginine 0.5%		D-glucose	1%	1.76 ± 0.01
			2%	1.60 ± 0.06
			5%	1.41 ± 0.01
	0%			2.78 ± 0.02
	0.5%			9.33 ± 0.04
L-arginine	1%	D-glucose 0.5%		0.86 ± 0.04
	2%			0.74 ± 0.03
	5%			0.70 ± 0.05
	0%			2.92 ± 0.03
	0.5%			2.93 ± 0.05
L-arginine hydrochloride	1%	D-glucose 0.5%		3.85 ± 0.07
	2%			4.35 ± 0.02
	5%			3.95 ± 0.04
	0%			3.18 ± 0.03
	0.5%			3.20 ± 0.08
L-arginine hydrochloride	1%	–		3.31 ± 0.05
	2%			3.60 ± 0.02
	5%			3.20 ± 0.08
		–	–	3.26 ± 0.07
		D-glucose	0.5%	9.31 ± 0.01
L-arginine 0.5%		D-galactose	0.5%	2.35 ± 0.04
		D-maltose	0.5%	1.96 ± 0.04
		D-sucrose	0.5%	2.49 ± 0.10
–	–			3.22 ± 0.06
L-arginine	0.5%			9.39 ± 0.03
L-tryptophan	0.5%	D-glucose 0.5%		1.93 ± 0.06
L-citrulline	0.5%			1.97 ± 0.01
L-glycine	0.5%			2.69 ± 0.02

[a]: with the presence of pGro7; [b]: produced from crude extracts.

optimal temperature was 50°C for the original ADI measured in *Pseudomonas putida* [22]. At the optimal 40°C, the specific activity of rADI reached 76 ± 0.03 U/mg, which was also higher than 58.8 U/mg of the original at 50°C.

We performed conversion experiments using the sodium phosphate buffer (500 mM, pH 6.0) containing 10% L-arginine. The L-citrulline products increased linearly at 30°C or 40°C. After 22 hrs of incubation, 20 g of L-arginine substrates were almost converted into L-citrulline by the crude enzyme from 25 ml cultures. The addition of the batch substrate or not, had no impact on the conversion (Figure 4). The above enzymatic characteristics suggested that rADI was a potential candidate for industrial conversion of L-arginine to L-citrulline.

Discussion

Previously, it had been observed that the improvement of the substrate L-arginine for correct refolding of denatured ADI proteins in the refolding solution [30]. However, L-arginine had no effect on the *in vitro* solubilization of rADI proteins in inclusion body forms. Thus, *in vivo* solubilization is important for potential industrial applications of rADI proteins. In the laboratories, the *E. coli* cells are routinely cultivated in LB medium with

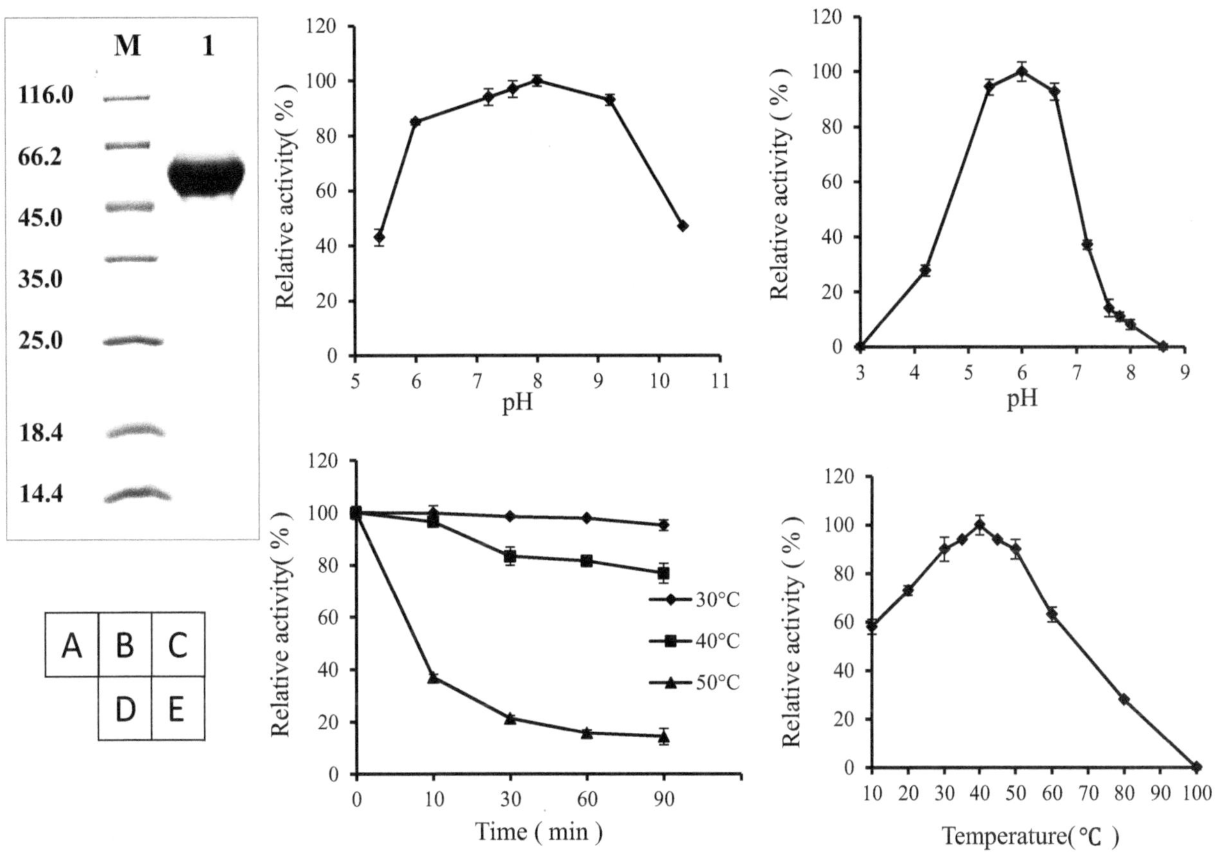

Figure 3 Enzymatic characteristics of the purified rADI protein. (A) SDS-PAGE analysis of the purified rADI proteins. The proteins were prepared by co-expression with pGro7 and under the conditions with the addition of 0.5% L-arginine and 0.5% D-glucose. Lanes: M, molecular mass marker protein; 1, rADI purified using Ni-NTA affinity chromatography. **(B)** The pH stability of the rADI after 12 hrs incubation at 4°C. **(C)** The optimal pH, assayed in buffers ranging from 3.0 to 8.6 at 40°C (Citric Acid-sodium citrate buffer for pH 3.0-5.4, Sodium phosphate buffer for pH 6.0-8.6). **(D)** Thermostability at 30°C (diamond), 40°C (square) and 50°C (triangle). **(E)** The optimal temperature, measured at various points from 10°C to 100°C in 500 mM Sodium phosphate buffer (pH 6.0).

tryptone and yeast extracts as the carbon and nitrogen sources. The present study showed that the combined addition of L-arginine and D-glucose in the LB medium significantly improved *in vivo* solubilization of rADI proteins, but the addition of L-arginine or D-glucose alone did not have the effect. L-arginine have to enter *E. coli* cells to play the *in vivo* solubilization functions. We suggested that the functions of L-arginine in solubilization of ADI were probably via the binding to the proteins, thus preventing the formation of inclusion bodies.

The role of D-glucose is still unknown during this process. There are three possibilities of its roles: (1) D-glucose helped L-arginine to transport through the cell membrane (2) D-glucose balanced the medium pH (3) D-glucose maintained the intracellular pH or metabolic processes. High concentrations of L-arginine caused alkalization of medium, which extensively inhibited the growth of cells. However, L-arginine hydrochloride did not change the medium pH significantly. The addition of the compound had no effect on the solubilization of the rADI proteins, even at low concentrations along with D-glucose. Thus, balancing medium pH or intracellular pH is not the function of D-glucose in the *in vivo* solubilization of rADI proteins.

Streptococcus lactis and *Oenococcus oeni* are able to utilize L-arginine as a source of energy via the L-arginine catabolism pathway [33,34]. Poolman *et al.* [33] reported that L-arginine entered these cells by an arginine-ornithine antiporter in *S. lactis*, which was initiated in the presence of a fermentable sugar. Only after initiated, L-arginine was able to subsequently enter cells via the ornithine and arginine concentration gradients formed during arginine metabolism [31,32]. It was suggested fermentable sugars provided the energy for the entrance of L-arginine into cells. Bioinformatics analysis showed that the genome of *E. coli* BL21(DE3) (GenBank accession no. CP001509) had some predicted proteins related to the arginine-ornithine antiporter, such as an arginine/ornithine antiporter transporter, an arginine/ornithine transporter subunit and an arginine/ornithine transport system ATPase [35]. We

Table 3 The effect of metal ions and reagents on the activity of rADI

Metal ions/ Reagents	Relative remaining activity (%) of rADI	
	1 mM	10 mM
Control	100.0 ± 2.9	100.0 ± 2.4
Li^+	99.6 ± 2.5	100.5 ± 0.8
Na^+	100.1 ± 3.5	102.5 ± 4.4
K^+	104.1 ± 2.5	100.1 ± 4.9
Mg^{2+}	96.4 ± 3.2	102.1 ± 1.5
Ca^{2+}	96.2 ± 3.9	91.2 ± 2.4
Mn^{2+}	97.4 ± 2.1	101.5 ± 4.2
Cr^{3+}	98.1 ± 2.3	81.5 ± 1.9
Co^{2+}	94.9 ± 2.8	73.7 ± 2.5
Ni^{2+}	100.6 ± 1.9	90.1 ± 0.1
Cu^{2+}	80.2 ± 1.2	33.6 ± 1.5
EDTA	100.3 ± 4.6	94.9 ± 1.8
SDS	22.2 ± 2.5	4.9 ± 1.6
DTT	71.0 ± 0.3	13.6 ± 1.2

supposed that the increased solubilization was associated with the *in vivo* interactions between L-arginine and newly synthesized rADI proteins during the process of refolding, which requires further determination.

Conclusions

Combined addition of L-arginine and D-glucose in the LB medium significantly improved *in vivo* solubility of rADI proteins but the addition of L-arginine or D-glucose alone did not have the effect. With the support of GroEL-GroES, the promotion of combined D-glucose and L-arginine was more significant. The studies present in this paper suggested a new strategy to increase the solubilization of overexpressed recombinant proteins in *E. coli* cells.

Methods

Strains and cultivation conditions

E. coli DH5α and BL21(DE3) strains (Table 4) were used as the hosts for cloning and expression of the ADI gene, respectively. The *E. coli* cells were routinely cultivated in Luria-Bertani (LB) medium (containing tryptone 10 g/L, yeast extract 5 g/L and NaCl 10 g/L).

Construction of recombinant plasmids

The ADI gene studied in this study is taken from *P. putida* (GenBank accession no. P41142). We optimized the gene sequence according to the codon frequency table of *E. coli* BL21(DE3) using the "codon randomization" strategy [23]. The artificial gene and its corresponding amino acid sequences (Figure 1) were deposited in GenBank under the accession no. KJ411883. After synthesized at Biosune (Shanghai, China), the gene was amplified using a specific primer pair (Forward: CATG*CCATGG*CTATGAGCGCCGAAAAACAGAAGTATG; Reverse: CCC*AAGCTT*GATAATCGATAGGGTCGCGAACAATG) to generate the digestion sites of endonucleases *Nco*I and *Hin*dIII (italics). After purification, the amplified fragment was digested using the restriction enzymes and ligated into plasmid pET-30a(+), named pET30a-ADI, which was confirmed by

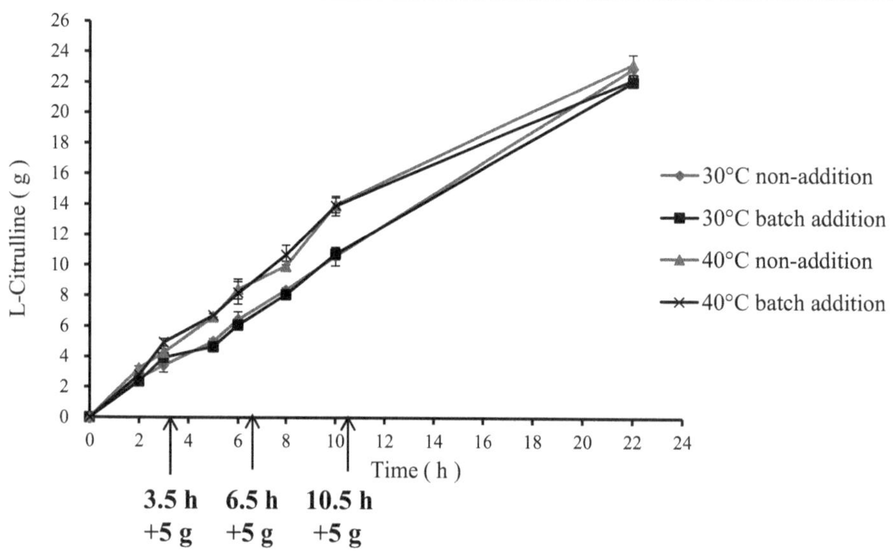

Figure 4 Conversation curves of L-arginine to L-citrulline by crude rADI enzymes. 20 g of L-arginine were added at the beginning (–◆- and -▲-) or batch-supplemented (–■- and -×–) from 5 g to 20 g at different time points (indicated by black arrows). The conversation experiments were performed at 30°C and 40°C.

Table 4 Bacterial strains and plasmids used in this study

Strains/plasmids	Genotype or description	Source
Strains		
E. coli DH5α	F^-,*sup*E44,ΔlacU169(φ*80lacZΔ*M15),*hsd*R17, *rec*A1,*end*A1, *gyr*A96, *thi*-1, *rel*A1	Life Technologies
E. coli BL21(DE3)	F^-, *omp*T, *hsd*S_B(r_B-m_B-), *gal*, *dcm*(DE3)	Life Technologies
Plasmids		
pG-KJE8	Encoding DnaK-DnaJ-GrpE / GroES-GroEL	Takara
pGro7	Encoding GroES-GroEL	Takara
pKJE7	Encoding DnaK-DnaJ-GrpE	Takara
pTf16	Encoding Tig	Takara
pColdTF	Expression vector	Takara
pET-30a (+)	Expression vector	Novagen
pGEX-6P-1	Expression vector	GE Healthcare
pBAD/gIIIA	Expression vector	Invitrogen

sequencing. The construction of recombinant plasmids, transformation and protein expression were performed according to the protocol described by Sambrook and Russell [36].

Co-expression with the chaperone teams

The plasmids pGro7, pG-KJE8, pTf16 and pKJE7 encode different chaperone teams (Table 4). The pET30a-ADI was co-expressed with these four plasmids respectively. *E. coli* BL21(DE3) cells harboring plasmid pET30a-ADI alone or together with a chaperone-encoding plasmid were cultivated in 5 ml of LB medium containing 40 μg/ml kanamycin and 20 μg/ml chloramphenicol for the selection of recombinants. After overnight shaking at 37°C, the culture was inoculated (1%) into fresh LB medium containing antibiotics and 0.04% L-arabinose (pG-KJE8 was induced by 5 μg/ml tetracycline and 0.04% L-arabinose). The culture was shaken at 37°C to induce the expression of chaperones. When the OD_{600} of the culture reached 0.6-0.8, isopropyl-β-D-thiogalactopyranoside (IPTG) was added to a final concentration of 0.1 mM to induce the expression of rADI. After additional 24 hrs of incubation at 16°C, the cells were harvested by centrifugation at 8,000 × *g* and 4°C for 10 min. The cell pellets were washed, resuspended in Tris-NaCl buffer (50 mM Tris, 150 mM NaCl, pH 8.0) and broken by sonication on ice. After centrifugation at 12,000 × *g* and 4°C for 30 min, the separated supernatant and cell debris were assayed enzymatic activities and in sodium dodecyl sulfate-polyacrylamide gel electrophoresis (SDS-PAGE, 10% gel concentration).

Cultivation-improved solubilization of rADI

The *E. coli* BL21(DE3) cells harboring the combined plasmids of pET30a-ADI and pGro7 were cultivated as described above. To assay the effects of cultivation on the solubility of rADI, different saccharides (D-glucose, D-sucrose, D-galactose or D-maltose) were added to a final concentration of 0.5% in the LB medium containing the antibiotics and L-arabinose. Together with the addition of IPTG, different amino acids (L-arginine, L-tryptophan, L-citrulline and L-glycine) were added to a final concentration of 0.5% when the culture OD_{600} reached 0.6-0.8. The cultures were incubated for additional 24 hrs at 16°C. *E. coli* BL21(DE3) harboring pET-30a(+) was used as negative control.

Purification of rADI

After harvested, cells were broken by sonication, and the supernatants were applied to the Ni-NTA affinity chromatography (GE Healthcare, Buckinghamshire, U.K.) to purify rADI. The protein concentration was determined using the Bradford method (Bio-Rad, Munich, Germany). Different fractions were collected and assayed enzymatic activities and in SDS-PAGE (10%).

Activity assays of rADI

The rADI activity was determined using a modified method of diacetyl monoxime thiosemicarbazide [17,37]. 40 mM L-arginine was dissolved in 500 mM sodium phosphate buffer (pH 6.0). After the addition of rADI, the mixture was incubated at 40°C for 10 min. Then, acid-ferric solution (containing 15% H_2SO_4, 60% H_3PO_4 and 0.005% $FeCl_3$) and DAM-TSC solution (containing 1% diacetyl monoxime and 0.06% thiosemicarbazide) were added to stop the reaction and the mixture was heated at 100°C for 10 min for coloration. The amount of L-citrulline was evaluated by measuring the optical density at 530 nm. The protein concentration was determined using Bradford-assay with bovine serum albumin as standard. One unit of the ADI activity was defined as the amount of enzymes to produce one μmol L-citrulline per minute under the assay conditions. All the assays were performed in triplicate. The total enzymatic activity was defined as units per milliliter of culture and the specific enzymatic activity was shown as units per milligram of purified enzyme.

Conversion rate of crude enzyme

The crude enzymes were prepared from sonication-broken cells using the above-described methods. 20 ml crude extracts from 100 ml culture broth were divided into four equal parts. The conversion of L-arginine to L-citrulline was assayed in 200 ml reaction buffer (500 mM sodium phosphate buffer, pH 6.0) at 30°C or 40°C. 20 g L-arginine was added at the beginning or batch-supplemented from 5 g to 20 g at different time points.

Competing interests

The authors declare that they have no competing interests.

Authors' contributions

YZL and YW designed the experiments and analyzed the data. YW performed all the experiments. YZL and YW wrote the paper. Both authors read and approved the final manuscript.

Acknowledgements

This work was financially supported by the National Natural Science Foundation of China (NSFC) Key Program (No. 31130004) and the NSFC for Distinguished Young Scholars (No. 30825001).

References

1. Georgiou G, Valax P: **Expression of correctly folded proteins in *Escherichia coli*.** *Curr Opin Biotechnol* 1996, **7**:190–197.
2. Baneyx F, Mujacic M: **Recombinant protein folding and misfolding in *Escherichia coli*.** *Nat Biotechnol* 2004, **22**:1399–1408.
3. Gasser B, Saloheimo M, Rinas U, Dragosits M, Rodriguez-Carmona E, Baumann K, Giuliani M, Parrilli E, Branduardi P, Lang C, Porro D, Ferrer P, Tutino ML, Mattanovich D, Villaverde A: **Protein folding and conformational stress in microbial cells producing recombinant proteins: a host comparative overview.** *Microb Cell Fact* 2008, **7**:11.
4. Frokjaer S, Otzen DE: **Protein drug stability: a formulation challenge.** *Nat Rev Drug Discov* 2005, **4**:298–306.
5. Schrodel A, de Marco A: **Characterization of the aggregates formed during recombinant protein expression in bacteria.** *BMC Biochem* 2005, **6**:10.
6. Platas G, Rodriguez-Carmona E, Garcia-Fruitos E, Cano-Garrido O, Villaverde A: **Co-production of GroELS discriminates between intrinsic and thermally-induced recombinant protein aggregation during substrate quality control.** *Microb Cell Fact* 2011, **10**:79.
7. Sorensen HP, Mortensen KK: **Advanced genetic strategies for recombinant protein expression in *Escherichia coli*.** *J Biotechnol* 2005, **115**:113–128.
8. de Marco A, Deuerling E, Mogk A, Tomoyasu T, Bukau B: **Chaperone-based procedure to increase yields of soluble recombinant proteins produced in E. coli.** *BMC Biotechnol* 2007, **7**:32.
9. Zuniga M, Perez G, Gonzalez-Candelas F: **Evolution of arginine deiminase (ADI) pathway genes.** *Mol Phylogenet Evol* 2002, **25**:429–444.
10. Ni Y, Schwaneberg U, Sun ZH: **Arginine deiminase, a potential anti-tumor drug.** *Cancer Lett* 2008, **261**:1–11.
11. Takaku H, Takase M, Abe S, Hayashi H, Miyazaki K: ***In vivo* anti-tumor activity of arginine deiminase purified from *Mycoplasma arginini*.** *Int J Cancer* 1992, **51**:244–249.
12. Misawa S, Aoshima M, Takaku H, Matsumoto M, Hayashi H: **High-level expression of *Mycoplasma* arginine deiminase in *Escherichia coli* and its efficient renaturation as an anti-tumor enzyme.** *J Biotechnol* 1994, **36**:145–155.
13. Kakimoto T, Shibatani T, Nishimura N, Chibata I: **Enzymatic production of L-citrulline by *Pseudomonas putida*.** *Appl Microbiol* 1971, **22**:992–999.
14. Miyazaki K, Takaku H, Umeda M, Fujita T, Huang WD, Kimura T, Yamashita J, Horio T: **Potent growth inhibition of human tumor cells in culture by arginine deiminase purified from a culture medium of a *Mycoplasma*-infected cell line.** *Cancer Res* 1990, **50**:4522–4527.
15. Takaku H, Misawa S, Hayashi H, Miyazaki K: **Chemical modification by polyethylene glycol of the anti-tumor enzyme arginine deiminase from *Mycoplasma arginini*.** *Jpn J Cancer Res* 1993, **84**:1195–1200.
16. Takaku H, Matsumoto M, Misawa S, Miyazaki K: **Anti-tumor activity of arginine deiminase from *Mycoplasma arginini* and its growth-inhibitory mechanism.** *Jpn J Cancer Res* 1995, **86**:840–846.
17. Liu YM, Sun ZH, Ni Y, Zheng P, Liu YP, Meng FJ: **Isolation and identification of an arginine deiminase producing strain *Pseudomonas plecoglossicida* CGMCC2039.** *World J Microbiol Biotechnol* 2008, **24**:2213–2219.
18. Kim JE, Jeong DW, Lee HJ: **Expression, purification, and characterization of arginine deiminase from *Lactococcus lactis ssp. lactis* ATCC 7962 in *Escherichia coli* BL21.** *Protein Expr Purif* 2007, **53**:9–15.
19. Burne RA, Parsons DT, Marquis RE: **Cloning and expression in *Escherichia coli* of the genes of the arginine deiminase system of *Streptococcus sanguis* NCTC 10904.** *Infect Immun* 1989, **57**:3540–3548.
20. Knodler LA, Schofield PJ, Gooley AA, Edwards MR: ***Giardia intestinalis*: purification and partial amino acid sequence of arginine deiminase.** *Exp Parasitol* 1997, **85**:77–80.
21. Knodler LA, Sekyere EO, Stewart TS, Schofield PJ, Edwards MR: **Cloning and expression of a prokaryotic enzyme, arginine deiminase, from a primitive eukaryote *Giardia intestinalis*.** *J Biol Chem* 1998, **273**:4470–4477.
22. Shibatani T, Kakimoto T, Chibata I: **Crystallization and properties of L-arginine deiminase of *Pseudomonas putida*.** *J Biol Chem* 1975, **250**:4580–4583.
23. Menzella HG: **Comparison of two codon optimization strategies to enhance recombinant protein production in *Escherichia coli*.** *Microb Cell Fact* 2011, **10**:15.
24. Donovan RS, Robinson CW, Glick BR: **Review: optimizing inducer and culture conditions for expression of foreign proteins under the control of the lac promoter.** *J Ind Microbiol* 1996, **16**:145–154.
25. Wen TN, Chen JL, Lee SH, Yang NS, Shyur LF: **A truncated *Fibrobacter succinogenes* 1,3-1,4-beta-d-glucanase with improved enzymatic activity and thermotolerance.** *Biochemistry (Mosc)* 2005, **44**:9197–9205.
26. Das K, Butler GH, Kwiatkowski V, Clark AD Jr, Yadav P, Arnold E: **Crystal structures of arginine deiminase with covalent reaction intermediates: mplications for catalytic mechanism.** *Structure* 2004, **12**:657–667.
27. Lu X, Galkin A, Herzberg O, Dunaway-Mariano D: **Arginine deiminase uses an active-site cysteine in nucleophilic catalysis of L-arginine hydrolysis.** *J Am Chem Soc* 2004, **126**:5374–5375.
28. Nishihara K, Kanemori M, Kitagawa M, Yanagi H, Yura T: **Chaperone coexpression plasmids: differential and synergistic roles of DnaK-DnaJ-GrpE and GroEL-GroES in assisting folding of an allergen of Japanese cedar pollen, Cryj2, in *Escherichia coli*.** *Appl Environ Microbiol* 1998, **64**:1694–1699.
29. Yan X, Hu S, Guan YX, Yao SJ: **Coexpression of chaperonin GroEL/GroES markedly enhanced soluble and functional expression of recombinant human interferon-gamma in *Escherichia coli*.** *Appl Microbiol Biotechnol* 2012, **93**:1065–1074.
30. Kozai M, Sasamori E, Fujihara M, Yamashita T, Taira H, Harasawa R: **Growth inhibition of human melanoma cells by a recombinant arginine deiminase expressed in *Escherichia coli*.** *J Vet Med Sci* 2009, **71**:1343–1347.
31. Liu SQ, Pilone GJ: **A review: arginine metabolism in wine lactic acid bacteria and its practical significance.** *J Appl Microbiol* 1998, **84**:315–327.
32. Liu SQ, Pritchard GG, Hardman MJ, Pilone GJ: **Arginine catabolism in wine lactic acid bacteria: is it via the arginine deiminase pathway or the arginase-urease pathway?** *J Appl Bacteriol* 1996, **81**:486–492.
33. Poolman B, Driessen AJ, Konings WN: **Regulation of arginine-ornithine exchange and the arginine deiminase pathway in *Streptococcus lactis*.** *J Bacteriol* 1987, **169**:5597–5604.
34. Tonon T, Lonvaud-Funel A: **Metabolism of arginine and its positive effect on growth and revival of *Oenococcus oeni*.** *J Appl Microbiol* 2000, **89**:526–531.
35. Jeong H, Barbe V, Lee CH, Vallenet D, Yu DS, Choi SH, Couloux A, Lee SW, Yoon SH, Cattolico L, Hur CG, Park HS, Segurens B, Kim SC, Oh TK, Lenski RE,

Studier FW, Daegelen P, Kim JF: **Genome sequences of *Escherichia coli* B strains REL606 and BL21(DE3).** *J Mol Biol* 2009, **394:**644–652.
36. Sambrook J, Russell DW, Sambrook J: *The Condensed Protocols from Molecular Cloning: A Laboratory Manual.* Cold Spring Harbor, N.Y.: Cold Spring Harbor Laboratory Press; 2006.
37. Ni Y, Liu Y, Schwaneberg U, Zhu L, Li N, Li L, Sun Z: **Rapid evolution of arginine deiminase for improved anti-tumor activity.** *Appl Microbiol Biotechnol* 2011, **90:**193–201.

Bactericidal activities of GM flax seedcake extract on pathogenic bacteria clinical strains

Magdalena Zuk[1,3*], Agata Dorotkiewicz-Jach[2], Zuzanna Drulis-Kawa[2], Malgorzata Arendt[1], Anna Kulma[1] and Jan Szopa[1,3]

Abstract

Background: The antibiotic resistance of pathogenic microorganisms is a worldwide problem. Each year several million people across the world acquire infections with bacteria that are antibiotic-resistant, which is costly in terms of human health. New antibiotics are extremely needed to overcome the current resistance problem.

Results: Transgenic flax plants overproducing compounds from phenylpropanoid pathway accumulate phenolic derivatives of potential antioxidative, and thus, antimicrobial activity. Alkali hydrolyzed seedcake extract containing coumaric acid, ferulic acid, caffeic acid, and lignan in high quantities was used as an assayed against pathogenic bacteria (commonly used model organisms and clinical strains). It was shown that the extract components had antibacterial activity, which might be useful as a prophylactic against bacterial infection. Bacteria topoisomerase II (gyrase) inhibition and genomic DNA disintegration are suggested to be the main reason for rendering antibacterial action.

Conclusions: The data obtained strongly suggest that the seedcake extract preparation is a suitable candidate for antimicrobial action with a broad spectrum and partial selectivity. Such preparation can be applied in cases where there is a risk of multibacterial infection and excellent answer on global increase in multidrug resistance in pathogenic bacteria.

Keywords: Antimicrobial compound, Phenolic acid, Flax, Alternative antibiotic, Flax seedcake

Background

The worldwide increase in the occurrence of multidrug resistance (MDR) in bacterial pathogens has become a serious problem in the clinical treatment of infections [1]. We are witnessing a high increase in the resistance levels of pathogenic bacteria and a sharp decrease in the efficacy of conventional antibacterial therapy, often across the whole spectrum of known therapeutics [2-4]. Even chemotherapeutics that are used as the last line of defense is less effective in an eradication of infections involving MDR strains. Bacteria can develop or acquire new mechanisms of resistance and share the information by horizontal transfer between species. Nowadays, over 90% of staphylococci, pneumococci and enterococci isolated from severe infections are resistant to antibiotics. Increasing prevalence of methicillin-resistant *Staphylococcus aureus* (MRSA), β-lactam- and macrolide-resistant pneumococci, and glycopeptide- and vancomycin-resistant enterococci have been reported [5]. Among the Gram-negative bacteria, the most serious treatment problems are caused by MDR *Pseudomonas aeruginosa* strains producing carbapenemases and MDR *Enterobacteriaceae* strains including *Escherichia coli* and *Klebsiella pneumoniae*, which produce extended beta-lactamases (ESBL) [6]. In the WHO (Word Health Organisation) report from World Health Day 2011 *Antimicrobial resistance: no action today, no cure tomorrow*, it was indicated that 150,000 people die each year as a result of only one very dangerous disease, multidrug-resistant tuberculosis (MDR-TB) [7]. The global increase in multidrug resistance in pathogenic bacteria has led to an increasing need for topical antimicrobial products that can be applied in cases of multibacterial infection. Many of the available agents have highly bactericidal activity, but their cytotoxicity can also interfere with human tissues [8]. Actually one of the great challenges of

* Correspondence: mzuk@ibmb.uni.wroc.pl
[1]Faculty of Biotechnology, Wrocław University, Przybyszewskiego 63/77, 51-148 Wroclaw, Poland
[3]Linum Fundation, Stabłowicka 147/149, 54-066 Wrocław, Poland
Full list of author information is available at the end of the article

modern bio-medical science is searching for compounds/agents that can be simultaneously effective as an unselective antibiotic and not indicate cytotoxic effect to human cells. Such preparations will be useful in the case of multibacterial and fungal infections.

Bioactive components isolated from plants are one alternative to commercially available traditional antibiotic. The rich source of bioactive metabolites may be flax (*Linum usitatissimum* L.), an annual plant that grows widely in the Mediterranean and temperate climate zones. It is not only a source of oil and fibers but also seedcake, the remains of the flax seeds after oil extraction. Seedcake has important dietary properties, and it is also used against many diseases, such as skin, respiratory tract and gastrointestinal tract diseases [9-11].

The health-promoting properties of flax can be further improved by genetic modification.

Findings presented in this paper indicated extensive spectrum antibacterial activity of preparation based on flax seedcake. The material for that set of experiments was derived from transgenic flax plants designated W92. They were obtained by plant transformation using the genes coding chalcone synthase (*CHS*), chalcone isomerase (*CHI*) and dihydroflavonol reductase (*DFR*), which all control the synthesis of antioxidative compounds generated via the phenylpropanoid pathway [12].

Significant increases in the amounts of various flavonoids (kaempferol, quercetin) were reported in the seeds of transgenic W92 flax. The levels of anthocyanins, proanthocyanidins and phenolic acids were also higher in the W92 flaxseeds than those of the control plants [13]. Ferulic, caffeic and p-coumaric acids and their glucoside derivatives were the most common phenolic acids present. Three main compounds from the phenylpropanoid pathway were identified in the fibers from W92 plants: vanillin, 4-hydrobenzoic acid and acetovanillone. UPLC analysis of fibers from W92 flax also revealed the presence of ferulic acid. The same compounds but in significantly lower amounts and different proportion between individual metabolites will be found in unmodified plants. Therefore, for further experiments on clinic bacterial stains, we used seedcake from W92 transgenic flax.

Detailed analysis of seedcake extract from transgenic flax indicates that such preparation is effective as an antibiotic wide spectrum of bactericidal activity, useful in the case of multibacterial and fungal infections is the main goal of this study. The data obtained strongly suggest that the seedcake extract a suitable candidate for antimicrobial action with a broad spectrum and partial selectivity. It is believed that this is the first report describing the potential of a product from GM flax for antimicrobial application.

Methods

Plant material

All of the presented experiments were performed on flax plants (*Linum usitatissimum L.*). Seeds from unmodified flax (cv. *Linola*) were obtained from the Flax and Hemp Collection of the Institute of Natural Fibers, Poland. Transgenic plants overproducing phenylpropanoid compounds were previously generated in laboratories of the University of Wroclaw [12] In brief, to construct the transgenic plants, we used a binary vector containing three cDNAs from *Petunia hybrida* encoding chalcone synthase [CHS, EMBL/GenBank database acc. no. *X04080*], chalcone isomerase [CHI, EMBL/GenBank database acc. no. *X14589*] and dihydroflavonol reductase [DFR, EMBL/GenBank database acc. no. *X15537*] in the sense orientation, under the control of the 35S promoter and OCS terminator. The vector was introduced into *Agrobacterium tumefaciens* and used for flax (cv. *Linola*) transformation. The transgenic plants were preselected via PCR using primers specific for the kanamycin resistance gene (*npt II*). The relevant selection of transgenic plants was performed by Northern blot analysis using radiolabeled cDNAs of all three transgenes (CHS, CHI, and DFR) as probes. The details on plant transformation, selection and transgenic plant analysis were described previously [12,14]. The reference specimens were collected in the University of Wroclaw Biotechnology Department in the form of tissue culture plants and seeds from all of the cultivation seasons.

In this study, we used the fifth generation transgenic plants W92 (transgenic line W92.40) and unmodified flax plants cv. Linola (as a control), both were field-grown in the 2011 growing season. The 600 m2 was seeded (sow density 600seeds/m2) and 12 kg of seeds, 15 kg stalks and finally 4,5 kg fiber for W92 as well as control plants were obtained. The seeds were harvested 4 months after sowing and used for oil pressing. The stems of plants were then retted by the dew method as described previously [15]. Briefly, the harvested plant stalks were spread evenly in field for at least 40 days. Combined action of water (dew), sun and naturally occurring in a soil bacteria and fungi caused degradation of the cell-wall polysaccharides and middle lamella, releasing the fibers from the stems. The retted stalks, called straw, were dried in open air and were stored for a short period (one week) to allow curing to occur, facilitating fibre removal. Final separation of the fibre is accomplished by a breaking process in which the brittle woody portion of the straw is broken, which removes the broken woody pieces (shaves) by beating or scraping.

Seedcake preparation

Flaxseeds (10 kg) were ground and transferred to an industrial warm gear oil press (Oil PressDD85G – IBG Monoforts

Oekotec GmbH& Co) to be cold pressed at a maximum of 40°C. After the cold pressing, 5 to 7% of the total fatty acids that are present in untreated seeds normally remain in the seedcakes as a residual oil fraction. In order to remove all of the fat from the seedcakes, the ground material (~100 μm grain) was defatted by threefold extraction with hot (65°C) hexane, after such treatment obtain 7.6 kg fully defatted seedcakes. Afterward, the seedcakes were dried and used for further analysis and for the preparation of the alkali hydrolyzed seedcake extract.

Determination of total phenolic content

The plant material (seeds, seedcakes and fibers) was crushed using a laboratory mill. The seeds and seedcakes were defatted using hot hexane before analysis. A 250-mg sample of each material was extracted three times with 800 μl of methanol containing 1% formic acid, and sonicated for 15 min. Next, the samples were centrifuged (10 min at 14000 g) and the clear supernatant was used for the analysis. The total phenolics content was determined by the optimized Folin-Ciocalteau method [16], referring to the calibration curve of gallic acid, which is a phenol compound used as a standard. A total of 100 μl of each sample solution was mixed with 0.2 ml of Folin-Ciocalteau reagent, 2 ml of H_2O, and 1 ml of 15% Na_2CO_3. The absorbance was measured with a model spectrophotometer at 765 nm after 2 hours incubation at room temperature. The total phenolics were estimated as gallic acid equivalent. The data were obtained from the average of three determinations.

Determination of total anthocyanin content via the pH-differential method [17] Samples of the ground and defatted seeds, seedcakes (15 mg) and fibers (15 mg) were extracted with 1 ml of methanol/HCl (95:5, v/v) in an ultrasonic bath for 30 min. The extract was centrifuged at 14000 g for 10 min. Two dilutions of the sample were performed: first, 100 μl of the supernatant was mixed with 900 μl of 0.025 M potassium chloride buffer, pH 1.0, and then 100 μl of supernatant was mixed with 900 μl of 0.4 M sodium acetate buffer, pH 4.5. The solution was allowed to stand at room temperature for 15 min, and then the absorbance at 520 nm and 700 nm was measured, which allowed for haze correction. The results were reported as cyanidin-3-*O*-glucoside equivalents.

Evaluation of proanthocyanin content

For the measurements, 15 mg samples of defatted seeds and seedcakes were used. Proanthocyanins were hydrolyzed with 1 ml of n-butanol/HCl (95:5, v/v) and 33 μl of 2% (w/v) $NH_4Fe(SO_4)_2$ $12H_2O$ in 2 M HCl for 40 min at 95°C. The extract was centrifuged at 14000 g for 10 min, and the supernatant was used for proanthocyanin content evaluation. Proanthocyanin detection was carried out by measuring absorption at 540 nm, and proanthocyanin content was expressed as catechin equivalents [18].

Measurement of phenolic acid and secoisolariciresinol diglucoside (SDG) contents

A 250 mg sample of defatted flax seeds or seedcakes was extracted three times with 1.5 ml of 80% methanol (v/v) for 10 min at 80°C. Prior to extraction, the seedcakes were finally defatted, again with hot hexane. The extracts were centrifuged and evaporated to near dryness at 40°C under a vacuum. The extract was then resuspended and subjected to alkaline hydrolysis (1 ml, 0.3 M aqueous sodium hydroxide) for 2 days at room temperature followed by neutralization using 2 M HCl to pH = 6.0. The extract (3× 1-3 μl injection) was analyzed on a Waters Acquity UPLC (Ultra Performance Liquid Chromatography) system with a 2996 PDA detector, using Acquity UPLC column BEH C18, 2.1100 mm, 1.7 μm. The mobile phase was A = acetonitrile and B = 20 mM ammonium formate, pH 3.0 in a gradient flow: 1 min, 10% A/90% B; 2–6 min gradient to 40% A/60% B, and 7 min gradient from 40% to100% A with a 0.4 ml/min flow rate [19]. Measurements were taken at 280 (SDG) and 320 nm (phenolic acids).

Extraction of phenylpropanoids from fibers

One gram of W92 flax fibers was ground in a Retch mill to a fine powder (100 μm grain) and extracted three times with10ml methanol. Extracts were pooled, evaporated under a vacuum and resuspended in 2 ml methanol. The remaining matter was hydrolyzed in 2 N NaOH at room temperature for 24 hours in order to release bound phenolics. Extracts were adjusted to pH 3.0 extracted three times with ethyl acetate, pH3.0 and then the extract was evaporated under a vacuum and resuspended in 2 ml of methanol.

UPLC analysis of phenolics in flax fibers

The components were analyzed using the Acquity UPLC system (Waters) equipped with an automated sample injector and PDA detector. A 10-μl sample was applied to an Acquity UPLC HSS- T3 column, 2.1 × 100 mm, 1.8 μm retaining better hydrophilic components. The mobile phase was passed through the column at a flow rate of 0.5 ml/min. The mobile phase consisted of 0.1% formic acid (A) and 100% methanol (B). For the first 2 min, isocratic elution was carried out using 100% of A, 2–5 min a linear gradient to 30% A/70% B, 5–5.5 min to 0% A/100% n B. In the final minute (5,5-6,5 min) the concentration of A was returned to 100%. An additional analysis of the very hydrophilic component was performed using a UPLC HILIC, 2.1 × 100 mm, 1.7 μm column. The mobile phase was passed through the column at a flow rate of 0.4 ml/min. The mobile phase consisted of 0.1% formic acid (A) and 100% acetonitrile (B). For the first 4 min 10% A/90% B was used, 4–8 min a linear gradient to 90% A/10% B. was applied, 8 - 9 min the gradient to 100% A was used. In the final minute, the concentration of eluting solvents was returned to 10% A /90% B [13].

Preparation of alkali hydrolyzed seedcake extract

Preparation was performed according modified method previously described by Zuk et al. [13] Defatted flax seedcakes (100 g) were extracted three times with 400 ml of 80% methanol (v/v) for 15 min at 80°C. The extract was centrifuged, the pellet was discarded and methanol from the supernatant fraction was evaporated at 40°C. The aqueous fraction of the extract was subjected to alkaline hydrolysis in a final concentration of 0.3 M sodium hydroxide for 2 days at room temperature. The hydrolyzed syrup was acidified with 2 M hydrochloric acid to pH 6.0. The solution was cooled down to 10°C then centrifuged with a high-speed centrifuge at 7000 rpm for 15 min to precipitate and remove water-soluble polysaccharides and proteins. After freeze-drying, the dry material was preserved at 4°C and used to prepare active water solutions (w/v) of 50, 30 and10mg/ml. The solutions were sterilized by filtration through an Acrodisc 0.22 μm filter (Gelman Sciences, Ann Arbor, MI).

Antioxidant capacity measurement *via* chemiluminescence method

The different seedcake extract solutions (from control and GM plants) or active seedcake preparation at a concentration of 9.5 mg/ml, which is the equivalent of 10 mM gallic acid in the total phenolic measurement, and a 10-mM solution of standard substances (ascorbic acid, chlorogenic acid, caffeic acid, SDG) were diluted in the range of 1000–15000 times with water, and directly analyzed according to the published method [20].This experiment was performed in a final volume of 250 μl on white microplates in a solution containing 0.1 M Tris–HCl buffer, pH 9.0, and 4 mM 2,2-azobis (2-amidinopropane) dihydrochloride (AAPH), freshly prepared. The luminol solution (100 mM) and diluted extract were automatically injected. The photons produced in the reaction were counted on an EG&G Berthold LB96P microplate luminometer at 30°C. The antioxidant potential (IC50) was defined as the amount of flax extract (mg FW) that inhibits luminol chemiluminescence by 50%.

Bacterial strains

Thirty-six Gram-negative clinical strains (*Pseudomonas aeruginosa* [n = 16], *Klebsiella pneumoniae* [n = 10] and *Escherichia coli* [n = 10]) and 15 Gram-positive clinical strains (*Staphylococcus aureus* [n = 5], *Staphylococcus epidermidis* [n = 5] and *Enterococcus faecalis* [n = 5]) were used to determine the antibacterial activity of seedcake extract. The tested strains were isolated from clinical samples from patients hospitalized in the Lower Silesian Centre of Pediatrics in Wroclaw, Poland. MDR isolates were included in the selection (Additional file 1: Table S1 and Table S2). As references, we used the strains *P. aeruginosa* ATCC 27853, *K.pneumoniae* ATCC 700603, *E. coli* ATCC 25922, *E.coli* ATCC 11229, *S. aureus* ATCC 29213 and *S. aureus* ATCC 6538 from the American Type Culture Collection (LGC Standards, Lomianki, Poland). The bacteria were stored at –70°C in Trypticase Soy Broth (Becton Dickinson and Company, Cockeysville, MD, USA) supplemented with 20% glycerol.

Determination of bacterial susceptibility

The antimicrobial activity of the seedcake extract was measured by determining the minimum inhibitory concentration (MIC). The MIC tests were performed by a broth microdilution method according to the European Committee on Antimicrobial Susceptibility Testing standards (EUCAST, www.eucast.org). Three different dilutions of W92 flax seedcake extract were prepared: 50, 30 and 10 mg/ml (for details see Table 1). The proper concentration of seedcake extract was prepared by mixing

Table 1 Content of phenylpropanoid compounds in three preparations of W92 seedcake extract (10 mg/ml,30 mg/ml and 50 mg/ml) dissolved in water

Compounds	Compounds content in dried extract (mg/gDW)	Preparations compounds content in:		
		50 mgDW/ml	30 mgDW/ml	10 mgDW/ml
Secoisolariciresinol diglucoside (SDG)	548.41 ± 9.56	27.4 ± 0.03	16.0 ± 0.02	5.5 ± 0.02
Ferulic acid	16.09 ± 0.63	0.803 ± 0.00	0.459 ± 0.04	0.178 ± 0.00
Ferulic acid glucoside	29.55 ± 0.31	1.476 ± 0.01	0.866 ± 0.10	0.302 ± 0.02
p-coumaric acid	4.94 ± 0.12	0.254 ± 0.01	0.143 ± 0.02	0.049 ± 0.00
Coumaric acid glucoside	23.83 ± 0.62	1.185 ± 0.00	0.743 ± 0.04	0.238 ± 0.01
Caffeic acid	0.32 ± 0.001	0.016 ± 0.00	0.008 ± 0.00	0.003 ± 0.00
Caffeic acid glucoside	38.18 ± 0.51	1.883 ± 0.02	1.093 ± 0.02	0.378 ± 0.01

The results are the mean values ± SD (n = 6).

with Mueller Hinton II Broth powder (MHB; Becton Dickinson and Company, Cockeysville, USA) and sterilized by filtration. The gentamicin (GM) and ampicillin (AM) for antibiotic susceptibility tests were obtained from Sigma-Aldrich Chemie GmbH (Steinheim, Germany). The serial concentration of antibiotics were prepared in MHB according to EUCAST standards [21]. For the experiments, bacterial strains were inoculated onto blood agar plates, incubated for 18 hours at 35°C, dissolved in PBS to an optical density equal to McFarland No. 0.5 and then diluted 1:10 (to approximate concentration 10^7 cells/ml). 10 μl samples of bacterial culture were added to a 96 microtitre plate containing 200 μl of the extract solution in MHB/ antibiotic solution in MHB. The final concentration of microorganisms was $5{\times}10^5$cfu/ml. The plates were then incubated for 18 hours at 35°C. The MIC was defined as the lowest concentration of seedcake extract at which no visible growth of bacteria was observed after 18 hours incubation. Positive controls (growth) consisted of bacteria in broth. Negative controls (sterility) consisted of uninoculated broth with which each of the extract dilutions was performed. Each assay was performed in triplicate and repeated three times on a different day to ensure reproducibility of the results.

Dynamics of microbial growth in the presence of seedcake extract

In vitro killing curves were determined in the presence of 10, 30 and 50 mg/ml seedcake extracts for the reference strains *P. aeruginosa* ATCC 27853, *K. pneumoniae* ATCC 700603, *E. coli* ATCC 11229 and *S. aureus* ATCC 6538. The control consisted of bacteria in MHB alone. The proper concentration of seedcake extract was prepared by mixing with MHB and sterilized by filtration. For the experiments, bacterial strains were inoculated onto blood agar plates, incubated for 18 hours at 35°C, dissolved in PBS to an optical density equal to McFarland No. 0.5 and then diluted 1:10 (to approximate concentration 10^7 cells/ml). 50 μl samples of bacterial culture were added to eppendorf tube containing 1000 μl of the extract solution in MHB. Samples of 10 μl were collected after 0, 2, 4, 8 and 24 hours of incubation, serially diluted and plated on Mueller Hinton II Agar (BioMerieux, France). The bacterial count (cfu/ml) was determined after 18 hours of incubation at 35°C. Each assay was performed in triplicate.

The inhibitory effect of standard substances on bacterial growth

For the experiments, *S. aureus* ATCC 6538 was inoculated onto blood agar plates, incubated for 18 hours at 35°C, and then diluted in PBS to an optical density equal to McFarland No. 0.5. A 10-μl sample of bacterial culture was diluted 1:10 (approximate concentration: 10^7 cells/ml) and added to the 96 microtitre plate containing the commercially available standard substances solution (ferulic acid, coumaric acid, caffeic acid, SDG) and seedcake extract in MHB. The final concentration of microorganisms was $5{\times}10^5$ cfu/ml. The plates were then incubated for 6 hours at 35°C. The growth inhibition effect (% of inhibition in relation to untreated bacteria) was calculated after 6hoursof incubation by measuring the OD600. Each assay was performed in triplicate.

The inhibitory effect of seedcake extract on bacteria gyrase activity

Reaction mixture (total volume 30 μl) consisted of 5× buffer for enzyme (NEB), sterile deionized water, gyrase from *E.coli* (NEB), substrate for gyrase (NEB) and particular amount of seedcake extract (to the final concentration 10 mg/ml; 30 mg/ml; 50 mg/ml and 75 mg/ml of dried extract respectively). Covalently closed substrate (pUC) in a relaxed DNA form was used. As negative control reaction mixture without gyrase or extract was used. As total inhibition control reaction mixture with novobiocin and without extract was used. Samples were incubated for 30 minutes at 37°C. Electrophoresis was run in 1×TBE for 2 hours in 0.8% agarose gel without ethidium bromide.

Pulse-field gel electrophoresis of *E.coli* DNA upon seedcake extract treatment

Bacteria cells (10^8 CFU/ml) incubated 1 hours with seedcake extract were centrifuged, pellet was resuspended in 200 μl PBS, embedded in 2% low melting agarose, then were treated first with lysozyme (10 μg/ml, 40 min at 30°C) and thereafter overnight with proteinase K (50 μg/ml at 56°C). Agarose plugs were washed 3 times with TE buffer and subjected to pulse-field electrophoresis. Gel was run in 0.5% TBE, t1 = 1 s, t2 = 50s with included angle 120°, at 12°C for 16.5 hours. After electrophoresis gel was stained with ethidium bromide and analysed under UV light (Gel Doc Imaging System, UVP Analitic).

Skin irritation test

An *in vitro* skin irritation test was performed according to the MTT Effective Time-50 (ET-50) protocol developed at MatTek Corporation with the use of the EpiDerm skin irritation test. This test allows the valuation of skin irritation due to cosmetic ingredients and ready-made products. 100 μl of three preparations of seedcake extracts (10 mg/ml, 30 mg/ml, 50 mg/ml) were put on the surface of the epidermis model. As a positive control (substance elicid skin irritation) the 1% Triton X-100 was used. After an incubation period of 2, 5 or 18 h, cell viability was assessed using the MTT colorimetric test. 50 μL of MTT stock solution (4 mg/mL) was added to each well to give a total reaction volume of 550 μL. After incubating for 4 h, the medium with MTT solution was removed from the plate. The formazan crystals in each well were dissolved in 500 μL of DMSO and incubated for 30 min with gentle shaking. The absorbance at 540 nm was read on an Asys UVM340 Microplate Reader (Biochrom, UK). The results were presented in% as a reference to the control (100%). Skin irritation potential is predicted if the remaining relative cell viability is below 50%. The experiment was performed in duplicate.

Statistical analysis

Pearson correlation quotient was calculated. The correlations were considered significantly significant at $p < 0,05$. All statistical calculations were performed using STATISTICA 7.1 software package (StatSoftPolska, Poland).

Results

Transgenic W92 flax (line W92.40) was previously obtained by simultaneous overexpression of three genes encoding key enzymes of flavonoid biosynthesis (CHS, CHI and DFR) under the control of the 35S promoter. Since the constitutive promoter was used for the modification, changes in compound contents were expected in the whole plant body. Increased amounts of many phenolic compounds (lignans, flavonoids, phenolic acids) were indeed detected in both the green parts and the seeds of W92 flax plants [13]. As a control, the level of phenolic compounds in unmodified plants was also measured.

Biochemical analysis of seedcakes and fibers from wild type and W92 plants

First, we identified the plant product that is most suitable for the isolation of antimicrobial compounds. There are two main products that are obtained from flax: seeds, the source of oil and seedcake, which is the material remaining after oil pressing; and fiber, which is derived from the plant stem. A detailed analysis of the phenylpropanoid contents in these products from control and transgenic plants is presented in Table 2. The comparison of phenylpropanoids content firstly revealed much higher amounts of these compounds in the seedcakes than in the fiber and secondly higher level of this compounds in raw products from genetically modified plants in comparison to control.

The total phenolic contents in seedcakes from transgenic W92 plants (46,6 mg/gFW) is significantly higher when in control- unmodified plants (16,2 mg/gFW). Secoisolariciresinol diglucoside (SDG) is the main phenylpropanoid compound identified in seedcakes, and its content –12,1 mg/g FW for control and 39.78 mg/g FW for W92plants is respectively about75% and 85% of the total phenolic content. The UPLC analysis also revealed the presence of SDG in the fibers but a much lower quantity (1.05 mg/gFW for transgenic plants). Other compounds with antioxidant, anti-inflammatory and antibacterial properties were also analyzed. Flavonols (kaempferol and quercetin), anthocyanins and proanthocyanidins were only detected and analyzed in the seeds and seedcakes; they were not detected in the fibers. We detected phenolic acids in control and transformed flax seeds, seedcakes and

Table 2 Biochemical analysis of seedcakes and fibers from wild type (WT) modified (W92) flax

Compounds	WT seedcakes (mg/gFW)	W92 seedcakes (mg/gFW)	P value transgene *vs.* control	WT fibers (mg/gFW)	W92 fibers (mg/gFW)	P value transgene *vs.* control
Anthocyanins	0.004 ± 0.000	0.013 ± 0.001	0.006	<0.001	<0.001	
Proanthocyanidin	0.021 ± 0.002	0.037 ± 0.003	0.02	<0.001	<0.001	
Secoisolariciresinol diglucoside (SDG)	12.112 ± 0.03	39.785 ± 0.0	0.005	0.348 ± 0.03	1.056 ± 0.05	0.01
Ferulic acid	0.57 ± 0.029	1.16 ± 0.035	0.01	0.082 ± 0.05	0,350 ± 0.08	0.002
Ferulic acid glucoside	1,488 ± 0.041	2.14 ± 0.053	0.02	0.29 ± 0.04	1.09 ± 0.05	0.002
Coumaric acid	0.340 ± 0.019	0.368 ± 0.016	0.04	<0.001	<0.001	
Coumaric acid glucoside	1.006 ± 0.030	1.721 ± 0.032	0.006	<0.001	<0.001	
Caffeic acid	0.006 ± 0.000	0.011 ± 0.001	0.06	<0.001	<0.001	
Caffeic acid glucoside	0.636 ± 0.009	1.336 ± 0.007	0.01	<0.001	<0.001	
p-Hydroxybenzoic acid	0.004 ± 0.001	0.011 ± 0.001	0.06	0,29 ± 0.01	0,350 ± 0.01	0.04
Flavonols	0.013 ± 0.001	0.022 ± 0.001	0.005	<0.001	<0.001	
Vanillin	<0.001	<0.001		0.193 ± 0.01	0.480 ± 0.01	0.02
Acetovanillone	<0.001	<0.001		0.073 ± 0.01	0.100 ± 0.003	0.01

The results are the mean values ± SD (n = 6).

fibers. The main phenolic acids in the seeds and seedcakes were ferulic acid, *p*-coumaric acid and caffeic acid and their glycoside derivatives [13]. Ferulic acid is also the most common phenolic acid in the fiber. However, the content of phenolic acids in the fibers was much lower than in the seedcakes. The compounds that were only identified in the fibers were vanillin and acetovanillone.

The general conclusion that might be drawn from the data is that seedcakes are the richest source of phenylpropanoid compounds in the plant and the level of this compounds in modified flax plants is significantly higher than in control plants. Therefore, for further experiments, we used seedcakes as a source of antimicrobial compounds.

Preparation of seedcake extract and study of antimicrobial activities on standard (reference) pathogenic bacterial strains

The seedcake extracts from control-wild type (WT) and W92 plants were prepared (dry material) as described in Methods. The content of particular constituents was verified by UPLC analysis and proportion between active compound are similar to observed while row seedcakes analysis (data not shown). Finally, dry extracts were dissolved in water to a final concentration (w/v) of 50, 30 and 10 mg/ml. The antioxidant and antibacterial properties of such obtained preparation were analysed, in a purpose, to find the best source to generate antimicrobial preparation based on flax. All preparation were analysed to measure total phenolic content (as was expected the highest in preparation from W92 plants) and antioxidant potential (W92 plants preparation indicate almost three time better antioxidant properties – lower IC-50 parameter than unmodified plants preparation). Finally the antimicrobial activity of seedcake preparations were tested on reference (usually used for antibiotics testing) bacterial strains. The low concentration of seedcake preparation from control plants do not show any antibacterial properties, only preparation 50 mg/ml indicate such properties (see Table 3) but this activity was predominantly lower (except E. coli) then observed for five time lower concentration of W92 based preparation. The comparable (50 mg/ml) concentration made from transgenic plant seedcakes kills all tested bacteria. As well lower (30 mg/ml) concentration of seedcake preparation based on W92 plants indicates high bactericidal activity reduction of bacterial from 87% to 100%. The obtained results to allow us to claim that seedcakes from W92 plants are the best source for antimicrobial preparation creation.

Antioxidant properties of seedcake extract

The process of microbial infection is usually started by oxygen free radical synthesis. Therefore, addition of external antioxidants might be helpful in arrest bacterial growth. Therefore, the antioxidant potential of the seedcake extract was measured using a chemiluminescence method, as described elsewhere [22]. The antioxidant potential was expressed as the IC_{50} value, which means the amount of extract inhibiting the oxidation of luminol by 50%. The antioxidant properties of the seedcake preparation were compared to those of well-known water-soluble antioxidants (ascorbic acid, chlorogenic acid, caffeic acid) and SDG standard solutions. The data are presented in Figure 1. It was found that the preparation made from seedcakes exhibits very high antioxidant potential: 13.5-fold higher than ascorbic acid and more than 4-fold higher than SDG. However, it was lower than that of the strong antioxidants caffeic acid (4-fold lower) and chlorogenic acid (6-fold lower). The very high antioxidative status of the preparation suggests that it may have great biomedical potential not only as an antimicrobial but also as an anti-inflammatory agent.

Antimicrobial properties of seedcake preparations

Next, the antimicrobial properties of seedcake extract on clinical strains of pathogenic bacteria were investigated. Three different concentrations of W92 flax seedcake

Table 3 The total phenolic content, antioxidant potential of dried seedcake extract preparation from wild type and modified flax plants and effect on reference bacterial strains

	WT seedcakes dried extract	W92 seedcakes dried extract		
	50 mgDW/ml preparation	50 mgDW/ml preparation	30 mgDW/ml preparation	10 mg/ml preparation
Total phenolics content [mg/ml]	24.26 ± 0.03	31.59 ± 0.05	18.45 ± 0.03	6.03 ± 0.06
IC-50 [μl]*	27.5 ± 0.04	10.56 ± 0.03	18.15 ± 0.03	26.36 ± 0.04
K. pneumoniae ATCC 700603**	64% ± 3.03	100% ± 0.00	98% ± 0.01	72% ± 0.93
S. aureus ATCC 29213**	52% ± 2.01	100% ± 0.01	93% ± 1.03	60% ± 0.63
P. aeruginosa ATCC 27853**	46% ± 3.50	100% ± 0.00	100% ± 0.02	85% ± 1.05
E. coli ATCC 25922**	75% ± 1.89	100% ± 0.23	87% ± 1.83	48% ± 1.82

The antimicrobial activity of seedcake preparations were tested on reference bacterial strains. The low concentration of seedcake preparation from control plants (10 mg/ml and 30 mg/ml) do not show any antibacterial properties (data not shown).

*The antioxidant potential (IC50) was defined as the amount (μl) of preparation that inhibits luminol chemiluminescence by 50%. **The presented results are the inhibition rate compared to the untreated bacteria (% of growth inhibition) after 6 h treatment. The results are the mean values ± SD (n = 5).

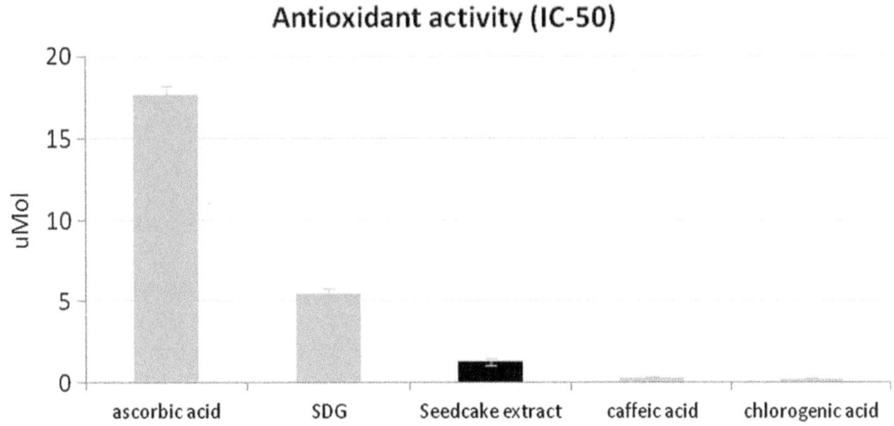

Figure 1 The antioxidant activities of seedcake extract and equimolar contents of standard antioxidants. The antioxidant activity was indicated as the IC-50 parameter. The results are the mean values ± SD (n = 6). The results are statistically significant (P < 0.05).

Table 4 Antibiotic resistance characteristics and MIC values of seedcake extracts tested on Gram-negative bacterial strains

Bacterial strains	Strain characteristics	MIC values (mg/mL)	Bacterial strains	Strain characteristics	MIC values (mg/mL)
K. pneumoniae ATCC 700603	reference strain	30	*P. aeruginosa* ATCC 27853	reference strain	30
K. pneumoniae 38	MDR; ESBL+	30	*P. aeruginosa* 9/5	MDR	10
K. pneumoniae 36	MDR; ESBL+	50	*P. aeruginosa* 12/3	MDR	10
K. pneumoniae 31	MDR; ESBL+	50	*P. aeruginosa* 14/3	MDR; carbapenemase+	10
K. pneumoniae 44	MDR; ESBL+	30	*P. aeruginosa* 15/3	MDR	10
K. pneumoniae 43	MDR; ESBL+	50	*P. aeruginosa* 49/3	S	30
K. pneumoniae 46	S	50	*P. aeruginosa* 82/3	MDR	30
K. pneumoniae 37	S	30	*P. aeruginosa* 113	MDR; carbapenemase+	30
K. pneumoniae 35	S	50	*P. aeruginosa* 249/P	MDR	30
K. pneumoniae 34	S	50	*P. aeruginosa* 3	MDR	30
K. pneumoniae 33	S	30	*P. aeruginosa* 12	MDR	30
E. coli ATCC 25922	reference strain	50	*P. aeruginosa* 14	MDR	30
E. coli 1426	S	50	*P. aeruginosa* 18	MDR	10
E. coli 1419	S	30	*P. aeruginosa* 20	MDR	10
E. coli 1417	S	30	*P. aeruginosa* 164	MDR	30
E. coli 1416	S	30	*P. aeruginosa* 0013	S	30
E. coli 1387	S	30	*P. aeruginosa* 0038	MDR	30
E. coli 1378	S	30			
E. coli 1303	S	50			
E. coli 1261	S	30			
E. coli 1288	S	50			
E. coli 1273	S	50			

The MIC tests were performed by a broth microdilution method according to the European Committee on Antimicrobial Susceptibility Testing standards (EUCAST, www.eucast.org). was estimated by three independed biological repetition.

MDR – acquired resistance to more than 3 groups of antibiotics; ESBL + – Extended-Spectrum Beta-Lactamases, S – sensitive, no acquired resistance.

preparation (50, 30 and 10 mg/ml; for details see Table 1) were prepared. The levels of particular active compounds in all tested dilutions and dry extract were measured and presented in Table 1. The content of phenolic acids, SDG and other compounds are proportional to values observed while row seedcake analysis and proportion of the particular compound do not change after freeze drying and dilution process. Therefore, such method of active solution preparation obtaining procedure can be used also in the future without fears of biological activity of achieved preparations.

The MIC values for control and clinical bacterial strains exposed to the seedcake preparations are presented in Tables 4 and 5. From the data obtained, it is clear that Gram-negative bacteria are more susceptible to the seedcake extract than the Gram-positive bacteria. *K. pneumoniae* and *E. coli* strains were inhibited by the 30 and 50 mg/ml concentrations of the extract. The lowest MIC values at the concentrations of 10 mg/ml and 30 mg/ml were noted for the *P. aeruginosa* strains regardless of the multidrug resistance feature of the clinical strains. Clinical Gram-positive isolates were less susceptible to the antibacterial activity of seedcake extract. The MIC was 50 mg/ml for clinical staphylococci and higher for clinical enterococci. To identify the bacteriostatic or bactericidal activity of seedcake extracts, killing curves for the reference strains were determined (Figure 2). In each analyzed case, an extract concentration of 10 mg/ml was insufficient to obtain an inhibitory or killing effect against the tested strains, and the growth parameters of the bacterial cultures under such conditions were similar to those of the control. For *P. aeruginosa* ATCC 27853 cultures, the extract concentration of 30 mg/ml inhibit bacterial growth, and after 24 hours of incubation, a cfu level equal to the initial one was detected (Figure 2A). Other bacterial cultures exposed to the same extract concentration exhibited an increase in cell number of one order of magnitude for *E. coli* ATCC 11229 and *S. aureus* ATCC6538, or two orders of magnitude for *K. pneumoniae* ATCC 700603 after 24 hours of incubation (Figure 2B-D). The 30 mg/ml concentration of seedcake extract was identified as bacteriostatic for all of the reference strains used for growth curve.

Gram-negative reference strains were much more susceptible to the highest concentration of extract, which reduced the bacterial count by around 3 orders of magnitude for *K. pneumoniae* to 5 for *E.coli* and 6 for *P. aeruginosa*. The *P. aeruginosa* cells appeared to be very sensitive to the bactericidal activity of seedcake extract compounds, because after 8 hours of incubation, the cfu/ml dropped from 10^6 to 10^2 and after one day of incubation, no living cell was detected in the culture (Figure 2A). The data suggest that the seedcake extract obtained from the transgenic flax plants is a suitable candidate for antimicrobial action with broad spectrum and partial selectivity.

Table 5 Antibiotic resistance characteristics and MIC values of seedcake extracts tested on Gram-positive bacterial strains

Bacterial strains	Strain characteristics	MIC value [mg/mL]
S. aureus ATCC 29213	Reference strain	50
S. aureus 1	S	50
S. aureus 2	S	50
S. aureus 3	S	>50
S. aureus 4	S	50
S. aureus 5	S	50
*S. epidermidis*1MRS	MRS	30
*S. epidermidis*2MRS	MRS	50
S. epidermidis 3MRS	MRS	50
S. epidermidis 4	S	50
*S. epidermidis*5MRS	MRS	>50
E. faecalis 1	S	>50
E. faecalis 2	S	>50
E. faecalis 3	S	>50
E. faecalis 4	S	>50
E. faecalis 5	S	>50

The MIC tests were performed by a broth microdilution method according to the European Committee on Antimicrobial Susceptibility Testing standards (EUCAST, www.eucast.org). was estimated by three independed biological repetition.
S – sensitive, no acquired resistance. MRS –methicillin-resistant staphylococci. The results are the mean values ± SD (n = 3). All results are statistically significant (P < 0.05).

Antimicrobial effect of standard substances

In order to verify which of the identified compounds in the seedcake extract have the greatest impact on its antibacterial action, we examined the effects on the reference strain *S. aureus* ATCC 6538 of the pure compounds that had been identified in the extract. In the experiment, the concentrations of pure compounds were close to the used concentrations of seedcake extract (sum of aglicons and glicosylated form of phenolic acids) and, with a concentration 1 or 2 mg/ml of the standard substances corresponding to 30 or 50 mg/ml of the seedcake preparation, respectively. The higher concentration of 10 mg/ml was also applied. The inhibition rate (% of growth inhibition) compared to the untreated control bacteria was measured. In addition, the inhibitory effect of the seedcake preparations used in earlier experiments with clinical strains was assessed. The data suggest that *p*-coumaric and ferulic acid make a considerable contribution to the antimicrobial properties of the seedcake extract. In the case of *p*-coumaric acid, 66.22% growth inhibition was observed at a concentration equal to the sum of its free form and its glycoside derivative in the preparation of 50 mg/ml. For ferulic acid, the growth inhibition rate was 63.5% for the same concentration level. A significantly lower inhibitory

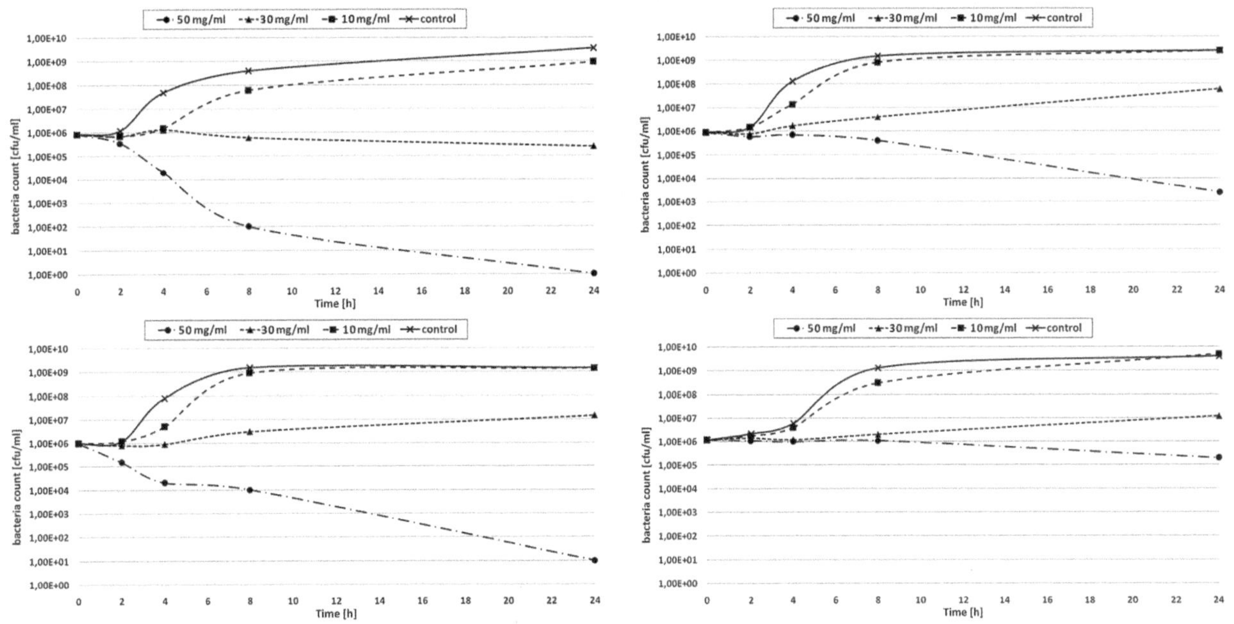

Figure 2 The antibacterial activity of seedcake extract on pathogenic bacteria. Growth curves of: *Pseudomonas aeruginosa* ATCC 27853, *Klebsiella pneumoniae* ATCC 700603, *Escherichia coli* ATCC 25922 and *Staphylococcus aureus* ATCC 29213 strains exposed to seedcake extract (preparation in a concentration of 10 mg/ml, 30 mg/ml and 50 mg/ml). The results are the mean values ± SD (n = 3). The results are statistically significant (P < 0.05).

effect was observed for caffeic acid (17.8%) and SDG (Table 6). It is proposed that to reach the antimicrobial effect of caffeic acid much higher concentrations need to be used than are present in the tested seedcake extracts, as indicated by the effect of the 10 mg/ml concentration. To find out if the antimicrobial effect of phenolic acids derives from the sum or synergy of their performance, simultaneous bacterial treatment of different combinations of mixed standard substances was performed. In all of the test situations, the results are significantly higher for the simultaneous effect of two, three or four substances than for a single component, and also higher than mathematical sum of the inhibitory effects observed for standard substances used in combination. Therefore, we suggest that the tested compounds exhibit a synergistic effect. The results observed for a single SDG addition and for mixtures including SDG shown that SDG has little or no influence on the antimicrobial properties of seedcake preparations. The efficacy of the seedcake extract and *p*-coumaric and ferulic acid were compared to the activity of ampicillin and gentamicin, which are respectively the inhibitors of peptidoglycan and protein synthesis

Table 6 The effect of pure compounds and seedcake extract on the growth of *Staphylococcus aureus* ATCC 29213 cultures

Compounds	1 mg/ml of each	2 mg/ml	10 mg/ml
p-coumaric acid	43.73% ± 0.2	66.22% ± 0.62	78.31% ± 1.2
Ferulic acid	29.99% ± 0.6	63.56% ± 0,89	78.31% ± 1.2
Caffeic acid	14.74% ± 0.5	17.82% ± 0,78	83.15% ± 1.2
SDG	4.94% ± 0.9	5.32% ± 1.0	7.41% ± 0.6
p-coumaric and ferulic acid	92.59% ± 2.3	100% ± 0.91	100% ± 0.0
p-coumaric, ferulic and caffeic acids	100% ± 0.6	100% ± 0.0	100% ± 0.0
SDG and ferulic acid.	30.35% ± 1.3	63.99% ± 0.92	82.51% ± 1.6
SDG, p-coumaric , ferulic and caffeic acids	100% ± 0.0	100% ± 0.0	100% ± 0.0
Seedcake extract	83.24% ± 1.8#	100% ± 0.2##	

The presented results are the inhibition rate compared to the control (% of growth inhibition). The results are the mean values ± SD (n = 5).
#30 mg/ml of extract contains 0.86 mg, 1.45 mg and 1.14 mg of p-coumaric, ferulic and caffeic acids, respectively, and 5.5 mg SDG, which is equal to 10 mg of dry extract, ##50 mg/ml of extract.

(see Additional file 2: Table S3). The enterococci group turned out to be weakly susceptible to*p*-coumaric and ferulic acids regardless of the antibiotic susceptibility profiles, so their highest concentrations (>0.25and >0.80 mg/ml, respectively) had no inhibitory effect on microbial growth. Ampicillin was efficient at 0.002 mg/ml. Staphylococci exhibited a high MIC level with phenolic acids but for several strains, a similar or even lower level than for ampicillin (0.25 mg/ml versus >0.51 mg/ml). *p*-coumaric acid was active against *S.epidermidis* MRS isolates at an equal concentration to gentamicin. The *E.coli* strains are antibiotic-susceptible bacteria, and the MICs of the phenolic acids were even 100 times higher than those for ampicillin and gentamicin for most of the tested cultures. *K. pneumoniae* isolates showed similar MICs for phenolic acids versus ampicillin. The flax compounds had the strongest impact on *Pseudomonas* growth. The ampicillin inhibitory concentration was over 10 times higher than of *p*-coumaric acid and similar to ferulic acid. For gentamicin-resistant *Pseudomonas* isolates, *p*-coumaric acid was 2–5 times more effective than the drug.

Potential mechanism of antimicrobial activity of seedcake extracts

In order to investigate the mechanism of flax phenolics action on bacterial cells we analysed the integrity of bacterial genomic DNA upon cell treatment with seedcake extract by pulse-field electrophoresis (Figure 3). Results demonstrated, that bacterial DNA was visibly degraded after extract treatment but typical pattern of apoptosis (occurrence of 50 kb or 500 kb domain) was not recorded. The lower than 5 kb products of genomic DNA degradation can be observed especially after treatment with 50 mg/ml dried seedcake extract. Nucleolytic activity however remained unchanged (data not shown). The *in.vitro* test on inpact of flax seedcake extract on bacterial topoisomerase II (gyrase) activity was performed (Figure 4). Samples treated with W92 seedcake extract showed partial (10 and 30 mg/ml) or total (50 and 75 mg/ml) inhibition of gyrase activity (Figure 4A). The similar experiment using dried seedcake extract from control plants was also preformed (Figure 4B) –the slight inhibition of gyrase activity was observed only after using the highest concentration of extract (50 mg/ml). The data presented in Figures 3 and 4 suggest that phenolics cause bacteria DNA disintegration and inhibits gyrase activity. Thus to the known effect of phenolics on bacteria cell membrane we may add new targets of their action and these are gyrase activity and genomic DNA disintegration.

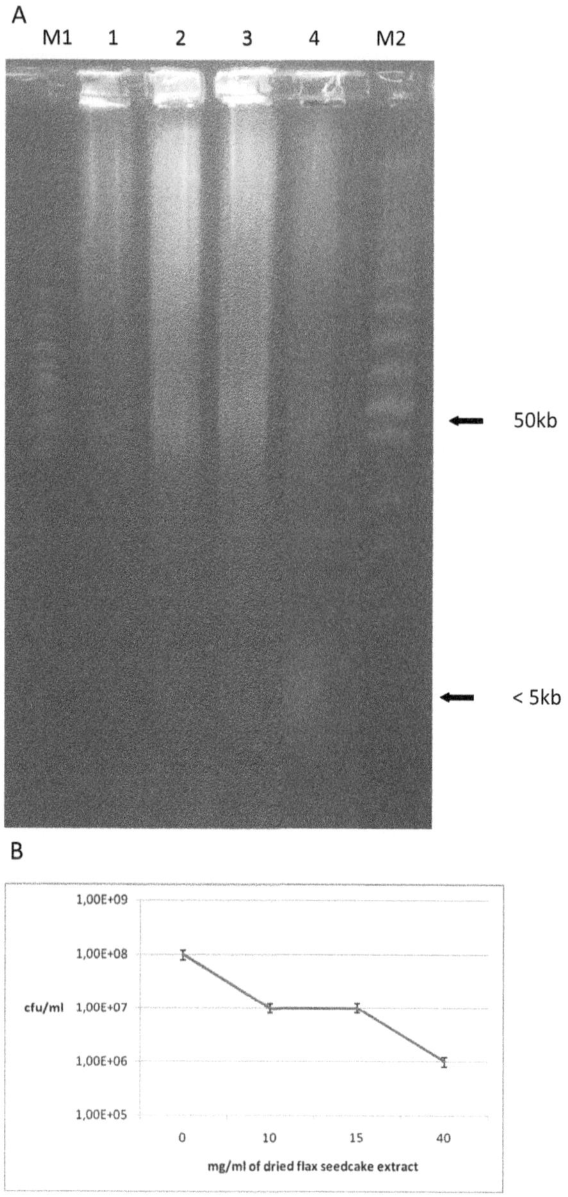

Figure 3 Pulsed-field electrophoresis of seedcake treated *Escherichia coli*. **A**. Line M1- MidRange Marker; line1-control (untreated bacteria); lane 2- bacteria incubated with 10 mg/ml dried seedcake extract; lane 3- bacteria incubated with 15 mg/ml dried seedcake extract; lane 4- bacteria incubated with 40 mg/ml dried seedcake extract; Lane M2- LowRange Marker. White arrow indicates ca. 50 kb, a size of typical apoptotic domain. **B**. Correlation between CFU number and seedcake extract concentration.

Evaluation of skin irritation caused by W92 seedcake preparations

To better assess the application potential of W92 seedcake preparations, the skin irritation test was performed with use of EpiDerm, a three-dimensional human skin model. This is a fully developed epidermis with a functional stratum corneum, (with keratinocytes in various stages of differentiation). It is commonly used to valuation the irritation potential of substances applied on the skin. This test can be applied as an alternative for evaluating dermal irritation prior to clinical tests. The W92 seedcake preparations were directly applied to the

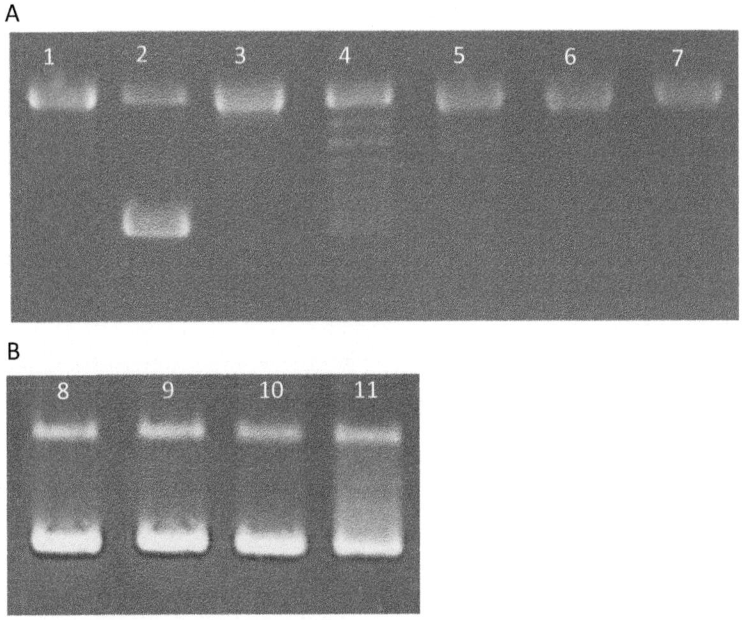

Figure 4 Gyrase activity assay after treatment of seedcake extract. Part **A**. Analysis of W92 seedcake extracts activity. Lane 1- negative control (without gyrase and extract); lane2- positive control (gyrase, without extract); lane 3- total inhibition control (novobiocin 25 μg/ml, gyrase, without extract); lane 4- reaction mixture with 10 mg/ml dried seedcake extract; lane 5- reaction mixture with 30 mg/ml dried seedcake extract; lane 6- reaction mixture with 50 mg/ml dried seedcake extract; lane 7- reaction mixture with 75 mg/ml dried seedcake extract. Part **B**. Analysis of control plant seedcake extracts activity. lane 8- positive control (gyrase, without extract), lane 9- reaction mixture with 10 mg/ml dried seedcake extract; lane 10- reaction mixture with 30 mg/ml dried seedcake extract; lane 11- reaction mixture with 50 mg/ml dried seedcake extract.

stratum corneum of this air-lifted, highly differentiated culture. The in vitro skin irritation evaluation was analysed 2, 5 and 18 h after application of the preparations. All analysed preparations doesn't indicate any irritating effects on the human skin model (Figure 5). The cell viability after the application of all seedcake preparation was between 94% to 104,6% after 2, 5 and 18 h treatment. As a positive control the 1% Triton X-100 was used –this solution indicate moderate irritation effect and cell viability varied between 34,8% to 25,4% depending on time of treatment. Obtained results prove that analysed preparations appeared to be safe for dermatological applications.

Discussion

Recently, many studies have focused on identifying potential antimicrobial agents from natural resources. The antimicrobial activity of plant extracts has been known for many years, as plants are known to produce useful antimicrobial phytochemicals [23]. Therefore, it is reasonable to try to identify a plant type that is a renewable source and provides sufficient amounts of these compounds. Plants with a certain tradition in their cultivation and with levels of antimicrobial compounds that can be increased by genetic engineering could be a sufficient source of antimicrobial preparations. Previously, a flax plant was transformed with three genes containing the cDNAs coding for key enzymes of the flavonoid biosynthesis pathway. This transgenic plant was analyzed to reveal the accumulation of flavonols and other phenolic compounds with potential antimicrobial activity. It was previously shown that the metabolite content of W92 plants, similarly to unmodified plants, varied by less than 10% from a year to year and that the stoichiometry between individual metabolites does not change either [24]. Therefore, these plants can be used as a stable source for the production of biotechnological preparations In this study, we analyzed the phenolic compound contents in extracts from seeds; seedcakes and fibers in more detail, and assessed their impact on pathogenic bacterial growth. Flavonoids are known to possess antifungal, antiviral and antibacterial activity. Moreover, several groups of flavonoids have demonstrated synergy in action within their groups and with existing chemotherapeutics [25]. It is thus hypothesized that transgenic plants overproducing flavonoids might prove to be a suitable source of antimicrobial compounds.

The seedcake extracts from the transgenic flax possess a unique composition of natural antioxidants [13]. Besides the presence of SDG, there are phenolic acids, such as ferulic, *p*-coumaric and caffeic acids and their glycosyl derivatives. Phenylpropanoids, particularly flavonoids and

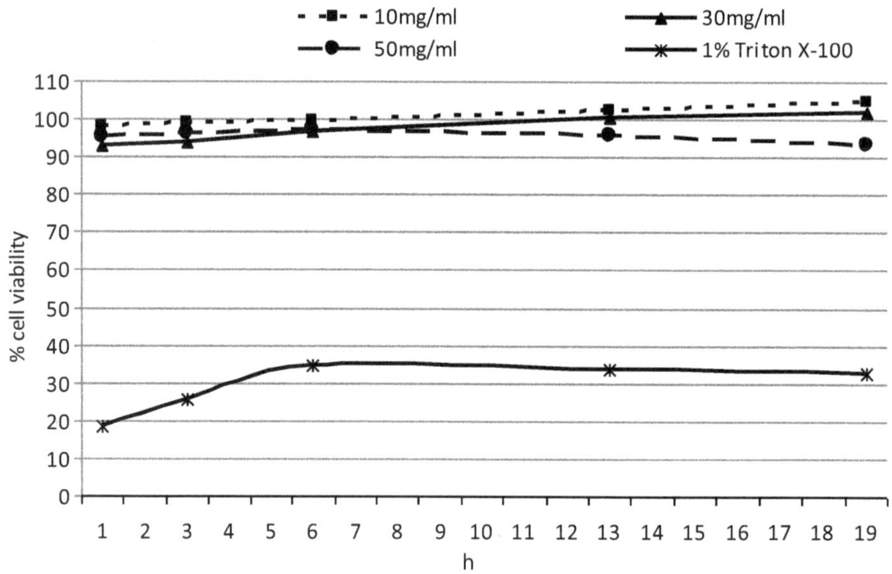

Figure 5 Skin irritation potential of seedcake preparations on reconstructed Human epidermal model EpiDerm. The irritation potential was determined for preparations 10 mg/ml, 30 mg/ml,50 mg/ml and 1% Triton X-100 as a positive control after 2, 5 and 18 h of treatment. The experiment was performed in duplicate.

phenolic acids, have been found to be effective antimicrobial agents against a wide range of microorganisms. Investigations into the mechanism of action of flavonoids revealed that they have multiple cellular targets rather than one specific site of action. *E.coli* cells treated with phenolics or anthocyanins exhibited local disintegration and irregularity in their outer membrane and leakage of cytoplasm [26]. Phenolics has the unique ability to perform redox reactions at the plasma membrane and sequester electrons from the respiratory processes. The partial hydrophobicity of phenolics enables them to bind to the outer membrane and to change the membrane fluidity. The irregularity in cell shape supports the theory of localized chemical interactions at the cell surface. Once the membrane has been penetrated, smaller phenolic compounds can enter the cell and disrupt its metabolism. It is reported that the partial lipophilic nature of phenolic acids enables them to cross the cell membrane by passive diffusion in their undissociated form, and consequently disturb the cell membrane structure [27]. They cause acidification of the cytoplasm, protein denaturation, and possibly also cell death [28]. It is also suggested that pH and the length of the alkyl chain determine the antimicrobial effect of phenolic acids [28].

The data obtained strongly suggest that mainly the phenolic acids are the most effective compounds in antimicrobial activity of seedcake extract. Although the exact mechanism of this action is as yet unknown, there are several suggestions in the literature on how phenylpropanoids compounds might affect bacteria growth. It is pointed out that the mechanism of action of phenolics revealed that they have multiple cellular targets rather than one specific site of action. Phenolics bind to the outer membrane and change the membrane fluidity which might results in the local disintegration of bacterial cell membrane and cytoplasm leakage [26,27]. They might cause acidification of the cytoplasm, protein denaturation, and possibly also cell death [28].

Detailed analysis of the impact of flax seedcake extract on bacteria cell revealed new targets of extract component action and these are genomic DNA and key enzyme involved in DNA replication. Treatment of bacteria cells with seedcake extract exhibit DNA disintegration and topoisomerase II (gyrase) inhibition. It is well known that bacterial gyrase is the target of several antibiotics or over antimicrobial agents that inhibit enzyme activity [29,30]. For example novobiocin blocks the binding of ATP to gyrase while nalidixic acid interferes with the breakage and resealing of DNA chains. It is thus suggested that seedcake extract might be used as alternative antibiotic to treat multibacterial infections.

We suggest that the structural diversity within plant phenolic compounds resulted in their impact on the growth of a broad spectrum of bacterial strains. According to experiments that included an assessment of the antibacterial activity of propolis, mixtures of phenolic compounds are effective against many bacterial strains [31]. This is perhaps also the reason why the mixture of phenolic acids and lignans in GM flax extract has far more effective antibacterial activity than a pure single compound. The suggestion is further supported by the result from the experiment described in this paper where a mixture of ferulic and p-coumaric acids was evaluated. The results suggest that the activity of an individual component changed in the presence of other compounds in the extracts [32,33]. The synergy in action of extract components might explain the high antimicrobial activity of the seedcake preparation.

The analysis of influence of W92 seedcake extract on a three-dimensional human skin model (EpiDerm), which is acommonly used to assess the irritation potential of dermally applied substances, indicate no irritation effect of tested preparation. This test can be applied as an alternative for evaluating dermal irritation prior to clinical tests.

Our results show that these preparations are safe for use on the skin that we suggest that seedcake preparations might be a good candidate for treatment of bacterial infected human skin (for example acne) and an alternative to antibiotic therapy of infected wounds. The use of such preparation for human therapy (especially for injection) should be certainly preceded by clinical tests.

Conclusion

The results of this study on genetically modified flax plants resulted in a potential new, pharmaceutical product containing biologically active compounds that exhibit antibacterial properties. Moreover, the natural products would be present in favorable proportions to counter bacterial infection according to the data. It should be pointed out that both the composition and component concentration of the product might be easily modified to give better specificity against different bacterial strains.

In this particular study, the seedcake extract was tested against sensitive and multidrug-resistant clinical bacterial strains. The obtained data are comparable to the inhibitory effect of the reference strains commonly used in antibiotic testing. The results showing the similar antibacterial activity of the flax seedcake preparation against different strains in a group of one species suggest a nonspecific mode of action on bacterial cells. An interesting aspect of seedcake antibacterial activity is the bactericidal efficiency, which was highest against *P. aeruginosa* among all the tested bacteria. These bacteria are known to be the most important nosocomial pathogens that develop various mechanisms of resistance to antibiotics and antiseptic chemical compounds. It should be emphasized that these strains were isolated from severe infections where the therapy options were very limited.

The presumed mechanism of seedcake extracts action may be related to interactions between the extract component and the bacterial membrane because Gram-negative strains exhibited higher susceptibility to this preparation. Gram-positive strains were more resistant than Gram-negatives, but at higher concentrations, the seedcake extracts still inhibited the growth of staphylococci and enterococci. The results of the treatment of the drug-resistant clinical bacterial strains indicate the great application potential of seedcake flax preparations. Therefore, we suggest that flax-derived natural products are a promising alternative to antibiotic therapy. Moreover, our results show that these preparations are safe for use on the skin. Such new application possibilities of flax and especially their biomedical relevance can contribute to the renewal of flax cultivation worldwide.

Additional files

Additional file 1: Table S1. Characteristics of Gram-negative strains used. **Table S2.** Characteristic of Gram-positive strains used.

Additional file 2: Table S3. The MIC values of seedcake extracts, standard substances and selected antibiotics tested on bacterial strains.

Competing interests

The authors declare that they have no competing interests.

Authors' contributions

MZ-carried out the biochemical analysis of seeds and seedcakes, analysis of antimicrobial effect of standard substances, drafted the manuscript, A D-J - carried out the microbiological analysis, Z D-K- interpretation of microbiological results and comparison to commercial antibiotics, MA –mechanism of antimicrobial activity analysis, AK – carried out fibre biochemical analysis. JS- conceived of the study and participate in its design and coordination. All authors read and approved the final manuscript.

Acknowledgements

This study was supported by grants 2012/06/A/NZ1/00006 and 2013/11/B/NZ9/00150 from the Ministry of Science and Higher Education.

Author details

[1]Faculty of Biotechnology, Wrocław University, Przybyszewskiego 63/77, 51-148 Wrocław, Poland. [2]Institute of Genetics and Microbiology, Wroclaw University, Przybyszewskiego 63/77, 51-148 Wroclaw, Poland. [3]Linum Fundation, Stabłowicka 147/149, 54-066 Wrocław, Poland.

References

1. Herati RS, Blumberg EE: **Losing ground: multidrug-resistant bacteria in solid-organ transplantation.** *Curr Opin Infect Dis* 2012, **25**(4):445–449.
2. Livermore DM: **Fourteen years in resistance.** *Int J Antimicrob Agents* 2012, **39**(4):283–294.
3. Pitout JD: **Infections with extended-spectrum beta-lactamase-producing enterobacteriaceae: changing epidemiology and drug treatment choices.** *Drugs* 2010, **3**:313–333.
4. Rossolini GM, Mantengoli E: **Antimicrobial resistance in Europe and its potential impact on empirical therapy.** *Clin Microbiol Infect* 2008, **14**(Suppl 6):2–8.
5. Johnson AP: **Antibiotic resistance among clinically important Gram-positive bacteria in the UK.** *J Hosp Infect* 1998, **40**:17–26.
6. EARSS: *European Antibiotic Resistance Surveillance System.* Available: http://www.who.int/world-health-day/2011/en/index.html, Accessed 15 October, 2012.
7. WHO: *Antimicrobial Resistance:no Action Today, no Cure Tomorrow.* 2011. Available: http://www.who.int/world-health-day/2011/en/index.html, Accessed 15 October, 2012.
8. Robles M, Toscano E, Cotta J, Lucena MI, Andrade RJ: **Antibiotic-induced liver toxicity: mechanisms, clinical features and causality assessment.** *Curr Drug Saf* 2010, **5**(3):212–222.
9. Prasad K: **Secoizolariciresinol diglucoside from flaxseed delays the development of type 2 diabetes in Zucker rat.** *J Lab Clin Med* 2001, **138**(1):32–39.

10. Christofidou-Solomidou M, Tyagi S, Tan KS, Hagan S, Pietrofesa R, Dukes F, Arguiri E, Heitjan DF, Solomides CC, Cengel KA: **Dietary flaxseed administered post thoracic radiation treatment improves survival and mitigates radiation-induced pneumonopathy in mice.** *BMC Cancer* 2011, **11**:269.
11. Miśta D, Kroliczewska B, Zawadzki W, Pecka E, Steininger M, Hull S, Zuk M, Szopa J: **The effect of Linola and W92/72 transgenic flax seeds on the rabbit caecal fermentation–in vitro study.** *Pol J Vet Sci* 2011, **14**(4):557–564.
12. Lorenc-Kukuła K, Wrobel-Kwiatkowska M, Starzycki M, Szopa J: **Engineering flax with increased flavonoid content and thus Fusarium resistance.** *PMPP* 2007, **70**(1–3):38–48.
13. Zuk M, Kulma A, Dymińska L, Szołtysek K, Prescha A, Hanuza J, Szopa J: **Flavonoid engineering of flax potentiate its biotechnological application.** *BMC Biotechnol* 2011, **11**(10):resc.
14. Wróbel-Kwiatkowska M, Lorenc-Kukula K, Starzycki M, Oszmiański J, Kępczyńska E, Szopa J: **Expression of −1,3-glucanase in flax causes increased resistance to fungi.** *Physiol Mol Plant Pathol* 2005, **65**:245–256.
15. Wróbel-Kwiatkowska M, Zuk M, Szopa J, Dymińska L, Mączka M, Hanuza J: **Poly-3-hydroxy butyric acid interaction with the transgenic flax fibers: FT-IR and Raman spectra of the composite extracted from a GM flax.** *Spectrochim Acta A Mol Biomol Spectrosc* 2009, **73**:286–294.
16. Blainski ALG, de Mello JC P: **Application and Analysis of the Folin Ciocalteu Method for the Determination of the Total Phenolic Content from Limonium Brasiliense L.** *Molecules* 2013, **18**:6852–6865.
17. Lee JRC, Wrolstad R: **Correlation of two anthocyanin quantification methods:HPLC and spectrophotometric methods.** *Food Chem* 2008, **110**:782–786.
18. Porter LJ, Hrstich LN, Chan BG: **The conversion of procyanidins and prodelphinidins to cyanidin and delphinidin.** *Phytochem* 1986, **25**:223–230.
19. Lorenc-Kukuła K, Zuk M, Kulma A, Czemplik M, Kostyn K, Skala J, Starzycki M, Szopa J: **Enginiering flax with the GT Family I Solanum sogerandinum gycosyltransferase SsGT1 Confers Increased Resistance to Fusarium Infection.** *J Agric Food Chem* 2009, **57**(15):6698–6705.
20. Lukaszewicz M, Matysiak-Kata I, Aksamit A, Oszmianski J, Szopa J: **14-3-3 Protein regulation of the antioxidant capacity of transgenic potato tubers.** *PlantSci* 2002, **163**:125–130.
21. EUCAST DISCUSSION DOCUMENT E.Dis 5.1: **Determination of minimum inhibitory concentrations (MICs) of antibacterial agents by broth dilution.** *Clin Microbiol Infect* 2003, **9**(8):1–7.
22. Lorenc- Kukuła K, Amarowicz R, Oszmiański J, Doermann P, Starzyki M, Skała J, Zuk M, Kulma A, Szopa J: **Pleiotropic effect of phenolic compounds content increases in transgenic flax plant.** *J Agric Food Chem* 2005, **53**:3685–3692.
23. Cowan M: **Plant products as antimicrobial agents.** *Clin Microbiol Rev* 1999, **12**(4):564–582.
24. Zuk M, Dyminska L, Kulma A, Boba A, Prescha A, Szopa J, Mączka MZA, Szołtysek K, Hanuza J: **IR and Raman studies of oil and seedcake extracts from natural and genetically modified flax seeds.** *Spectrochim Acta A Mol Biomol Spectrosc* 2011, **78**(3):1080–1089.
25. Cushni T, Lamb A: **Antimicrobial activity of flavonoids.** *Int J Antimicrob Agents* 2005, **26**:343–356.
26. Lacombe A, Wu VC, Tyler S, Edwards K: **Antimicrobial action of the American cranberry constituents; phenolics, anthocyanins, and organic acids, against Escherichia coli O157:H7.** *Int J Food Microbiol* 2010, **139**(1-2):102–107.
27. Campos FM, Couto JA, Figueiredo AR, Toth LV, Rangel AO, Hogg TA: **Cell membrane damage induced by phenolic acids on wine lactic acid bacteria.** *Int J Food Microbiol* 2009, **135**(2):144–151.
28. Sa'nchez-Maldonado AF, Schieber A, Ganzle MG: **Structure–function relationships of the antibacterial activity of phenolic acids and their metabolism by lactic acid bacteria.** *J Appl Microbiol* 2011, **111**(5):1176–1184.
29. Heddle JG, Zamble DB, Holfelder F, Miller DA, Wenzell LM, Walsh C, Maxwell A: **The antibiotic microcin B17 is a DNA gyrase poison: characterisation of the mode of inhibition.** *J Mol Biol* 2001, **307**(5):1223–1234.
30. Mustaev A, Malik M, Zhao X, Kurepina N, Luan G, Oppegard LM, Hiasa H, Marks KR, Kerns RJ, Berger JM, Drlica K: **Fluoroquinolone-Gyrase-DNA Complexes: two modes of drug binding.** *J Biol Chem* 2014, **289**(18):12300–12312.
31. Kosalec I, Pepeljnjak S, Bakmaz M, Vladimir-Knezevic S: **Flavonoid analysis and antimicrobial activity of commercially available propolis products.** *Acta Pharm* 2005, **55**:423–430.
32. Liao F, Wang L, Yang L-B, Zhang L, Peng X, Sun MX: **Antisense oligodeoxynucleotide inhibition as an alternative and convenient method for gene function analysis in pollen tubes.** *PLoS One* 2013, **8**(3):e59112.
33. Borchers AT, Keen CL, Gerstiwin ME: **Mushrooms, tumors, and immunity: an update.** *Exp Biol Med* 2004, **229**:393–406.

10

Recombinant fusion protein of cholera toxin B subunit with YVAD secreted by *Lactobacillus casei* inhibits lipopolysaccharide-induced caspase-1 activation and subsequent IL-1 beta secretion in Caco-2 cells

Yukihiro Hiramatsu[1], Masatatsu Yamamoto[2], Tomomitsu Satho[1*], Keiichi Irie[1], Akiko Kai[1], Saori Uyeda[1], Yuki Fukumitsu[1], Akihisa Toda[3], Takeshi Miyata[4], Fumio Miake[1], Takeshi Arakawa[5,6] and Nobuhiro Kashige[1]

Abstract

Background: *Lactobacillus* species are used as bacterial vectors to deliver functional peptides to the intestine because they are delivered live to the intestine, colonize the mucosal surface, and continue to produce the desired protein. Previously, we generated a recombinant *Lactobacillus casei* secreting the cholera toxin B subunit (CTB), which can translocate into intestinal epithelial cells (IECs) through GM1 ganglioside. Recombinant fusion proteins of CTB with functional peptides have been used as carriers for the delivery of these peptides to IECs because of the high cell permeation capacity of recombinant CTB (rCTB). However, there have been no reports of rCTB fused with peptides expressed or secreted by *Lactobacillus* species. In this study, we constructed *L. casei* secreting a recombinant fusion protein of CTB with YVAD (rCTB–YVAD). YVAD is a tetrapeptide (tyrosine–valine–alanine–aspartic acid) that specifically inhibits caspase-1, which catalyzes the production of interleukin (IL)-1β, an inflammatory cytokine, from its inactive precursor. Here, we examined whether rCTB–YVAD secreted by *L. casei* binds to GM1 ganglioside and inhibits caspase-1 activation in Caco-2 cells used as a model of IECs.

Results: We constructed the rCTB–YVAD secretion vector pSCTB–YVAD by modifying the rCTB secretion vector pSCTB. *L. casei* secreting rCTB–YVAD was generated by transformation with pSCTB–YVAD. Both the culture supernatant of pSCTB–YVAD-transformed *L. casei* and purified rCTB–YVAD bound to GM1 ganglioside, as did the culture supernatant of pSCTB-transformed *L. casei* and purified rCTB. Interestingly, although both purified rCTB–YVAD and rCTB translocated into Caco-2 cells, regardless of lipopolysaccharide (LPS), only purified rCTB–YVAD but not rCTB inhibited LPS-induced caspase-1 activation and subsequent IL-1β secretion in Caco-2 cells, without affecting cell viability.

Conclusions: The rCTB protein fused to a functional peptide secreted by *L. casei* can bind to GM1 ganglioside, like rCTB, and recombinant YVAD secreted by *L. casei* may exert anti-inflammatory effects in the intestine. Therefore, rCTB secreted by *L. casei* has potential utility as a vector for the delivery of YVAD to IECs.

Keywords: Caspase-1, Cholera toxin B subunit, GM1 ganglioside, Interleukin-1β, *Lactobacillus casei*, YVAD

* Correspondence: satho@cis.fukuoka-u.ac.jp
[1]Faculty of Pharmaceutical Sciences, Fukuoka University, 8-19-1, Nanakuma, Jonan-ku, Fukuoka, Japan
Full list of author information is available at the end of the article

Background

Lactic acid bacteria (LAB) are not pathogenic and their classification is "generally recognized as safe". Over the past several decades, LAB have been used in foods and medicines because they confer beneficial effects on the health of the host. Moreover, after their administration, LAB are delivered live to the intestine, colonizing the mucosal surface and exerting various effects [1]. Therefore, LAB that produce heterologous proteins have been used as bacterial vectors for the delivery of functional proteins to the intestine. Many studies using recombinant DNA technology have used *Lactobacillus* species, which are present in large numbers in the human gut and are resistant to gastric and bile acids [2]. These live recombinant lactobacilli colonize the intestinal mucosal surface and produce the desired protein [3]. Although *Escherichia coli* has generally been used for the production of heterologous proteins, coliform lipopolysaccharide (LPS) contamination always poses a problem. In contrast to *E. coli*, *Lactobacillus* species are gram-positive bacteria and consequently do not contain LPS. Therefore, we selected *Lactobacillus* species for the secretion of functional heterologous proteins.

Cholera toxin (CT) is an enterotoxin produced by *Vibrio cholerae*, which is composed of a toxic A subunit (CTA) and nontoxic B subunit (CTB). CT gains entry to intestinal epithelial cells (IECs) when CTB binds to GM1 ganglioside, a cell-surface receptor present on mammalian cells. CTB alone can translocate into IECs through the GM1 ganglioside without toxicity [4]. Many groups have reported that recombinant CTB (rCTB) expressed in various bacteria, yeasts, and plants also binds to GM1 ganglioside. Previously, we constructed a recombinant *Lactobacillus casei* that secretes CTB, and showed that the rCTB secreted by *L. casei* has GM1-ganglioside-binding activity similar to that of CT from *V. cholerae* [5]. Recombinant fusion proteins of CTB with functional proteins and peptides, such as vaccine antigens [6] and the insulin B chain peptide [7], have been used as carriers to deliver these proteins and peptides to IECs, because they also bind to GM1 ganglioside. However, it has not been determined whether recombinant fusion proteins of CTB with functional proteins or peptides expressed by *Lactobacillus* species bind GM1 ganglioside and translocate into IECs.

The synthetic tetrapeptide composed of tyrosine, valine, alanine, and aspartic acid (YVAD) is a specific inhibitor of caspase-1 [8]. Caspase-1 catalyzes the production of interleukin (IL)-1β, an inflammatory cytokine, from its precursor (pro-IL-1β), and its overexpression in and secretion from IECs exacerbates intestinal inflammation [9,10]. Caspase-1 is also produced as an inactive precursor, pro-caspase-1, which is activated by inflammatory stimuli, such as LPS and mature caspase-1 itself [11,12]. Therefore, YVAD has anti-inflammatory properties, acting as a decoy substrate for caspase-1 instead of pro-IL-1β and pro-caspase-1. However, recombinant bacteria expressing or secreting YVAD have not been reported because it is difficult to express and secrete recombinant low-molecular-weight peptides in bacteria. Furthermore, for YVAD to inhibit caspase-1 activation and subsequent IL-1β secretion, it must be translocated into IECs. However, the cell permeation capacity of YVAD is low because of its strong polarity [13]. Here, we investigated whether fusing rCTB to YVAD would allow the secretion of recombinant YVAD from *L. casei* and facilitate the translocation of YVAD into IECs.

In this study, we constructed *L. casei* that secretes a recombinant fusion protein of CTB with YVAD (rCTB–YVAD) and confirmed that rCTB–YVAD secreted by *L. casei* binds to GM1 ganglioside, translocates into human epithelial colorectal adenocarcinoma Caco-2 cells used as a model of IECs, and inhibits the activation of caspase-1 and subsequent IL-1β secretion from Caco-2 cells.

Results and discussion

Secretion of rCTB-YVAD by *L. casei* transformed with pSCTB-YVAD

Recombinant fusion proteins of CTB with functional peptides expressed in various bacteria [7,14], yeasts [6], and plants [15] have been reported to bind GM1 ganglioside. However, there have been no reports of recombinant fusion proteins of CTB with functional peptides expressed in *Lactobacillus* species. Liljegvist *et al.* reported that the fusion of the serum albumin binding region (BB, approximately 25 kDa) to the C-terminus of CTB had no effect on the GM1-ganglioside-binding activity of CTB, whereas this activity was abolished when BB was fused to the N-terminus of CTB [16]. These observations demonstrate the importance of fusing the functional peptide to the C-terminus of CTB to retain its GM1-ganglioside-binding activity. In contrast, Dertzbaugh and Cox reported that CTB can bind to nickel ions without a His-tag because of a specific histidine residue within its sequence [17]. However, we recently showed that the purification of rCTB with a His-tag from the culture supernatant of *L. casei* was 20 times more efficient than the purification of rCTB without a His-tag [5]. The GM1-ganglioside-binding activity of rCTB with a His-tag, secreted by *L. casei*, was similar to those of rCTB without a His-tag and native CT from *V. cholerae* [5]. Therefore, to avoid any steric effects on the structure and GM1-ganglioside-binding activity of rCTB secreted by *L. casei*, we fused YVAD and the His-tag to the C-terminus of CTB. The rCTB–YVAD secretion vector pSCTB–YVAD includes the promoter region of the lactate dehydrogenase (LDH) gene, the secretory signal sequence (SS) and terminator region (Term) of the β-*N*-acetylglucosaminidase gene from *L. casei*, the CTB-coding sequence from *V. cholerae*, the YVAD-coding sequence (tatgttgctgat; this nucleotide sequence was determined by reference to the codon usage of *L. casei* ATCC 334 and ATCC 393), and the His-tag-coding sequence in the *Lactobacillus–E. coli* shuttle vector pHIL253 [5] (Figure 1). Protein bands of about 12–14 kDa were detected in the supernatant of *L. casei* transformed with pSCTB–YVAD or pSCTB, when immunoblotted with an antibody directed against CT (Figure 2A). These results indicate that rCTB–YVAD and rCTB, with predicted molecular weights of 13,231 and 12,783, respectively, were secreted by *L. casei* transformed with pSCTB–YVAD and pSCTB, respectively. Furthermore, the supernatant of *L. casei* transformed with pSCTB–YVAD showed strong binding activity to GM1 ganglioside compared with that of pHIL253 (0.78 ± 0.04 *vs* 0.06 ± 0.01 at OD_{405}, respectively, $P < 0.01$; Figure 2B). This GM1-ganglioside-binding activity was similar to that of the supernatant of *L. casei* transformed with pSCTB (0.61 ± 0.08 at OD_{405}). These observations indicate that rCTB–YVAD secreted by *L. casei* transformed with pSCTB–YVAD has binding activity for GM1 ganglioside similar to that of rCTB secreted by *L. casei* transformed with pSCTB.

Purification of rCTB–YVAD secreted by pSCTB–YVAD-transformed *L. casei*

To analyze the inhibitory effect of rCTB–YVAD secreted by *L. casei* transformed with pSCTB–YVAD, it was necessary to coincubate Caco-2 cells with *L. casei* secreting rCTB–YVAD. However, wild-type *L. casei* and *L. casei* transformed with pHIL253, pSCTB–YVAD, or pSCTB do not grow in minimum essential medium (MEM), which is used to culture Caco-2 cells, but grow in MRS/K (see Additional file 1: Figure S1A). Furthermore, neither rCTB–YVAD nor rCTB was detected in the supernatant of *L. casei* transformed with pSCTB–YVAD or pSCTB, respectively, when grown in MEM, but both were detected when the transformants were grown in MRS/K (see Additional file 1: Figure S1B). Therefore, to analyze the inhibitory effect of rCTB–YVAD on LPS-induced caspase-1 activation and subsequent IL-1β secretion, we purified rCTB–YVAD from the culture supernatant of *L. casei* transformed with pSCTB–YVAD using its His-tag and a nickel-bound affinity resin. The purified rCTB–YVAD migrated as a single band on SDS-PAGE when visualized with Coomassie Brilliant Blue (CBB) staining (Figure 3A). Furthermore, purified rCTB–YVAD was detected by immunoblotting with an antibody directed against CT (Figure 3B) and bound more strongly to GM1 ganglioside

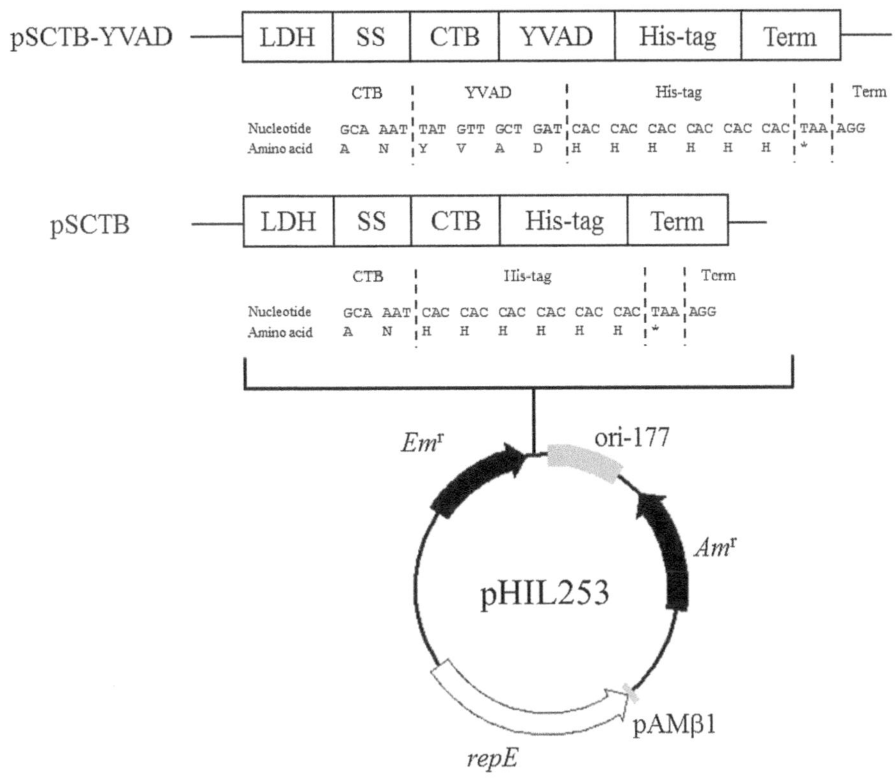

Figure 1 Construction of rCTB–YVAD secretion vector pSCTB–YVAD. pSCTB–YVAD includes the promoter of the *LDH* gene, SS of the β-*N*-acetylglucosaminidase gene, the CTB–YVAD-coding sequence, His-tag-coding sequence, and the Term of the β-*N*-acetylglucosaminidase gene in the *Lactobacillus–E. coli* shuttle vector pHIL253. The nucleotide and amino acid sequences around the CTB/YVAD fusion site are shown.

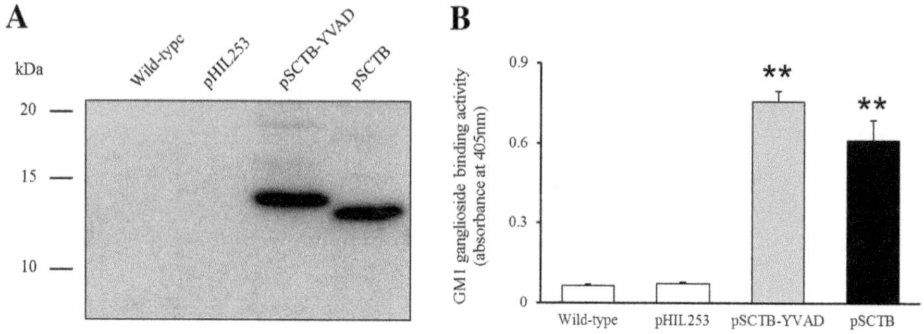

Figure 2 Secretion and GM1-ganglioside-binding activity of rCTB-YVAD. Wild-type *L. casei* and *L. casei* transformed with pHIL253, pSCTB-YVAD, or pSCTB were grown at 30°C. The culture supernatants were collected and concentrated 10-fold. Immunoblotting with an antibody directed against CT was performed to analyze the secretion of rCTB-YVAD or rCTB into the supernatant. Data are representative of three separate experiments **(A)**. The GM1-ganglioside-binding activity in the supernatant was analyzed with GM1-ELISA. Data represent the means ± SEM of three separate experiments performed in triplicate **(B)**. $^{**}P < 0.01$ compared with pHIL253.

than did PBS (0.64 ± 0.03 *vs* 0.08 ± 0.01 at OD_{405}, respectively, $P < 0.01$; Figure 3C), similar to the supernatant of *L. casei* transformed with pSCTB–YVAD (Figure 2B). The GM1-ganglioside-binding activity of rCTB–YVAD was similar to that of rCTB (0.56 ± 0.06 at OD_{405}; Figure 3C). Purified rCTB–YVAD and rCTB were used in further experiments. About 1 mg of rCTB–YVAD protein was obtained from one liter of culture supernatant from *L. casei* transformed with pSCTB–YVAD. This purification efficiency was similar to that for rCTB purified from the culture supernatant of pSCTB-transformed *L. casei*.

Viability of Caco-2 cells after rCTB-YVAD and rCTB treatment

There were no significant differences in the viability of Caco-2 cells treated with PBS, rCTB-YVAD, or rCTB in the presence of LPS (Figure 4). Therefore, rCTB–YVAD and rCTB had no effect on Caco-2 cell viability in the presence of LPS. Consequently, rCTB–YVAD and rCTB were used at concentrations of 50 μM in subsequent experiments.

Translocation of rCTB-YVAD and rCTB into Caco-2 cells

The translocation of rCTB–YVAD is a prerequisite for the inhibition of LPS-induced caspase-1 activation and subsequent IL-1β secretion. Therefore, we examined whether rCTB–YVAD translocates into Caco-2 cells. rCTB–YVAD was detected within Caco-2 cells in the absence or presence of LPS (Figure 5). Similarly, rCTB was detected within Caco-2 cells. These results indicate that both rCTB–YVAD and rCTB are

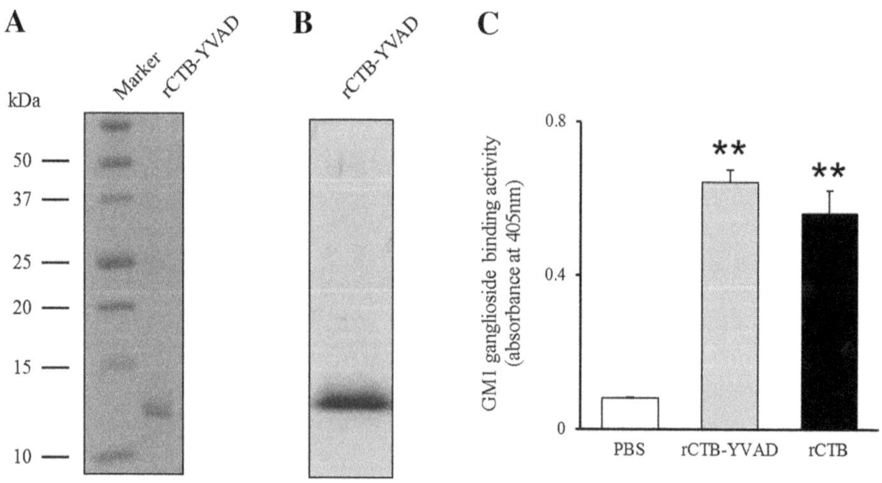

Figure 3 Purification of rCTB-YVAD from the culture supernatant of *L. casei* transformed with pSCTB-YVAD. rCTB-YVAD was purified using its His-tag with a nickel-bound affinity resin. Purification was confirmed by SDS-PAGE followed by CBB staining **(A)** and immunoblotting with an antibody directed against CT **(B)**. Data are representative of three separate experiments. The GM1-ganglioside-binding activities of purified rCTB-YVAD and rCTB were analyzed by GM1-ELISA. Data represent the means ± SEM of three separate experiments performed in triplicate **(C)**. $^{**}P < 0.01$ compared with PBS.

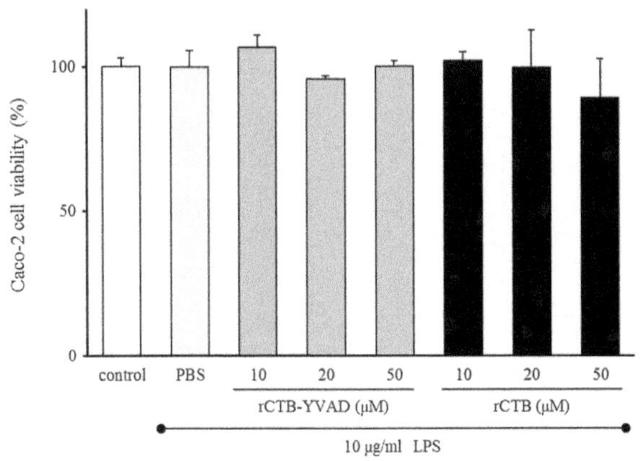

Figure 4 Viability of Caco-2 cells treated with rCTB–YVAD or rCTB. Caco-2 cells were treated with three concentrations of rCTB–YVAD or rCTB (10, 20, or 50 μM) for 48 h in the presence of 10 μg/ml LPS. Caco-2 cell viability was measured with the WST-1 method and by comparing the absorbance of untreated control cells with that of PBS-, rCTB–YVAD-, and rCTB-treated cells. Data represent the means ± SEM of three separate experiments performed in triplicate.

translocated into Caco-2 cells, regardless of LPS treatment.

The cell permeation of functional peptides, such as vaccine antigens and the insulin B chain peptide, is increased by their fusion with CTB [6,7], because CTB translocates easily into IECs by binding to GM1 ganglioside [4]. For YVAD to inhibit caspase-1 activation and subsequent IL-1β secretion, it is necessary for YVAD to translocate into Caco-2 cells. Our results suggest that the fusion of CTB to YVAD contributed to the translocation of YVAD into Caco-2 cells. The translocation of rCTB–YVAD into the cells, regardless of the presence or absence of LPS, suggests that rCTB–YVAD translocates into Caco-2 cells through GM1 ganglioside, which is constantly expressed, regardless of the inflammatory status.

Inhibitory effect of rCTB–YVAD on LPS-induced caspase-1 activation and subsequent IL-1β secretion in Caco-2 cells

The level of caspase-1 activity was significantly increased in the lysate of LPS-treated Caco-2 cells compared with that in the untreated control cells ($P < 0.01$). rCTB–YVAD significantly inhibited LPS-induced caspase-1 activation compared with that in cells treated with LPS + PBS (0.24 ± 0.01 *vs* 0.41 ± 0.03 at OD_{405}, respectively, $P < 0.01$; Figure 6A). This observation suggests that rCTB-YVAD acts as a decoy substrate for caspase-1. The level of IL-1β secretion increased in Caco-2 cells after LPS treatment compared with that in the untreated controls ($P < 0.01$). Treatment with rCTB–YVAD significantly reduced LPS-induced IL-1β secretion compared with that in cells treated with LPS + PBS (28.5 ± 3.6 *vs* 86.5 ± 4.7 pg/ml, respectively, $P < 0.01$; Figure 6B). In contrast, treatment with rCTB had no significant effect on LPS-induced caspase-1 activation (0.42 ± 0.02 at OD_{405}) or IL-1β secretion (73.3 ± 5.0 pg/ml) compared with those in cells treated with LPS + PBS (Figure 6). These results indicate that rCTB had no effect on caspase-1 activation or subsequent IL-1β secretion, and that the inhibitory effect of YVAD on capase-1 was not abolished by the fusion of YVAD to CTB.

The expression and secretion of YVAD by bacteria have been limited by its low molecular weight. Therefore, we constructed recombinant *L. casei* secreting YVAD as a fusion protein with CTB and showed that YVAD secreted by *L. casei* inhibits caspase-1 activation and subsequent IL-1β secretion. The results of this study indicate that YVAD secreted by bacteria exerts an anti-inflammatory effect. Meng *et al.* also reported that recombinant green

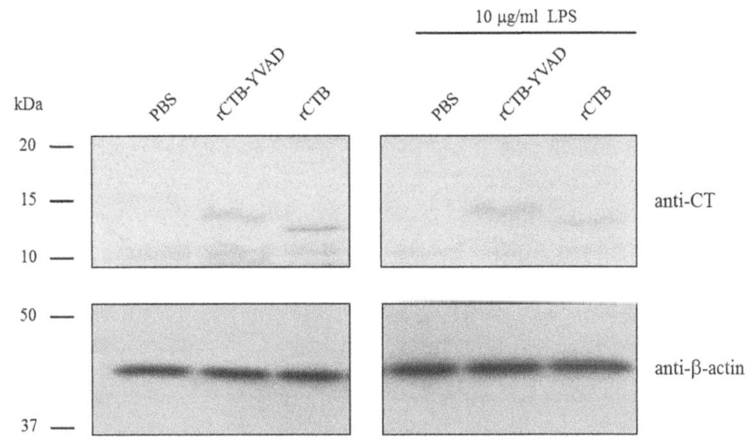

Figure 5 Translocation of rCTB–YVAD or rCTB into Caco-2 cells. Caco-2 cells were treated with 50 μM rCTB–YVAD or rCTB for 6 h in the absence or presence of 10 μg/ml LPS. The translocation of rCTB–YVAD or rCTB into Caco-2 cells was detected with immunoblotting using an antibody directed against CT. β-actin was used as the internal control. Data are representative of three separate experiments.

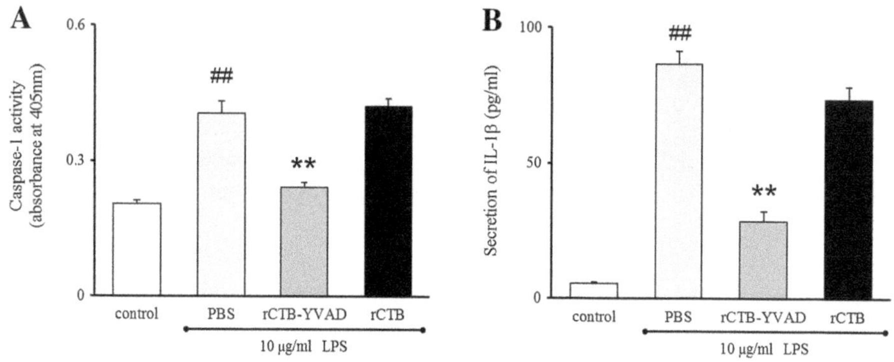

Figure 6 Inhibitory effects of rCTB–YVAD on LPS-induced caspase-1 activation and IL-1β secretion in Caco-2 cells. Caco-2 cells were treated with 10 μg/ml LPS for 12 h. The cell lysates were incubated at 30°C for 2 h in the absence or presence of 50 μM rCTB-YVAD or rCTB. Caspase-1 activity was determined with a colorimetric assay **(A)**. Caco-2 cells were treated with 50 μM rCTB–YVAD or rCTB in the presence of 10 μg/ml LPS. The concentrations of IL-1β in the supernatant were determined with an ELISA **(B)**. Data are the means ± SEM of three separate experiments performed in triplicate. $^{\#\#}P < 0.01$ compared with the control, $^{**}P < 0.01$ compared with LPS + PBS treatment.

fluorescent protein (GFP) fused to the C-terminus of CTB expressed in silkworms emitted green fluorescence similar to that emitted by recombinant GFP alone [18]. That report and the results of the present study suggest that the effects of functional peptides or proteins are not abolished by their fusion to the C-terminus of CTB.

Conclusions

We constructed an *L. casei* that secretes a recombinant CTB protein fused to YVAD. Although rCTB–YVAD bound GM1 ganglioside and translocated into Caco-2 cells, like rCTB, rCTB–YVAD but not rCTB inhibited LPS-induced caspase-1 activation and subsequent IL-1β secretion without affecting cell viability. These results indicate not only that a recombinant fusion protein of CTB with a functional peptide secreted by *L. casei* has GM1-ganglioside-binding activity, but also that recombinant YVAD secreted by *L. casei* exerts an anti-inflammatory effect. The results of this study suggest that rCTB secreted by *L. casei* has potential utility as a system for the delivery of YVAD into IECs. However, we were unable to examine the anti-inflammatory effects of rCTB–YVAD-secreting *L. casei* in this study because *L. casei* could not grow or secrete rCTB–YVAD in MEM. We confirmed that it is difficult to completely mimic the intestinal environment in *in vitro* experiments. Therefore, further studies, such as an *in vivo* study of the ingestion of rCTB–YVAD secreting *L. casei*, are required to examine the effects of rCTB–YVAD-secreting *L. casei*. If successful, such a study would confirm that rCTB-secreting *L. casei* has potential utility as a delivery system for functional peptides into the intestine.

Methods

Bacterial strains and culture conditions

The strains used in this study were *L. casei* ATCC 27092 and *E. coli* DH5α. *L. casei* was grown at 37°C in MRS broth to produce the recombinant strain or at 30°C in MRS/K (MRS with 0.2 M potassium phosphate buffer) to produce the cell culture supernatant [5]. Erythromycin (5 μg/ml) was added to MRS or MRS/K to select the recombinant strain. *E. coli* was grown at 37°C in LB medium with or without ampicillin (100 μg/ml).

Construction of the rCTB–YVAD secretion vector

PCR was performed using the plasmid pSCTB [5] as the template DNA, KOD-Plus- DNA polymerase (Toyobo, Osaka, Japan), and primers containing the YVAD-coding sequence fused to the C-terminus of the *CTB* gene: sense, 5′-caccaccaccaccaccactaaaggccttc-3′; anti-sense, 5′-atcagcaacataatttgccatactaattgc-3′. Initial denaturation was at 94°C for 2 min, followed by 30 cycles of denaturation at 94°C for 15 sec, annealing at 50°C for 30 sec, and extension at 68°C for 7 min. The DNA fragment of about 6,600 bp was separated by agarose gel electrophoresis and extracted from the gel with the QIAquick Gel Extraction Kit (Qiagen, Valencia, CA). The extracted DNA was phosphorylated at the 5′ end with T4 polynucleotide kinase (TaKaRa Bio, Shiga, Japan) and self-ligated with T4 DNA ligase (TaKaRa Bio). The rCTB–YVAD secretion vector pSCTB–YVAD was confirmed by sequencing, and then introduced into *L. casei* by electroporation, as described previously [5].

Secretion of rCTB–YVAD by *L. casei*

Overnight cultures of *L. casei* transformed with pSCTB–YVAD or pSCTB were inoculated into MRS/K medium to an optical density at 600 nm (OD_{600}) of 0.05, and the cells were grown at 30°C until they reached an OD_{600} of 2.0. The culture supernatants were collected by centrifugation (7,000 × *g*, 5 min, 4°C) and concentrated 10-fold with Amicon Ultra Centrifugal Filter Units (10 kDa; Merck Millipore, Tokyo, Japan). The secretion and specific GM1-ganglioside-binding activities of rCTB–YVAD and rCTB in the concentrated supernatant were confirmed with immunoblotting using an antibody directed against CT or a GM1 enzyme-linked immunosorbent assay (GM1-ELISA), respectively.

Immunoblotting

The concentrated supernatants of *L. casei* (20 μl/lane), rCTB–YVAD (100 ng/lane), and cellular protein extracts of Caco-2 cells (50 μg/lane) were subjected to immunoblotting analysis. They were separated with SDS-PAGE (10–18%) and transferred to polyvinylidene difluoride membranes (GE Healthcare, Buckinghamshire, UK). Antibodies directed against CT (Sigma-Aldrich, St. Louis, MO) and β-actin (Cell Signaling Technology, Boston, MA) were used as the primary antibodies. Alkaline phosphatase (AP)-labeled anti-rabbit IgG antibody (Cell Signaling Technology) was used as the secondary antibody, and binding was detected with a chemiluminescent substrate of AP (CDP-Star Reagent; Biolabs, Beverly, MA).

GM1-ELISA

A GM1-ELISA was performed to determine specific GM1-ganglioside-binding activities. Briefly, 96-well microtiter plates (Sumitomo Bakelite Co., Ltd., Tokyo, Japan) were coated with 5 μg/ml monosialoganglioside GM1 (Sigma-Aldrich) diluted in bicarbonate buffer (15 mM Na_2CO_3, 35 mM $NaHCO_3$, pH 9.6), and incubated overnight at 4°C. After incubation, the plates were washed three times with PBS supplemented with 0.05% Tween-20 (PBS-T), and then blocked with PBS containing 1% bovine serum albumin (Nacalai Tesque, Kyoto, Japan) at 37°C for 2 h. After the plates were washed, the concentrated supernatant of *L. casei* (100 μl/well), rCTB–YVAD (50 ng/well), or rCTB (50 ng/well) was applied to the wells and incubated at 37°C for 2 h. The plates were incubated at 37°C for 2 h with antibody directed against CT and AP-labeled anti-rabbit IgG antibody used as the primary and secondary antibodies, respectively. The plates were incubated with AP substrate (Sigma-Aldrich) at 37°C for 20 min, and the OD_{405} was then measured with a microplate reader (ImmunoMini Nj-2300; Nunc, Rochester, NY).

Purification of rCTB–YVAD secreted by *L. casei*

rCTB–YVAD from the culture supernatant of *L. casei* transformed with pSCTB–YVAD was purified using the His-tag and an affinity resin containing bound nickel ions. The culture supernatant of *L. casei* transformed with pSCTB–YVAD was collected by centrifugation (12,000 × *g*, 30 min, 4°C) after growth in MRS/K medium at 30°C until the OD_{600} was 2.0. After the supernatant was filtered at 0.22 μm, imidazole was added to a final concentration of 20 mM, and the culture supernatant was then adjusted to pH 7.0. Nickel resin (Ni Sepharose High Performance; GE Healthcare) was added to the culture supernatant and then mixed gently overnight at 4°C. The open column was filled with resin, and then washed with wash buffer (10 mM sodium phosphate, 500 mM NaCl, 20 mM imidazole, pH 7.4). rCTB–YVAD bound with nickel resin was eluted with elution buffer (10 mM sodium phosphate, 500 mM NaCl, 500 mM imidazole, pH 7.4). The eluted rCTB–YVAD was concentrated and the buffer replaced with PBS using Amicon Ultra Centrifugal Filter Units (10 kDa). The protein concentration of rCTB–YVAD was determined with Coomassie Protein Assay Reagent (Pierce, Perbio Science, Bonn, Germany). The purity of rCTB–YVAD was confirmed with SDS-PAGE on 17% polyacrylamide gel, followed by CBB staining and immunoblotting using an antibody directed against CT. The GM1-ganglioside-binding activities of purified rCTB–YVAD and rCTB were confirmed with GM1-ELISA.

Caco-2 cell viability assay

Caco-2 cells were cultured as described previously [19]. Aliquots of 5×10^3 Caco-2 cells were plated in each well of a 96-well plate (Nunc). The cells were treated with three concentrations of rCTB–YVAD or rCTB (10, 20, or 50 μM) in the presence of 10 μg/ml LPS from *E. coli* O55:B5 (Sigma-Aldrich). After incubation for 48 h, 20 μl of WST-1 Cell Proliferation Reagent (TaKaRa Bio) was added to each well. After 2 h, the OD_{450} and OD_{630} were measured with a microplate reader. Cell viability was calculated as (OD_{450} − OD_{630} of treated cells/OD_{450} − OD_{630} of untreated control cells) × 100%.

Detection of translocated rCTB–YVAD and rCTB in Caco-2 cells

Aliquots of 7×10^5 Caco-2 cells were plated in each well of six-well plates (Nunc). Cells were treated with 50 μM rCTB–YVAD or rCTB in the absence or presence of 10 μg/ml LPS. After incubation for 6 h, cellular protein extracts were prepared with PRO-PREP Protein Extraction Solution (iNtRON Biotechnology, Kyungki-Do, South Korea), according to the manufacturer's protocol. The protein concentrations of the cellular protein extracts were determined with Coomassie Protein Assay Reagent. Intracellular rCTB–YVAD and rCTB were detected by immunoblotting with an antibody directed against CT. Equal loading was confirmed with an antibody directed against β-actin.

Inhibitory effect on caspase-1 activity

Caspase-1 activity was determined with a modification of a previously described method [20,21]. Aliquots of 1×10^7 Caco-2 cells were plated in 90 mm plastic culture dishes (Nunc) and treated with or without 10 μg/ml LPS for 12 h. The cells were washed with PBS and resuspended in buffer W (20 mM HEPES, 10 mM KCl, 1.5 mM $MgCl_2$, 1 mM EGTA, 1 mM EDTA, pH 7.4) supplemented with 10 mM DTT and 1 mM phenylmethylsulfonyl fluoride. The cells were incubated at 4°C for 15 min and disrupted by 20 passages through a 23G needle. The lysates were then centrifuged at 12,000 × *g* for 5 min at 4°C and the supernatants collected. The protein concentrations of the lysates were determined with Coomassie Protein Assay Reagent. Aliquots of 10 μg/μl lysate were incubated at 30°C for 2 h in the absence or presence of 50 μM rCTB–YVAD or rCTB. The caspase-1 activity in the 10-fold-diluted lysate was determined with a Caspase 1 Assay Kit, Colorimetric (Calbiochem, La Jolla, CA).

Measurement of IL-1β by ELISA

Aliquots of 7×10^5 Caco-2 cells were plated in each well of six-well plates. The cells were treated with 50 μM rCTB–YVAD and rCTB in the presence of 10 μg/ml LPS for 48 h. The cell supernatants were centrifuged at 15,000 × *g* for 5 min at 4°C and stored at −80°C until IL-1β analysis. The concentrations of IL-1β in the cell supernatants were determined with a human IL-1β ELISA Kit (R&D Systems, Abingdon, UK).

Statistical analyses

Data are presented as means ± SEM. Statistical analyses were performed with Origin Pro 8.1 (OriginLab, Northampton, MA). Differences were analyzed with one-way ANOVA followed by Tukey's test. In all analyses, $P < 0.05$ was deemed to indicate significance.

Additional file

Additional file 1: Figure S1. Culture of *L. casei* and its secretion of rCTB–YVAD or rCTB in MRS/K or MEM. The data confirm that *L. casei* does not grow or secrete rCTB–YVAD when cultured in MEM.

Abbreviations

BB: Serum albumin binding region; CBB: Coomassie Brilliant Blue; CT: Cholera toxin; CTA: Cholera toxin A subunit; CTB: Cholera toxin B subunit; GFP: Green fluorescent protein; GM1-ELISA: GM1 enzyme-linked immunosorbent assay; His-tag: Six-histidine-residue tag; IECs: Intestinal epithelial cells; IL-1β: Interleukin-1β; LAB: Lactic acid bacteria; LDH: Lactate dehydrogenase; LPS: Lipopolysaccharide; pro-caspase-1: Precursor of caspase-1; pro-IL-1β: Precursor of IL-1β; rCTB: Recombinant CTB; rCTB–YVAD: Recombinant fusion protein of CTB with YVAD; SS: Secretory signal sequence; Term: Terminator region; YVAD: Tetrapeptide composed of tyrosine, valine, alanine, and aspartic acid.

Competing interests

The authors declare that they have no competing interests.

Authors' contributions

YH, MY, TS, and NK designed the research; YH, AK, SU, and YF acquired the data; YH analyzed the data; YH, TS, and KI drafted the manuscript. AT, TM, FM, and TA clarified the manuscript. All authors read and approved the final manuscript.

Acknowledgments

This research was supported by the following grants: JSPS KAKENHI (grant numbers 24658099 and 24659060); JST Project to Develop Innovative Seeds (grant number 12–046); and the Central Research Institute from Fukuoka University (grant numbers 126012 and 127007).

Author details

[1]Faculty of Pharmaceutical Sciences, Fukuoka University, 8-19-1, Nanakuma, Jonan-ku, Fukuoka, Japan. [2]Department of Molecular Oncology, Graduate School of Medical and Dental Sciences, Kagoshima University, 8-35-1, Sakuragaoka, Kagoshima, Japan. [3]Daiichi College of Pharmaceutical Sciences, 22-1, Tamagawa-cho, Minami-ku, Fukuoka, Japan. [4]Division of Molecular Functions of Food, Department of Biochemistry and Biotechnology, Kagoshima University, 1-21-24, Korimoto, Kagoshima, Japan. [5]Molecular Microbiology Group, Department of Tropical Infectious Diseases, COMB, Tropical Biosphere Research Center, University of the Ryukyus, 1 Senbaru, Nishihara, Okinawa, Japan. [6]Division of Host Defense and Vaccinology, Department of Microbiology, Graduate School of Medicine, University of the Ryukyus, 207 Uehara, Nishihara, Okinawa, Japan.

References

1. del Rio B, Dattwyler RJ, Aroso M, Neves V, Meirelles L, Seegers JF, Gomes-Solecki M: **Oral immunization with recombinant *Lactobacillus plantarum* induced a protective immune response in mice with Lyme disease.** *Clin Vaccine Immunol* 2008, **15:**1429–1435.
2. Jung SW, Kim WJ, Lee KG, Kim CW, Noh WS: **Fermentation characteristics of exopolysaccharide-producing lactic acid bacteria from sourdough and assessment of the isolates for industrial potential.** *J Microbiol Biotechnol* 2008, **18:**1266–1273.
3. Pouwels PH, Lee RJ, Shaw M, den Bak-Glashouwer MJ H, Tielen FD, Smit E, Martinez B, Jore J, Conway PL: **Lactic acid bacteria as antigen delivery vehicles for oral immunization purposes.** *Int J Food Microbiol* 1998, **41:**155–167.
4. Fishman PH: **Role of membrane ganglioside in the binding and action of bacterial toxins.** *J Membr Biol* 1982, **69:**85–97.
5. Okuno T, Kashige N, Satho T, Irie K, Hiramatsu Y, Sharmin T, Fukmits Y, Uyeda S, Harakuni T, Miyata T, Arakawa T, Imoto M, Toda A, Nakashima Y, Miake F: **Expression and secretion of cholera toxin B subunit in lactobacilli.** *Biol Pharm Bull* 2013, **36:**952–958.
6. Harakuni T, Sugawa H, Komesu A, Tadano M, Arakawa T: **Heteropentameric cholera toxin B subunit chimeric molecules genetically fused to a vaccine antigen induce systemic and mucosal immune responses: a potential new strategy to taeget recombinant vaccine antigens to mucosal immune systems.** *Infect Immun* 2005, **73:**5654–5665.
7. Yuki Y, Hara-Yakoyama C, Guadiz AA, Udaka S, Kiyono H, Chatterjee S: **Production of a recombinant cholera toxin B subunit-insulin B chain peptide hybrid protein by *Brevibacillus choshinensis* expression system as a nasal vaccine against autoimmune disease.** *Biotechnol Bioeng* 2005, **92:**803–809.
8. Garcia-Calvo M, Peterson EP, Leiting B, Ruel R, Nicholson DW, Thornberry NA: **Inhibition of human caspases by peptide-based and macromolecular inhibitors.** *J Biol Chem* 1998, **273:**32608–32613.
9. Thornberry NA, Bull HG, Calaycay JR, Chapman KT, Howard AD, Kostura MJ, Miller DK, Molineaux SM, Weidner JR, Aunins J, Elliston KO, Ayala JM, Casano FJ, Chin J, Ding GJF, Egger LA, Gaffney EP, Limjuco G, Palyha OC, Raju SM, Rolando AM, Salley JP, Yamin TT, Lee TD, Shively JE, MacCross M, Mumford RA, Schmidt JA, Tocci MJ: **A novel heterodimetric cysteine protease is required for interleukin-1 beta processing in monocytes.** *Nature* 1992, **356:**768–774.
10. Mahida YR, Wu K, Jewell DP: **Enhanced production of interleukin 1-beta by mononuclear cells isolated from mucosa with active ulcerative colitis of Crohn's disease.** *Gut* 1989, **30:**835–838.
11. Schumann RR, Belka C, Reuter D, Lamping N, Kirschning CJ, Weber JR, Pfeil D: **Lipopolysaccharide activates caspase-1 (interleukin-1-converting enzyme) in cultured monocytic and eondothelial cells.** *Blood* 1998, **91:**577–584.
12. Ramage P, Cheneval D, Chvei M, Graff P, Hemmig R, Heng R, Kocher HP, Mackenzie A, Memmert K, Revesz L, Wishart W: **Expression, refolding, and**

autocatalytic proteolytic processing of the interleukin-1 beta-converting enzyme precursor. *J Biol Chem* 1995, **270**:9378–9383.

13. Charrier JD, Durrant SJ, Studley J, Lawes L, Weber P: **Synthesis and evauation of novel prodrugs of caspase inhibitors.** *Bioorg Med Chem Lett* 2012, **22**:485–488.
14. Sadeghi H, Bregenholt S, Wegmann D, Petersen JS, Holmgren J, Lebens M: **Genetic fusion of human insulin B-chain to the B subunit of cholera toxin enhances *in vitro* antigen presentation and induction of bystander suppression *in vivo*.** *Immunology* 2002, **106**:237–245.
15. Kim TG, Kim HY, Yang MS: **Cholera toxin B subunit-domain III of dengue virus envelope glycoprotein E fusion protein production in transgenic plants.** *Protein Expr Purif* 2010, **74**:236–241.
16. Liljegvist S, Ståhl S, Andréoni C, Binz H, Uhlén M, Murby M: **Fusions to the cholera toxin B subunit: influence on pentamerization and GM1 binding.** *J Immunol Methods* 1997, **210**:125–135.
17. Dertzbaugh MT, Cox LM: **The affinity of cholera toxin for Ni^{2+} ion.** *Protein Eng* 1998, **11**:577–581.
18. Meng Q, Wang W, Shi X, Jin Y, Zhang Y: **Protection against autoimmune diabetes by silkworm-produced GFP-tagged CTB-insulin fusion protein.** *Clin Dev Immunol* 2011, **2011**:831704.
19. Hiramatsu Y, Satho T, Irie K, Shiimura S, Okuno T, Sharmin T, Uyeda S, Fukumitsu Y, Nakashima Y, Miake F, Kashige N: **Differences in TLR9-dependent inhibitory effects of H_2O_2-induced IL-8 secretion and NF-kappa B/I kappa B-alpha system activation by genomic DNA from five *Lactobacillus* species.** *Microbes Infect* 2013, **15**:96–104.
20. Martinon F, Burns K, Tschopp J: **The inflammasome: a molecular platform triggering acivation of inflammatory caspases and processing of proIL-beta.** *Mol Cell* 2002, **10**:417–426.
21. Yamamoto M, Yaginuma K, Tsutsui H, Sagara J, Guan X, Seki E, Yasuda K, Yamamoto M, Akira S, Nakanishi K, Noda T, Taniguchi S: **ASC is essential for LPS-induced activation of procaspase-1 independently of TLR-associated signal adaptor molecules.** *Genes Cells* 2004, **9**:1055–1067.

11

Biosynthesis of *trans*-4-hydroxyproline by recombinant strains of *Corynebacterium glutamicum* and *Escherichia coli*

Yulan Yi[1], Huakai Sheng[1], Zhimin Li[1,2*] and Qin Ye[1]

Abstract

Background: *Trans*-4-hydroxy-L-proline (*trans*-Hyp), one of the hydroxyproline (Hyp) isomers, is a useful chiral building block in the production of many pharmaceuticals. Although there are some natural biosynthetic pathways of *trans*-Hyp existing in microorganisms, the yield is still too low to be scaled up for industrial applications. Until now the production of *trans*-Hyp is mainly from the acid hydrolysis of collagen. Due to the increasing environmental concerns on those severe chemical processes and complicated downstream separation, it is essential to explore some environment-friendly processes such as constructing new recombinant strains to develop efficient process for *trans*-Hyp production.

Result: In this study, the genes of *trans*-proline 4-hydroxylase (*trans*-P4H) from diverse resources were cloned and expressed in *Corynebacterium glutamicum* and *Escherichia coli*, respectively. The *trans*-Hyp production by these recombinant strains was investigated. The results showed that all the genes from different resources had been expressed actively. Both the recombinant *C. glutamicum* and *E. coli* strains could produce *trans*-Hyp in the absence of proline and 2-oxoglutarate.

Conclusions: The whole cell microbial systems for *trans*-Hyp production have been successfully constructed by introducing *trans*-P4H into *C. glutamicum* and *E. coli*. Although the highest yield was obtained in recombinant *E. coli*, using recombinant *C. glutamicum* strains to produce *trans*-Hyp was a new attempt.

Keywords: *Trans*-4-hydroxy-L-proline, Recombinant *Corynebacterium glutamicum*, Recombinant *Escherichia coli*, Proline 4-hydroxylases

Background

Hydroxyproline (Hyp) is a specific amino acid component of collagen. The amount of Hyps varies from 80 to 100 residues per 1000 residues in mammalian collagen, which can be used to estimate collagen content and act as an important indicator to collagen quality [1]. There are five naturally occurring Hyps, including three diastereomers of 4-hydroxyproline and two diastereomers of 3-hydroxyproline. Among them, *trans*-4-hydroxy-L-proline is the most abundant component in the constitution of collagen and can enhance the procollagen synthesis. Its derivative N-acetyl *trans*-4-hydroxyproline (oxaceprol) is an atypical inhibitor of inflammation and useful for the treatment of diseases affecting the connective tissues such as osteoarthritis [2]. *Trans*-Hyp has been widely used in medicine, biochemistry, food, cosmetic and other aspects of industry [3]. Additionally, *trans*-Hyp has also been found in the composition of some secondary metabolites such as actinomycins and echinocandins [4].

Trans-Hyp is manufactured industrially most by acid hydrolysis of mammalian collagen because of its rich amount in the collagen. However, it obviously results in many environmental issues and brings great difficulties into the down stream processing [5]. There are several identified pathways of hydroxyproline biosynthesis. In animal tissue, 4-hydroxyproline is catalyzed by prolyl 4-hydroxylase, which takes peptidyl proline as substrate rather than free proline [6]. 4-hydroxy-2-oxoglutaric acid can be enzymatically transformed to hydroxyproline [7].

* Correspondence: lizm@ecust.edu.cn
[1]State Key Laboratory of Bioreactor Engineering, East China University of Science and Technology, 130 Meilong Road, Shanghai 200237, China
[2]Shanghai Collaborative Innovation Center for Biomanufacturing Technology, Shanghai, China

Some bacteria or fungi have been found to form hydroxyproline via fermentation directly [8]. Although the titer of product is low, these findings show the possibility of utilizing biological processes to produce *trans*-Hyp.

The proline 4-hydroxylases (P4Hs) have been identified from several microbial strains, which can catalyze the hydroxylation of L-proline at the 4-position to produce *trans*-Hyp in the presence of 2-oxoglutarate, oxygen and ferrous ion [9-11]. P4Hs have an optimum pH range of 6.0 to 7.5 and temperature range of 30°C to 40°C. Its activity is inhibited by metal ions such as Zn^{2+} and Cu^{2+}. Lawrence et al. have studied the effect of co-substrates on the hydroxylation of L-proline by P4H and pointed that 2-oxoglutarate was essential for proline hydroxylation since the replacement of 2-oxoglutarate with 2-oxopentanoate, 2-oxoadipate, pyruvate or 2-oxomalonate (all at 0.5 mM) led to no detectable hydroxylation of L-proline [10]. Although 2-oxoglutarate as the oxygen donator is required for hydroxylation of L-proline to 4-hydroxy-L-proline in vitro, it is unnecessary to add extra 2-oxoglutarate in vivo in the production of 4-hydroxy-L-proline by recombinant strains. 2-oxoglutarate is a key metabolic intermediate in the tricarboxylic acid cycle (TCA cycle) in *Escherichia coli* strain, which can result in the formation of hydroxyproline from glucose and proline directly [12]. Shibasaki et al. have analyzed the possible metabolic pathways of 2-oxoglutatate. They concluded that 2-oxoglutatate can be supplied either through the action of proline dehydrogenase (PutA) from L-proline or through the action of isocitric dehydrogenase (Icd) from glucose. The addition of L-proline to a glucose-containing minimal medium had a positive effect on both the proline 4-hydroxylase activity and production level [13]. But the availability of intracellular proline may still be limited because the biosynthesis of proline in wild type *E. coli* is strictly regulated to very low level. Thus, the precursor and co-factor in the microbial production of hydroxyproline need to be considered simultaneously.

Corynebacterium glutamicum is one of the most important industrial microorganisms and widely used in amino acids, vitamins and nucleic acids production [14]. Leuchtenberger et al. has summarized the commercial application of *C. glutamicum* to the fermentative production of amino acids [15]. Lee et al. has reported a novel glutamate and proline producing method through the utilization of phenol in *C. glutamicum* [16]. Masaaki Wachi reported a strategy for optimizing the industrial production of amino acids by reinforcing the export systems of *C. glutamicum* [17]. The metabolic pathways of amino acids are sophisticated and controlled tightly in *C. glutamicum.* But *C. glutamicum* as the platform of amino acid production has been studied in details and there are lots of molecular tools used for its genetic modifications, which contribute to *C. glutamicum* as one of the most popular host systems [18,19]. Additionally, Ikeda et al. [20] and Kalinowski et al. [21] have completed genome sequencing of *C. glutamicum* ATCC13032, which made *C. glutamicum* into a new era of system biology. To overproduce amino acids by *C. glutamicum,* not only the modification of biosynthetic pathway and regulation mechanism but also the transportation of amino acids plays significant roles in the final yield of a particular amino acid [22,23].

To produce *trans*-Hyp by microbial fermentation, it is essential to contain both the proline pathway and subsequent hydroxylation activity in the microbial cells. Considering the metabolic pathway in *C. glutamicum* or *E. coli* (Figure 1), proline and 2-oxoglutarate can be produced from glucose. NADPH and ATP can also be regenerated during the glucose metabolism. Although there are some efforts being conducted in *E. coli,* for example, Hyp was produced from glucose using a recombinant *E. coli* by introducing a proline 4-hydroxylase and the mutated *proB* (encodingγ-glutamyl kinase) gene encoding the feedback resistant enzyme [24]. However, the development of recombinant *C. glutamicum* for hydroxyproline secretion has never been reported. In this study, genes of *trans*-P4H from diverse sources were screened and cloned into different *C. glutamicum* strains. The recombinant *E. coli* expressed different genes were also investigated. Meanwhile, a surprisingly high yield in shake flasks by recombinant *E. coli* without extra addition of L-proline was achieved.

Results and discussion

Construction of *trans*-Hyp producing recombinant strains

There are several genes being speculated as putative L-proline 4-hydroxylase gene in the database, including genes in *Pseudomonas stutzeri, Janthinobacterium* sp., *Bordetella bronchiseptica RB50, Bradyrhizobium japonicum, Achromobacter xylosoxidans C54* and *Dactylosporangium.* sp. Using PCR, we cloned and obtained the putative genes of P4H from *P. stutzer* and *B. bronchiseptica RB50,* named *p4hP* and *p4hB.* They were ligated to the corresponding plasmids after digestion and converted to *C. glutamicum* and *E. coli,* respectively. The length of *p4hP* was 918 bps while *p4hB* was 924 bps. These sequences were 100% identical to the reported genes in NCBI. The gene of *trans*-P4H from *Dactylosporangium sp.* (*p4hD*) had been expressed in *E. coli* successfully and can transform L-proline with good enzymatic properties [11-13,24-26]. The length of *p4hD* was 816 bps encoding a 272-amino-acid polypeptide with the molecular weight of 29,715 daltons [11,26]. In this study, *p4hD* was applied with some modifications on the nuclear bases. The original gene sequence of *p4hD* was analyzed (http://www.kazusa.or.jp/codon/) and the results showed there were some rare codons for both *C. glutamicum* and

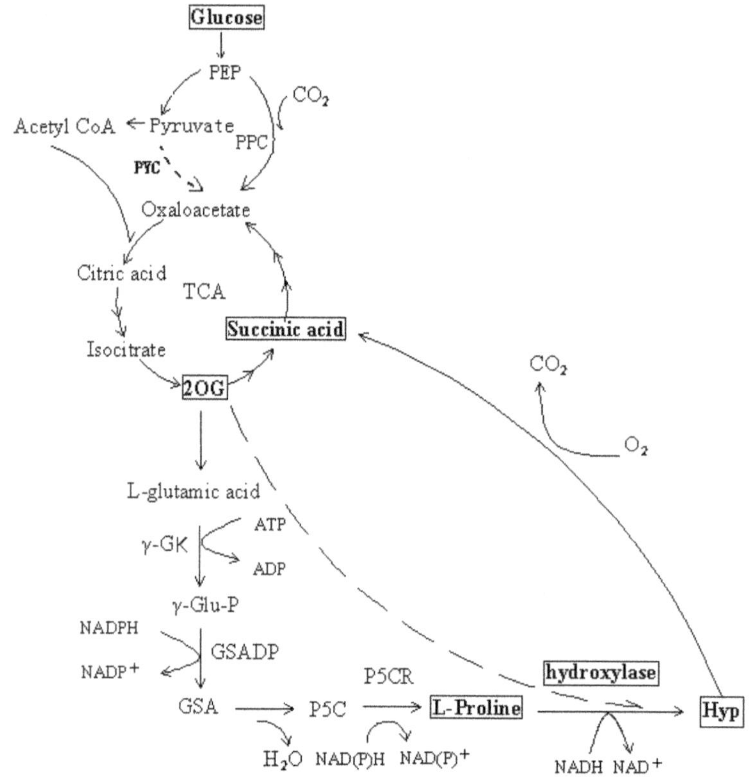

Figure 1 The metabolism of *trans*-4-hydroxyproline in recombinant bacteria. 2OG: 2 - 2-Ketoglutaric acid; γ-GK: γ- Glutamyl kinase; γ-Glu-P: γ- Glutamyl phosphate; GSADH; Glutamyl phosphate reductase; GSA: Glutamyl semialdehyde; P5C: Pyrroline - 5 - carboxylic acid; P5CR: P5C reductase; Hyp: Hydroxyproline.

E. coli. It has been reported that rare codons are strongly associated with low level of protein expression [27]. Codon optimization for heterologous protein expression has often been shown to drastically increase protein expression [28]. Thus, the rare codons of *p4hD* gene were substituted for those used with high frequency in *C. glutamicum* and the GC content was adjusted from 73% to 61% through synonymous conversion, which was close to that of *C glutamicum*. The modified gene of *p4hD* was synthesized according to the above modifications (Additional file 1).

The expression of P4H is one of the important aspects on the construction of trans-Hyp biosynthetic pathway. Figure 2 shows the SDS-PAGE of *trans*-P4Hs expressed in recombinant *C. glutamicum* and *E. coli.* All the recombinant *trans*-P4Hs were expressed as soluble proteins without inclusion bodies. It was obviously that the recombinant *trans*-P4Hs in *E. coli* were expressed much more than those in *C. glutamicum* (Figure 2). Many factors have influences on the expression of foreign proteins including promoters, the host-vector system and cultural conditions etc. [29,30]. Since it was the first to express *trans*-P4Hs in *C. glutamicum*, more comprehensive studies such as promoter selection and culture condition optimization will be considered in our future work.

Comparison of P4H activities

Oxygenases are widely applied in industry since they can catalyze the highly specific oxyfunctionalization of unactivated C-H bonds under mild conditions, especially transferring molecular oxygen to a substrate [31]. P4H belongs to a family of 2-oxoacid-dependent dioxygenase, which is a monomeric protein and utilizes the monomeric rather than polymeric substrates [10]. The activities of *trans*-P4Hs using recombinant whole cells in this study were measured (Table 1). Our data indicated that the expressed protein level and enzymatic activity was higher at 30°C. The results also showed that the plasmids were very stable as the plasmid stabilities of recombinant *E. coli* and *C. glutamicum* strains were all more than 98% at the end of fermentation.

The recombinant cells with expressing of different genes showed different levels of catalytic activities toward L-proline. The activity of *trans*-P4H expressed by *E. coli* BL21/ pET28a-*p4hD* was the highest among all the constructed recombinant strains. The new cloned and expressed genes from *P. stutzeri* and *B. bronchiseptica* also showed interested activities. As for different host strains, *E. coli* represented better than *C. glutamicum*, which may be related to the performance of corresponding

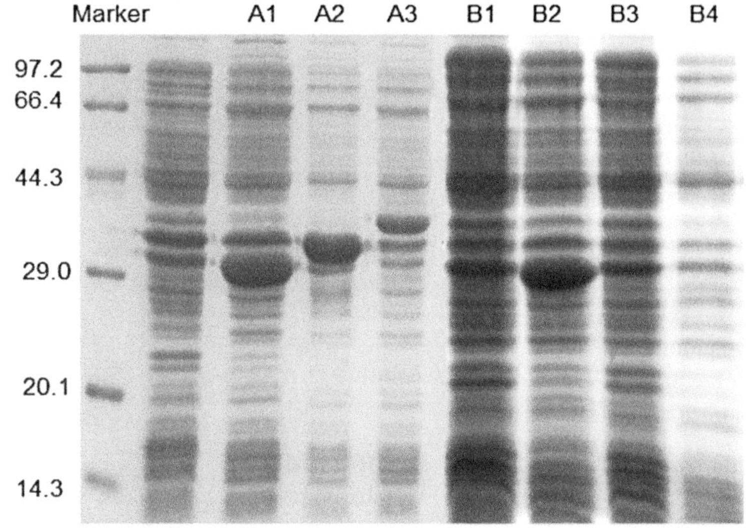

Figure 2 Expression of *trans*-P4Hs in different strains. A1: *E. coli* BL21/pET28a-*p4hD*; A2: *E. coli* BL21/pET28a-*p4hP*; A3: *E. coli* BL21/pET28a-*p4hB*. B1: *C. glutamicum* ATCC15940/pECXK99E-*p4hD*; B2: *C. glutamicum* ATCC21355/pECXK99E-*p4hD*; B3: *C. glutamicum* ATCC21157/pECXK99E-*p4hD*; B4: *C. glutamicum* 49-1/ pECXK99E- *p4hD*.

plasmid. Four L-proline producing strains of *C. glutamicum* were used as expression host strains and the resulted recombinant strains showed different enzymatic activities. The highest specific enzymatic activity among *C. glutamicum* strains was 40.7 U/mg · wet cell by *C. glutamicum* ATCC13032/pEC-XK99E-*p4hB*. However, the specific enzymatic activity of recombinant *E. coli*/pET28a -*p4hD* was up to 60.4 U/mg · wet cell. The growth of three recombinant *E. coli* strains was similar. But there was significant difference among the recombinant *C. glutamicum* strains. The recombinant *C. glutamicum* strains with higher specific enzymatic activities grew less than those with lower specific enzymatic activities. Additionally, the enzymatic activity of *E. coli* BL21 /pET28a -*p4hD* was similar to that of *E. coli* W1485/pWFH1 and higher than that of *E. coli* BL21/pET24-*p4h*1 of [12,13]. The *p4hD* in *E. coli* W1485/ pWFH1 was the original one in *Dactylosporangium* sp., while *p4hD* in *E. coli* BL21/pET24-*p4h*1 was modified. Although the codon optimization in this study was designed for *C. glutamicum*, the results indicated that it was also successfully in *E. coli*.

Trans-Hyp production in flasks

The production of *trans*-Hyp by different recombinant *C. glutamicum* and *E. coli* strains was also shown in Table 1. The yields of *trans*-Hyp by these recombinant strains depended both on the enzymatic activity of P4H and cell growth. *E. coli* BL21/ pET28a-*p4hD* had the

Table 1 Comparison of *trans*-P4Hs activities and *trans*-Hyp production by different recombinant *C. glutamicum* and *E. coli* strains

Strains	Specific activities (U/mg · wet cell weight)	*Trans*-Hyp (g/L)	OD600 (OD620)*
C. glutamicum ATCC13032/pEC-XK99E- *p4hD*	37.4 ± 1.4	0.072 ± 0.001	5.5 ± 0.7
C. glutamicum ATCC13032/pEC-XK99E- *p4hP*	20.7 ± 1.1	0.106 ± 0.002	7.3 ± 0.5
C. glutamicum ATCC13032/pEC-XK99E- *p4hB*	40.7 ± 0.8	0.079 ± 0.016	5.4 ± 0.03
C. glutamicum ATCC15940/pEC-XK99E- *p4hD*	12.9 ± 0.5	0.103 ± 0.001	14.0 ± 0.2
C. glutamicum ATCC21355/pEC-XK99E- *p4hD*	35.9 ± 0.1	0.087 ± 0.005	6.6 ± 0.2
C. glutamicum ATCC21157/pEC-XK99E- *p4hD*	12.3 ± 0.9	0.112 ± 0.004	13.3 ± 0.1
C. glutamicum 49-1/pEC-XK99E- *p4hD*	12.4 ± 0.6	0.113 ± 0.001	13.8 ± 0.5
E. coli BL21/pET-28a -*p4hD*	60.4 ± 1.8	0.470 ± 0.028	6.5 ± 0.2
E. coli BL21/pET-28a -*p4hP*	22.2 ± 0.5	0.126 ± 0.007	7.3 ± 0.05
E. coli BL21/pET-28a -*p4hB*	50.0 ± 2.2	0.115 ± 0.006	6.9 ± 0.1

*Optical density at 600 nm for *E. coli* and 620 nm for *C. glutamicum*.

highest yield, which was coincided of its specific enzymatic activity. Although the recombinant *E. coli* strains grew similarly in the production medium, there was significant difference in the production of *trans*-Hyp which did not keep the same level with the specific enzymatic activities. The productions of *trans*-Hyp by recombinant *C. glutamicum* strains were also much less than that of *E. coli* BL21/pET28a-*p4hD*. It was due to both the less expression of *trans*-P4H and less cell growth in *C. glutamicum*. The L-proline production of four *C. glutamicum* strains was also less than 1 g/L. There was little difference of *trans*-Hyp production among the recombinant strains of *C. glutamicum* with same gene *p4hD*, despite that some strains had better enzymatic performance and proline production.

Using the recombinant strains to directly synthesize *trans*-Hyp from glucose via fermentation was achieved since both the enzymes and precursors needed in the process were available. The over expressed foreign *trans*-P4Hs catalyzed the hydroxylation of L-proline at the *trans*-4 position, while 2– ketoglutarate was supplied by glucose through TCA cycle and then oxidatively decarboxylated to succinate (Figure 1). It was reported that proline was demanded in the production of Hyp by recombinant *E. coli* only with *p4h* gene. The carbon in proline added during the fermentation only flowed into amino acids synthesized from TCA cycle intermediates and not into gluconeogenesis [13]. However, the accumulated Hyp was at a relative high level even without the addition of proline in this study. It could be understood that *Corynebacterium* had the powerful biosynthetic pathway of proline [32,33]. The biosynthetic pathway of proline has also been identified in *E. coli*, which may contribute to the synthesis of *trans*-Hyp by recombinant *E. coli* strains. The modification of proline pathway in *E. coli* enhanced the yield of Hyp, whereas the formation of Hyp can also relieve the feedback inhibition of proline [24]. The amount of proline (0-4 mM) promoted the production of *trans*-Hyp. However, continuously addition of L-proline didn't improve the production yield significantly (Table 2). In this study, the time of cultivation was significantly less than those reported in the literatures, which might attribute to the different media used and also indicated there was great potential with optimization. In fact, 2.28 g/L of *trans*-Hyp was produced by recombinant *E. coli* without adding L-proline in flasks with a little modification of media and 6.72 g/L was achieved by supplement only 4 mM L-proline.

Table 2 Hyp production under different L-proline supplementation

Supplementary addition of L- proline (mM)	0	1	2	4	8	12
Hyp (g/L)	2.28	2.81	3.25	6.72	5.56	6.32

In order to further increase the biosynthesis of *trans*-Hyp by recombinant *C. glutamicum* and *E. coli*, alternative approaches should be considered as well. In *E. coli*, the degradation of proline should be overcome. Although the *trans*-Hyp production by a *putA* mutant of *E. coli* was not improved furthermore, the yield based on the proline utilized was enhanced greatly. In both *C. glutamicum* and *E. coli*, the expression of recombinant P4H as one of the oxygenases is involved in the physiological metabolism of host cells including the cofactor, co-substrate and oxygen. Moreover, without a powerful proline synthetic pathway in *E. coli*, the availability and transportation of substrate will limit the transformation seriously.

Conclusions

In this study, two new and a modified *trans*-P4Hs were expressed in *C. glutamicum* and *E. coli* successfully. Different amount of *trans*-Hyp were produced by these recombinant strains detected. Although the yield in recombinant *C. glutamicum* was less than that in recombinant *E. coli*, *C. glutamicum* as a native proline producing strain was worthy of further optimizing. This is the first report of producing *trans*-Hyp by introducing L-proline 4-hydroxylase into *C. glutamicum*.

Methods

Strains and plasmids

The bacterial strains and plasmids used in this study are listed in Table 3. *E. coli* BL21(DE3) and five *C. glutamicum* strains among which four were reported to produce L-proline were used as hosts. Plasmids pET-28a and pEC-XK99E were applied as the vectors respectively. *Trans*- 4-hydroxy-L-proline was purchased from Sigma-Aldrich Trading Co., Ltd.

Construction of recombinant strains

The full sequences of *p4h* encoding predicted hydroxylases were amplified from the strains' genomic DNA as followed in Table 4. The primers used in this study for gene cloning and plasmid construction were also as listed in Table 4, which were incorporated *Sal*I/*EcoR*I restriction sites for recombinant *E. coli* and *Sal*I/*Xba*I for recombinant *C. glutamicum*. The PCR products were digested with restriction enzymes above and inserted into the vector pET-28a and pEC-XK99E respectively, which resulted into the expression plasmid pET-28a-*p4h* and pEC-XK99E-*p4h*. The inserted fragment was sequenced and verified the identity to the anticipated sequence. Recombinant plasmid was transferred into *E. coli* competent cells by chemical $CaCl_2$ method. The preparation of *C. glutamicum* competent cell and electro transformation of exogenous gene were conducted according to the method in reference [34].

Table 3 Strains and plasmids used in this study

Strains & plasmids	Properties	Source/reference
E. coli		
JM109	Cloning host	Our laboratory
BL21(DE3)	*F-, ompT, hsdS(rBB-mB-), gal, dcm (DE3)*	Our laboratory
C. glutamicum		
ATCC13032	Wild-type	Our laboratory
ATCC 15940	L-Proline production	Our laboratory
ATCC 21355	L-Proline production	[27]
ATCC 21157	L-Proline production	[27]
49-1 Plasmids	L-Proline production	Our laboratory
pET-28a	His_4-tag, T7 promoter, Kan^r	Our laboratory
pEC-XK99E	*E. coli - C. glutamicum* shuttle expression vector, Kan^r	Our laboratory
pET-28a-*p4hP*	pET-28a containing the *p4h* gene from *P. stutzeri*	This study
pEC-XK99E-*p4hP*	pEC-XK99E containing the *p4h* gene from *P. stutzeri*	This study
pET-28a-*p4hB*	pET-28a containing the *p4h* gene from *B. bronchiseptica*	This study
pEC-XK99E-*p4hB*	pEC-XK99E containing the *p4h* gene from *B. bronchiseptica*	This study
pET-28a- *p4hD*	pET-28a containing the *p4h* gene from *Dactylosporangium* sp.	This study
pEC-XK99E-*p4hD*	pEC-XK99E containing the *p4h* gene from *Dactylosporangium* sp.	This study

^{r} indicates resistant.

Media

Luria broth (LB) medium, tryptone 10 g/L; yeast extract 5 g/L; solium chloride 10 g/L, was used for seed cultivation of *E. coli* strains. LBG medium containing 1% glucose additionally was used for *C. glutamicum* seed cultivation.

The medium (MEC) for batch culture of *E. coli* in shake flasks contained: glucose 10 g/L, glycerol 5 g/L, $CO(NH_2)_2$ 10 g/L, yeast extract 10 g/L, K_2HPO_4 1 g/L, NaCl 2 g/L, $MgSO_4 \cdot 7H_20$ 0.2 g/L, $FeSO_4 \cdot 7H_2O$ 1 mM, $MnSO_4 \cdot 4H_2O$ 10 mg/L, $ZnSO_4 \cdot 7H_2O$ 10 mg/L, VB_1 200 ug/L.

The medium (MCG) for batch culture of *C. glutamicum* in shake flasks contained: glucose 10 g/L; glycerol 5 g/L; $CO(NH_2)_2$ 10 g/L; corn syrup 15 g/L; K_2HPO_4 1 g/L; NaCl 2 g/L; $MgSO_4 \cdot 7H_20$ 0.2 g/L; $FeSO_4 \cdot 7H_2O$ 1 mM; $MnSO_4 \cdot 4H_2O$ 10 mg/L; $ZnSO_4 \cdot 7H_2O$ 10 mg/L; VB_1 200 mg/L; ethyl alcohol absolute 1.5%.

Cultivations

The seed culture of *E. coli* strains was prepared by transferring 1 ml of glycerol stock to 30 ml of LB medium in a 250-ml flask, which was incubated overnight at 37°C and 220 rpm. Then 6% of seed culture was inoculated into 30 ml of MEC medium in a 250 ml flask and incubated at 37°C and 220 rpm for about 36 h. The initial pH of the

Table 4 Primers used in this study for gene cloning and plasmid construction

Primer name	Sequences (5' → 3')	Source
p4hP –pET-28a-S	TGTAGAATTCATGATCTCACCTGCGCA	*P. stutzeri*
p4hP –pET-28a-A	ATATAAGCTTCTAGCTGCCGACCAGCTTC	
p4hP -pEC-XK99E-S	TAATGAATTCGTGAACCCTATGCAAGC	*P. stutzeri*
p4hP -pEC-XK99E -A	ATAGTCTAGATCAGAGATACTGTTGCGG	
p4hB–pET-28a-S	TGATGAATTCCTAGCGCTCGACCAGTTT	*B. bronchiseptica*
p4hB –pET-28a-A	CTGCAAGCTTATGATTTCACCTGCTCAGG	
p4hB -pEC-XK99E-S	TATAGAATTCATGATCTCACCTGCCCAG	*B. bronchiseptica*
p4hB -pEC-XK99E -A	TGTCTCTAGATTATTCCACCAGCTTCAG	
p4hD –pET-28a-S	TATAGAGCTCATGCTGACTCCGACCGA	*Dactylosporangium* sp.
p4hD –pET-28a-A	GATCAAGCTTTTAAACTGGCTGGGCAAG	
p4hD -pEC-XK99E-S	TATAGAGCTCATGCTGACTCCGACCGA	*Dactylosporangium* sp.
p4hD -pEC-XK99E-A	GATCTCTAGATTAAACTGGCTGGGCAAG	

medium was adjusted to 7.4. Induction (0.5 mM IPTG) was performed when the optical density was around 0.5, and the growth temperature was reduced from initial 37°C to 30°C. The cultivation of *C. glutamicum* was similar to that of *E. coli* expect that they were conducted at 30°C in the whole process using MCG medium. Experiments were performed in parallel on the same media without any induction.

Trans-P4H activities

The intracelleluar *trans*-P4H activities were measured by the whole-cell reaction procedures. After 8 hours induction in fermentation medium, cells were harvested by centrifugation at 12000× g for 20 min. The harvested cells were resuspended in the reaction mixture as followed. Each reaction mixture contained 80 mM 2-[N- morpholino] ethanesulfonic acid (MES) buffer (pH 6.5), 4 mM L-proline, 8 mM 2 - ketoglutarate, 2 mM FeSO4, 4 mM L-ascorbic acid The final cell concentration was about 100 g wet weight/L. The reaction mixtures were incubated at 35°C for 10 min and then cellular activity was inactivated completely by heat treatment at 100°C and 5 minutes. The amount of *trans*-4-hydroxy-L-proline in the supernatant of each mixture after centrifugation was determined. The amount of the enzyme which forms 1 nmol of Hyp in one minute was defined as 1 U.

Analytical methods

The cell concentration was determined by measuring the optical density of appropriately diluted sample at 600 nm (*E. coli*, OD_{600}) and 620 nm (*C. glutamicum* OD_{620}) with a UV-visible spectroscopy system (Xinmao, Shanghai, China). Hyp was oxidized by Chloramine T and analyzed using spectrophotometric determination [35].

Statistical analysis

All measurements for growth, *trans*-Hyp production and *trans*-P4H activity were performed in triplicate, and the data were averaged and presented as the mean ± standard deviation.

Additional file

Additional file 1: Supplement.

Competing interests

The authors declare that they have no competing interests.

Authors' contributions

YY participated in the design and conducted the experiments in the study, and drafted the manuscript. HS participated in the analysis of proline and hydroxyproline. ZL conceived the study and helped to draft the manuscript. QY discussed the experiments and manuscript. All authors read and approved the final manuscript.

Acknowledgements

This study was supported by the National High Technology Research and Development Program of China (Grant No's. 2012AA022104 & 2012AA021205), the Fundamental Research Funds for the Central Universities (Grant No's. 222201313007 & 222201314048), the National Special Fund for State Key Laboratory of Bioreactor Engineering (No.2060204).

References

1. Kuttan R, Radhakrishnan AN: **Biochemistry of the hydroxyprolines.** *Adv Enzymol* 1973, **37**:273–347.
2. Mai Hoa BT, Hibi T, Nasuno R, Matsuo G, Sasano Y, Takagi H: **Production of N-acetyl cis-4-hydroxy-L-proline by the yeast N-acetyltransferase Mpr1.** *J Biosci Bioeng* 2012, **114**(2):160–165.
3. Remuzon P: ***Trans* 4-Hydroxy-L-proline, a novel and versatile chiral starting block.** *Tetrahedron* 1996, **52**:13803–13835.
4. Wichmann CF, Liesch JM, Schwartz RE: **L-671,329, a new antifungal agent. II. Structure determination.** *Antibiot* 1989, **42**(2):168–173.
5. Izumi Y, Chibata I, Itah T: **Herstellung und Verwendung von Aminosäuren.** *Angew Chem* 1978, **90**(3):187–194.
6. Gorres KL, Raines RT: **Prolyl 4-hydroxylase.** *Crit Rev Biochem Mol Biol* 2010, **45**:106–124.
7. Hashimoto S, Katsumata R, Ochiai K: **Process for producing optically active 4-hydroxy-2-ketoglutaric acid using microorganisms.** *US Patent.* 5,607,848 [P]. 1997-3-4.
8. Serizawa N, Matsuoka T, Hosoya T, Furuya K: **Fermentation production of trans-4-hydroxy-L-proline by Clonostachys cylindrospora.** *Biosci Biotech Biochem* 1995, **59**(3):555–557.
9. Onishi M, Okumura R, Okamoto R, Ishikura T: **Proline hydroxylation by cell free extract of a streptomycete.** *Biochem Biophys Res Commun* 1984, **120**(1):45–51.
10. Lawrence CC, Sobey WJ, Field RA, Baldwin JE, Schofield CJ: **Purification and initial characterization of proline 4-hydroxylase from *Streptomyces griseoviridus* P8648: a 2-oxoacid, ferrous-dependent dioxygenase involved in etamycin biosynthesis.** *Biochem J* 1996, **313**(1):185–191.
11. Baldwin JE, Field RA, Lawrence CC, Lee V, Robinson JK, Schofield CJ: **Substrate specificity of proline 4-hydroxylase: chemical and enzymatic synthesis of 2*S*, 3*R*, 4*S* –epoxyproline.** *Tetrahedron Lett* 1994, **35**(26):4649–4652.
12. Shibasaki T, Mori H, Ozaki A: **Enzymatic production of trans-4-Hydroxy -L-proline by region- and stereospecific hydroxylation of L-proline.** *Biosci Biotechno Biochem* 2000, **64**(4):746–750.
13. Falcioni F, Blamk LM, Oliver F, Andreas K, Bruno B, Schmid A: **Proline availability regulates proline-4-hydroxylase synthesis and substrate uptake in proline-hydroxylating recombinant *Escherichia coli*.** *Appl Environ Microbiol* 2013, **79**:3091–3100.
14. Eggeling L, Bott M: **Handbook of *Corynebacterium glutamicum*.** In Boca Raton: CRC Press; 2005:1–616.
15. Wolfgang L, Klaus H, Karlheinz D: **Biotechnological production of amino acids and derivatives: current status and prospects.** *Appl Microbiol Biotechnol* 2005, **69**:1–8.
16. Soo YL, Yang HK, Jiho M: **Conversion of phenol to glutamate and proline in *Corynebacterium glutamicum* is regulated by transcriptional regulator ArgR.** *Appl Microbiol Biotechnol* 2010, **85**:713–720.
17. Masaaki W: **Amino Acid Exporters in *Corynebacterium glutamicum*.** *Microbiol Monographs* 2013, **23**:335–349.
18. Jiang LY, Chen SG, Zhang YY, Liu JZ: **Metabolic evolution of Coryne- bacterium glutamicum for increased production of L-ornithine.** *BMC Biotechnol* 2013, **13**:47.
19. Meiswinkela TM, Rittmannb D, Lindnera SN, Wendischa VF: **Crude glycerol-based production of amino acids and putrescine by *Corynebacterium glutamicum*.** *Bioresour Technol* 2013, **145**:254–258.
20. Ikeda M, Nakagawa SL: **The *Corynebacterium glutamicum* genome: features and impacts on biotechnological processes.** *Appl Microbiol Biotechnol* 2003, **62**(2/3):99–109.
21. Kalinowski J, Bathe B, Bartels D, Bischoff N, Bott M, Burkovski A, Dusch N, Eggeling L, Eikmanns BJ, Gaigalat L, Goesmann A, Hartmann M, Huthmacher K, Kramer R, Linke B, McHardy AC, Meyer F, Mockel B, Pfefferle W, Puhler A, Rey DA, Ruchert C, Rupp O, Sahm H, Wendisch VF, Wiegrabe I, Tauch A: **The complete *Corynebacterium glutamicum* ATCC 13032 genome sequence and its impact on the production of L-aspartate-derived amino acids and vitamins.** *J Biotechnol* 2003, **104**(1/3):5–25.

22. Becker J, Wittmann C: **Systems and synthetic metabolic engineering for amino acid production–the heartbeat of industrial strain development.** *Curr Opin Biotechnol* 2012, **23**(5):718–726.
23. Ikeda M, Katsumata R: **Transport of aromatic amino acids and its influence on overproduction of the amino acids in *Corynebacterium glutamicum*.** *J Ferment Bioeng* 1994, **78**(6):420–425.
24. Shibasaki T, Hashimoto S, Mori H, Ozaki A: **Construction of a novel hydroxyproline-producing recombinant Escherichia coli by introducing a proline 4-hydroxylase gene.** *J Biosci Bioeng* 2000, **90**(5):522–525.
25. Shibasaki T, Mori H, Chiba S, Ozaki A: **Microbial Proline 4-Hydroxylase Screening and Gene Cloning.** *Appl Environ Microbiol* 1999, **65**(9):4028–4031.
26. Klein C, Hüttel W: **A simple procedure for selective hydroxylation of l-Proline and l-Pipecolic acid with recombinantly expressed proline hydroxylases.** *Adv Synth Catal* 2011, **353**(8):1375–1383.
27. Hayes CS, Bose B, Sauer RT: **Stop codons preceded by rare arginine codons are efficient determinants of ssrA tagging in *Escherichia coli*.** *Proc Natl Acad Sci U S A* 2002, **99**(6):3440–3445.
28. Gustafsson C, Govindarajan S, Minshull J: **Codon bias and heterologous protein expression.** *Trends Biotechnol* 2004, **22**:346–353.
29. Ahn SJ, Seo JS, Kim MS, Jeon SJ, Kim NY, Jang JH, Kim KH, Hong YK, Chung JK, Lee HH: **Cloning, site-directed mutagenesis and expression of cathepsin L-like cysteine protease from Uronema marinum (Ciliophora: Scuticociliatida).** *Mol Biochem Parasitol* 2007, **156**(2):191–198.
30. Retnoningrum DS, Pramesti HT, Santika PY, Valerius O, Asjarie S, Suciati T: **Codon optimization for high level expression of human bone morphogenetic protein- 2 in *Escherichia coli*.** *Protein Expr Purif* 2012, **84**:188–194.
31. Blank LM, Ebert BE, Buehler K, Bühler B: **Redox Biocatalysis and Metabolism: Molecular Mechanisms and Metabolic Network Analysis.** *Antioxid Redox 527 Signal* 2010, **13**:349–394.
32. Tsuchida T, Kubota K, Yoshinage F: **Improvement of L-proline production by sulfaguanidine resistant mutants derived from L-glutamic acid-producing bacteria.** *Agric Biol Chem* 1986, **50**(9):2201–2207.
33. Nakanishi T, Hagino H: **Process for producing L-proline by fermentation.** *US Patent* 1984, **4**:444,885.
34. Yu BQ, Shen W, Zhuge J: **An improved method for integrative electro transformation of *Corynebacterium glutamicum* with xenogeneic DNA.** *China Biotechnol* 2005, **25**(2):78–81
35. Yin MW, Nan YM, Wang XM: **Improvement of the spectrophotometric method for the determination of hydroxyproline.** *J Henan Med Univ* 1994, **29**:74–77.

12

Simultaneously improving xylose fermentation and tolerance to lignocellulosic inhibitors through evolutionary engineering of recombinant *Saccharomyces cerevisiae* harbouring xylose isomerase

Justin Smith, Eugéne van Rensburg and Johann F Görgens*

Abstract

Background: Yeasts tolerant to toxic inhibitors from steam-pretreated lignocellulose with xylose co-fermentation capability represent an appealing approach for 2nd generation ethanol production. Whereas rational engineering, mutagenesis and evolutionary engineering are established techniques for either improved xylose utilisation or enhancing yeast tolerance, this report focuses on the simultaneous enhancement of these attributes through mutagenesis and evolutionary engineering of *Saccharomyces cerevisiae* harbouring xylose isomerase in anoxic chemostat culture using non-detoxified pretreatment liquor from triticale straw.

Results: Following ethyl methanesulfonate (EMS) mutagenesis, *Saccharomyces cerevisiae* strain $D5A^{+}$ (ATCC 200062 strain platform), harbouring the xylose isomerase (XI) gene for pentose co-fermentation was grown in anoxic chemostat culture for 100 generations at a dilution rate of 0.10 h^{-1} in a medium consisting of 60% (v/v) non-detoxified hydrolysate liquor from steam-pretreated triticale straw, supplemented with 20 g/L xylose as carbon source. In semi-aerobic batch cultures in the same medium, the isolated strain $D5A^{+H}$ exhibited a slightly lower maximum specific growth rate ($\mu_{max} = 0.12 \pm 0.01$ h^{-1}) than strain TMB3400, with no ethanol production observed by the latter strain. Strain $D5A^{+H}$ also exhibited a shorter lag phase (4 h vs. 30 h) and complete removal of HMF, furfural and acetic acid from the fermentation broth within 24 h, reaching an ethanol concentration of 1.54 g/L at a yield ($Y_{p/s}$) of 0.06 g/g xylose and a specific productivity of 2.08 g/gh. Evolutionary engineering profoundly affected the yeast metabolism, given that parental strain $D5A^{+}$ exhibited an oxidative metabolism on xylose prior to strain development.

Conclusions: Physiological adaptations confirm improvements in the resistance to and conversion of inhibitors from pretreatment liquor with simultaneous enhancement of xylose to ethanol fermentation. These data support the sequential application of random mutagenesis followed by continuous culture under simultaneous selective pressure from inhibitors and xylose as primary carbon source.

Keywords: *Saccharomyces cerevisiae*, Yeast hardening, Evolutionary engineering, Random mutagenesis, Triticale hydrolysate, EMS, Lignocellulose

* Correspondence: jgorgens@sun.ac.za
Department of Process Engineering, Stellenbosch University, Private Bag X1, Matieland 7602, South Africa

Background

Saccharomyces cerevisiae remains the preferred microbe for producing ethanol from pretreated lignocellulose, given its general robustness, high ethanol production rates and ethanol tolerance [1-3]. Critical interventions required to enhance the efficiency of this yeast for commercial 2nd generation ethanol production include (i) introducing capability to ferment xylose [1,4,5] and (ii) enhancing tolerance to toxic by-products from steam pretreatment [1,6-9].

Two approaches can be followed for introducing xylose utilising capability in *S. cerevisiae*, namely the cloning of xylose reductase (XR) and xylitol dehydrogenase (XDH) or cloning xylose isomerase (XI), usually in conjunction with xylulokinase (XK) to convert xylulose to xylulose-5-phosphate. These rational metabolic engineering approaches were exemplified in the respective benchmark studies of Wahlbom *et al.* [10] and Kuyper *et al.* [11]. Other examples where rational approaches were followed for improved xylose metabolism include studies by Gorsich *et al.* [12] and Toivari *et al.* [13]. Enhancing yeast tolerance, or "hardening" requires a more comprehensive intervention employing rational approaches to genome modification, random mutagenesis and directed evolutionary engineering under selective pressure (see excellent reviews by Sauer [14] and Nevoigt [15]). It should be noted, however, that these techniques can also be employed to enhance xylose utilisation in the absence of hydrolysate inhibitors (see below), which suggests technique overlap in achieving both improved xylose utilisation and inhibitor tolerance. Furthermore, rational approaches are often challenging due to the complexity of the yeast genome, requiring an in-depth understanding of the metabolome and its response to dynamic inputs, making the investigator reliant on a variety of "-omics" technologies [15]. Rational approaches were also occasionally found insufficient to instil required capabilities in the yeast phenotype [16]. As such, random mutagenesis and selection followed by evolutionary engineering is often the methodological sequence of choice, where the former could lead to phenotypes with enhanced capabilities for either xylose utilisation or inhibitor tolerance but without prior knowledge of specific metabolic pathways. The latter allows for selection under process-relevant conditions [14], especially where inhibitors from lignocellulosic pretreatment are present.

In the literature, strain development predominantly focused on the XR/XDH/XK system where strains with the TMB prefix, produced by Swedish researchers, featured quite prominently. *S. cerevisiae* TMB3400 produced by Wahlbom *et al.* [10] was later subjected to evolutionary engineering using furfural in sequential batch culture [17], and in another study to sequential batch and continuous culture at a dilution rate (D) of 0.05 h^{-1} using spruce hydrolysate, subsequent to UV mutagenesis [18]. Similarly, *S. cerevisiae* strain TMB3001 [19] also harbouring XR/XDH/XK was subjected to evolutionary engineering in continuous culture (D = 0.05 h^{-1}) in the absence of inhibitors to enhance xylose fermentation [20], as well as in continuous culture (D = 0.1 h^{-1}) in the presence of hydrolysate [21,22]. Other strains of *S. cerevisiae* were also subjected to either mutagenesis [23] or evolutionary engineering [24,25] or both [26] (all in the absence of inhibitors), with a few instances where other yeast species were also investigated for enhanced xylose utilisation, including *Scheffersomyces* (*Pichia*) *stipitis* [27] and *Pachysolen tannophilus* [28]. Amartey and Jeffries [29] described enhanced tolerance by a *Pichia* strain after sequential subculture in different concentrations of corn cob hydrolysate, after overliming with $Ca(OH)_2$ and in an innovative approach, Almario *et al.* [30] used visualising evolution in real-time (VERT) to assess molecular mechanisms associated with short term adaptation [31] in batch culture.

Noteworthy studies pertaining to strain development harbouring the xylose isomerase (XI) only appeared after the successful expression of XI from *Piromyces* in *S. cerevisiae* [11] followed by evolutionary engineering in oxygen-limited chemostat culture at D = 0.06 h^{-1} for 2 000 hours in the absence of hydrolysate. Similar work was published by Wisselink *et al.* [32] and Zhou *et al.* [33] using sequential batch and continuous culture. In these studies, cassettes containing several enzymes from the pentose phosphate pathway and for arabinose utilisation were transformed into *S. cerevisiae*. However, in none of these studies were hydrolysates used as additional selective pressure criterion for evolutionary engineering of the xylose fermenting yeast in continuous culture. In fact, to the knowledge of the authors, there appeared only one study where close to the full range of strain development techniques was applied to an XI-containing strain [16]. Using *S. cerevisiae* Ethanol Red as platform, the by now classical approach commenced with rational metabolic engineering through transformation with a cassette containing the XI gene from *Clostridium phytofermentans*, followed by random mutagenesis using EMS with hydrolysate and xylose as selective criteria. Gene shuffling preceded evolutionary engineering carried out in sequential batch culture without hydrolysate but at very high concentrations of xylose (40 g/L). Using a mating approach in a follow-up study [34], this strain designated GS1.11-26 was mated with another tolerant yeast strain after screening 580 yeast strains using dilute acid-pretreated spruce hydrolysate. Three resulting strains, designated as GSF335, GSF767 and GSE16 revealed a marked improved phenotype compared to Ethanol Red, producing up to 23% (v/v) more ethanol.

The purpose of the present study was to improve the inhibitor tolerance and simultaneously enhance xylose fermentation to ethanol of a recombinant xylose-utilising *S. cerevisiae* strain D5A$^+$ harbouring the *xylA* gene from

Bacteriodes thetaiotaomicron producing XI. "Hardening" was accomplished through a combination of random mutagenesis with EMS and long-term evolutionary engineering in chemostat culture using xylose as carbon source and liquor from steam-pretreated triticale straw as selective criteria at both steps (mutagenesis and chemostat culture). The degree of hardening achieved was evaluated through comparison of the fermentative performance of the hardened yeast to the original parental strain, during exposure to pretreatment liquor supplemented with either glucose or xylose in batch culture. Two additional non-recombinant *S. cerevisiae* strains MEL2 and MH1000, as well as *S. cerevisiae* strain TMB3400, were included for comparison. Finally, the fermentative performance of the hardened yeast under SSF conditions was assessed using pressed steam-pretreated sweet sorghum bagasse as substrate.

Results

Chemical characterisation of steam pretreated triticale straw and sweet sorghum bagasse

The chemical composition of the liquor from steam-pretreated triticale straw (used in continuous culture) and sweet sorghum bagasse (used in SSF experiments) are shown in Table 1. The transition between the two feedstocks was required due to limited material availability. Xylose was the most abundant sugar in both the triticale and sorghum pretreatment liquor fractions. This result supported the general observation that predominantly hemicellulose is solubilised during steam pretreatment [35,36] although it should be noted that no acidic catalyst was used during pretreatment in the present study. After steam pretreatment the WIS fractions were first subjected to complete acid hydrolysis before chemical analysis. Glucose was the predominant sugar in the hydrolysed sweet sorghum WIS fraction, which was expected since the WIS fraction of the pretreated material was enriched for both lignin and cellulose, as previously documented [37,38]. Acetic acid was the most abundant by-product from steam-pretreatment, followed by furfural and HMF, in both the triticale and sweet sorghum pretreatment liquor fractions. Although the acetic acid concentration was significantly greater (17%) in the sorghum pretreatment liquor, the sugar degradation products furfural and HMF were 1.5 and 1.8 times more concentrated in the triticale pretreatment liquor. Phenolic compounds and their respective degradation products were not quantified in this study. Given that HMF, furfural and acetic acid are the main by-products responsible for inhibition of microbial metabolism [39], specific emphasis was placed on the concentration of these by-products as an indicator of pretreatment liquor toxicity.

Table 1 Chemical composition of steam-pretreated triticale straw and sweet sorghum bagasse

	Triticale pretreatment liquor (g/L)	Sweet sorghum pretreatment liquor (g/L)	Sweet sorghum water-insoluble solids (g/100 g)
By-products			
Acetic acid	3.34 ± 0.10	4.01 ± 0.30	-
Formic acid	0.94 ± 0.10	0.72 ± 0.20	-
HMF	0.44 ± 0.01	0.23 ± 0.02	-
Furfural	1.63 ± 0.05	1.08 ± 0.05	-
Sugars			
Glucose	0.71 ± 0.05	0.27 ± 0.03	51.48 ± 0.3
Xylose	3.24 ± 0.50	2.70 ± 0.06	12.29 ± 0.23
Cellobiose	0.16 ± 0.01	0.59 ± 0.01	10.49 ± 1.22
Arabinose	0.24 ± 0.02	0.67 ± 0.09	0.46 ± 0.10

Conditions of steam pretreatment for triticale straw and sweet sorghum bagasse were 200°C for 20 min and 180°C for 5 min, respectively. The chemical composition of the water insoluble solid fraction was determined after washing and subsequent to acid hydrolysis. The composition of pretreated triticale straw WIS was excluded from the table since this fraction was not utilised in this study. Standard deviations from duplicate analyses are also shown.

Random mutagenesis

Yeast cells that survived chemical mutagenesis after exposure to 2 and 3 µl EMS/mL for 1 h were pooled and the growth compared to that of the parental D5A$^+$ strain in a chemically-defined medium supplemented with diluted (67% v/v) and undiluted triticale pretreatment liquor with 20 g/L xylose as carbon source (Figure 1). Xylose in combination with the pretreatment liquor was essential for selecting mutants displaying functional xylose utilisation together with increased inhibitor tolerance. As evident from Figure 1, mutants of strain D5A$^+$ reached substantially greater total cell counts (±2-fold greater) than the parental D5A$^+$ strain in the diluted liquor, whereas the undiluted liquor resulted in decreased cell counts, possibly resulting from cell lysis due to the harshness of the medium. The dark medium colour precluded the use of turbidity to quantify culture growth, necessitating the use of serial dilutions and plate counts. Based on this data, a triticale hydrolysate concentration of 67% (v/v) was regarded as the maximum sub-lethal concentration to which the yeast could be exposed.

Evolutionary engineering using continuous culture

Mutated cells were subjected to evolutionary engineering using continuous culture at an initial dilution rate of 0.05 h^{-1} for one week (Table 2). This low initial dilution rate would allow adaptation of cells to a relatively high pretreatment liquor concentration of 50% (v/v) while maintaining selective pressure, albeit at pretreatment liquor levels lower than the maximum sub-lethal concentration. Plate counts were used to confirm that the culture did not wash out of the reactor, upon which the dilution rate was incrementally adjusted over a further

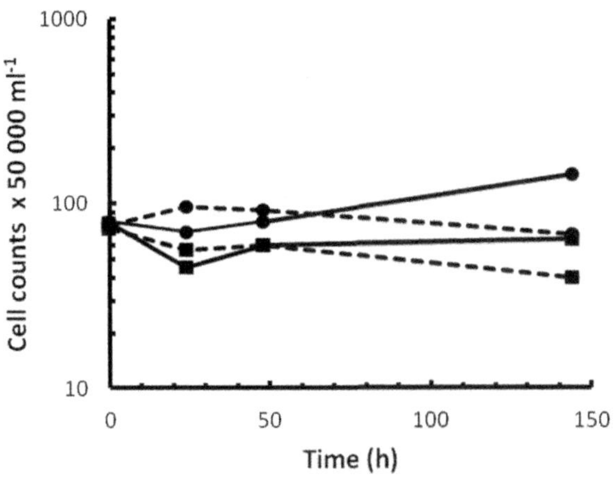

Figure 1 Comparison of parental and mutated strain in batch culture. Batch growth profiles for the *S. cerevisiae* parental strain D5A$^+$ (squares) and mutant strain D5A$^+$ (circles) in water-diluted (67% v/v, solid line) and undiluted (broken line) liquor from steam-pretreated triticale supplemented with 20 g xylose/L in shake flask cultures. Culture growth was quantified as total cell counts using a counting chamber. Data represents the average values from duplicate counts.

one week period to a dilution rate of 0.1 h^{-1}. This dilution rate was maintained at a hydrolysate concentration of 50% (v/v) for 49 generations (two weeks), after which the hydrolysate concentration was increased to 60% (v/v) for a further 100 generations (four weeks). Substantial washout of the culture occurred at dilution rates greater than 0.125 h^{-1}, which approximated the μ_{max} value of 0.12 h^{-1}, recorded in batch culture (see below). Although attempts were made to attain steady state at a dilution rate close to 0.125 h^{-1}, the proximity of this dilution rate to the critical dilution rate resulted in large variation in cell counts over a two week period at a hydrolysate concentration of 60% (v/v, Table 2). The absence of total wash-out could possibly be attributed to substantial wall growth after continuous growth for several weeks. Due to constraints in pretreatment liquor availability, no further increases in the concentration above 60% (v/v) were attempted. The structure of the population was not characterised, where a single colony was isolated after two rounds of subculture on YPX agar (in the absence of hydrolysate inhibitors) and used for subsequent work. This strategy was based on the premise that after 220 generations (Table 2) under stringent selective conditions (high inhibitor concentration, xylose as primary carbon source), the culture would be predominantly comprised of strains with improved performance and hence, a greater probability of isolating a more resistant strain. As the results showed, this was indeed the case. Developing an appropriate screening procedure of isolates with beneficial phenotypes is an ongoing activity in our group.

Analysis of strain performance in semi-aerobic batch culture

The strain isolated from continuous culture after 100 generations was designated as strain D5A^{+H} and was grown in batch culture on either xylose or glucose to assess its ability to utilise these carbon sources in the presence of pretreatment liquor. Culture growth was also compared to three other strains of *S. cerevisiae* namely strains MEL2 [40], MH1000 [41] and TMB3400 [10], also supplemented with liquor from steam-pretreated triticale. Strain D5A$^+$ (parental recombinant strain not subjected to mutagenesis) was included as control on both carbon sources in the presence of liquor. As additional control, the parental D5A$^+$ strain was also grown in the defined medium without hydrolysate, with either xylose or glucose as carbon source. In all other experiments the different strains were grown at a pretreatment liquor concentration of 50% (v/v). Fermentation profiles of the recombinant xylose-utilising yeast grown on xylose are shown in Figure 2, with the fermentation profiles on glucose shown in Figure 3. Tables 3 and 4 provide a summary of fermentation parameters calculated from duplicate batch experiments.

In the presence of 50% pretreatment liquor with 20 g/L xylose as carbon source, strain D5A^{+H} displayed a marked improvement in terms of culture growth compared to the control strain as evident from a more than 10-fold increase in biomass concentration (Figure 2E). Strain D5A^{+H} also

Table 2 Cell counts at steady state

Dilution rate (h^{-1})	CFU/mL	Pretreatment liquor (%, v/v)	Time (weeks)	Generations
Inoculation	$1.8 \times 10^6 \pm 0.42 \times 10^6$	50	0	-
0.05	$3.5 \times 10^7 \pm 7.8 \times 10^6$	50	1	13
0.1	$38 \times 10^6 \pm 4.8 \times 10^6$	50	2	49
0.1	$36 \times 10^6 \pm 6.4 \times 10^6$	60	4	97
0.125	$41 \times 10^6 \pm 17 \times 10^6$	60	2	61
Total generations				220

Steady state biomass concentrations, expressed as plate counts (CFUs/mL) for the mutated strain D5A$^+$ at increasing dilution rates with the corresponding time at which the chemostat was operated at each dilution rate. The biomass concentration at the time of inoculation is shown for comparison. Standard deviations represent triplicate determinations corresponding to measurements at three consecutive residence times.

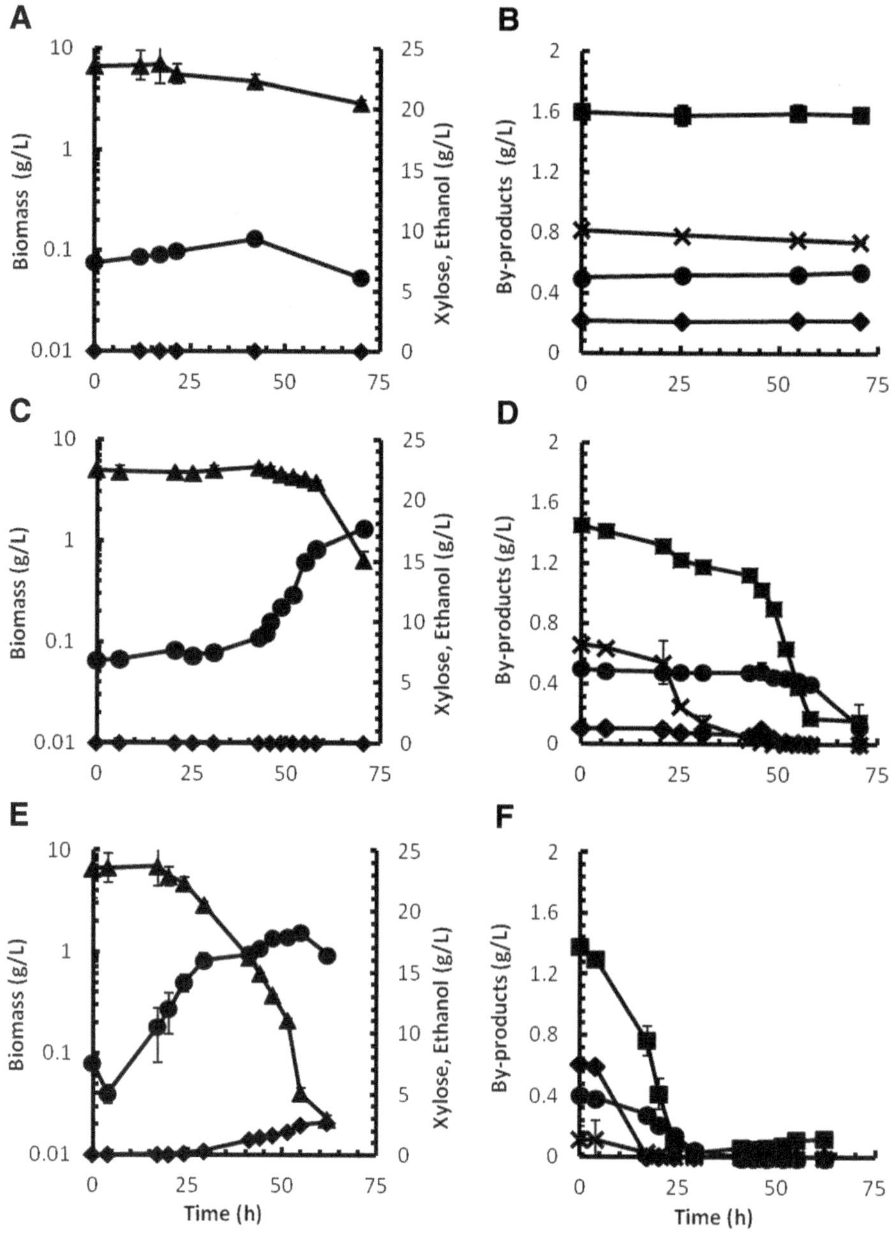

Figure 2 Batch culture profiles after hardening with xylose as carbon source. Batch culture profiles for the recombinant xylose-utilising *S. cerevisiae* strains D5A$^+$ (parental strain, **A** and **B**), TMB3400 **(C and D)** and D5A^{+H} (hardened strain, **E** and **F**) in a growth medium supplemented with 50% (v/v) pretreatment liquor from steam-pretreated triticale and 20 g xylose/L in shake flask cultures. Figure symbols left-hand column: biomass (circles), ethanol (diamonds), xylose (triangles). Figure symbols right-hand column: acetic acid (squares), formic acid (circles), HMF (diamonds) and furfural (crosses). Error bars represent standard deviations from duplicate experiments.

exhibited a significantly shorter lag phase of less than 4 h compared to that of strain TMB3400 (Figure 2C), which suggested that strain D5A^{+H} exhibited a greater level of tolerance towards the inhibitors in the growth medium. However, there was no significant difference between the μ_{max} values of strains D5A^{+H} and TMB3400 (Table 3). The performance of strain D5A^{+H} stood in stark contrast to the poor growth and xylose utilisation of strain D5A$^+$ (parental strain, not subjected to hardening, Figure 2A) in the presence of inhibitors. Furthermore, in the absence of inhibitors, the parental D5A$^+$ strain produced no ethanol, whereas an ethanol concentration of up to 1.54 g/L at an ethanol yield ($Y_{p/s}$) of 0.06 g/g xylose was recorded for strain D5A^{+H} in the presence of inhibitors. Although low levels of glucose was present in the hydrolysate from pretreatment (0.71 g/L, Table 1), this low concentration was insufficient to support an ethanol titre of 1.54 g/L (Table 3), pointing to improved ethanol fermentation from

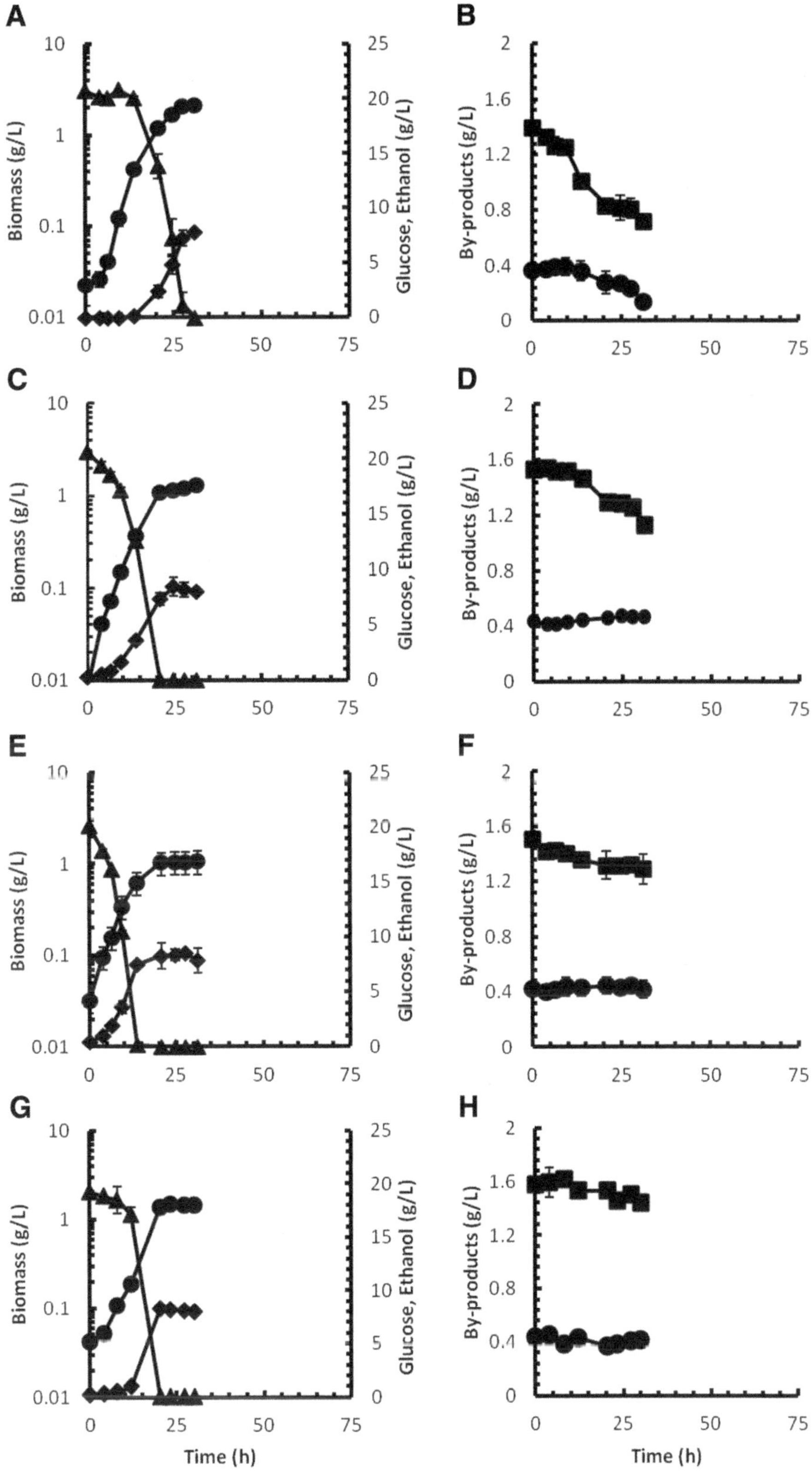

Figure 3 (See legend on next page.)

(See figure on previous page.)
Figure 3 Batch culture profiles after hardening with glucose as carbon source. Batch cultivation profiles for *S. cerevisiae* strains D5A^{+H} **(A and B)**, TMB3400 **(C and D)**, MH1000 **(E and F)** and MEL2 **(G and H)** in 50% (v/v) triticale pretreatment liquor supplemented with 20 g glucose/L in shake flask cultures. Figure symbols: biomass (circles), glucose (triangles), ethanol (diamonds), acetic acid (squares), formic acid (circles). Error bars represent standard deviations from duplicate experiments.

xylose by this strain subsequent to mutagenesis and evolutionary engineering.

Xylose consumption by strain D5A^{+H} closely followed culture growth attesting to its use as primary carbon source. On the other hand, marked xylose consumption by strain TMB3400 only occurred towards the end of the cultivation period (Figure 2C), with the magnitude of xylose consumption apparently not strongly correlated with the increase in biomass concentration. In fact, the data revealed that strain D5A^{+H} consumed 33% more xylose supplied to the culture than that consumed by strain TMB3400 (Table 3). This level of consumption was also reflected in the more than 7-fold greater biomass yield ($Y_{x/s}$) on xylose by the latter strain (Table 3), which suggested that a carbon source other than xylose was preferentially utilised by strain TMB3400. Conversely, the low $Y_{x/s}$ for strain D5A^{+H} might be attributable to ethanol production. Whereas no ethanol production was recorded for strain TMB3400 (Figure 2C) nor for strain D5A^{+} in the absence of liquor (Table 3), an ethanol concentration of 1.54 g/L was recorded for strain D5A^{+H} (Figure 2E), which pointed to increased efficiency in the conversion of xylose to ethanol, especially considering that the $Y_{x/s}$ and $Y_{p/s}$ values (biomass and ethanol on xylose) for strain D5A^{+H} were relatively similar (Table 3).

Table 3 Growth parameters with xylose as carbon source

Parameter	No liquor	Supplemented with liquor		
	D5A^{+}	D5A^{+}	D5A^{+H}	TMB3400
μ_{max} (h^{-1})	0.37 ± 0.002	N/D	0.12 ± 0.01	0.14 ± 0.01
Biomass[†] (g/L)	2.08 ± 0.01	N/D	0.82 ± 0.1	0.84 ± 0.01
Ethanol[†] (g/L)	N/D	N/D	1.54 ± 0.14	N/D
Glycerol[†] (g/L)	0.15 ± 0.01	N/D	2.7 ± 0.27	0.19 ± 0.05
$Y_{x/s}$[‡] (g/g)	0.55 ± 0.023	N/D	0.07 ± 0.001	0.51 ± 0.06
$Y_{p/s}$[§] (g/g)	N/D	N/D	0.06 ± 0.001	N/D
$Y_{p/x}$[¶] (g/g)	N/D	N/D	1.9 ± 0.03	N/D
q_s[*] (g/gh)	0.17 ± 0.003	N/D	2.08 ± 0.05	0.23 ± 0.02
q_p[††] (g/gh)	N/D	N/D	0.02 ± 0.003	N/D

Yield coefficients and specific rates of growth, substrate consumption and ethanol production for strains D5A^{+}, D5A^{+H} and TMB3400 with 20 g xylose/L as primary carbon source in batch culture with and without with 50% (v/v) hydrolysate liquor from steam-pretreated triticale. Standard deviations from duplicate fermentations are shown.
N/D None detected.
[†]Biomass and product concentrations at the end of the exponential growth phase.
[‡]$Y_{x/s}$, biomass yield on xylose, calculated from the slope of the biomass concentration plotted as a function of the residual xylose concentration.
[§]$Y_{p/s}$, ethanol yield on xylose, calculated from the slope of the ethanol concentration plotted as a function of the residual xylose concentration.
[¶]$Y_{p/x}$, ethanol yield on biomass, calculated from the slope of the ethanol concentration plotted as a function of the biomass concentration.
[*]q_s, specific rate of substrate utilisation in g xylose/g biomass per hour during the mid-exponential phase.
[††]q_p, specific rate of product formation in g ethanol/g biomass per hour during the mid-exponential phase.

The by-product time profiles for strains D5A^{+} TMB3400 and D5A^{+H} are shown in Figure 2B, D and F, respectively. The largest decrease in by-products coincided with the exponential growth phase of strains D5A^{+H} and TMB3400. This inverse correlation was especially evident in the case of strain TMB3400 where a distinct biphasic profile was evident and could be related to the longer lag phase of this culture, which lasted for up to 42 h. The shorter lag phase of strain D5A^{+H} coincided with earlier commencement of by-product utilisation and/or detoxification. No significant decrease in either organic acids or furaldehydes was observed in the D5A^{+} control culture, attributable to the absence of culture growth (Figure 2A and Table 3).

Using glucose as carbon source, similar growth profiles were recorded for strains D5A^{+H} (Figure 3A) and TMB3400 (Figure 3C) in the presence of 50% (v/v) pretreatment liquor, which corresponded to that of the two reference strains, MH1000 and MEL2. The μ_{max} for strain D5A^{+H} in the presence of inhibitors was marginally greater than that of the other strains (Table 4), although a short lag phase of 4 h was apparent for this strain compared to the reference strains. The μ_{max} for the parental D5A^{+} strain was more than 20-fold lower than that recorded for strain D5A^{+H} (Table 4). Although the $Y_{x/s}$ between the strains varied (Table 4), the $Y_{x/s}$ value of strain D5A^{+H} was significantly greater than that of the other strains and could be related to the significantly greater biomass concentration recorded at the end of the exponential growth phase (Figure 3A, Table 4) and was corroborated by the significantly lower ethanol yield on glucose ($Y_{p/s}$). In other words, strain D5A^{+H} clearly produced less ethanol relative to biomass ($Y_{p/x}$), although the greater biomass concentration of this strain resulted in an ethanol titre similar to that of the other strains at the end of the exponential growth phase. Due to a severe lag phase in excess of 48 h recorded for the D5A^{+} parental strain, the data for this strain was omitted from Figure 3. The by-product profiles displayed in Figure 3B, D, F and H showed a substantial decrease in the acetic acid concentration for strains D5A^{+H} (50%), TMB3400 (27%), MH1000 (20%) and MEL2 (<10%), respectively. No clear pattern of formic acid removal was

Table 4 Growth parameters with glucose as carbon source

Parameter	No liquor	Supplemented with liquor				
	D5A^{+}	D5A^{+}	D5A^{+H}	TMB3400	MEL2	MH1000
μ_{max} (h^{-1})	0.56 ± 0.01	0.03 ± 0.01	0.29 ± 0.02	0.22 ± 0.00	0.21 ± 0.01	0.23 ± 0.01
Biomass† (g/L)	1.42 ± 0.06	1.18 ± 0.05	2.13 ± 0.04	1.32 ± 0.02	1.46 ± 0.02	1.07 ± 0.02
Ethanol† (g/L)	7.77 ± 0.15	8.86 ± 0.30	7.76 ± 0.13	8.48 ± 0.14	8.28 ± 0.02	8.55 ± 0.50
Glycerol† (g/L)	0.67 ± 0.01	0.44 ± 0.00	0.51 ± 0.01	0.46 ± 0.01	0.39 ± 0.07	0.43 ± 0.06
$Y_{x/s}$‡ (g/g)	0.08 ± 0.00	0.06 ± 0.01	0.1 ± 0.00	0.06 ± 0.01	0.08 ± 0.00	0.05 ± 0.00
$Y_{p/s}$§ (g/g)	0.42 ± 0.01	0.43 ± 0.03	0.37 ± 0.01	0.42 ± 0.01	0.43 ± 0.00	0.44 ± 0.02
$Y_{p/x}$¶ (g/g)	5.18 ± 0.03	5.37 ± 0.2	3.58 ± 0.2	6.16 ± 0.7	6.08 ± 0.06	6.85 ± 0.3
q_s* (g/gh)	3.66 ± 0.07	1.6 ± 0.14	1.1 ± 0.1	3.03 ± 0.1	4.64 ± 0.38	6.4 ± 0.1
q_p†† (g/gh)	2.25 ± 0.00	0.69 ± 0.00	0.41 ± 0.00	0.54 ± 0.03	0.85 ± 0.03	1.8 ± 0.02

Yield coefficients and specific rates of growth, substrate consumption and ethanol production for strains D5A^{+}, D5A^{+H}, TMB3400, MH1000 and MEL2 with 20 g glucose/L as primary carbon source in batch culture with and without with 50% (v/v) hydrolysate liquor from steam-pretreated triticale. Standard deviations from duplicate fermentations are shown.

†Biomass and product concentrations at the end of the exponential growth phase.

‡$Y_{x/s}$, biomass yield on glucose, calculated from the slope of the biomass concentration plotted as a function of the residual glucose concentration.

§$Y_{p/s}$, ethanol yield on glucose, calculated from the slope of the ethanol concentration plotted as a function of the residual glucose concentration.

¶$Y_{p/x}$, ethanol yield on biomass, calculated from the slope of the ethanol concentration plotted as a function of the biomass concentration.

*q_s, specific rate of substrate utilisation in g glucose/g biomass per hour during the mid-exponential phase.

††q_p, specific rate of product formation in g ethanol/g biomass per hour during the mid-exponential phase.

distinguishable for strains TMB3400, MH1000 and MEL2, whereas the formic acid concentration in the culture of strain D5A^{+H} decreased by more than 65% over the cultivation period.

Supplying either glucose or xylose as the primary carbon source distinctly affected the yeast metabolism, evident from the absolute differences in the fermentation parameters presented in Tables 3 and 4. The most prominent differences between strains D5A^{+} and TMB 3400 are highlighted here. In the absence of pretreatment liquor with xylose as carbon source, the μ_{max} of strain D5A^{+} was 34% lower than with glucose, whereas the biomass yield coefficient was almost 7-fold greater. Ethanol production was absent with xylose as carbon source, whereas comparatively high levels of ethanol was produced on glucose. A similar response was evident for strain TMB3400 where the μ_{max} value on xylose was 36% lower than on glucose, with a substantially lower $Y_{x/s}$ value on glucose (>8-fold difference), and complete absence of ethanol formation. Although these perturbations were less pronounced in the case of strain D5A^{+H}, the most prominent difference was the ability of the latter strain to produce ethanol when xylose was the sole source of carbon. However, xylose was not as effectively utilised as a fermentable carbon source compared to glucose ($Y_{EtoH/Glc} >> Y_{EtoH/Xyl}$).

Strain performance during simultaneous saccharification and fermentation

Fermentation profiles for strains D5A^{+H} and TMB3400 during simultaneous saccharification and fermentation (SSF) are presented in Figure 4 using the water-insoluble solids (WIS) from steam-pretreated sweet sorghum bagasse as carbon source. Although it would have been preferable to continue with triticale solids we were restricted to using limited amounts of sorghum solids, whereas most hydrolysate from triticale was consumed during continuous culture experiments. Given that this section of the work was to provide a first order assessment of strain performance in SSF in the presence of inhibitors, we opted for an approach where triticale hydrolysate was used for hardening and sorghum was used for SSF work, in light of the relatively similar inhibitor concentrations in the hydrolysate from both feedstocks. Although the pretreatment liquor was separated from the WIS using a mechanical press, the solids were not washed prior to addition to the growth medium, which implied that a portion of the inhibitory by-products from the pretreatment liquor were carried into the SSF medium. A fed-batch fermentation strategy was employed to reduce broth viscosity and minimise exposure of the culture to high inhibitor levels, where feed times are indicated by the arrows.

On sweet sorghum bagasse solids, strain D5A^{+H} produced 45% more ethanol than strain TMB3400 reaching a maximum of 19.22 g/L after 150 h. Although this concentration equated to a very low 50% of the theoretical maximum, accounting for both glucose and xylose in the WIS and assuming a sugar to ethanol conversion factor of 0.51 g/g, this data served to illustrate the difference in the performance of the yeasts, since much room for SSF optimisation remains. For both strains, ethanol increased at the greatest rate before the third solids feed at 48 h, followed by a substantial decrease in the ethanol production rate. The profiles of residual glucose and xylose concentrations also differed distinctly between the two strains. For strain TMB3400, residual glucose levels

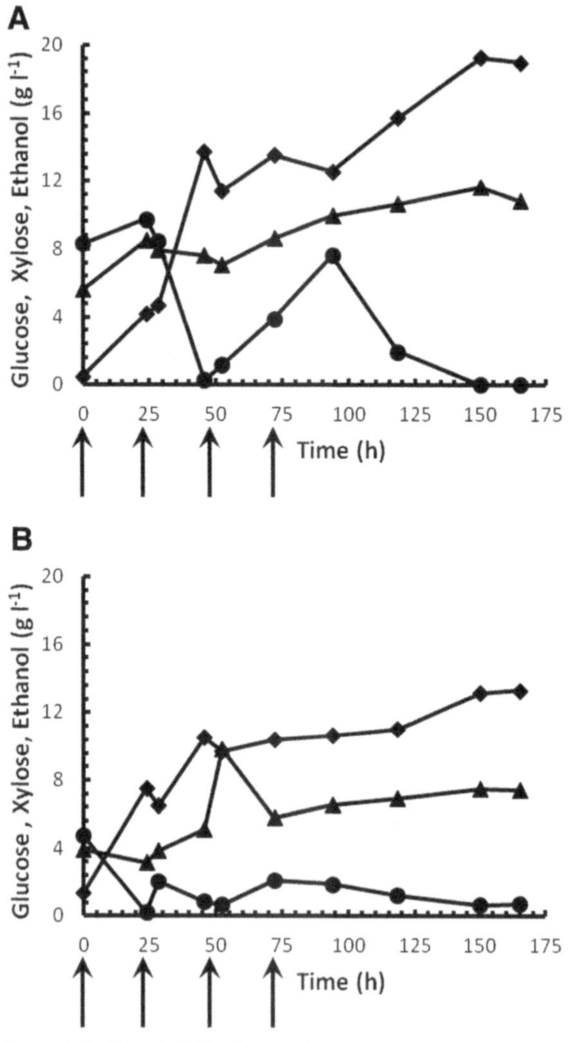

Figure 4 Fed-batch SSF culture using steam-pretreated sweet sorghum bagasse. Single fed-batch SSF experiments of pressed solids fed-batch using *S. cerevisiae* strains $D5A^{+H}$ **(A)** and TMB 3400 **(B)**. Figure symbols: ethanol (diamonds), glucose (circles) and xylose (triangles). Arrows indicate the addition of substrate in 2.5% (g/v) increments at 0, 24, 48 and 72 h.

remained consistently below 2 g/L throughout the fermentation, in spite of the addition of substrate at 24, 48 and 72 hrs. In the case of strain $D5A^{+H}$, glucose accumulated to levels above 7.5 g/L directly after the addition of substrate at 24 and 48 hrs, after which consumption again commenced. The accumulation in residual glucose in the culture of strain $D5A^{+H}$, and to a lesser extent with strain TMB3400, coincided with the levelling off of ethanol production. With strain $D5A^{+H}$ a further increase in ethanol production again coincided with an increase in residual glucose consumption which occurred after approx. 96 h, whereas a less pronounced increase in ethanol production was evident for strain TMB3400 after 118 h. No residual glucose was detected in the culture of strain $D5A^{+H}$ after 165 h, whilst a residual glucose concentration of 0.73 g/L was recorded in the culture of strain TMB 3400. This observation suggested strong recovery by strain $D5A^{+H}$ from introduced toxic by-products after the third feed, especially in light of the final ethanol titres recorded in this culture, whereas the behaviour of strain TMB3400 remained more consistent throughout the cultivation. Xylose accumulated throughout the fermentation of both strains TMB3400 and $D5A^{+H}$, although the final residual xylose concentration was more than 35% (7.5 g/L) lower for strain TMB3400 when compared to strain $D5A^{+H}$ (11.6 g/L).

Discussion

Two key strain development techniques, namely random mutagenesis and evolutionary engineering were sequentially used to harden a recombinant *S. cerevisiae* strain $D5A^{+}$ against inhibitors from steam-pretreated triticale straw. Simultaneously, selective pressure through xylose as carbon source was imposed during both steps to enhance the metabolism of this pentose sugar via the XI pathway. Xylose isomerase represents the advantage of avoiding co-factor imbalances in the microbial metabolism at low oxygen tension, as opposed to the XR/XDH pathway, where XR was found to prefer NADPH as co-factor [11,24,42]. Furthermore, during evolutionary engineering, Koppram *et al.* [18] observed decreased xylose consumption by strain TMB3400 (XR/XDH system), apparently due to competition for reductive power between XR and furaldehyde. An expected outcome from the present study would thus be improved xylose fermentation to ethanol in semi-aerobic culture in the presence of pretreatment liquor from XI expression.

Given the complexity of the yeast metabolism on xylose as carbon source in the presence of inhibitors, random mutagenesis presented a more practical approach than rational metabolic engineering [15] to simultaneously enhance xylose fermentation and inhibitor tolerance. As such, we resorted to chemical mutagenesis using EMS where an expected outcome would be at least improved growth in the presence of hydrolysate inhibitors, using xylose as sole source of carbon compared to the parental strain as control. Evolutionary engineering for at least 100 generations at a high dilution rate (close to μ_{max}) with xylose as sole source of carbon and in the presence of inhibitors from non-detoxified liquor would have the expected outcome of enhanced tolerance evidenced by an improved μ_{max} value, and ethanol yield and productivity compared to known benchmark strains. Therefore, there is a clear correlation between the developmental steps employed and the outcomes expected from the study, namely (i) xylose fermentation from metabolic engineering using XI, (ii) enhanced inhibitor tolerance and conversion from mutagenesis and evolutionary engineering, and (iii) a superior phenotype for ethanol production from xylose or glucose also from mutagenesis and evolutionary engineering.

Hardening methodology

The high dilution rate in our chemostat culture supplemented with hydrolysate would allow for strong selective pressure in favour of yeast cells with improved xylose carbon flow through the metabolism. In this way cells with high growth rate would be selected. Most studies dealing with chemostat-based evolutionary engineering maintained continuous cultures at very low dilution rates in the region of $D = 0.02$ to $0.05\ h^{-1}$ [18,20,43]. Whereas low dilution rates are normally used to select for substrate affinity [14,44], the purpose in our study was to select for high growth rate, given the negative effect that inhibitors have on the culture μ_{max} [8,45]. This approach corresponds to the more frequent use of sequential batch culture for evolutionary engineering, where the culture would grow at μ_{max} [11,16-18,22,26,29,46,47]. Theoretically, evolved strains with the ability to maintain a higher growth rate should become dominant in a culture at a dilution rate close to μ_{max}. An expected outcome would be an improvement in ethanol productivity, although this improvement was difficult to quantify in our work due to the absence of ethanol production by strain TMB3400 and the parental strain in xylose batch cultures (Table 3). The benchmark strains outperformed strain D5A^{+H} in terms of ethanol productivity in glucose batch cultures (Table 4).

Strain performance with xylose as carbon source

An oxidative metabolism implies a greater biomass yield than would be the case with a respiro-fermentative metabolism and hence, a concomitant decrease in the ethanol yield [48]. This principle was patently evident from the data presented in Tables 3 and 4 where the $Y_{x/s}$ value of the parental D5A^{+} strain on glucose without hydrolysate supplementation was more than six-fold lower than when xylose was supplied as carbon source in the chemically defined medium, which implied a predominantly oxidative metabolism with xylose as carbon source. These observations could, on the one hand, support the notion that *S. cerevisiae* fails to recognise xylose as a fermentable carbon source [49]. On the other hand, the strain D5A^{+H} produced 1.54 g/L ethanol in the presence of 50% (v/v) pretreatment liquor, whereas no measurable ethanol was produced by strain TMB3400 or by the parental D5A^{+} strain on xylose without pretreatment liquor (Table 3). Furthermore, strain D5A^{+} exhibited strong ethanol production from glucose at a yield of 0.42 g/g glucose (82% of the theoretical maximum), which still makes it a prime candidate for 2nd generation ethanol production from cellulose. It is for the latter reason that the recombinant strain was subjected to mutagenesis and evolutionary engineering in an attempt to enhance xylose fermentation to ethanol, which was indeed accomplished. This result suggested that the mutagenesis treatment and/or evolutionary engineering profoundly affected the metabolism for funnelling carbon from xylose into the EMP pathway for ethanol production. The reprogramming of these pathways was described by Liu [50] and could possibly have been responsible for our observed result. Complete genomic analysis, coupled with transcriptomic analysis is earmarked for the future to elucidate the nature of this modification.

The negative impact of inhibitors on the ethanol productivity and yield [6,51] as well as on the biomass yield due to interactions between furaldehydes and acetic acid [52] is well described. Previous reports also illustrated that yeast has a diminished ability to deal with the toxic effects exerted by lignocellulosic inhibitors when grown on xylose as carbon source [53]. This theoretical basis might explain why the parental strain D5A^{+} in the present study was unable to grow and retain viable cell numbers on xylose in the presence of 50% (v/v) pretreatment liquor (Figure 2A). However, given this complete absence of growth of the un-adapted parental strain in the presence of inhibitors, the evolutionary engineering approach followed proved effective to introduce tolerance into the D5A yeast platform, as evident from the comparable μ_{max} values of strains D5A^{+H} and TMB3400 (Table 3). The fact that the μ_{max} of $0.12\ h^{-1}$ of the adapted strain D5A^{+H} on xylose in the presence of pretreatment liquor (corresponding to the dilution rate close to the washout dilution rate in chemostat culture with 60% (v/v) pretreatment liquor) was 68% lower than that of the parental strain without hydrolysate clearly illustrated the toxic effect of the inhibitors on this yeast, in spite of the level of tolerance that was introduced into the yeast (Table 3).

Notably the μ_{max} value of strain TMB3400 recorded in the present study on xylose in the presence of pretreatment liquor was comparable to data published by Wahlbom *et al.* [10] for that strain in aerobic bioreactor cultures without pretreatment inhibitors. Therefore, compared to strain D5A^{+H} on xylose as carbon source, the presence of inhibitors appeared to have a less drastic effect on strain TMB3400, possibly attributable to the presence of XR. Furfural was shown to serve as an effective electron acceptor for XR, as evident from decreased xylitol formation, resulting in an improved ethanol yield [42], although no ethanol production by strain TMB3400 on xylose was observed in the present study (Table 3). Furthermore, strain D5A^{+H} exhibited a much shorter lag phase than strain TMB3400, where the lag phase for the latter strain lasted up to 30 h (Figure 2A, C and E). A long lag phase in the presence of inhibitors is a known occurrence, attributable to the time required for the cells to carry out detoxification [54,55]. Therefore, in spite of the inhibitory effect on strain D5A^{+H}, the similarity of the growth rates between the two strains and the shorter lag phase pointed to our D5A^{+H} strain as a possible candidate for further development for industrial fermentations.

The *in situ* conversion of HMF and furfural by strain D5A^{+H} was clearly superior compared to the parental D5A^{+} strain and TMB 3400 (Figure 2B, D and F). Hence, HMF and furfural removal commenced immediately after inoculation due to a comparatively shorter lag phase, and similar to that reported by Palmqvist and Hahn-Hägerdal [8]. Previous reports identified a range of oxido-reductases as the primary enzymes responsible for the *in situ* reduction of HMF and furfural to their less toxic alcohol derivatives [56], such as the native reductases ADH6 and ADH7 [31,57-59]. One may, therefore, speculate on enhanced catalytic activity and/or higher degree of expression by one or more of these reductases brought on by mutagenesis and evolutionary engineering of strain D5A^{+H}. Less than 12% of the xylose consumed by strain D5A^{+H} was converted to ethanol (Table 3), which implies accumulation of another metabolite, since the total recovered carbon from biomass, ethanol and glycerol was only 60%. Xylitol production has been observed in recombinant *S. cerevisiae* strains expressing XI and was ascribed to the endogenous activity of non-specific NADPH-dependent aldo-ketoreductases expressed from the *GRE3* gene [60,61]. Xylitol was not quantified during these fermentations, and future investigations will verify whether xylitol can account for these discrepancies in the carbon mass balance and transcriptomic profiling as well as enzyme activity data is required to validate this hypothesis.

The ability of strain D5A^{+H} to attain high growth rates and a shorter lag phase in spite of the presence of weak organic acids may point to a decreased sensitivity towards anion accumulation. The more abundant organic acids, i.e. acetic and formic acid, also play a pivotal role in contributing to pretreatment liquor toxicity [8]. However these acids are typically not reduced in a typically enzyme-mediated manner, as is the case for HMF and furfural [62], but rather accumulate intracellularly in dissociated form due to the neutrality of the cytosol [8]. The detoxification profiles show acetic and formic acid (Figure 2B, D and F) decreasing over time, with the fastest overall removal of these weak acids apparent for strain D5A^{+H}. This observation could probably be related to earlier onset of cell growth. The greater biomass concentration probably resulted in faster diffusion of weak acids down its concentration gradient into the cytoplasm of the yeast cells. Although a large proportion of these anions were probably still secreted, this efflux was probably masked by increasing biomass and repetition of the cycle.

Strain performance with glucose as carbon source

Compared to the parental D5A^{+} strain in the absence of pretreatment liquor, the influence of inhibitors on the cultures with glucose as carbon source was clearly evident from μ_{max} values that were up to 2.7-fold lower (strain MEL2) and q_p values that were up to 5.5-fold lower (strain D5A^{+H} Table 4). This data clearly illustrated the dramatic negative effect of inhibitors on productivity rates, whereas the biomass and ethanol yield was affected to a much lesser extent and corresponded to data reviewed by Liu [50]. Although strain D5A^{+H} surpassed strains TMB 3400, MH1000 and MEL2 in terms of biomass yield and growth rate, this better performance did not correspond to higher ethanol productivity or yield during growth on glucose (Table 4). The comparatively higher $Y_{x/s}$ value of strain D5A^{+H} on glucose could possibly be viewed as a secondary effect of mutagenesis and long term selection, since this trait was not observed in the parental yeast in the absence of pretreatment liquor. On the other hand, strains MH1000 and MEL2, which had comparatively lower μ_{max} values, produced ethanol at a significantly greater rate than strains D5A^{+H} or TMB3400. This result is not surprising, given the robust nature of these yeasts, specifically strain MH1000, which was purposefully included due to its high ethanol productivity [41]. The by-product detoxification profiles showed that both acetic and formic acid (Figure 3B, D, F and H) decreased over time for all the strains. However, once again strain D5A^{+H} outperformed the other strains since approximately 20% more acetic acid was removed than was the case with strain TMB3400. This observation supports our previous suggestion of a decreased sensitivity to these acids was brought about through mutagenesis and evolutionary engineering.

Practical implementation through simultaneous saccharification and fermentation

Whereas our results clearly demonstrated a degree of inhibitor tolerance by strain D5A^{+H}, it is also of industrial importance to evaluate its effectiveness during SSF using steam-treated material. Therefore the ability of this strain was assessed along with that of the benchmark strain TMB3400 (Figure 4). A fed-batch approach was followed, to allow for a higher solids content of 10% (w/v), whilst minimising yeast inhibition [63], and viscosity and mass transfer effects. The fed-batch strategy should also further assist in xylose metabolism by limiting the concentration of residual glucose [63,64]. Strain D5A^{+H} performed substantially better than strain TMB3400 in SSF culture with pressed WIS, producing 46% more ethanol at a combined (glucose and xylose) theoretical yield of 50% vs. 34% for strain TMB3400. The data recorded for strain D5A^{+H} was slightly below that reported in the literature where ethanol yields from steam-pretreated sorghum ranged between 60-64% of theoretical maximum [65,66]. Although xylose was poorly utilised by strain D5A^{+H} in SSF culture, as evident from accumulation of residual xylose (Figure 4A), it is possible that the greater ethanol titre produced by strain D5A^{+H} could in part have stemmed from xylose, given the level of ethanol produced by this strain in batch cultures supplemented with hydrolysate and xylose (Table 3). Conversely, a possible explanation for the comparatively poor degree of xylose utilisation observed for the D5A^{+H} strain during SSF could be attributed to residual glucose levels that remained high for a large portion of the total fermentation time, which may have resulted in the inhibition of xylose consumption. High concentrations of glucose are known to inhibit xylose transport in yeast, due to competition between the two substrates for the hexose transporters which have an inherently lower affinity for xylose [67].

Conclusions

The adaptation of yeast cultures for the production of ethanol from pretreated lignocellulosic substrates is of critical importance due to the toxic nature of the broth after pretreatment. Whereas other methods such as recursive breeding and genome shuffling may also be employed to beneficiate multi-allelic traits such as ethanol productivity and inhibitor tolerance [68], the experimental plan followed in this study yielded promising results. Therefore, we conclude that mutagenesis in combination with long term evolutionary engineering was successfully applied to introduce a greater level of tolerance in *S. cerevisiae* D5A^{+H}, together with improved xylose fermentation. To the knowledge of the authors, this is one of very few studies where EMS treatment followed by evolutionary engineering at high hydrolysate concentrations with xylose as the sole source of carbon was successfully applied to a strain of *S. cerevisiae* harbouring the XI pathway for xylose consumption, resulting in comparatively superior performance to a known yeast benchmark. The physiological changes achieved in this study included an improved ability to ferment xylose in the presence of steam pretreated triticale straw liquor in batch culture. Future studies will include a second phase of long-term adaptation with emphasis on selection for increased ethanolic fermentation of xylose in the presence of pretreatment liquor, by keeping the fermentation broth strictly anaerobic. Additionally, genome sequencing as well as transcriptomic profiling of D5A^{+H} will be focused upon to elucidate the nature of the molecular mechanisms of tolerance inferred through the experimental approaches presented in this study, followed by directed mutagenesis at selected targets using a rational approach.

Methods

Triticale straw and sweet sorghum bagasse steam pretreatment

Triticale straw was kindly provided by the Department of Genetics at Stellenbosch University and sweet sorghum bagasse was obtained from the University of Kwazulu Natal both from South Africa. All biomass was steam-pretreated in a steam gun unit consisting of a 19 litre reactor vessel and a cyclone tank for material collection. Triticale straw biomass was soaked in water from reverse osmosis for a 16 h period in a 1:1 mass ratio after which excess water was removed from the material in a spin dryer for 5 min at 4000 rpm. Triticale straw was treated for 20 min at 200°C, whereas dry (i.e. not pre-soaked) sweet sorghum bagasse was treated for 5 min at 190°C. For both feedstocks, a solids loading of 500 g (dry weight) was used per steam gun run. Water insoluble solids (WIS) were removed from the slurry through compression in a 50 ton mechanical press at 25 kPa. The liquid and WIS fractions of the treated material were collected separately and stored at −20°C.

Strains and maintenance

Saccharomyces cerevisiae strain D5A^{+} was used for performing random mutagenesis and evolutionary engineering. Strain D5A^{+} is a metabolically engineered xylose-utilising variant of the NREL D5A strain (ATCC 200062) and this recombinant strain as well as all other strains is deposited in the yeast culture collection of the Dept. Microbiology at Stellenbosch University, South Africa. Strain D5A^{+} contains a chromosomal integration of the putative xylose isomerase (*xylA*) gene from *Bacteroides thetaiotaomicron* [69]. The synthetic XI, which showed 83.1% identity with the *Piromyces* XI, was expressed from the multi-copy integrative vector pBKD1 together with the *XKS1* gene expressing xylulokinase from *S. cerevisiae* (R. den Haan, Stellenbosch University, unpublished results). *Saccharomyces cerevisiae* TMB3400, employed as benchmark in this study, is a chemically induced mutant of the xylose utilising TMB 3001 strain, which is a CEN.PK derivative expressing xylose reductase (XR) and xylitol dehydrogenase (XDH) from the chromosomally integrated *Pichia stipitis* genes *XYL1* and *XYL2*, and over-expresses the homologous xylulokinase enzyme [10]. Two non-recombinant industrial isolates were also included, namely *S. cerevisiae* MEL2, which is a wild-type yeast isolated from grape marcs [40] and *S. cerevisiae* MH1000, which is a robust in-house yeast with a high fermentative capacity [41]. Strains where transferred from −80°C freezer stock cultures and routinely cultured on YPD (strains MEL2 and MH1000) and YPX agar plates (strains D5A^{+} and TMB3400), containing (per litre): 10 g yeast extract, 20 g bacteriological peptone, 20 g of either glucose or xylose and 15 g agar. All chemicals and reagents where obtained from Sigma Aldrich (Manheim, Germany).

The growth medium used for the cultivation of all yeast was based on the mineral medium described previously [70] and contained (per litre): 20 g yeast extract 0.5 g citric acid 3.4 g KH_2PO_4, 7.5 g $(NH_4)_2SO_4$, 0.8 g $MgSO_4{\cdot}7H_2O$,

0.05 g $CaCl_2{\bullet}2H_2O$ and 1 mL of a trace element solution, containing (per litre): 4.5 mg $ZnSO_4{\bullet}7H_2O$, 0.3 mg $CoCl_2{\bullet}6H_2O$, 1.5 mg $MnSO_4{\bullet}H2O$, 0.3 mg $CuSO_4{\bullet}5H_2O$, 3 mg $FeSO_4{\bullet}7H2O$, 0.4 mg $Na_2MoO_4{\bullet}2H_2O$, 1 mg H_3BO_3 and 0.1 mg KI. All chemicals and reagents where obtained from Sigma Aldrich. Xylose and glucose were autoclaved separately at a concentration of 200 g/L and added aseptically after sterilisation to a final concentration of 20 g/L. To minimise evaporation of volatile by-products such as furfural, which precluded the use of autoclaving for sterilisation, pretreatment liquor was filter sterilised using a 0.45 μm nylon filter (Pall Corporation, NY, USA) and added aseptically to the medium. During continuous culture, the growth medium was supplemented with the anaerobic growth factors ergosterol and Tween 80 (Sigma Aldrich). Prior to addition to the growth medium, ergosterol was dissolved in 5 mL pure ethanol and subsequently added to this mixture. Final concentrations of ergosterol and Tween 80 in the growth medium were 0.01 and 0.42 g/L, respectively. Due to the sheer volume of hydrolysate required for continuous culture, the hydrolysate was not sterilised prior to feeding the culture.

Random mutagenesis and primary selection culture

For mutagenesis the recombinant *S. cerevisiae* strain $D5A^+$ was grown in batch culture using 50 mL of the chemically defined medium as described above, in a 250 mL Erlenmeyer flask supplemented with 20 g/L xylose as carbon source and incubated on an orbital shaker (New Brunswick Scientific, Edison, NJ, USA) at 30°C and 150 rpm. Flasks were fitted with foil-capped cotton plugs. Cells were harvested during the late exponential growth phase by centrifugation at 8 000 rpm for 5 min and re-suspended to a final concentration of 3.1×10^6 CFUs/mL in phosphate-buffered saline (PBS) consisting of (per litre): 8 g NaCl, 0.2 g KCl, 1.44 g Na_2HPO_4, 0.24 g KH_2PO_4 and adjusted to pH 7.4 using 3 mol/L KOH. Chemical mutagenesis was carried out with ethyl methanesulfonate (EMS, Sigma Aldrich) according to the method described by Winston [71] using a 1 h incubation time on a rotating platform. EMS-treated cells were plated onto YPX agar, containing 20 g/L of xylose, and incubated at 30°C for 72 h to allow for colony growth of cells surviving the treatment. Untreated cells were plated on separate YPX plates as control. The percentage survival of mutants was calculated using the equation $[CFU_{treated\ plate}/CFU_{untreated\ control}] \times 100$. In all cases cells were diluted using PBS. Relative to the control, the percentage survival of the mutated cells in the presence of 2, 3 and 6 μL EMS were 26 ± 2.3%, 21 ± 4.4% and 7.4 ± 1.2, respectively.

Colonies of surviving yeast treated with 2 and 3 μL EMS were washed from YPX plates with PBS, pooled and used as inoculum for selection of resistant mutants in medium containing pretreatment liquor. Two separate selection cultures at different pretreatment liquor concentrations where grown in batch culture in 500 mL Erlenmeyer flasks containing 100 mL medium. Whereas the one selection culture consisted out of the chemically defined medium (see above) supplemented with pretreatment liquor at a concentration of 67% (v/v), the other selection culture consisted of undiluted pretreatment liquor in which the dry powder of the medium components were dissolved. The growth of the mutated cells were followed over a six day incubation period at 30°C in a rotary shaker at 150 rpm and compared to that of the parental $D5A^+$ strain which was grown under the exact same conditions and served as a control. Due to the dark coloration of the medium, cell growth was quantified with cell counts using an Improved Neubauer counting chamber/heamocytometer. For both the mutant and control strain, 20 g/L xylose was supplemented as the primary source of carbon.

Continuous culture

Mutants that survived selection in the growth medium supplemented with 67% (v/v) pretreatment liquor were harvested by centrifugation after a six day cultivation period suspended in 50 mL PBS to a concentration of 72×10^7 cells/mL and inoculated into a bioreactor for continuous culture. All fermentations were conducted in computer-controlled glass bioreactors (Sartorius Stedum BBI, Göttingen, Germany) with a total volume of 5000 mL and a working volume of 2000 mL. The bioreactor was equipped with a marine impeller, an exhaust gas reflux cooler perfused with water at 5°C, a pH electrode (Mettler Toledo, Halstead, UK) and a polarographic dissolved oxygen probe (Hamilton, Bonaduz, GR, Switzerland). Cultures were grown at 30°C and pH 5.0 by automatic titration with 3 mol KOH/L. Prior to inoculation of the reactor, the system was flushed with nitrogen to minimise oxygen levels at the onset of fermentation. To ensure anoxic conditions, no air was supplied to the reactor vessel during the six week cultivation which was operated at a stirrer speed of 200 rpm. Although neither air impermeable tubing nor a nitrogen overlay in the headspace of the culture was used, no measurable dissolved oxygen was observed throughout the cultivations. The culture was grown as batch culture for the first 48 h after inoculation before the feeding of the growth medium with hydrolysate commenced using a peristaltic pump (Watson Marlow Ltd., Falmouth, UK) adjusted to a flow rate of 50 mL/h to obtain a dilution rate of 0.05 h^{-1}. To increase selective pressure, the dilution rate was gradually increased to a maximum of 0.125 h^{-1}, which was in the vicinity of μ_{max} values determined in batch culture. At a dilution rate of 0.1 h^{-1}, the hydrolysate concentration was increased from the original concentration of 50% (v/v) to a final concentration of 60% (v/v). The biomass concentration was estimated by total cell counts using an Improved Neubauer counting chamber. The large quantities of hydrolysate required for continuous culture (in excess of a total of 200 litres) precluded the use of filter sterilisation prior to addition to the bioreactor. Hence, the fermentation was performed under non-sterile conditions. Culture steady state was defined as a deviation in biomass concentration (cell counts/mL) of approximately 20% of the mean measured over three residence times. Although this variation is larger than what is usually acceptable, we allowed for a greater degree of error due to the harsh conditions impacting cell growth, the use of plate counts (not optical density), and limited hydrolysate availability, given a chemostat culture operated at a 2 L working volume. The total number of yeast generations was calculated by dividing the fermentation time by the cell doubling time at a given dilution rate.

Strain isolation from continuous culture

At the end of six weeks of continuous cultivation the remaining yeast culture was harvested and subsequently purified through two rounds of sub-culturing on YPX agar. Culture purity was confirmed using microscopy. To verify that the isolated yeast was indeed a sub-strain of the parental recombinant D5A$^+$ strain the isolated mutant was screened for resistance against the fungicide genetecin. A genetecin resistance gene was previously cloned into the parental strain for use as a selection marker during its genetic manipulation [69]. The adapted mutant (designated D5A^{+H}) was found to be genetecin resistant, confirming that the isolated mutant yeast was related to the original parental D5A$^+$ strain.

Analysis of D5A^{+H} performance and benchmark comparison

Batch cultivations where performed in 500 mL Erlenmeyer flasks containing 100 mL growth medium consisting of the chemically defined medium (see above) supplemented with 50% (v/v) triticale pretreatment liquor. Glucose was supplied as the primary carbon source (20 g/L) for growth experiments using strains MH1000 and MEL2 whereas either glucose or xylose was supplied as a source of carbon (final concentration of 20 g/L) for the strains D5A$^+$, D5A^{+H} and TMB 3400. The inoculum of each culture was prepared by growing the yeast for up to 36 h in a preconditioning medium consisting of the chemically defined medium (see above) supplemented with 20% (v/v) triticale hydrolysate. Batch cultures were inoculated at an initial OD value of 0.2, measured at 600 nm and incubated at 30°C on an orbital shaker at 150 rpm. Flasks were assumed to be semi-aerobic since cotton plugs were capped with aluminium foil.

Simultaneous Saccharification and Fermentation (SSF)

All SSF experiments were conducted in the same bioreactors fitted with Rushton impellers and chemically defined medium as described above without supplemented carbon, implying that all carbon supplied to the culture was liberated by enzymatic hydrolysis from the steam-pretreated sweet sorghum bagasse. Nitrogen was sparged at a rate of 100 mL/min (0.05 vvm) to minimise oxygen availability. Due to the different temperature optima between enzymatic hydrolysis (45°C) and fermentation (30°C) a three hour pre-saccharification step at 45°C always preceded the SSF experiments. Yeast biomass was inoculated into the SSF medium to a final concentration of 5 g/L (wet mass). The inoculum for the SSF experiments was prepared using the same pre-conditioning method described in the previous section. Solids where loaded in 2.5% (w/v) increments at times 0, 24, 48 and 72 h to yield a final WIS solids loading of 10% (w/v) based on the total volume of the culture after the final WIS addition. A single dosage of cellulase was introduced at the onset of the pre-saccharification phase through addition of 0.667 mL Cellic® CTec2 (Novozymes, Denmark) per gram WIS. This volume was based on the same volume of Spezyme (Genencor, Palo Alto, CA, USA) that corresponded to approximately 15 FPU/g WIS, which was in excess of the required amount to achieve efficient hydrolysis of a 10% (w/v) solids loading of cellulosic material [72].

Analytical methods

For the estimation of yeast biomass in g/L optical density measurements (absorbance measured at 600 nm) was correlated to yeast dry weight (g/L) through individual OD vs. dry weight standard curves for all strains used in this study. Dry biomass measurements were taken after fermentation samples were washed and vacuum filtered through a 0.22 µm Whatman filter paper and dried for a minimum of 12 h at 105°C. The chemical composition of the hydrolysate and water insoluble solids was determined according to the standard NREL procedures (LAP - 001 to LAP - 005 LAP - 010 and LAP - 017). Sugars, organic acids, ethanol and glycerol where separated on a Water Breeze (Waters Corporation, Milford, MA) high performance liquid chromatograph, fitted with an Aminex HPX-87H resin column (Bio-Rad Laboratories, Hercules, CA) and an H cartridge guard column (Bio-Rad). The column was operated at 45°C with 0.5 g/L H_2SO_4 as the mobile phase at a flow rate of 0.6 mL/min and detection was performed using a Waters 2410 refractive-index detector. HMF and furfural where separated on an Aminex HPX-87P column (Biorad). The column was eluted at 80°C with 0.5 g/L H_2SO_4 as the mobile phase at a flow rate of 0.6 mL/min and detection was performed using a Waters UV detector adjusted to 230 nm. Integration of the area underneath the peak of an eluted compound was relayed to concentration through the use of individual standard curves, based on standards injected at a range of concentrations prior to each run.

Calculations

Maximum specific growth rates (μ_{max}) where determined by linear regression of the slope of the natural logarithm of the biomass concentration plotted as a function of time using a minimum of four data points. The biomass ($Y_{x/s}$) and product ($Y_{p/s}$) yield coefficients were calculated by linear regression of the slope when biomass and ethanol where plotted as a function of the substrate concentration (either glucose or xylose) using a minimum of four data points recorded during the exponential growth phase. The specific rates of glucose xylose consumption (q_s) and ethanol production (q_p) were taken as the slope of the regression line fitted through the metabolite concentration vs. time profile normalised to the amount of biomass present at the midpoint of the exponential growth phase. Data was analysed on Microsoft Excel and imported into Design Expert v.1.6 (Stat-Ease Inc., MN, USA) for performing a one way analysis of variance (ANOVA), where a P-value of less than 0.05 was considered as statistically significant.

Abbreviations

EMS: Ethyl methyl sulfonate; CFU: Colony forming unit; HMF: Hydroxymethyl furfural; SSF: Simultaneous saccharification and fermentation; WIS: Water-insoluble solids; XDH: Xylitol dehydrogenase; XI: Xylose isomerase; XK: Xylulokinase; XR: Xylose reductase.

Competing interests

The authors declare have no competing interests to declare.

Authors' contributions

JS designed the experiments and carried out experimental work. EVR contributed to the experimental design, provided guidance for experimental work, especially continuous culture, and critically read and formatted the document. JFG provided critical inputs into the study in terms of experimental approaches and the content of the document. All authors read and approved the final manuscript.

Acknowledgements

The authors gratefully acknowledge Dr. L. Favaro Prof. S. Casella and Prof. M. Basaglia from Department of Agronomy Food Natural resources Animals and Environment (DAFNAE) University of Padova, Agripolis, Viale dell'Università 16, 35020 Legnaro (PD), Italy for providing *S. cerevisiae* strain MEL2, used as benchmark in this study. The Technology Innovation Agency (TIA), an entity of the Department of Science and Technology (DST) of the Republic of South Africa, is acknowledged for funding support of this study and its authors.

References

1. Hahn-Hägerdal B, Wahlbom C, Gárdonyi M, van Zyl W, Cordero Otero R, Jönsson L: **Metabolic engineering of *Saccharomyces cerevisiae* for xylose utilization.** *Adv Biochem Eng Biotechnol* 2001, **73**:53–84.
2. Hahn-Hägerdal B, Pamment N: **Microbial pentose metabolism.** *Appl Biochem Biotechnol* 2004, **116**:1207–1209.
3. Lynd LR, Weimer PJ, Van Zyl WH, Pretorius IS: **Microbial cellulose utilization: Fundamentals and biotechnology.** *Microbiol Mol Biol Rev* 2002, **66**:506–577.
4. Stephanopoulos G, Aristidou AA, Nielsen J: *Metabolic engineering: principles and methodologies.* San Diego: Academic; 1998.
5. von Sivers M, Zacchi G, Olsson L, Hahn-Hägerdal B: **Cost analysis of ethanol production from willow using recombinant *Escherichia coli.*** *Biotechnol Prog* 1994, **10**:555–560.
6. Almeida J, Modig T, Petersson A, Hahn-Hägerdal B, Lidén G, Gorwa-Grauslund M: **Increased tolerance and conversion of inhibitors in lignocellulosic hydrolysates by *Saccharomyces cerevisiae.*** *J Chem Technol Biotechnol* 2007, **82**:340–349.
7. Larsson S, Reimann A, Nilvebrant NO, Jönsson LJ: **Comparison of different methods for the detoxification of lignocellulose hydrolyzates of spruce.** *Appl Biochem Biotechnol* 1999, **77**:91–103.
8. Palmqvist E, Hahn-Hägerdal B: **Fermentation of lignocellulosic hydrolysates. II: inhibitors and mechanisms of inhibition.** *Bioresour Technol* 2000, **74**:25–33.
9. Parajó J, Domínguez H, Domínguez J: **Biotechnological production of xylitol. Part 3: operation in culture media made from lignocellulose hydrolysates.** *Bioresour Technol* 1998, **66**:25–40.
10. Wahlbom C, van Zyl W, Jönsson L, Hahn-Hägerdal B, Otero R: **Generation of the improved recombinant xylose–utilizing *Saccharomyces cerevisiae* TMB 3400 by random mutagenesis and physiological comparison with *Pichia stipitis* CBS 6054.** *FEMS Yeast Res* 2003, **3**:319–326.
11. Kuyper M, Winkler AA, Dijken JP, Pronk JT: **Minimal metabolic engineering of *Saccharomyces cerevisiae* for efficient anaerobic xylose fermentation: a proof of principle.** *FEMS Yeast Res* 2004, **4**:655–664.
12. Gorsich SW, Dien BS, Nichols NN, Slininger PJ, Liu ZL, Skory CD: **Tolerance to furfural-induced stress is associated with pentose phosphate pathway genes *ZWF1, GND1, RPE1,* and *TKL1* in *Saccharomyces cerevisiae.*** *Appl Microbiol Biotechnol* 2006, **71**:339–349.
13. Toivari MH, Salusjarvi LS, Ruohonen L, Penttila M: **Endogenous xylose pathway in *Saccharomyces cerevisiae.*** *Appl Environ Microbiol* 2004, **70**:3681–3686.
14. Sauer U: **Evolutionary engineering of industrially important microbial phenotypes.** *Adv Biochem Eng Biotechnol* 2001, **73**:129–169.
15. Nevoigt E: **Progress in metabolic engineering of *Saccharomyces cerevisiae.*** *Microbiol Mol Biol Rev* 2008, **72**:379–412.
16. Demeke MM, Dietz H, Li Y, Foulquié-Moreno MR, Mutturi S, Deprez S, Den Abt T, Bonini BM, Lidén G, Dumortier F, Verplaetse A, Boles E, Thevelein JM: **Development of a D-xylose fermenting and inhibitor tolerant industrial *Saccharomyces cerevisiae* strain with high performance in lignocellulose hydrolysates using metabolic and evolutionary engineering.** *Biotechnol Biofuels* 2013, **6**:89.
17. Heer D, Sauer U: **Identification of furfural as a key toxin in lignocellulosic hydrolysates and evolution of a tolerant yeast strain.** *Microb Biotechnol* 2008, **1**:497–506.
18. Koppram R, Albers E, Olsson L: **Evolutionary engineering strategies to enhance tolerance of xylose utilizing recombinant yeast to inhibitors derived from spruce biomass.** *Biotechnol Biofuels* 2012, **5**:32.
19. Eliasson A, Christensson C, Wahlbom CF, Hahn-Hägerdal B: **Anaerobic xylose fermentation by recombinant *Saccharomyces cerevisiae* carrying *XYL1, XYL2,* and *XKS1* in mineral medium chemostat cultures.** *Appl Environ Microbiol* 2000, **66**:3381–3386.
20. Sonderegger M, Sauer U: **Evolutionary engineering of *Saccharomyces cerevisiae* for anaerobic growth on xylose.** *Appl Environ Microbiol* 2003, **69**:1990–1998.
21. Martín C, Marcet M, Almazán O, Jönsson LJ: **Adaptation of a recombinant xylose-utilizing *Saccharomyces cerevisiae* strain to a sugarcane bagasse hydrolysate with high content of fermentation inhibitors.** *Bioresour Technol* 2007, **98**:1767–1773.
22. Tomás-Pejó E, Ballesteros M, Oliva J, Olsson L: **Adaptation of the xylose fermenting yeast *Saccharomyces cerevisiae* F12 for improving ethanol production in different fed-batch SSF processes.** *J Ind Microbiol Biotechnol* 2010, **37**:1211–1220.
23. Tantirungkij M, Izuishi T, Seki T, Yoshida T: **Fed-batch fermentation of xylose by a fast-growing mutant of xylose-assimilating recombinant *Saccharomyces cerevisiae.*** *Appl Microbiol Biotechnol* 1994, **41**:8–12.
24. Pitkänen JP, Rintala E, Aristidou A, Ruohonen L, Penttilä M: **Xylose chemostat isolates of *Saccharomyces cerevisiae* show altered metabolite and enzyme**

levels compared with xylose, glucose, and ethanol metabolism of the original strain. *Appl Microbiol Biotechnol* 2005, **67**:827–837.
25. Zha J, Shen M, Hu M, Song H, Yuan Y: **Enhanced expression of genes involved in initial xylose metabolism and the oxidative pentose phosphate pathway in the improved xylose-utilizing *Saccharomyces cerevisiae* through evolutionary engineering.** *J Ind Microbiol Biotechnol* 2014, **41**:27–39.
26. Liu E, Hu Y: **Construction of a xylose-fermenting *Saccharomyces cerevisiae* strain by combined approaches of genetic engineering, chemical mutagenesis and evolutionary adaptation.** *Biochem Eng J* 2010, **48**:204–210.
27. Laplace JM, Delgenes JP, Moletta R, Navarro JM: **Fermentation of lignocellulosic sugars to ethanol: Selection of mutants of *Pichia stipitis* affected for D-glucose utilization.** *Enzyme Microb Technol* 1992, **14**:644–648.
28. James AP, Zahab DM, Mahmourides G, Maleszka R, Schneider H: **Genetic and biochemical characterization of mutations affecting the ability of the yeast *Pachysolen tannophilus* to metabolize D-xylose.** *Appl Environ Microbiol* 1989, **55**:2871–2876.
29. Amartey S, Jeffries T: **An improvement in *Pichia stipitis* fermentation of acid-hydrolysed hemicellulose achieved by overliming (calcium hydroxide treatment) and strain adaptation.** *World J Microbiol Biotechnol* 1996, **12**:281–283.
30. Almario MP, Reyes LH, Kao KC: **Evolutionary engineering of *Saccharomyces cerevisiae* for enhanced tolerance to hydrolysates of lignocellulosic biomass.** *Biotechnol Bioeng* 2013, **110**:2616–2623.
31. Almeida J, Bertilsson M, Gorwa-Grauslund M, Gorsich S, Lidén G: **Metabolic effects of furaldehydes and impacts on biotechnological processes.** *Appl Microbiol Biotechnol* 2009, **82**:625–638.
32. Wisselink HW, Toirkens MJ, Wu Q, Pronk JT, van Maris AJ: **Novel evolutionary engineering approach for accelerated utilization of glucose, xylose, and arabinose mixtures by engineered *Saccharomyces cerevisiae* strains.** *Appl Environ Microbiol* 2009, **75**:907–914.
33. Zhou H, Cheng J, Wang BL, Fink GR, Stephanopoulos G: **Xylose isomerase overexpression along with engineering of the pentose phosphate pathway and evolutionary engineering enable rapid xylose utilization and ethanol production by *Saccharomyces cerevisiae*.** *Metab Eng* 2012, **14**:611–622.
34. Demeke MM, Dumortier F, Li Y, Broeckx T, Foulquie-Moreno R, Thevelein JM: **Combining inhibitor tolerance and D-xylose fermentation in industrial *Saccharomyces cerevisiae* for efficient lignocellulose-based bioethanol production.** *Biotechnol Biofuels* 2013, **6**:120.
35. Bobleter O: **Hydrothermal degradation of polymers derived from plants.** *Prog Polym Sci* 1994, **19**:797–841.
36. Garrote G, Dominguez H, Parajo JC: **Hydrothermal processing of lignocellulosic materials.** *Eur J Wood Prod* 1999, **57**:191–202.
37. Gupta R, Khasa YP, Kuhad RC: **Evaluation of pretreatment methods in improving the enzymatic saccharification of cellulosic materials.** *Carbohydr Polym* 2011, **84**:1103–1109.
38. Samuel R, Foston M, Jaing N, Cao S, Allison L, Studer M, Wyman C, Ragauskas AJ: **HSQC (heteronuclear single quantum coherence) ^{13}C-^{1}H correlation spectra of whole biomass in perdeuterated pyridinium chloride–DMSO system: an effective tool for evaluating pretreatment.** *Fuel* 2011, **90**:2836–2842.
39. Taylor MP, Mulako I, Tuffin M, Cowan D: **Understanding physiological responses to pre–treatment inhibitors in ethanologenic fermentations.** *Biotechnol J* 2012, **7**:1169–1181.
40. Favaro L, Basaglia M, van Zyl W, Casella S: **Using an efficient fermenting yeast enhances ethanol production from unfiltered wheat bran hydrolysates.** *Appl Energy* 2013, **102**:170–178.
41. van Zyl J, van Rensburg E, van Zyl W, Harms T, Lynd L: **A kinetic model for simultaneous saccharification and fermentation of avicel with *Saccharomyces cerevisiae*.** *Biotechnol Bioeng* 2011, **108**:924–933.
42. Wahlbom C, Hahn-Hägerdal B: **Furfural, 5-hydroxymethyl furfural, and acetoin act as external electron acceptors during anaerobic fermentation of xylose in recombinant *Saccharomyces cerevisiae*.** *Biotechnol Bioeng* 2002, **78**:172–178.
43. Kuyper M, Hartog MM, Toirkens MJ, Almering MJ, Winkler AA, Dijken JP, Pronk JT: **Metabolic engineering of a xylose-isomerase-expressing *Saccharomyces cerevisiae* strain for rapid anaerobic xylose fermentation.** *FEMS Yeast Res* 2005, **5**:399–409.
44. Harder W, Kuenen JG, Matin A: **A review. Microbial selection in continuous culture.** *J Appl Bacteriol* 1977, **43**:1–24.
45. Taherzadeh MJ, Gustafsson L, Niklasson C, Lidén G: **Physiological effects of 5-hydroxymethylfurfural on *Saccharomyces cerevisiae*.** *Appl Microbiol Biotechnol* 2000, **53**:701–708.
46. Keller FA, Bates D, Ruiz R, Nguyen Q: **Yeast adaptation on softwood prehydrolysate.** *Appl Biochem Biotechnol* 1998, **70**:137–148.
47. Liu ZL, Slininger PJ, Gorsich SW: **Enhanced biotransformation of furfural and hydroxymethylfurfural by newly developed ethanologenic yeast strains.** *Appl Biochem Biotechnol* 2005, **121-124**:451–460.
48. Jeffries T, Shi N-Q: **Genetic engineering for improved xylose fermentation by yeasts.** *Adv Biochem Eng Biotechnol* 1999, **65**:117–161.
49. Souto-Maior AM, Runquist D, Hahn-Hägerdal B: **Crabtree-negative characteristics of recombinant xylose-utilizing *Saccharomyces cerevisiae*.** *J Biotechnol* 2009, **143**:119–123.
50. Liu ZL: **Molecular mechanisms of yeast tolerance and in situ detoxification of lignocellulose hydrolysates.** *Appl Microbiol Biotechnol* 2011, **90**:809–825.
51. Klinke HB, Thomsen AB, Ahring BK: **Inhibition of ethanol-producing yeast and bacteria by degradation products produced during pre-treatment of biomass.** *Appl Microbiol Biotechnol* 2004, **66**:10–26.
52. Palmqvist E, Grage H, Meinander NQ, Hahn-Hägerdal B: **Main and interaction effects of acetic acid, furfural, and ρ–hydroxybenzoic acid on growth and ethanol productivity of yeasts.** *Biotechnol Bioeng* 1999, **63**:46–55.
53. Helle S, Cameron D, Lam J, White B, Duff S: **Effect of inhibitory compounds found in biomass hydrolysates on growth and xylose fermentation by a genetically engineered strain of *S. cerevisiae*.** *Enzyme Microb Technol* 2003, **33**:786–792.
54. Brandberg T, Franzén CJ, Gustafsson L: **The fermentation performance of nine strains of *Saccharomyces cerevisiae* in batch and fed-batch cultures in dilute-acid wood hydrolysate.** *J Biosci Bioeng* 2004, **98**:122–125.
55. Liu ZL, Slininger PJ, Dien BS, Berhow MA, Kurtzman CP, Gorsich SW: **Adaptive response of yeasts to furfural and 5-hydroxymethylfurfural and new chemical evidence for HMF conversion to 2,5-bis-hydroxymethylfuran.** *J Ind Microbiol Biotechnol* 2004, **31**:345–352.
56. Larroy C, Fernández M, González E, Parás X, Biosca JA: **Characterization of the *Saccharomyces cerevisiae YMR318C* (*ADH6*) gene product as a broad specificity NADPH-dependent alcohol dehydrogenase: relevance in aldehyde reduction.** *Biochem J* 2002, **361**:163–172.
57. Larroy C, Pares X, Biosca JA: **Characterization of a *Saccharomyces cerevisiae* NADP (H)-dependent alcohol dehydrogenase (ADHVII), a member of the cinnamyl alcohol dehydrogenase family.** *Eur J Biochem* 2002, **269**:5738–5745.
58. Liu ZL, Moon J, Andersh BJ, Slininger PJ, Weber S: **Multiple gene-mediated NAD (P) H-dependent aldehyde reduction is a mechanism of in situ detoxification of furfural and 5-hydroxymethylfurfural by *Saccharomyces cerevisiae*.** *Appl Microbiol Biotechnol* 2008, **81**:743–753.
59. Petersson A, Almeida J, Modig T, Karhumaa K, Hahn-Hägerdal B, Gorwa-Grauslund M, Lidén G: **A 5-hydroxymethyl furfural reducing enzyme encoded by the *Saccharomyces cerevisiae ADH6* gene conveys HMF tolerance.** *Yeast* 2006, **23**:455–464.
60. Kuhn A, van Zyl C, van Tonder A, Prior BA: **Purification and partial characterization of an aldo-keto reductase from *Saccharomyces cerevisiae*.** *Appl Environ Microbiol* 1995, **61**:1580–1585.
61. Walfridsson M, Bao X, Anderlund M, Lilius G, Bülow L, Hahn-Hägerdal B: **Ethanolic fermentation of xylose with *Saccharomyces cerevisiae* harboring the *Thermus thermophilus xylA* gene, which expresses an active xylose (glucose) isomerase.** *Appl Environ Microbiol* 1996, **62**:4648–4651.
62. Taherzadeh MJ, Karimi K: **Fermentation inhibitors in ethanol processes and different strategies to reduce their effects.** In *Biofuels: Alternative feedstocks and conversion processes.* Edited by Pandey A, Larroche C, Ricke S, Dussap C, Gnansounou E. Oxford: Academic, 2011.287–311.
63. Olofsson K, Rudolf A, Lidén G: **Designing simultaneous saccharification and fermentation for improved xylose conversion by a recombinant strain of *Saccharomyces cerevisiae*.** *J Biotechnol* 2008, **134**:112–120.
64. Öhgren K, Bengtsson O, Gorwa-Grauslund M, Galbe M, Hahn-Hägerdal B, Zacchi G: **Simultaneous saccharification and co-fermentation of glucose and xylose in steam-pretreated corn stover at high fiber content with *Saccharomyces cerevisiae* TMB3400.** *J Biotechnol* 2006, **126**:488–498.
65. Ballesteros M, Oliva J, Negro M, Manzanares P, Ballesteros I: **Ethanol from lignocellulosic materials by a simultaneous saccharification and**

fermentation process (SFS) with *Kluyveromyces marxianus* CECT 10875. *Process Biochem* 2004, **39**:1843–1848.
66. Shen F, Hu J, Zhong Y, Liu ML, Saddler JN, Liu R: **Ethanol production from steam-pretreated sweet sorghum bagasse with high substrate consistency enzymatic hydrolysis.** *Biomass Bioenergy* 2012, **41**:157–164.
67. Sedlak M, Ho N: **Production of ethanol from cellulosic biomass hydrolysates using genetically engineered *Saccharomyces* yeast capable of cofermenting glucose and xylose.** *Appl Biochem Biotechnol* 2004, **113–116**:403–416.
68. Yu L, Pei X, Lei T, Wang Y, Feng Y: **Genome shuffling enhanced L-lactic acid production by improving glucose tolerance of *Lactobacillus rhamnosus*.** *J Biotechnol* 2008, **134**:154–159.
69. de Villiers G: *Development of recombinant Saccharomyces cerevisiae for improved D-xylose utilisation*, MSc thesis. Stellenbosch: Stellenbosch University, Microbiology Department; 2006.
70. Verduyn C, Postma E, Scheffers WA, van Dijken JP: **Effect of benzoic acid on metabolic fluxes in yeasts: a continuous-culture study on the regulation of respiration and alcoholic fermentation.** *Yeast* 1992, **8**:501–517.
71. Winston F: **EMS and UV mutagenesis in yeast.** *Curr Protoc Mol Biol* 2008, **82**:13.3B.1–13.3B.5.
72. Cannella D, Hsieh CW, Felby C, Jørgensen H: **Production and effect of aldonic acids during enzymatic hydrolysis of lignocellulose at high dry matter content.** *Biotechnol Biofuels* 2012, **5**:26.

An efficient method for stable protein targeting in grasses (Poaceae): a case study in *Puccinellia tenuiflora*

Yuanyuan Bu[1†], Mengqing Zhao[1†], Bo Sun[1], Xinxin Zhang[1], Tetsuo Takano[2] and Shenkui Liu[1*]

Abstract

Background: An efficient transformation method is lacking for most non-model plant species to test gene function. Therefore, subcellular localization of proteins of interest from non-model plants is mainly carried out through transient transformation in homologous cells or in heterologous cells from model species such as *Arabidopsis*. Although analysis of expression patterns in model organisms like yeast and *Arabidopsis* can provide important clues about protein localization, these heterologous systems may not always faithfully reflect the native subcellular distribution in other species. On the other hand, transient expression in protoplasts from species of interest has limited ability for detailed sub-cellular localization analysis (e.g., those involving subcellular fractionation or sectioning and immunodetection), as it results in heterogeneous populations comprised of both transformed and untransformed cells.

Results: We have developed a simple and reliable method for stable transformation of plant cell suspensions that are suitable for protein subcellular localization analyses in the non-model monocotyledonous plant *Puccinellia tenuiflora*. Optimization of protocols for obtaining suspension-cultured cells followed by *Agrobacterium*-mediated genetic transformation allowed us to establish stably transformed cell lines, which could be maintained indefinitely in axenic culture supplied with the proper antibiotic. As a case study, protoplasts of transgenic cell lines stably transformed with an ammonium transporter-green fluorescent protein (PutAMT1;1-GFP) fusion were successfully used for subcellular localization analyses in *P. tenuiflora*.

Conclusions: We present a reliable method for the generation of stably transformed *P. tenuiflora* cell lines, which, being available in virtually unlimited amounts, can be conveniently used for any type of protein subcellular localization analysis required. Given its simplicity, the method can be used as reference for other non-model plant species lacking efficient regeneration protocols.

Keywords: Non-model plant, Suspension-cultured cells, Green fluorescent protein (GFP), *Agrobacterium*, Subcellular localization

Background

The subcellular localization of plant proteins is highly correlated with their functions, and as such, is a particularly relevant aspect of functional studies. In many cases, subcellular localization can be predicted *in silico* based on the primary protein sequence by exploiting the conservation of signal peptides and motifs responsible for protein targeting to different cell compartments [1]. Despite recent advances in improving the accuracy of algorithms for prediction [2,3], experimental validation of predicted localization is still the golden standard to obtain reliable functional information.

Transient gene expression assays allow for rapid and high-throughput analyses of plant genes, and thus have become widely used for characterization studies of gene function. Accordingly, several methods for transient gene expression have been developed, such as polyethylene glycol-mediated protoplast transfection [4], biolistic bombardment [5], and *Agrobacterium*-mediated transient assays

* Correspondence: shenkuiliu@nefu.edu.cn
†Equal contributors
[1]Key Laboratory of Saline-Alkali Vegetation Ecology Restoration in Oil Field (SAVER), Ministry of Education, Alkali Soil Natural Environmental Science Center (ASNESC), Northeast Forestry University, Hexing Road No. 26, Xiangfang District, Harbin City, Heilongjiang Province 150040, China
Full list of author information is available at the end of the article

[6]. In addition, stable plant transgenic lines (particularly *Arabidopsis*) expressing epitope-tagged or otherwise modified genes offer advantages in terms of ensuring a sustainable supply of plant material with homologous protein expression, the potential for mutant complementation, as well as the ability to conduct a global-scale examination throughout all tissues and cell types. Although the commonly used floral dip procedure can also be used toward this end, this procedure is more time consuming and laborious as it requires the maintenance and growth of the transgenic plants to maturity, which can take up to several weeks or longer.

All these methods have specific advantages and disadvantages depending on the experimental question and setup at hand. In general, however, heterologous expression systems, either transient or stable, have the major drawback of being prone to experimental artifacts. There are several challenges associated with the above-described method for determining protein subcellular localization in non-model plants, including the difficulty in obtaining transgenic plant lines and to express a protein and determine its subcellular localization in the original plant cells, because an efficient regeneration system has not yet been established for several wild plant species. To date, gene function analysis of wild plant species has been conducted using the model plant *Arabidopsis* or in yeast [7-9]; however, because these are heterologous systems, even when they successfully express certain wild plant genes, the localization of the encoded proteins might not be representative of localization in the native species [10]. Therefore, in order to best study the subcellular localization of wild plant species proteins, an effective homologous expression system needs to be established. While stable transformation of gametes or regeneration of whole plants from transformed somatic cells is routinely achieved in model plant species (e.g., by floral dipping in *Arabidopsis thaliana*), for most non-model species, the lack of suitable protocols and the phenomenon of regeneration recalcitrance make the obtainment of whole-plant transformants an arduous task. To circumvent this problem, we here propose a method for protein subcellular localization analyses that does not rely on obtaining whole-plant transformation, but instead focuses on the *Agrobacterium*-mediated stable transformation of cells in suspension-cultures (Figure 1). Plant suspension-cultured cells, a proven physiological and versatile cell system, are widely used in plant biology as a convenient tool for high-throughput functional analysis, and have also been successfully used for genetic analyses [11-16]. Here, we present a reliable and highly efficient method for stable gene expression in *Puccinellia tenuiflora* suspension-cultured cells by *Agrobacterium*-mediated transformation. We applied this method to express an endoplasmic reticulum-green fluorescent protein (ER-GFP) fusion protein in *P. tenuiflora*, a grass halophyte with extreme tolerance to alkaline soils (pH ≈ 10).

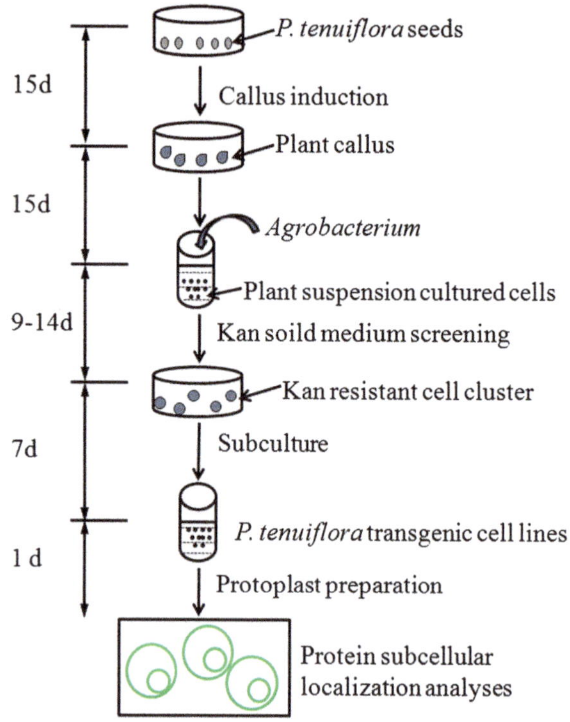

Figure 1 Overview of the method for *Agrobacterium*-mediated stable transformation of *P. tenuiflora* suspension-cultured cells. The time (in days) required for each step of the procedure is indicated to the left side of the figure.

Methods

Materials and preparations

Wild *P. tenuiflora* seeds and *Agrobacterium tumefaciens* harboring the plasmid *pBI121-ER-GFP* were provided from the Alkali Soil Natural Environment Science Center of Northeast Forestry University, Anda practice base, Harbin, China. This plasmid contains 499 amino acids from the *PutAMT1;1* ammonium transporter from *P. tenuiflora*, which has been previously demonstrated to be localized at the nucleus periphery in correspondence with the ER and the plasma membrane [17]. *P. tenuiflora* plants were collected from an area of alkaline soil in Northeast China (Heilongjiang Province). No specific permissions were required for these activities, as sampling did not involve any endangered or protected species.

Callus induction and subculture training

P. tenuiflora seeds were soaked in distilled water at 4°C for 1–2 d. After the seeds were dried and soaked in 75% alcohol for 1 min, they were rinsed 3–4 times with sterile water, then with 10% sodium hypochlorite for 30 min, and finally with sterile water 3–5 times. The calli were induced on the disinfected dry seeds after inoculation in the induction medium, which comprised Murashige and Skoog (MS) medium containing 0.5 g/L proline, 0.5 mg/L glutamic acid, 30 g/L sucrose, 2.0 mg/L 2,4-dichlorophenoxyacetic acid (2,4-D), and 0.8% agar (pH 5.8) at 25°C under illumination of 80 μmol/($m^2 \cdot s$) for 12 h/d. Calli were obtained after 2 weeks. Weakly yellow, loose calli were selected for subculture training.

Establishment of the system of *P. tenuiflora* suspension-cultured cell system

The selected calli were cultured at 22°C with shaking at 120 rpm in the dark. After culturing for 1, 3, 5, 7, 9, 11, 13, 15, 17, 19, or 21 d, the evenly suspended cells were harvested. Simultaneously, for subculture training, 10 mL of suspension-cultured cells (dry weight approximately 40 mg) were inoculated in liquid MS medium with different nutrient conditions: 1) different types of nitrogen nutrients, including 1/2 MS (i.e., half concentration of regular MS), MSN (i.e., half the ammonium nitrate amount of MS), MSK (i.e., half the potassium nitrate amount), MSE (i.e., additional 0.5 mg/L glutamic acid), MSP (i.e., additional 0.5 g/L proline), MSPE (i.e., 0.5 mg/L glutamic acid and 0.5 g/L proline), MSCH (i.e., 0.6 g/L casein acid hydrolysate), and MSPECH (i.e., 0.5 mg/L glutamic acid, 0.5 g/L proline, and 0.6 g/L casein acid hydrolysate); all the different media contained 2.0 mg/L 2,4-D, and 30 g/L sucrose (pH = 5.8); 2) different concentrations of sucrose, including 2.0 mg/L 2,4-D, and 0, 10, 20, 25, 30, 35, 40, or 50 g/L sucrose at pH = 5.8; 3) the same hormone at different concentrations: MSPECH + 0, 1.0, 2.0, 3.0, or 4.0 mg/L 2, 4-D and 30 g/L sucrose at pH = 5.8; 4) different pH: MSPECH + 2.0 mg/L 2, 4-D and 30 g/L sucrose at pH = 3.8, 4.8, 5.8, 6.8, 7.8, 8.8, 9.8, or 10.8. All media were tested at 22 ± 1°C with shaking at 120 rpm in the dark. After culturing for 15 d, the suspension cultures (50 mL) were centrifuged at 2000 rpm for 5 min, then dried at 70°C for 24 h and weighed; this process was repeated three times.

Effect of acetosyringone (AS) on transformation

Three-day-old *P. tenuiflora* suspension-cultured cells were used to test the effects of AS on the transformation process. AS (100 mg/L) was added to the culture medium to create our subculture, while the cells in the culture medium without AS were used as the control.

Isolation of protoplasts from *P. tenuiflora* suspension-cultured cells

Five-day-old *P. tenuiflora* suspension-cultured cells (5 mL) were centrifuged at 800 rpm for 5 min, and the precipitated cells were washed and centrifuged again at 800 rpm for 5 min. The newly precipitated cells were weighed and incubated in 0.05 mM MES (pH 5.8) containing 2% cellulose enzyme, 1% pectinase, 0.2 M $CaCl_2 \cdot 2H_2O$, and 0.6 M mannitol for 3–4 h at 28°C with gentle shaking. After the enzymatic digestion, the cells were centrifuged at 100 rpm for 5 min and the supernatant was discarded. The cells were resuspended in 2 mM MES (pH 5.7) containing 154 mM NaCl, 125 mM $CaCl_2$, and 5 mM KCl for 30 min. Protoplasts were harvested by centrifugation at 1000 rpm for 5 min, washed three times using the resuspension solution, and then solubilized in the solution containing 0.16 M mannitol and 0.02 M $CaCl_2 \cdot 2H_2O$.

Transformation of *P. tenuiflora* suspension-cultured cells

Two-week-old suspension-cultured *P. tenuiflora* cells (5 mL) were inoculated into 20 mL of freshly modified MS medium containing 100 mg/L AS. After 3 d, the cells were co-cultured with *Agrobacterium* at optical density at 600 nm (OD_{600}) of 0.5, 1.0, 1.5, or 2.0 with shaking at 120 rpm and maintained at 22°C in the dark for 3 d. Then, the cells were filtered over nylon nets and washed with 200 mg/L cefotaxime (Cef) in sterile water. After being transferred to the liquid medium for 24 h, 1 mL suspension-cultured cells was inoculated into the modified MS medium containing 2.0 mg/L 2,4-D, 30 g/L sucrose (pH = 5.8), additional 0, 25, 50, 75, or 150 mg/L kanamycin (Kan), and 0, 100, 200, 300, 400, or 500 mg/L Cef at 25°C in the dark. Subsequently, the cluster of cells was transferred into 50 mL of liquid medium containing Kan and Cef and incubated at 22°C with shaking at 120 rpm in the dark for 9–14 d to obtain the transgenic suspension-cultured cells. Finally, the cells were harvested and dried at 70°C for 24 h. The dry weight was measured three times.

GFP fluorescence observation

Green fluorescence of *P. tenuiflora* transgenic cells was observed using a stereomicroscope (Olympus, SZX9, Japan). The detection of the GFP signals was carried out using a laser-scanning confocal imaging system (Olympus Fluoview, FV500, Japan).

Results and discussion

Most of the currently available methods for protein subcellular localization can be readily applied to the majority of model plant species such as *Arabidopsis* and tobacco, which could, in principle, lend themselves as heterologous recipients for localization studies of proteins from non-model plant species [18,19]. However, the relatively high failure rates reported in studies adopting such heterologous approaches (e.g., for *Arabidopsis* proteins expressed in tobacco) indicate that homologous expression systems are preferable whenever possible [10,20]. The lack of established regeneration protocols for the large majority of non-model plant species hinders the application of whole-plant stable transformation methods for protein subcellular localization studies, which rely mainly on transient transformation approaches. The latter methods, however, are limited with respect to the level of detail to which localization can be assessed, as subcellular fractionation or sectioning and immunodetection methods cannot be applied, thus seriously preventing analyses of functional characterization of genes in non-model species. Therefore, adaptation of methods for the stable transformation of cell cultures that are currently applied mainly for functional studies [21,22] to protein subcellullar localization analyses would provide two main advantages for cases dealing with non-model species: (1) the faithful representation of native localization patterns that are typical of homologous systems, and (2) the versatility, robustness, and virtually unlimited amount of material available for characterizing stable transformation methods. As a case study for the development of such an approach, we chose to use *P. tenuiflora*, a non-model halophite grass belonging to the Poaceae family that has been the subject of several previous studies owing to its ability to grow in soils with extremely high salinity and pH [23-26]. Given that *P. tenuiflora* is recalcitrant to regeneration, it can be conveniently used as a representative case for the proof of concept of the novel method.

Establishment of suspension cell lines of *P. tenuiflora*

To establish *P. tenuiflora* suspension-cultured cells, a callus was induced by dedifferentiation of mature *P. tenuiflora* seeds. In brief, among the calli with different morphologies induced on the seeds of *P. tenuiflora*, we chose a yellow, loose callus (Figure 2A) with a clear nucleus and without visible intercellular aggregations (Figure 2B) for subculture. Unlike the standard method used for filtering rice [21], dispersed cells were obtained by slight grinding of the callus. The obtained cell clusters were mixed with liquid culture medium and cultured for 15 d. The culture conditions were optimized to produce the suspension-cultured cell system of *P. tenuiflora* as described below.

Optimization of the liquid medium

We optimized the liquid medium in terms of concentration of sucrose (Figure 3A), amounts of 2,4-D (Figure 3B), pH (Figure 3C), and types of nitrogen nutrients (Figure 3D) added in a total volume of 50 mL of MS medium. After 15 d of cultivation, the dry weight of the harvested cells was measured. The dry weight of *P. tenuiflora* suspension-cultured cells was found to be highest in MSPECH medium (pH = 5.8) containing 30 g/L sucrose and 2 mg/L of 2,4-D (Figure 3), which was thus chosen as the optimum medium for the culture of *P. tenuiflora* suspension-cultured cells.

Growth performance of P. tenuiflora suspension-cultured cells

In order to understand the growth dynamics of *P. tenuiflora* suspension-cultured cells, we characterized their dry weight in terms of culturing time under the optimized conditions. Figure 4A shows the growth curve of the *P. tenuiflora* suspension-cultured cells at different growth stages. Suspension-cultured cells (10 mL) were inoculated into 50 mL of MS medium, and the dry weight of the suspension was determined to be approximately 40 mg. After 5 d of cultivation, the growth rate increased and stabilized at ~15 d of cultivation, after which the cells started to show signs of aging, thus prejudicing further tests. Furthermore, as shown in Figure 4B, the cells were well dispersed, and the small cell clusters presented inclusions and uniform appearance during the 5–15 d of cultivation. Our results showed that during the 5–15 d of cultivation, the cultured cells were stable and reached a high dry weight, which were suitable conditions to perform *Agrobacterium* transformation.

P. tenuiflora suspension-cultured cells are a stable genetic transformation system

The *Agrobacterium*-mediated transformation was used to establish the genetic transformation system of the

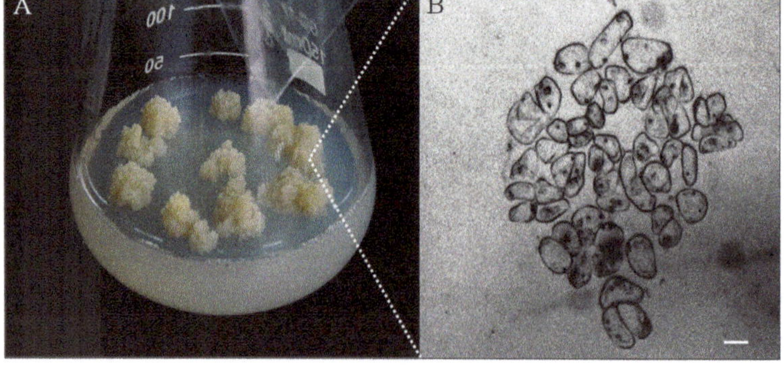

Figure 2 Callus induction in *P. tenuiflora*. (A) A friable and yellow callus of *P. tenuiflora* suspension cultures. **(B)** Magnified image of the single cells, obtained using a microscope, showing no visible intercellular aggregations and clear nuclei. Scale bar = 10 μm.

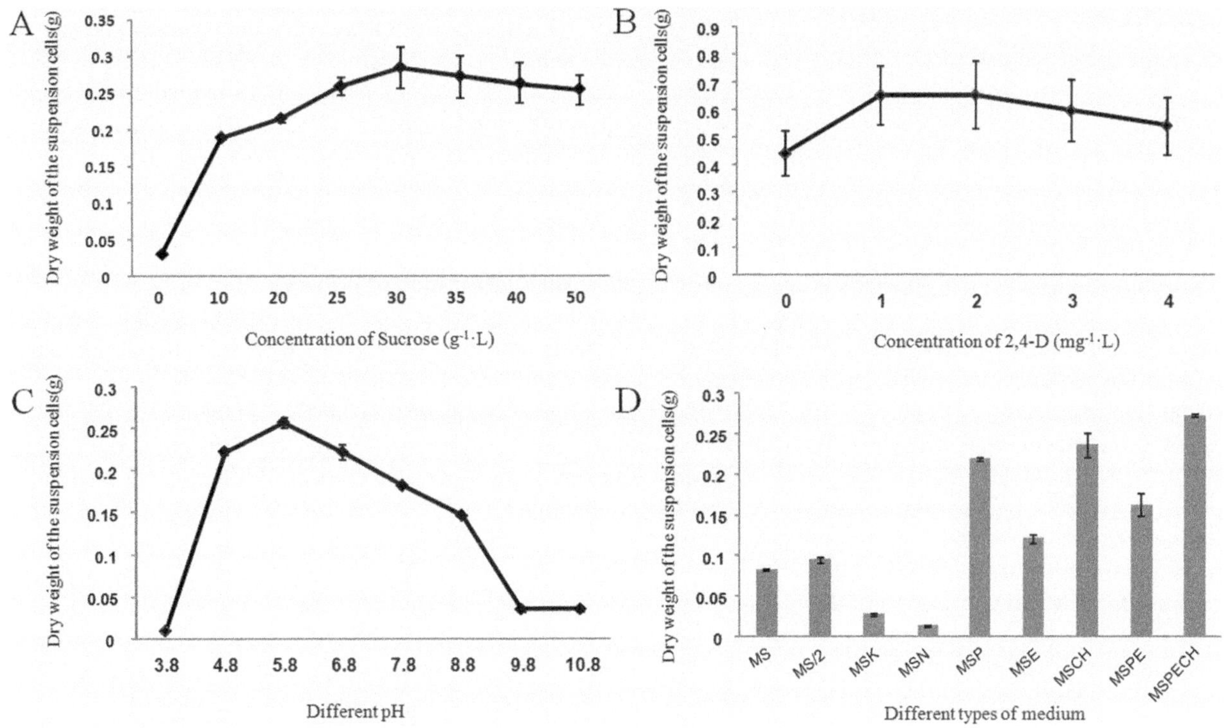

Figure 3 Liquid culture medium optimization for *P. tenuiflora* suspension-cultured cells. Dry weight of the harvested *P. tenuiflora* suspension-cultured cells at different concentrations of sucrose **(A)** and 2,4-dichlorophenoxyacetic acid (2,4-D) **(B)**, different pH **(C)**, and medium types containing different nitrogen nutrients **(D)**. The values represent the mean ± standard errors (n = 3).

P. tenuiflora suspension-cultured cells. The *pBI121-ER-GFP* plasmid was used to transform the *P. tenuiflora* suspension-cultured cells. The transgenic cells were screened on a solid medium containing the appropriate marker. To ensure high transformation efficiency, the *Agrobacterium* concentration, AS effect, co-culture time during transformation, and working concentrations of Kan and Cef were optimized as described below.

Agrobacterium concentration

Considering that *Agrobacterium* concentration can affect the transformation efficiency, four different concentrations of *Agrobacterium* were tested: OD_{600} = 0.5, 1.0, 1.5, and 2.0. Although Kan-resistant cell clusters grew at *Agrobacterium* concentrations of an OD_{600} of 0.5, 1.0, and 1.5, the growth of the cell clusters was negatively affected at higher concentration (OD_{600} = 2.0). Therefore, we suggest

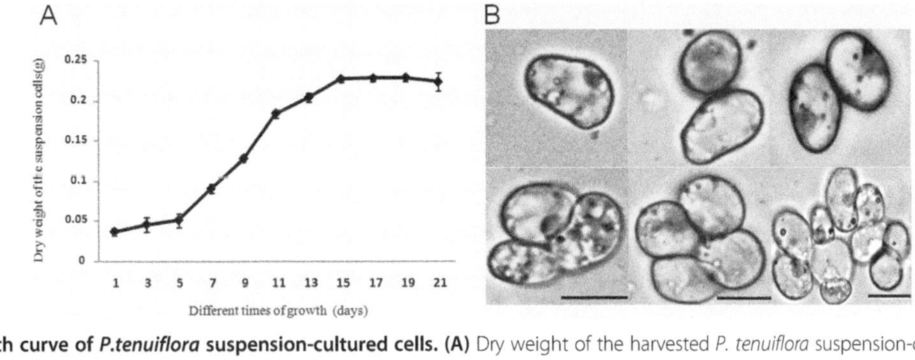

Figure 4 Growth curve of *P.tenuiflora* suspension-cultured cells. (A) Dry weight of the harvested *P. tenuiflora* suspension-cultured cells at different growth stages. **(B)** Image of *P. tenuiflora* suspension-cultured cells obtained using a microscope. Scale bar = 10 μm. The values represent the mean ± standard errors (n = 3).

that *Agrobacterium* concentration for genetic transformation should be kept as low as possible (an *Agrobacterium* concentration corresponding to $OD_{600} = 0.3$ was suitable for genetic transformation; data not shown). In addition, the length of co-culture during the transformation was also important for obtaining high transformation efficiency, as the bacteria were observed to be completely removed from the suspension cells after 3 d of co-culture (data not shown). Therefore, 3 d of co-culture was chosen as the optimum culturing time in order to obtain high transformation efficiency. Pandey et al. reported that optimal β-glucuronidase expression was observed in cumin embryos co-cultivated with an *Agrobacterium* cell suspension at an OD_{600} of 0.6 for 72 h [27], while the optimal transformation efficiency of suspension-cultured *Glycyrrhiza inflata* Batalin cells was achieved using an *Agrobacterium* suspension of an OD_{600} of 0.4 over 24 h of co-cultivation [22]. These results suggested that co-cultivating the lowest concentration of *Agrobacterium* possible with plant suspension-cultured cells for approximately 3 d could provide optimal conditions for achieving highly efficient transformation.

Optimization of the concentrations of Kan, Cef, and AS

The working concentrations of Kan and Cef were examined. Figure 5A shows that the dry weight of the suspension-cultured cells significantly decreased with increasing concentrations of Kan, and that the working Kan concentration of ~75 mg/L was most appropriate for this study. Moreover, the dry weight of the suspension-cultured cells slightly decreased with increasing concentrations of Cef. However, since the growth inhibition at 200 mg/L Cef was only slightly reduced compared to that at 300 mg/L Cef (Figure 5B), the optimum concentrations of Kan and Cef were determined to be 75 mg/L and 200 mg/L, respectively.

AS is a commonly used agent affecting *Agrobacterium*-mediated plant genetic transformation. To test the AS-mediated effects on the *Agrobacterium*-mediated genetic transformation of *P. tenuiflora* suspension-cultured cells, 3-d-old AS-pretreated suspension cells were co-cultured with *Agrobacterium* at $OD_{600} = 0.5$ for 3 d; non-pretreated cells were used as the control. A slight increase in the number of cell clusters from the AS-treated suspension cells on the solid medium was observed compared to that in the control (data not shown), suggesting that AS pretreatment could improve the efficiency of infection of *P. tenuiflora* suspension-cultured cells.

Subcellular localization of proteins

In order to carry out a proof of concept study for the use of stably transformed *P. tenuiflora* suspension-cultured cells in the analysis of the subcellular localization of proteins, the plasmid *pBI121-ER-GFP*, encoding an ER-GFP fusion protein was used for genetic transformation. First, the calli were inoculated into MSPECH culture medium containing 2 mg/L 2,4-D and 30 g/L sucrose (pH 5.8) for 15 d. After 3 d, AS-pretreated cells were co-cultured with *Agrobacterium* at $OD_{600} \approx 0.3$ for 3 d, and the transgenic cell lines were selected on solid MSPECH medium containing 75 mg/L Kan and 200 mg/L Cef by observing them with a stereomicroscope (Olympus, SZX9, Japan). Green fluorescent cells were manually selected for subculture to obtain highly pure *P. tenuiflora* transgenic cell lines according to previously published protoplast isolation methods, with slight modifications for improvement [28-30]. After 5 d of cultivation, the GFP signals of the transgenic protoplasts were mainly concentrated on the plasma membrane and at the ER in association with the nuclear periphery, which is consistent with the subcellular localization pattern previously established for the PutAMT1;1 protein [17] (Figure 6A-B). Given the availability of stably transformed *P. tenuiflora* cell lines expressing the PutAMT1;1-GFP fusion protein, we plan to carry out further studies using subcellular fractionation and immunogold electron microscopy to determine the localization of this protein in its native cellular environment.

Taken together, the results presented herein indicate that the *P. tenuiflora* suspension-cultured cell system

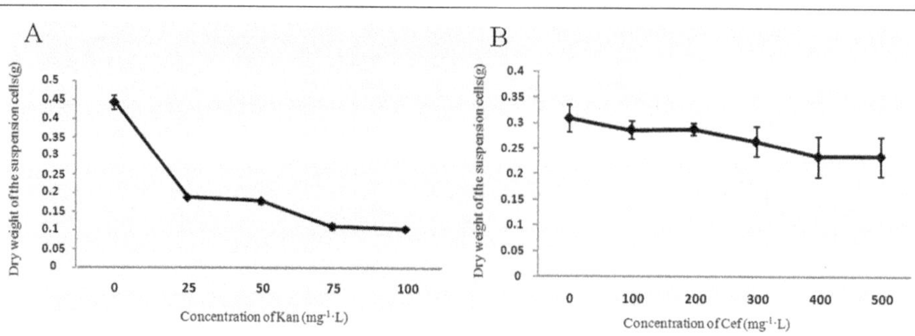

Figure 5 Test of the sensitivity of *P.tenuiflora* suspension-cultured cells against different concentrations of kanamycin (Kan) and cefotaxime (Cef). Dry weight of the harvested *P. tenuiflora* suspension-cultured cells at different concentrations of Kan **(A)** and Cef **(B).** The values represent the mean ± standard errors (n = 3).

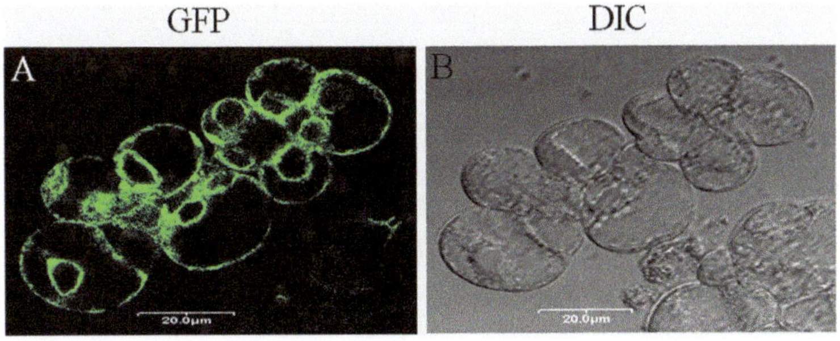

Figure 6 Subcellular localization of an endoplasmic reticulum-green fluorescent protein (ER-GFP) fusion protein in *P. tenuiflora* suspension-cultured cells. (A) GFP, **(B)** Differential interference contrast (DIC). All images were created using a laser-scanning confocal imaging system (Olympus Fluoview, FV500, Japan). GFP fluorescence was excited using an argon laser (488 nm). Scale bar = 20 μm.

could be successfully established and employed for protein subcellular localization analysis. Although previous studies reported that this approach may not be suitable for subcellular localization and other fluorescence-based analyses [31,32], we demonstrate that this optimized expression system based on *P. tenuiflora* suspension-cultured cells proved to be simple, reliable, and stable. Nevertheless, like all transformation methods relying on cells isolated from single organs without intervening regeneration of whole plants, this method can only be applied to examine the localization of proteins in cell compartments/structures that are present in the recipient cells. Therefore, the method is not suitable for studies requiring localization to the cell walls or plasmodesmata, for example, or to track protein movements in the context of organized tissues or organs. We note, however, that application of this method to other cell types deriving from different tissues such as the leaf, root, and inflorescence requires further validation to best ascertain the scope of applications of the method. In comparison to transient assays, our method is more labor-intensive and time-consuming, and thus does not lend itself readily to application in high-throughput studies [10]. However, the method does have an important advantage of providing homogeneous populations of transformed cells in virtually unlimited amounts for extended periods of time that can be used in all localization assays, which cannot otherwise be carried out on transiently transformed protoplasts owing to their complexity or particular technical requirements (e.g., cell fractionation, embedding and sectioning, immunolocalization [33,34]). Furthermore, we expect that our stable transformation method could be suitable for dynamically documenting protein re-localization through the different phases of the cell cycle or in response to environmental cues, although this was not directly tested in the present study. Most importantly, the fact that this method does not rely on whole plant regeneration makes it applicable to any non-model plant species lacking suitable regeneration protocols, which is a substantial advantage. As the method is based on simple, species-specific optimization steps of protocols that were previously developed for functional analyses in a relatively distantly related monocot species (rice), we expect that, with minor modifications, the method could be applied to several other non-model species from the Poaceae family.

Conclusions

We have developed a rapid and stable suspension-cultured cell system for non-model plants (i.e., those lacking suitable regeneration protocols) based on an *Agrobacterium*-mediated approach. This method was successfully used for determining ER-GFP fusion protein subcellular localization in *P. tenuiflora*. Our method is simple and rapid, and is applicable to evaluating the localization of proteins only in cell compartments/structures that are present in the recipient cells.

Abbreviations

GFP: Green fluorescent protein; ER: Endoplasmic reticulum; MS: Murashige and Skoog; 2,4-days: 2,4-dichlorophenoxyacetic acid; 1/2 MS: Half concentration of regular MS; MSN: Contains 1/2 amount of ammonium nitrate of MS; MSK: Contains 1/2 amount of potassium nitrate; MSE: Contains additional 0.5 mg/L glutamic acid; MSP: Contains additional 0.5 g/L proline; MSPE: Contains 0.5 mg/L glutamic acid and 0.5 g/L proline; MSCH: Contains 0.6 g/L casein acid hydrolysate; MSPECH: Contains 0.5 mg/L glutamic acid, 0.5 g/L proline, and 0.6 g/L casein acid hydrolysate; AS: Acetosyringone; Kan: Kanamycin; Cef: Cefotaxime; GUS: β-glucuronidase.

Competing interests

The authors declare that they have no competing interests.

Authors' contributions

YB, MZ and SL designed the study. YB and MZ performed the experiments and drafted the manuscript. BS conducted the statistical analysis. XZ prepared vectors. SL and TT supervised the study and critically revised the manuscript. All authors read and approved the final manuscript.

Acknowledgements

This work was supported by the Program for Changjiang Scholars and Innovative Research Team in University of China (PCSIRT), State Forestry Administration 948 Program of PR China (No. 2008429) and the Heilongjiang Provincial Program for Distinguished Young Scholars (JC200609) awarded to Shenkui Liu. Further support was provided by the Fundamental Research Funds for the Central Universities (No. DL13BAX13) awarded to Yuan-yuan Bu. The funders had no role in study design. We thank the Editage Corporation for providing the English editing service.

Author details

[1]Key Laboratory of Saline-Alkali Vegetation Ecology Restoration in Oil Field (SAVER), Ministry of Education, Alkali Soil Natural Environmental Science Center (ASNESC), Northeast Forestry University, Hexing Road No. 26, Xiangfang District, Harbin City, Heilongjiang Province 150040, China. [2]Asian Natural Environmental Science Center (ASNESC), The University of Tokyo, Nishitokyo, Tokyo 188-0002, Japan.

References

1. Bodén M, Hawkins J: **Prediction of subcellular localization using sequence-biased recurrent networks.** *Bioinformatics* 2005, **21**(10):2279–2286.
2. Adelfio A, Volpato V, Pollastri G: **SCLpredT: Ab initio and homology-based prediction of subcellular localization by N-to-1 neural networks.** *Springerplus* 2013, **2**:502.
3. Liu L, Zhang Z, Mei Q, Chen M: **PSI: a comprehensive and integrative approach for accurate plant subcellular localization prediction.** *PLoS One* 2013, **8**(10):e75826.
4. Sheen J: **Signal transduction in maize and Arabidopsis mesophyll protoplasts.** *Plant Physiol* 2001, **127**:1466–1475.
5. Christou P: **Strategies for variety-independent genetic transformation of important cereals, legumes and woody species utilizing particle bombardment.** *Euphytica* 1995, **85**:13–27.
6. Yang Y, Li R, Qi M: **In vivo analysis of plant promoters and transcription factors by agroinfiltration of tobacco leaves.** *Plant J* 2000, **22**:543–551.
7. Ardie SW, Liu S, Takano T: **Expression of the AKT1-type K(+) channel gene from Puccinellia tenuiflora, PutAKT1, enhances salt tolerance in Arabidopsis.** *Plant Cell Rep* 2010, **29**(8):865–874.
8. Ardie SW, Xie L, Takahashi R, Liu S, Takano T: **Cloning of a high-affinity K+ transporter gene PutHKT2;1 from *Puccinellia tenuiflora* and its functional comparison with OsHKT2;1 from rice in yeast and Arabidopsis.** *J Exp Bot* 2009, **60**(12):3491–3502.
9. Liu H, Zhang X, Takano T, Liu S: **Characterization of a PutCAX1 gene from Puccinellia tenuiflora that confers Ca^{2+} and Ba^{2+} tolerance in yeast.** *Biochem Biophys Res Commun* 2009, **383**(4):392–396.
10. Marion J, Bach L, Bellec Y, Meyer C, Gissot L, Faure JD: **Systematic analysis of protein subcellular localization and interaction using high-throughput transient transformation of Arabidopsis seedlings.** *Plant J* 2008, **56**(1):169–179.
11. Ogawa Y, Sakurai N, Oikawa A, Kai K, Morishita Y, Mori K, Moriya K, Fujii F, Aoki K, Suzuki H, Ohta D, Saito K, Shibata D: **High-throughput cryopreservation of plant cell cultures for functional genomics.** *Plant Cell Physiol* 2012, **53**(5):943–952.
12. Callard D, Axelos M, Mazzolini L: **Novel molecular markers for late phases of the growth cycle of Arabidopsis thaliana cell-suspension cultures are expressed during organ senescence.** *Plant Physiol* 1996, **112**(2):705–715.
13. Mitsukawa N, Okumura S, Shirano Y, Sato S, Kato T, Harashima S, Shibata D: **Overexpression of an Arabidopsis thaliana high-affinity phosphate transporter gene in tobacco cultured cells enhances cell growth under phosphate-limited conditions.** *Proc Natl Acad Sci U S A* 1997, **94**(13):7098–7102.
14. Uno Y, Furihata T, Abe H, Yoshida R, Shinozaki K, Yamaguchi-Shinozaki K: **Arabidopsis basic leucine-zipper transcription factors involved in an abscisic acid-dependent signal transduction pathway under drought and high-salinity conditions.** *Proc Natl Acad Sci U S A* 2000, **97**(21):11632–11637.
15. Takahashi S, Katagiri T, Hirayama T, Yamaguchi-Shinozaki K, Shinozaki K: **Hyperosmotic stress induced a rapid and transient increase in inositol 1,4,5-trisphosphate independent of abscisic acid in Arabidopsis cell culture.** *Plant Cell Physiol* 2001, **42**(2):214–222.
16. Stolc V, Samanta MP, Tongprasit W, Sethi H, Liang S, Nelson DC, Hegeman A, Nelson C, Rancour D, Bednarek S, Ulrich EL, Zhao Q, Wrobel RL, Newman CS, Fox BG, Phillips GN Jr, Marklev JL, Sussman MR: **Identification of transcribed sequences in Arabidopsis thaliana by using high-resolution genome tiling arrays.** *Proc Natl Acad Sci U S A* 2005, **102**(12):4453–4458.
17. Bu Y, Sun B, Zhou A, Zhang X, Lee I, Liu S: **Identification and Characterization of a PutAMT1;1 Gene from *Puccinellia tenuiflora*.** *PLoS One* 2013, **8**(12):e83111.
18. Wang M, Yuan F, Hao H, Zhang Y, Zhao H, Guo A, Hu J, Zhou X, Xie CG: **BolOST1, an ortholog of Open Stomata 1 with alternative splicing products in Brassica oleracea, positively modulates drought responses in plants.** *Biochem Biophys Res Commun* 2013, **442**(3–4):214–220.
19. Sasaki K, Christov NK, Tsuda S, Imai R: **Identification of a novel LEA protein involved in freezing tolerance in wheat.** *Plant Cell Physiol* 2013, in press.
20. Campanoni P, Sutter JU, Davis CS, Littlejohn GR, Blatt MR: **A generalized method for transfecting root epidermis uncovers endosomal dynamics in Arabidopsis root hairs.** *Plant J* 2007, **51**:322–330.
21. Ozawa K, Takaiwa F: **Highly efficient Agrobacterium-mediated transformation of suspension-cultured cell clusters of rice (*Oryza sativa* L.).** *Plant Sci* 2010, **179**:333–337.
22. Li Y, Li S, Dong Y, Zhang Y, Fu C, Yu L: **Stable transformation of suspension-cultured *Glycyrrhiza inflata* batalin cells with agrobacterium tumefaciens.** *Z Naturforsch C* 2012, **67**(11–12):603–610.
23. Kobayashi S, Abe N, Yoshida KT, Liu S, Takano T: **Molecular cloning and characterization of plasma membrane- and vacuolar-type Na^+/H^+ antiporters of an alkaline-salt-tolerant monocot. Puccinellia tenuiflora.** *J Plant Res* 2012, **125**(4):587–594.
24. Wang SM, Zhao GQ, Gao YS, Tang ZC, Zhang CL: **Puccinellia tenuiflora exhibits stronger selectivity for K^+ over Na^+ than wheat.** *J Plant Nutr* 2004, **27**:1841–1857.
25. Peng YH, Zhu YF, Mao YQ, Wang SM, Su WA, Tang ZC: **Alkali grass resists salt stress through high [K^+] and an endodermis barrier to Na^+.** *J Exp Bot* 2004, **55**:939–949.
26. Wang CM, Zhang JL, Liu XS, Li Z, Wu GQ, Cai JY, Flowers TJ, Wang SM: **Puccinellia tenuiflora maintains a low Na + level under salinity by limiting unidirectional Na + influx resulting in a high selectivity for K + over Na+.** *Plant Cell Environ* 2009, **32**:486–496.
27. Pandey S, Mishra A, Patel MK, Jha B: **An efficient method for agrobacterium-mediated genetic transformation and plant regeneration in cumin (*cuminum cyminum* L.).** *Appl Biochem Biotechnol* 2013, **171**(1):1–9.
28. Schirawski J, Planchais S, Haenni AL: **An improved protocol for the preparation of protoplasts from an established Arabidopsis thaliana cell suspension culture and infection with RNA of turnip yellow mosaic tymovirus: a simple and eliable method.** *J Virol Methods* 2000, **86**(1):85–94.
29. Lee LY, Fang MJ, Kuang LY, Gelvin SB: **Vectors for multi-color bimolecular fluorescence complementation to investigate protein-protein interactions in living plant cells.** *Plant Methods* 2008, **4**:24.
30. Wu FH, Shen SC, Lee LY, Lee SH, Chan MT, Lin CS: **Tape-Arabidopsis sandwich-A simpler Arabidopsis protoplast isolation method.** *Plant Methods* 2009, **5**:16.
31. Zhang Y, Su J, Duan S, Ao Y, Dai J, Liu J, Wang P, Li Y, Liu B, Feng D, Wang J, Wang H: **A highly efficient rice green tissue protoplast system for transient gene expression and studying light/chloroplast-related processes.** *Plant Methods* 2011, **7**(1):30.
32. Li JF, Park E, von Arnim AG, Nebenfuhr A: **The FAST technique: a simplified Agrobacterium-based transformation method for transient gene expression analysis in seedlings of Arabidopsis and other plant species.** *Plant Methods* 2009, **5**:6.
33. Szymanski WG, Kierszniowska S, Schulze WX: **Metabolic labeling and membrane fractionation for comparative proteomic analysis of Arabidopsis thaliana suspension cell cultures.** *J Vis Exp* 2013, **79**:e50535.
34. Marie-Laure F–G, Sophie P, Loïc F, Véronique G, Azeddine D: **An improved chemical fixation method suitable for immunogold localization of green fluorescent protein in the Golgi apparatus of tobacco bright yellow (BY-2) cells.** *J Histochem Cytochem* 2003, **51**(7):931–940.

14

Immobilization of carboxypeptidase from *Sulfolobus solfataricus* on magnetic nanoparticles improves enzyme stability and functionality in organic media

Silvia Sommaruga[1], Elisabetta Galbiati[1], Jesus Peñaranda-Avila[1], Chiara Brambilla[1], Paolo Tortora[1*], Miriam Colombo[1] and Davide Prosperi[1,2]

Abstract

Background: Superparamagnetic iron oxide nanoparticles (MNP) offer several advantages for applications in biomedical and biotechnological research. In particular, MNP-based immobilization of enzymes allows high surface-to-volume ratio, good dispersibility, easy separation of enzymes from the reaction mixture, and reuse by applying an external magnetic field. In a biotechnological perspective, extremophilic enzymes hold great promise as they often can be used under non-conventional harsh conditions, which may result in substrate transformations that are not achievable with normal enzymes. This prompted us to investigate the effect of MNP bioconjugation on the catalytic properties of a thermostable carboxypeptidase from the hyperthermophilic archaeon *Sulfolobus solfataricus* (CP*Sso*), which exhibits catalytic properties that are useful in synthetic processes.

Results: CPSso was immobilized onto silica-coated iron oxide nanoparticles via NiNTA-His tag site-directed conjugation. Following the immobilization, CP*Sso* acquired distinctly higher long-term stability at room temperature compared to the free native enzyme, which, in contrast, underwent extensive inactivation after 72 h incubation, thus suggesting a potential utilization of this enzyme under low energy consumption. Moreover, CP*Sso* conjugation also resulted in a significantly higher stability in organic solvents at 40°C, which made it possible to synthesize N-blocked amino acids in remarkably higher yields compared to those of free enzyme.

Conclusions: The nanobioconjugate of CPSso immobilized on silica-coated magnetic nanoparticles exhibited enhanced stability in aqueous media at room temperature as well as in different organic solvents. The improved stability in ethanol paves the way to possible applications of immobilized CP*Sso*, in particular as a biocatalyst for the synthesis of N-blocked amino acids. Another potential application might be amino acid racemate resolution, a critical and expensive step in chemical synthesis.

Keywords: Carboxypeptidase, Magnetic nanoparticles, His-tag immobilization, Enzyme stability, Hyperthermophilic microorganisms

* Correspondence: paolo.tortora@unimib.it
[1]Dipartimento di Biotecnologie e Bioscienze, Università di Milano-Bicocca, Piazza della Scienza 2, Milano 20126, Italy
Full list of author information is available at the end of the article

Background

Due to their unique magnetic properties, superparamagnetic iron oxide nanoparticles (MNP) represent a well-known and effective nanosystem for applications in biomedical and biotechnological research. In fact, MNP are being developed as drug carriers [1,2], imaging agents [1,3], analytical probes [4,5] and recyclable support for enzyme immobilization [6,7].

Compared with chemical catalysis, enzyme-mediated reactions allow for more specific chemo-, regio-, and stereoselectivity in organic synthesis [8]. However, long-term stability and recyclability of enzymes have been considered the main limitations to their extensive utilization [9].

Recently, several nanoparticles have been employed to improve traditional enzyme immobilization methods in order to enhance loading, activity and stability of enzymes and to reduce the biocatalyst costs in industrial biotechnology [10,11]. In particular, MNP-based immobilization of enzymes presents several advantages, including (i) high surface-to-volume ratio offered by nanosize magnetic beads, (ii) good dispersibility, (iii) easy separation of enzymes from the reaction mixture, and (iv) reuse by applying an external magnetic field [12]. One of the crucial points in protein immobilization on nanoscale solid surfaces is that structural modifications may occur, which could affect protein activity and stability to different extents depending on the protein and the conjugation strategy [13]. For this reason, there has been an increasing interest in developing new reliable approaches for the immobilization of enzymes on magnetic nanoparticles [14-16]. However, although great efforts have been made for this purpose, the actual effect of immobilization on enzyme functionality is still poorly understood. Furthermore, no such studies involving enzymes from extremophile microorganisms have been carried out so far.

Extremophilic enzymes hold great promise in industrial biotechnology, as they can often be used under non-conventional harsh conditions, which may result in substrate transformations that are not achievable with normal enzymes [17].

In this work, we investigated the effect of MNP bioconjugation on the catalytic properties of a thermostable carboxypeptidase from *Sulfolobus solfataricus* (CP*Sso*). CP*Sso* is a heat- and pressure-resistant zinc-metalloprotease consisting of four identical 43 kDa subunits [17-20]. The catalytic and kinetic mechanisms of CP*Sso* have been well established and were confirmed by a 3D model that was developed and validated in the past years [21,22]. CP*Sso* exhibits nonconventional catalytic properties that are useful in several synthetic processes. First, it removes any amino acid from the C-terminus of short peptides, with the sole exception of proline, and also hydrolyzes N-blocked amino acids, thus acting as an aminoacylase [22]. Second, despite its remarkable thermophilicity, it maintains a significant fraction of its maximal activity even at room temperature. Finally, CP*Sso* maintains a significant activity in solvent mixtures even at high content of organic fraction [23]. These peculiar properties highlight the biotechnological potential of this enzyme, in particular to achieve the synthesis of N-blocked amino acid in organic media.

Results

Synthesis of Ni^{2+}-functionalized silica-coated magnetic nanoparticles (NiNTASiMNP)

High-quality magnetite nanocrystals were synthesized in organic solvents and transferred to water phase using tetramethylammonium hydroxide (TMAOH) as previously described [24]. Next, the water-soluble iron oxide nanoparticles were individually coated with a 10 nm thick silica shell by reaction with tetraethoxysilane (TEOS) in aqueous ammonia and further reacted with the chelating agent ICPTES-NTA and $NiCl_2$, [25] resulting in fully armed Ni^{2+} functionalized nanoparticle (NiNTASiMNP). The overall synthesis procedure is summarized in Figure 1. In Figure 2, TEM images show that the final size of SiMNP was 30 ± 3 nm with a magnetic core of 10 ± 1 nm, indicating a homogeneous sample of silica-coated magnetic nanoparticles. Hydrodynamic diameter of SiMNP in ethanol was 80.4 ± 2.3 nm, as determined by dynamic light scattering, which remarkably increased to 128.3 ± 3.7 nm as a consequence of NTA functionalization. Such a difference accounts for a transient loss in nanoparticle stability forming small aggregates after reaction with ICPTES-NTA. However, stable nanoparticle dispersion was recovered after Ni^{2+} chelation, resulting in hydrodynamic size of 89.0 ± 2.8 nm. Zeta-potential (ζ) of SiMNP, NTA-SiMNP and NiNTASiMNP in water was also investigated. SiMNP and NTASiMNP were negatively charged (−19.11 ± 1.70 mV and −24.42 ± 1.03 mV, respectively), the second being more negative due to the presence of NTA carboxylic groups. The capture of Ni^{2+} ions by the external carboxylic groups of NTA resulted in a more positive charge in NiNTASiMNP (−12.00 ± 1.03 mV). The amount of Ni^{2+} ions captured by NTASiMNP was quantified by inductively coupled plasma optical emission spectrometry (ICP-OES), resulting in 41 ng Ni per mg nanoparticles.

Immobilization of CP*Sso* on NiNTASiMNP

CP*Sso* protein containing 6 × His tag was produced in *Escherichia coli* and purified by Ni-chelate affinity chromatography, as previously described [21]. Subsequent Ni^{2+} capture by NTA groups of NTASiMNP promoted the active Ni^{2+}-NTA affinity-oriented immobilization of His-tagged CP*Sso* (Figure 1). The conjugation reaction was performed at room temperature by incubating NiNTASiMNP (1 mg) with purified CP*Sso* (200 μg). The

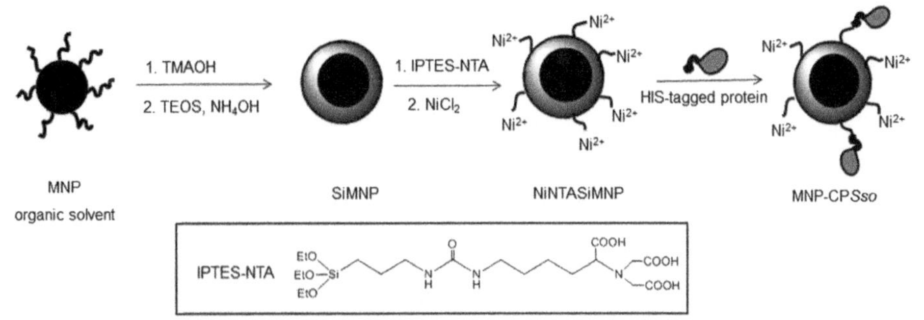

Figure 1 **Procedure of CP*Sso* immobilization on NiNTASiMNP.**

resulting MNP-CP*Sso* was isolated from unreacted CP*Sso* by centrifugation and washed three times. By measuring the amount of protein found in the supernatants after the enzyme binding process, we determined an amount of CP*Sso* immobilized on nanoparticles of 150 μg per mg of MNP-CP*Sso*. The average CP*Sso* loading was estimated to be about 2 molecules per nanoparticle. To check that the conjugation reaction occurred specifically to Ni^{2+} and His tag, CP*Sso* was incubated in parallel with non-functionalized nanoparticles (NTASiMNP). Then, reaction mixtures were washed, and bound and unbound CP*Sso* determined enzymatically. Non-specifically bound activity was only 4.5% of that bound through the NiNTA linker (Additional file 1: Table S1). These data confirm that CP*SSo* immobilization is mediated by the interaction between Ni^{2+} and His tag. After immobilization, CP*Sso* substantially retained its activity, as shown by comparison of equal amount of free and bound CP*Sso* (data not shown).

Assessment of the catalytic stability of MNP-CP*Sso*

The catalytic stability of free and conjugated CP*Sso* was analyzed by monitoring the inactivation profiles of the samples incubated in aqueous solution at three different temperatures (25°C, 40°C and 75°C, respectively) and, subsequently, in the presence of organic solvents, *i.e.* dimethyl formamide (DMF) or ethanol, at different water ratios.

As expected, CP*Sso* was fully active after 30 min incubation at 75°C in aqueous buffer, while MNP-CP*Sso* proved to be less stable (Figure 3a). No significant differences were observed in the stability of free and conjugated enzyme at 40°C (Figure 3b). Functionality profiles obtained at 25°C showed that free CP*Sso* gradually lost its activity resulting in a complete inactivation after 96 h. In such a case, however, enzyme conjugation led to a substantial increase in stability, up to 85% of initial activity being retained after 96 h (Figure 3c). A tentative interpretation of these results is presented in the section Discussion.

The inactivation profiles obtained at 40°C and different concentrations of DMF indicate that CP*Sso* gradually lost its activity by increasing the organic component in the solvent mixture, while the nanobioconiugate retained 80% of residual activity even in the presence of 80% DMF (Figure 3d-f). The results obtained in the presence of ethanol at 40°C and various concentrations are of particular interest. The inactivation profiles show that CP*Sso* had a residual activity of 50% after 6 h, which decreased to 20% after 24 h incubation (Figure 3g-i). However, MNP-CP*Sso* revealed a significantly improved stability in ethanol at the different tested concentrations compared with free CP*Sso*, up to 80-90% of residual activity after 6 h, and 70% after 24 h incubation in 80% ethanol being retained.

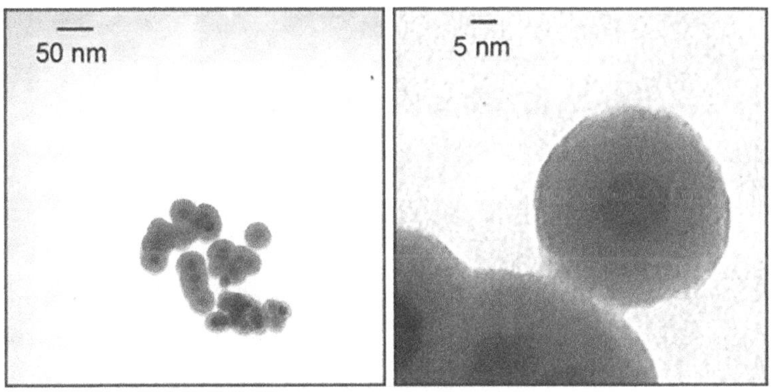

Figure 2 **TEM images of synthesized SiMNP at two different magnifications.**

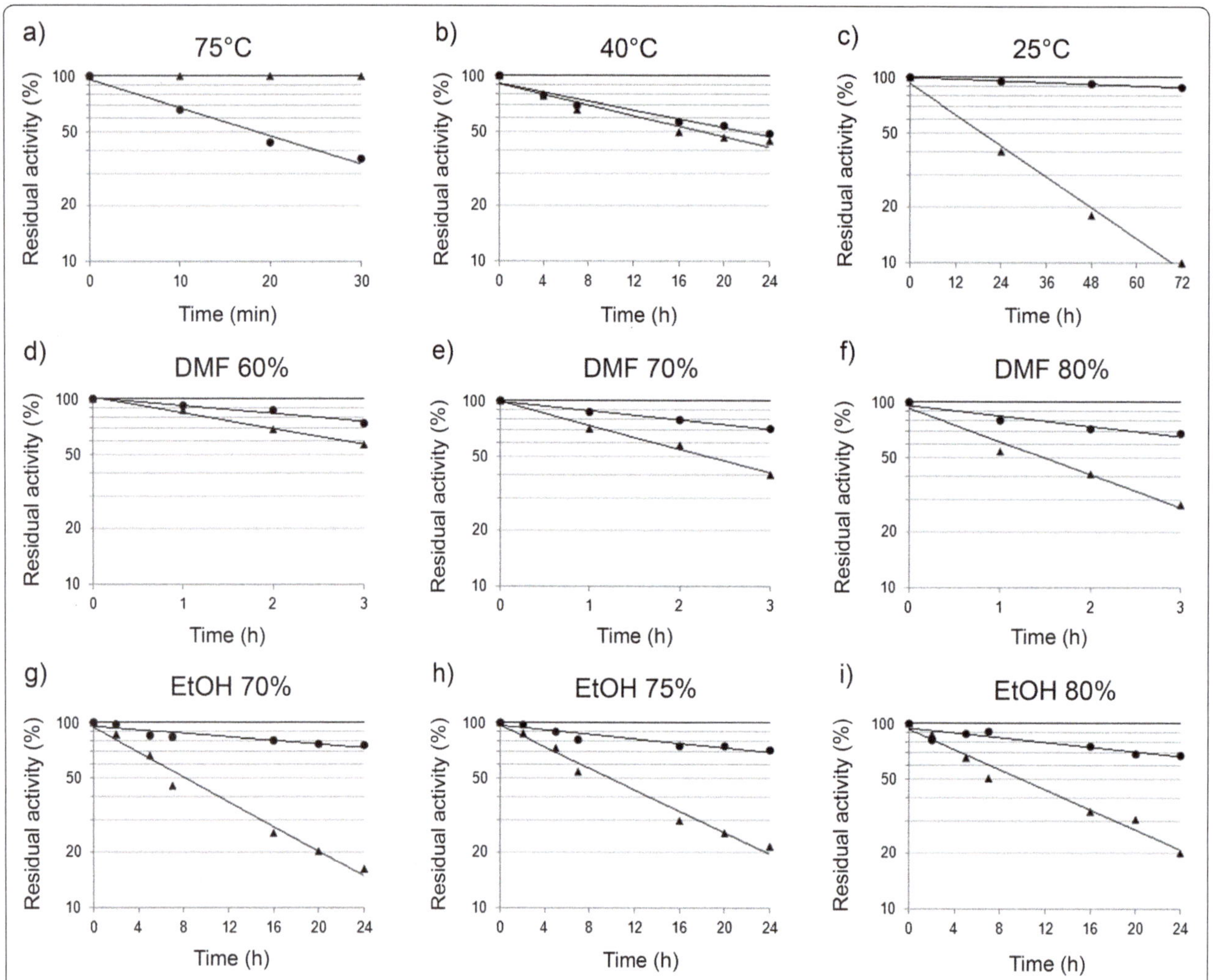

Figure 3 Inactivation profiles of free (▲) and conjugated (•) CP*Sso* in aqueous medium (a-c) at 75°C, 40°C and 25°C or in the presence of the indicated DMF (d-f) or ethanol (g-i) concentrations at 40°C. The enzyme was incubated under shaking at a concentration of 0.6 mg mL^{-1}, in 50 mM potassium Mes pH 6.5.

Synthesis of N-blocked amino acids promoted by MNP-CP*Sso*

Preliminary experiments were carried out in order to select the water-cosolvent system and the chemical-physical parameters most suitable to promote the synthesis of *N*-blocked amino acids by CP*Sso*, using arginine and benzoate as a model system (Figure 4). Our results show that the best rate of benzoyl-arginine (BA) synthesis was obtained in 70% ethanol at pH 6.5 and 40°C after 6 h of incubation, using 10 U mL^{-1} CP*Sso*, 0.4 M benzoate and

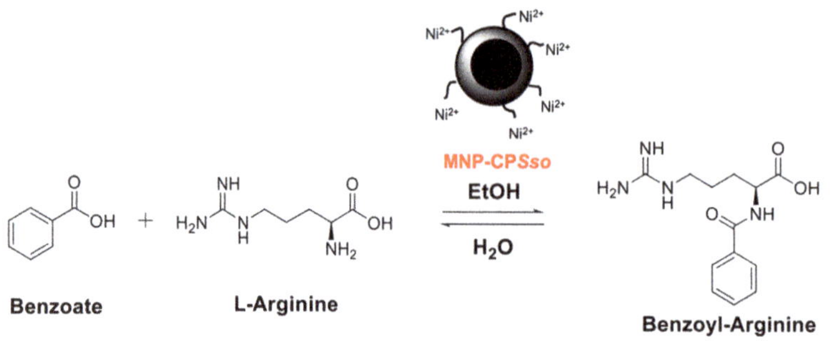

Figure 4 MNP-CP*Sso*-catalyzed synthesis of an N-blocked amino acid.

0.1 M arginine (data not shown). Next, BA synthesis was performed using both free CP*Sso* and MNP-CP*Sso* under the aforementioned optimal conditions. By taking advantage of the remarkable enzyme stability in organic solvent, higher ethanol concentration (75% and 80%) and a longer incubation time (24 h), were also tested. The results obtained after a 6-h incubation (Figure 5) show a similar synthetic rate for free and conjugated enzyme in 70% ethanol. However, at higher ethanol concentrations, only MNP-CP*Sso* could synthesize appreciable amounts of BA, although in much lower yields. Remarkably, when the synthesis was extended to 24 h, higher yields were observed, with the sole exception of unbound CP*Sso* in 70% ethanol. In particular, at 70-75% ethanol, far better yields were attained when using MNP-CPSso (Figure 5). Under all adopted conditions, no detectable enzyme activity was found in the reaction mixture after removal of the MNPs (data not shown).

The activity of NiNTASiMNP was retained without significant drop in functionality for three-times recycled uses in comparison with the initial activity, comparable with other systems of similar efficiency [26].

Discussion

In our ongoing effort in designing innovative approaches for the orientation-controlled conjugation of proteins on metal-based nanoparticles for biotechnological and biomedical applications [5,27-31], we have recently investigated the modulation of the structural properties of the immobilized protein from a biophysical point of view [13]. In the present paper, we report the results of a study conducted on a hyperthermophilic enzyme, namely CP*Sso*, aimed at determining the effect of its immobilization on superparamagnetic nanoparticles on the biological activity of the enzyme.

The nanobioconjugate was obtained by affinity-oriented immobilization of His-tagged CP*Sso* on silica-coated magnetic nanoparticles functionalized with Ni-NTA groups (NiNTASiMNP). The procedure for the synthesis of NiNTASiMNP is illustrated in Figure 1 and the resulting functionalized Fe_3O_4@silica core-shell nanoparticles gave a stable dispersion in aqueous environment, as determined by dynamic light scattering analysis.

An important requirement for protein immobilization is that the matrix should provide a biocompatible and inert environment, *i.e.* it should not interfere with the native structure of the protein, which would thereby compromise its biological activity. Thus, we measured the enzymatic activity of free and immobilized CP*Sso* and confirmed that its activity was preserved after conjugation. Next, we evaluated the protein activities in an aqueous milieu to assess the effect of conjugation on the stability of the hyperthermophilic enzyme in dependence of the temperature. Noteworthy, in this respect, is the substantial stability at room temperature of MNP-conjugated CP*Sso*, unlike the free enzyme, which should make it possible to perform transformations without the use of energy for heating or cooling the reaction environment, otherwise needed for maintaining enzyme stability. In a theoretical perspective, this might be rationalized on the basis of the following considerations. As free enzyme undergoes inactivation at 40°C and 25°C, but not at 75°C, in agreement with our previous report [22], this fits with the "cold denaturation" pattern, which is believed to be a general phenomenon in protein chemistry [32]. In the case of free CP*Sso*, optimum stability temperature might be at, or close to 75°C, as significant inactivation was previously observed at 85°C [33]. In contrast, the conjugated enzyme was essentially stable at 25°C and underwent progressive destabilization on increasing temperature, suggestive of a substantial downshift of the stability optimum. Based on enzyme activity decay constants determined at the three temperatures, we assessed an activation enthalpy ($\Delta H^{\ddagger}$) of enzyme inactivation around 150 $kJ \cdot mol^{-1}$ for the conjugated enzyme, well below values close to 500 $kJ \cdot mol^{-1}$ determined for the free form [33,22]. As the enzyme retains its activity after conjugation, it must retain its overall structure as well. Thus, the dramatic drop in $\Delta H^{\ddagger}$ of bound with respect to free enzyme, is suggestive of major conformational constraints of the

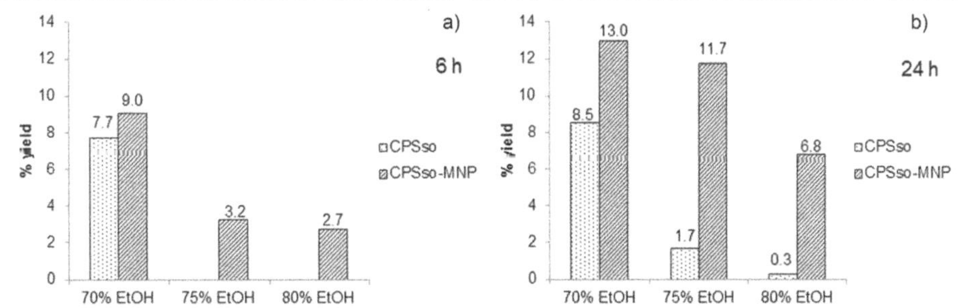

Figure 5 Synthesis of benzoyl-arginine catalysed by free and conjugated CP*Sso* in the presence of different ethanol concentration. 0.4 M benzoate, 0.1 M arginine were incubated with 5 U mL^{-1} CP*Sso* at 40°C in 50 mM potassium Mes, pH 6.5, and ethanol at the indicated concentrations. After a 6 h **(a)** or 24 h **(b)** incubation, the amount of benzoyl-arginine synthesized was determined spectrophotometrically.

transition state as a result of immobilization. This is confirmed by an activation entropy value ($\Delta S^{\ddagger}$) close to zero (as determined from the data in Figure 3a to c), to be compared with 1.1 $kJ \cdot mol^{-1} K^{-1}$ of the free enzyme [33,22].

It is well-known that enzyme immobilization generally results in increased stability toward the denaturing effect of water-soluble organic solvents [34]. Also, it is believed that such polar solvents first strip water from the surface of the protein, then strongly compete for intramolecular hydrogen bonds, with resulting protein unfolding [35]. In the case of CP*Sso*, the conjugation quite likely limits its conformational flexibility, as above outlined, which may partially prevent the solvent from penetrating the protein interior. The stronger effect of DMF with respect to ethanol fits well with its higher polarity and "denaturing capacity", as previously defined [36]. Remarkably, we found that at 40°C the conjugated enzyme was even more stable in 80% ethanol than in water.

Improved stability of MNP-CP*Sso* in organic solvent is relevant to possible industrial applications of the enzyme as a biocatalyst in synthetic reactions carried out in organic environment. In particular, also thanks to its broad substrate specificity, CP*Sso* could be an ideal candidate as a biocatalyst for the synthesis of *N*-blocked amino acids in water-cosolvent mixtures following the thermodynamic method. Actually, we previously found that the enzyme is capable of both hydrolyzing N-blocked amino acids in aqueous solution [20], and catalyzing the synthetic reaction in organic solvents [33], irrespective of the amino acid involved (with the sole exception of proline). The presence of organic solvents in solution is in fact deemed necessary to shift the equilibrium toward the formation of the peptide bond, due to their generally low dielectric constant and the consequent prevalence of undissociated forms of carboxylic and amino groups, which are indeed the reactive species in amide linkage formation [37]. It is worthwhile to be mentioned, in this respect, that at 70-75% ethanol, MNP-CP*Sso* catalyzed significant synthesis of benzoyl-arginine (BA) (Figure 5), which is promising for possible industrial applications, whereas the free enzyme's performance was very poor under the same conditions. The lower levels of BA produced in 75% and 80% compared to 70% ethanol, are likely due to enzyme inhibition by the organic solvent. In fact, it is well known that the kinetic parameters, k_{cat} and K_m, are strongly affected by organic solvents [38].

Conclusions

We developed a nanobioconjugate of CP*Sso* immobilized on silica-coated magnetic nanoparticles, which exhibited enhanced stability in aqueous media at room temperature as well as in different organic solvents. The improved stability in ethanol paves the way to possible applications of immobilized CP*Sso*; in particular, our results show that the enzyme could be employed as a biocatalyst for the synthesis of *N*-blocked amino acids. Another potential application might be amino acid racemate resolution, a critical and expensive step in chemical synthesis. This could be accomplished by chemically synthesizing *N*-blocked racemic amino acid mixtures, followed by CP*Sso*-catalyzed selective hydrolysis of the blocking group from the L-enantiomer.

Methods

Chemicals and Instrumentation

All chemicals were purchased from Sigma-Aldrich (St. Louis, MO) and used as received. Water was deionized and ultrafiltered by a MilliQ apparatus from Millipore Corporation (Billerica, MA) before use. Transmission electron microscopy (TEM) images were obtained by a Zeiss EM-109 microscope operating at 80 kV, available at the "Centro di Microscopia Elettronica per le Nanotecnologie applicate alla medicina" (CMENA, University of Milan). Dynamic light scattering (DLS) measurements were performed at 90° with a 90 Plus Particle Size Analyzer from Brookhaven Instruments Corporation (Holtsville, NY), working at 15 mW of a solid state laser (λ = 661 nm). Zeta-potential (ζ) measurements were performed on the same instrument, equipped with a couple of AQ-809 electrodes, and analyses were processed by a ZetaPlus software. Viscosity and refractive index of pure water were used to characterize the solvent. Nanoparticles were dispersed in the solvent and sonicated in a S15H Elmasonic apparatus (Elma, Singen, Germany) before analysis. Final sample concentration used for measurements was typically of 5 $\mu g\ mL^{-1}$. For ICP-OES analysis, 2 ml of the NiNTASiMNP dispersion (5 $mg\ mL^{-1}$) was measured in triplicate with Optima 7000 DV ICP-OES (Perkin Elmer).

Synthesis of NiNTASiMNP

Magnetic nanoparticles (**MNP**) were synthesized in organic solvent by solvothermal decomposition as described in our previous work [30]. The 8 nm iron oxide nanoparticles coated with oleic acid dissolved in chloroform (20 mg, 6.7 mL) were treated with an aqueous solution of tetramethylammonium hydroxide (TMAOH, 0.6 g in 20 ml) overnight under vigorous stirring at room temperature [27,30,39]. The organic solution containing the surfactant was removed by centrifugation at 1000 rpm (3 min). The aqueous solution of TMAOH-coated MNP (20 mg) was concentrated in Amicon Ultra 100.000 MWCO (Millipore) (3000 rpm, 5 min). The resulting nanoparticle dispersion (1.2 mL) was added to a mixture of EtOH (3 mL) and aqueous ammonia (75 µL, 28% by weight), then, a solution of tetraethyl orthosilicate (TEOS) in ethanol (45 µL in 500 µL EtOH) was added dropwise and stirred overnight in order to obtain individual-particle silica coating. The precipitate was collected by magnet, washed three times with EtOH and dried under vacuum. The resulting nanoparticles were dispersed in EtOH (33 mL) (**SiMNP**).

For the synthesis of ICPTES-NTA, a solution of ε-(*N*-benzyloxycarbonyl)lysine (**1**) (1.2 g, 4.2 mmol) in 1.5 M NaOH (15 mL) was added dropwise to a solution of bromoacetic acid (2.5 g, 18 mmol) in 1.5 M NaOH (9 mL) at 0°C over 2 h. The mixture was allowed to warm up to 18°C, stirred at this temperature overnight, then at 50°C for 2 h. The cooled reaction mixture was treated dropwise with 1 M HCl (24 mL). The white precipitate was filtered off, washed with 0.1 M HCl (12 mL) and water (2 × 12 mL), then dried under vacuum and then oven-dried at 90°C to give the triacid derivative (**2**) [40].

The product (2 g) was dissolved in 50 mL of MeOH/H_2O (20:1) under sonication to dissolve the product and, after the addition of a spatula tip of 5% Pd/C, hydrogenated at room temperature and atmospheric pressure. Total hydrogenation of compound was achieved after 90 min as monitored by TLC in CH_3CN/H_2O (4:1), visualized by charring with a solution of $(NH_4)_6Mo_7O_{24}$ (21 g), $CeSO_4$ (1 g) and 96% H_2SO_4 (31 mL) in 500 mL of deionized H_2O, followed by heating. The catalyst was filtered off, and solvent removed in vacuo. The resulting precipitate was redissolved in H_2O (20 mL) and EtOH (90 mL) was added. The product crystallizes at 0°C. The crystals were filtered off and dried in vacuo yielding 1.02 g of **3**. Isocyanatopropyltriethoxysilane (ICPTES, 247.4 mg, 0.37 mmol) was dissolved in $CHCl_3$ (5 mL) and **3** (0.37 mmol) was then added, followed by a solution of triethylamine (3.34 mmol) in MeOH (10 mL). The resultant solution was stirred overnight. Finally, the solvent was dried under vacuum, yielding 265 mg of **4** (Additional file 1: Scheme S1) [25].

An ICPTES-NTA solution (20 mg in 20 mL EtOH) was added to the SiMNP aqueous dispersion (33 mL, EtOH) and the mixture was heated at 80°C for 3 h under stirring. **NTASiMNP** were recovered after centrifugation and washed twice with EtOH to remove the excess of unreacted ICPTES-NTA. In the last step, $NiCl_2 \bullet 6H_2O$ (95 mg in 4 mL H_2O) was added to the dispersion of NTASiMNP (20 mg in 6 mL H_2O). The pH of the solution was adjusted to pH 8.4 with 0.1 N NaOH and the dispersion was sonicated for 30 min. The nanoparticles were washed twice with water by centrifuging at 2500 rpm. The collected washed particles (**NiNTASiMNP**) were dissolved in water and stored at the final concentration of 5 mg mL^{-1}. This procedure is briefly outlined in Figure 1, first two steps.

Affinity purification of histidine-tagged CP*Sso*

Wild-type and mutated histidine-tagged CP*Sso* were purified by Ni-chelate affinity chromatography. *E. coli* cells were grown under shaking at 37°C in LB medium (3 L) containing 0.1% glucose, 50 mg mL^{-1} ampicillin, 25 mg mL^{-1}choramphenicol until A_{600} reached 0.8. Induction was then carried out for 3.5 h with 0.4 mM isopropyl β-D-thiogalactopyranoside. Cells were harvested, washed with 50 mM Tris–HCl, pH 8.0, 0.1 M NaCl, 10 mM $ZnCl_2$, and resuspended in 3 volumes of 50 mM Tris–HCl, pH 8.0, 1 mM PMSF, 5 mM 2-mercaptoethanol, complete EDTA-free protease inhibitor (1 tablet/50 mL; Roche), 13 units mL^{-1} bovine pancreas DNase (Sigma-Aldrich, St. Louis, MO). Cells were disrupted by sonication and centrifuged at 40000 × g for 30 min. After addition of 10% (by vol.) glycerol and 10 mM imidazole, the supernatant was loaded at a flow rate of 1 mL min^{-1} onto a Ni-NTA Superflow (Qiagen) column (2 mL bed volume) pre-equilibrated with 50 mM potassium Mes, pH 6.0, 10 mM imidazole, 5 mM 2-mercaptoethanol, 10% (by vol.) glycerol. The column was washed with 10 volumes of the same buffer and the enzyme eluted with a 20-mL continuous linear imidazole gradient, 10 mM to 1 M, in the same buffer, and 2-mL fractions were collected (Additional file 1: Figure S1). Fractions displaying the highest purity (at least 95%, as assessed in SDS-PAGE) [41] were concentrated in Amicon Ultra 100.000 MWCO (Millipore) and buffer was exchanged to 50 mM potassium Mes, pH 6.0.

CP*Sso* activity and protein assays

CP*Sso* activity was assayed continuously using benzoyl-arginine (Bz-Arg) or furylacryloyl-phenylalanine (Fur-Phe) as substrates. The assay mixture contained 50 mM potassium Mes pH 6.5, 0.1 mM Bz-Arg or 0.2 mM Fur-Phe, in a total volume of 1.5 mL. After a 5-min preincubation at 60°C the enzyme was added, and the decrease in absorbance monitored at the same temperature and 239 nm or 330 nm. At these wavelengths, the decrease of the molar absorption coefficient was 6240 M^{-1} cm^{-1} for Bz-Arg and 5750 M^{-1} cm^{-1} for Fur-Phe, respectively. One unit of enzyme activity is defined as the amount that hydrolyzes 1 μmol min^{-1} of substrate under the standard assay conditions. Protein content was determined using the Coomassie Plus Protein Assay Reagent from Pierce and bovine plasma immunoglobulin G as the standard protein.

CP*Sso* conjugation to NiNTASiMNP

In a plastic tube, NiNTASiMNP (1 mg) were incubated with purified CP*Sso* (200 μg) in 50 mM potassium Mes pH 6.5 in a final volume of 5 mL and the mixture was stirred on an orbital shaker for 15 min at 4°C. CP*Sso*-MNP were isolated from unreacted CP*Sso* by centrifugation at 5000 × g for 3 min and the supernatant was discarded. Nanoparticles were washed three times with 50 mM potassium Mes pH 6.5 (1 ml) and stored in 50 mM potassium Mes pH 6.5 at 4°C. By measuring the amount of protein found in the supernatant after the enzyme binding process, we determined an amount of CP*Sso* immobilized on nanoparticles of 150 μg per mg of MNP-CP*Sso*. The average number of CP*Sso* loaded on MNP-CP*Sso* was estimated to be about 2 molecules per nanoparticle. This procedure is briefly outlined in Figure 1, last step.

Assessment of CP*Sso* stability in aqueous medium

Free and conjugated CP*Sso* were incubated at the desired temperatures in a thermoshaker under stirring and at a concentration of 60 μg mL^{-1}, in 50 mM potassium Mes, pH 6.5. At different times, protein aliquots (about 30 mU) were taken from the mixture and added to 1.5 mL of degassed 50 mM potassium Mes, pH 6.5, 0.1 mM Bz-Arg, already thermostated in a cuvette at 60° C. The reaction was monitored at 239 nm as previously described.

Assessment of CP*Sso* stability in dimethylformamide

Free and conjugated CP*Sso* were incubated at 40°C in a thermoshaker and at a concentration of 60 μgmL^{-1}, in 50 mM potassium Mes, pH 6.5 and the indicated DMF concentrations. At different times, protein aliquots (about 30 mU) were taken from the mixture and added to 1.5 mL of degassed 50 mM potassium Mes, pH 6.5, 0.2 mM Fur-Phe, already thermostated in a cuvette at 60°C. The reaction was monitored at 330 nm as previously described.

Assessment of CP*Sso* stability in ethanol

Free and conjugated CP*Sso* were incubated at 40°C in a thermoshaker and at a concentration of 60 μg mL^{-1}, in 50 mM potassium Mes, pH 6.5, and the indicated ethanol concentrations. At different times, protein aliquots (about 30 mU) were taken from the mixture and added to 1.5 mL of degassed 50 mM potassium Mes, pH 6.5, 0.1 mM Bz-Arg, already thermostated in a cuvette at 60°C. The reaction was monitored at 239 nm as previously described.

Synthesis of N-blocked amino acids

A solution of 50 mM potassium Mes, pH 6.5, including reaction substrates (sodium benzoate and the amino acid) and organic solvent at the desired concentration was prepared. Final volume was 40 μL including enzyme and nanoparticle. Control samples were prepared with NiNTASiMNP and without CP*Sso*. Samples and controls were incubated in a thermoshaker at the desired temperature for 6 or 24 h. At the end of the respective incubation times, each sample was boiled at 100°C for 15 min to inactivate the enzyme.

Determination of the products of synthesis

After boiling the samples, quantitative determination of *N*-blocked amino acid produced was carried out by total hydrolysis catalyzed by CP*Sso*. Aliquots were taken from the samples, added to 2.5 mL of degassed 50 mM potassium Mes, pH 6.5, at an appropriate dilution factor (around 1:1000) so that the initial absorbance was not greater than 1.4 absorbance units. Then, the mixture was thermostated for 10 min at 40°C and the assay performed by monitoring the decrease in absorbance at 40°C in the presence of 20–30 mU of CP*Sso* at 242 nm. The decrease in extinction coefficients (Δε) for Bz-amino acids was 6058 M^{-1} cm^{-1}.

Additional file

Additional file 1: Scheme S1. Synthesis of ICPTES-NTA. **Figure S1.** Progress of *CPSso* purification by Ni-chelate chromatography, as monitored by SDS-PAGE (12% gel). M: molecular weight markers with the respective molecular weights (kDa); CE: crude extract; FT: column flow through; W: column wash; Fr1 to Fr10: individual fractions eluted by an imidazole gradient. For other details, see Materials and Methods. **Table S1.** CPSso binds to MNPs via NiNTA functional groups. The enzyme (ca. 1000 mU) was incubated in the presence of 1 mg of MNP (NiNTA-SiMNP or NTASiMNP) for 15 min at 4°C under gentle shaking. Then, the mixes were centrifuged and the supernatant (containing the unbound enzyme) removed. Next, the MNPs were washed twice with 50 mM potassium MES, 6.5. Finally, they were resuspended in the same buffer and bound activity determined. Activities in the other fractions (unbound, wash1 and wash 2) were also determined.

Abbreviations

CP*Sso*: *Sulfolobus solfataricus* carboxypeptidase; DMF: Dimethylformamide; ICPTES: Isocyanatopropyltriethoxysilane; IPTES-NTA: 15-carboxy-16-(carboxymethyl)-4,4-diethoxy-9-oxo-3-oxa-8,10,16-triaza-4-silaoctadecan-18-oic acid-Ni^{2+}; potassium Mes: 2-(N-Morpholino)ethanesulfonic acid potassium salt; MNP: Superparamagnetic iron oxide nanoparticles; NiNTA: Nichel nitriloacetic acid; NiNTASiMNP: Ni^{2+}-functionalized silica-coated magnetic nanoparticles; SiMNP: Silica-coated magnetic nanoparticles; TEM: Transmission electron microscopy; TEOS: Tetraethylorthosilicate; TMAOH: Tetramethylammonium hydroxide.

Competing interests

The authors declare that they no competing interests.

Authors' contributions

Conceived and designed the experiments: DP PT SS. Performed the experiments: CB EG JPA MC SS. Analyzed the data: DP MC PT SS. Wrote the paper: DP SS PT. All authors read and approved the final manuscript.

Acknowledgments

This work was supported by Regione Lombardia (NanoMeDia Project) and Fondazione Regionale per la Ricerca Biomedica, FRRB (Early System Nanotechnology).

Author details

[1]Dipartimento di Biotecnologie e Bioscienze, Università di Milano-Bicocca, Piazza della Scienza 2, Milano 20126, Italy. [2]Istituto di Scienze e Tecnologie Molecolari, CNR, via Fantoli 16/15, Milano 20138, Italy.

References

1. Colombo M, Carregal-Romero S, Casula MF, Gutiérrez L, Morales MP, Böhm IB, Heverhagen JT, Prosperi D, Parak WJ: **Biological applications of magnetic nanoparticles.** *Chem Soc Rev* 2012, **41**:4306–4334.
2. Sun C, Lee JSH, Zhang M: **Magnetic nanoparticles in MR imaging and drug delivery.** *Adv Drug Delivery Rev* 2008, **60**:1252–1265.
3. Weissleder R, Pittet MJ: **Imaging in the era of molecular oncology.** *Nature* 2008, **452**:580–589.
4. Lee H, Sun E, Ham D, Weissleder R: **Chip-NMR biosensor for detection and molecular analysis of cells.** *Nat Med* 2008, **14**:869–874.

5. Polito L, Monti D, Caneva E, Delnevo E, Russo G, Prosperi D: **One-step bioengineering of magnetic nanoparticles via a surface diazo transfer/azide-alkyne click reaction sequence.** *Chem Commun* 2008, **5**:621–623.
6. Wang W, Xu Y, Wang DIC, Li Z: **Recyclable nanobiocatalyst for enantioselectivesulfoxidation: facile fabrication and high performance of chloroperoxidase-coated magnetic nanoparticles with iron oxide core and polymer shell.** *J Am Chem Soc* 2009, **131**:12892–12893.
7. Dyal A, Loos K, Noto M, Chang SW, Spagnoli C, Shafi KVPM, Ulman A, Cowman M, Gross RA: **Activity of Candida rugosa lipase immobilized on γ-Fe_2O_3 magnetic nanoparticles.** *J Am Chem Soc* 2003, **125**:1684–1685.
8. Koeller KM, Wong CH: **Enzymes for chemical synthesis.** *Nature* 2001, **409**:232–240.
9. Schmid A, Dordick JS, Hauer B, Kiener A, Wubbolt M, Witholt B: **Industrial biocatalysis today and tomorrow.** *Nature* 2011, **409**:258–267.
10. Yiu HHP, Keane MA: **Enzyme–magnetic nanoparticle hybrids: new effective catalysts for the production of high value chemicals.** *J Chem Technol Biotechnol* 2012, **87**:583–594.
11. Abad JM, Mertens SFL, Pita M, Fernandez VM, Schiffrin DJ: **Functionalization of thioctic acid-capped gold nanoparticles for specific immobilization of histidine-tagged proteins.** *J Am Chem Soc* 2005, **127**:5689–5694.
12. Johnson PA, Park HJ, Driscoll AJ: **Enzyme nanoparticle fabrication: magnetic nanoparticle synthesis and enzyme immobilization.** *Methods Mol Biol* 2011, **679**:183–191.
13. Occhipinti E, Verderio P, Natalello A, Galbiati E, Colombo M, Mazzucchelli S, Tortora P, Doglia SM, Prosperi D: **Investigating the structural biofunctionality of antibodies conjugated to magnetic nanoparticles.** *Nanoscale* 2011, **3**:387–390.
14. Li D, Teoh WY, Gooding JJ, Selomulya C, Amal R: **Functionalization strategies for protease immobilization on magnetic nanoparticles.** *Adv Funct Mater* 2010, **20**:1767–1777.
15. Yu CC, Kuo YY, Liang CF, Chien WT, Wu HT, Chang TC, Jan FD, Lin CC: **Site-specific immobilization of enzymes on magnetic nanoparticles and their use in organic synthesis.** *Bioconjugate Chem* 2012, **23**:714–724.
16. Johnson A, Zawadzka A, Deobald L, Crawford R, Paszczynski A: **Novel method for immobilization of enzymes to magnetic nanoparticles.** *J Nanoparticle Res* 2008, **10**:1009–1025.
17. Colombo S, Toietta G, Zecca L, Vanoni M, Tortora P: **Molecular cloning, nucleotide sequence and expression of a carboxypeptidase-encoding gene from the archaebacterium *Sulfolobus solfataricus*.** *J Bacteriol* 1995, **177**:5561–5565.
18. Bec N, Villa A, Tortora P, Mozhaev VV, Balny C, Lange R: **Enhanced stability of carboxypeptidase from Sulfolobus solfataricus at high pressure.** *Biotechnol Lett* 1996, **18**:483–48832.
19. Tortora P, Vanoni M: ***Sulfolobus* carboxipeptidase.** In *Handbook of proteolytic enzymes.* 2nd edition. Edited by Barrett AJ, Rawlings ND, Woessner F. London. Elsevier; 2004:953–955.
20. Occhipinti E, Bec N, Gambirasio B, Baietta G, Martelli PL, Casadio R, Balny C, Lange R, Tortora P: **Pressure and temperature as tools for investigating the role of individual non-covalent interactions in enzymatic reactions. *Sulfolobus solfataricus* carboxypeptidase as a model enzyme.** *Biochim Biophys Acta* 2006, **1764**:563–572.
21. Occhipinti E, Martelli PL, Spinozzi F, Corsi F, Formantici C, Molteni L, Amenitsch H, Mariani P, Tortora P, Casadio R: **3D structure of *Sulfolobus solfataricus* carboxypeptidase developed by molecular modeling is confirmed by site-directed mutagenesis and small-angle X-ray scattering.** *Biophys J* 2003, **85**:1165–1175.
22. Sommaruga S, De Palma A, Mauri PL, Trisciani M, Basilico F, Martelli PL, Casadio R, Tortora P, Occhipinti E: **A combined approach of mass spectrometry, molecular modeling, and site-directed mutagenesis highlights key structural features responsible for the thermostability of *Sulfolobus solfataricus* carboxypeptidase.** *Proteins* 2008, **71**:1843–1852.
23. Colombo S, D'Auria S, Fusi P, Zecca L, Raia CA, Tortora P: **Purification and characterization of a thermostable carboxypeptidase from the extreme thermophilic archaebacterium *Sulfolobus solfataricus*.** *Eur J Biochem* 1992, **206**:349–357.
24. Corsi F, Fiandra L, De Palma C, Colombo M, Mazzucchelli S, Verderio P, Allevi R, Tosoni A, Nebuloni M, Clementi E, Prosperi D: **HER2 expression in breast cancer cells is downregulated upon active targeting by antibody-engineered multifunctional nanoparticles in mice.** *ACS Nano* 2011, **5**:6383–6393.
25. Mazzucchelli S, Verderio P, Sommaruga S, Colombo M, Salvade A, Corsi F, Galeffi P, Tortora P, Prosperi D: **Multiple presentation of Scfv800E6 on silica nanospheres enhances targeting efficiency toward HER-2 receptor in breast cancer cells.** *Bioconjugate Chem* 2011, **22**:2296–2303.
26. Hwang ET, Gang H, Chung J, Gu MB: **Carbonic anhydrase assisted calcium carbonate crystalline composites as a biocatalyst.** *Green Chemistry* 2012, **8**:2216–2220.
27. Mazzucchelli S, Colombo M, De Palma C, Salvadè A, Verderio P, Coghi MD, Clementi E, Tortora P, Corsi F, Prosperi D: **Synthesis of single-domain protein A-engineered magnetic nanoparticles: toward a universal strategy to site-specific labeling of antibodies for targeted detection of tumor cells.** *ACS Nano* 2010, **4**:5693–5702.
28. Colombo M, Mazzucchelli S, Collico V, Avvakumova S, Pandolfi L, Corsi F, Porta F, Prosperi D: **Protein-assisted one-pot synthesis and biofunctionalization of spherical gold nanoparticles for selective targeting of cancer cells.** *Angew Chem Int Ed* 2012, **51**:9272–9275.
29. Colombo M, Mazzucchelli S, Montenegro JM, Galbiati E, Corsi F, Parak WJ, Prosperi D: **Protein oriented ligation on nanoparticles exploiting O6-alkylguanine-DNA transferase (SNAP) genetically encoded fusion.** *Small* 2012, **8**:1492–1497.
30. Colombo M, Sommaruga S, Mazzucchelli S, Polito L, Verderio P, Galeffi P, Corsi F, Tortora P, Prosperi D: **Site-specific conjugation of scFv antibodies to nanoparticles by bioorthogonal strain-promoted alkyne-nitronecycloaddition.** *Angew Chem Int Ed* 2012, **51**:496–499.
31. Baietta G: *La carbossipeptidasi dall'archaeon Sulfolobus solfataricus: un enzima termostabile che promuove la sintesi del legame peptidico in solvente organico.* Master's Degree Thesis: University of Milan; 2000.
32. Dias CL, Ala-Nissila T, Wong-ekkabut J, Vattulainen I, Grant M, Karttunen M: **The hydrophobic effect and its role in cold denaturation.** *Cryobiology* 2010, **60**:91–99.
33. Villa A, Zecca L, Fusi P, Colombo S, Tedeschi G, Tortora P: **Structural features responsible for kinetic thermal stability of a carboxypeptidase from the archaebacterium Sulfolobus solfataricus.** *Biochem J* 1993, **295**:827–831.
34. Olofsson L, Söderberg P, Nicholls IA: **Influence of water miscible organic solvents on alpha-chymotrypsin in solution and immobilized on Eupergit CM.** *Biotechnol Lett* 2006, **28**:929–935.
35. Mattos C, Ringe D: **Proteins in organic solvent.** *Curr Opin Struct Biol* 2001, **11**:761–764.
36. Khmelnitsky YL, Mozhaev VV, Belova AB, Sergeeva MV, Martinek K: **Denaturation capacity: a new quantitative criterion for selection of organic solvents as reaction media in biocatalysis.** *Eur J Biochem* 1991, **198**:31–41.
37. Bongers J, Heimer EP: **Recent applications of enzymatic peptide synthesis.** *Peptides* 1994, **15**:183–193.
38. Carrea G, Riva S: **Properties and synthetic applications of enzymes in organic solvents.** *Angew Chem Int Ed* 2000, **39**:2226–2254.
39. Liong M, Shao H, Haun JB, Lee H, Weissleder R: **Carboxymethylated polyvinyl alcohol stabilizes doped ferrofluids for biological applications.** *Adv Mater* 2010, **22**:5168–5172.
40. Schmitt L, Dietrich C, Tampe R: **Synthesis and characterization of chelator-lipids for reversible immobilization of engineered proteins at self-assembled lipid interfaces.** *J Am Chem Soc* 1994, **116**:8485–8492.
41. Laemmli UK: **Cleavage of structural proteins during the assembly of the head of bacteriophage T4.** *Nature* 1970, **227**:680–685.

A novel *Streptomyces* spp. integration vector derived from the *S. venezuelae* phage, SV1

Bahgat Fayed[1], Ellen Younger[2], Gabrielle Taylor[1] and Margaret C M Smith[1*]

Abstract

Background: Integrating vectors based on the *int/attP* loci of temperate phages are convenient and used widely, particularly for cloning genes in *Streptomyces* spp.

Results: We have constructed and tested a novel integrating vector based on *g27*, encoding integrase, and *attP* site from the phage, SV1. This plasmid, pBF3 integrates efficiently in *S. coelicolor* and *S. lividans* but surprisingly fails to generate stable integrants in *S. venezuelae*, the natural host for phage SV1.

Conclusion: pBF3 promises to be a useful addition to the range of integrating vectors currently available for *Streptomyces* molecular genetics.

Keywords: *Streptomyces*, Cloning, Integration vector, Serine integrase, Bacteriophage, SV1

Background

Bacteria in the genus *Streptomyces* are a prolific source of natural products, many of which are used in the clinic as antibiotic, anticancer, immune-modulatory or other therapeutic agents. Furthermore these soil bacteria have an unusual life style; vegetative growth is mycelial and when nutrients become scarce a sporulation cycle initiates [1]. The phages that infect these bacteria have been exploited in the development of vectors for genetic engineering of *Streptomyces* and closely related genera, in particular in the study of natural product pathways [2].

The development of integrating vectors that integrate via site-specific recombination between a site on the plasmid vector, the *attP* site and a site in the bacterial chromosome, the *attB* site have been widely adopted by researchers wishing to genetically manipulate *Streptomyces* genes [3]. The *int/attP* site from the integrating plasmid, pSAM2, was first exploited in a novel vector that could integrate into the endogenous *attB* site in the *Streptomyces* genome [4]. The advantage of the integration vectors over freely replicating plasmid vectors are the very low copy number (usually single or two copies integrated in tandem), the ease of construction of plasmids, which can be done in *E. coli*, and the simple method of plasmid transfer into *Streptomyces* via conjugation [5].

The idea of using phage integrases by the research group at Eli Lilley, led to the development of integrating vectors encoding the *int/attP* locus from the *Streptomyces* phage ϕC31 [6-8]. The recombination event that leads to phage integration is a conservative reciprocal DNA cleavage and rejoining mechanism occurring at the centre of the *attP* and *attB* sites producing the integrated plasmid flanked by hybrid *attP/B* sites called *attL* and *attR* [7]. Phage integrases are known to be highly directional, with tight control over integration versus excision. Integration (or *attB x attP*) is the default reaction for phage integrases whilst excision (*attL x attR*) requires activation by a recombination directionality factor (RDF) or Xis [9,10]. Consequently the integration vectors based on the ϕC31 *int/attP* system and lacking any other phage genes are 100% stable in most *Streptomyces* species. The ϕC31 integrating vectors integrated with higher efficiency and were more stable than the pSAM2-derived integration vectors and are now widely adopted by researchers in *Streptomyces* genetics. The use of the ϕC31 integration system is also being widely adopted for genome engineering in eukaryotes, in particular in tissue culture and model organisms such as the mouse and *Drosophila* [11].

In 2003 the *int/attP* locus from the phage ϕBT1 was used to generate an alternative suite of phage-derived integration vectors for *Streptomyces* [12]. These vectors were demonstrated to be completely orthogonal to the ϕC31 derived plasmids and plasmids derived from the

* Correspondence: maggie.smith@york.ac.uk
[1]Department of Biology, University of York, York YO10 5DD, UK
Full list of author information is available at the end of the article

two phage *int/attP* loci could be used in combination without loss of integrating efficiency. The ϕBT1 *int/attP* integration vectors are also widely used in the *Streptomyces* community. Recently vectors based on phage TG1 have been developed for use in *Streptomyces avermitilis* [13].

Here we present a new integrating vector derived from *S. venezuelae* phage SV1 [14,15]. We demonstrate the efficient integration of the plasmid into several *Streptomyces* spp. although, surprisingly, we failed to obtain stable integrants in *S. venezuelae*.

Results and discussion

The genome of phage SV1 was sequenced previously and, consistent with the temperate nature of the phage, *g27* encodes an integrase [15]. Gp27 is a serine integrase, whose closest homologue in the database is from *Streptomyces prunicolor* (WP_019054986.1; 54% identity). SV1 is only distantly related (between 11 and 13% identical) to ϕC31, TG1 and ϕBT1 integrases so the SV1 integration system encoded by SV1 *g27/attP* is therefore very likely to be another orthologous system to ϕC31 and ϕBT1 integration systems.

The DNA upstream and downstream of SV1 *g27* was studied for a likely *attP* site. Precedent dictates that *attP* is normally upstream or downstream of the integrase gene but in some mobile genetic elements, such as *SCCmec*, can be located quite distal from their cognate integrase genes [11,16]. In SV1 the *attP* site is unlikely to be upstream of *g27* as the upstream gene, *g28*, overlaps with *g27* by the sequence 5′ATGA, which couples the start codon (ATG) of *g27* with the stop codon of *g28* (TGA). Downstream of *g27* is a non-coding region of 342 bp before the start of the downstream gene, *g26*. The *attP* sites for the serine integrases commonly comprise a perfect inverted repeat flanking a spacer of at least 20 bp [11]. Within the *g26-g27* intergenic region in SV1 there are two perfect inverted repeats (IRs); the IR distal to *g27* has a spacer of 5 bp and the IR proximal to *g27* has a 22 bp spacer. Moreover the IR proximal to *g27* has a 10 bp perfect inverted repeat so, together with the spacer DNA, the length of this DNA element is 42 bp which is the same length as the ϕC31 *attP* site. The length *attP* sites used by other serine integrases is between 42 and 69 bp [11,17]. The *attP* site in SV1 is therefore likely to be located between nucleotides 20504 and 20545 and is downstream of the integrase gene, *g27*.

As the SV1 *g27* is likely to be expressed as part of an operon, we decided to test its integrating properties by swapping the ϕC31 *int* ORF expressed from the *tcp830* promoter in pEY25 for the SV1 *g27/attP* locus (Figure 1). Primers were designed to clone the SV1 *g27/attP* locus 20487 to 22295 comprising the putative *attP* site and the *g27* ORF in both orientations downstream of the *tcp830* promoter in pEY25, to generate pMS98 and pBF1 (Figure 1). Both plasmids were introduced into *E. coli* ET12567 (pUZ8002) and used in conjugation reactions with *S. coelicolor* J1929. Surprisingly the numbers of exconjugants were not significantly different for pBF1 (in which the SV1 *g27* is co-directional with the *tcp830* promoter) and pMS98 (in which the SV1 *g27* is oriented towards *tcp830*) and this occurred with or without addition of anhydrotetracycline (Table 1). In fact exconjugants containing pBF1 tended to overproduce the two pigments actinorhodin and undecylprodigiosin indicating a possible stress response, perhaps due to overexpression of SV1 integrase. We assume that there are fortuitous sequences upstream of the *g27* gene in pMS98 that resemble both promoter and ribosome binding sites for integrase expression after conjugation to *Streptomyces*. The low frequency of exconjugants from *E. coli* containing pEY25 in this experiment is likely to be due to the absence of the ϕC31 *attP* site (Figure 1).

To validate the integration of the plasmids via SV1 *g27/attP* site into the *S. coelicolor* genome, we have identified the integration site of SV1 phage using an inverse PCR technique. Genomic DNA from an exconjugant of *S. coelicolor* J1929 containing the integrated pMS98 was cut with a restriction enzyme for which there is no recognition site within pMS98 and self-ligated. Primers reading outwards from the SV1 plasmid into the *S. coelicolor J1929* genome were then used to generate a PCR product (Figure 2). The plasmid pMS98 was found to have integrated into SCO4383 encoding a putative 4-Coumarate-CoA Ligase, a key enzyme in the phenylpropanoid pathway that, at least in plants, is important in secondary metabolism pathways for flavonoids and monolignols [18]. Based on the DNA sequence of this PCR product, we constructed two further primers against the integration region (SC04383) to amplify *attB*, *attR*, and *attL* (using *S. coelicolor* DNA, and *S. coelicolor* J1929:pMS98 genomic DNA as templates) (Figure 3). The resulting DNA sequences confirmed that the SV1 *attB* site is within SCO4383 and the *attP* site is located, as predicted, downstream of the SV1 *g27* gene (Figure 4). The *attP* and the *attB* have similar features to those observed in the other serine integrase attachment sites; they both contain inverted repeats but these are different in the *attB* and *attP* sequences, and there is an identical sequence of 4 bp in the centre of both substrates where the recombination occurs (Figure 4) [11,12].

Plasmid pMS98 was modified to remove unnecessary DNA, the *attR* site from ϕC31, and the *tcp830* promoter to generate pBF3 (Figure 1). This novel integration vector has unique XhoI, XbaI, KpnI and PvuII sites for cloning. To test whether pBF3 could integrate into a range of *Streptomyces* genomes it was introduced by conjugation into *S. coelicolor, S. lividans, S. venezuelae, S. avermitilis* and *S. albus*. Conjugation frequencies of

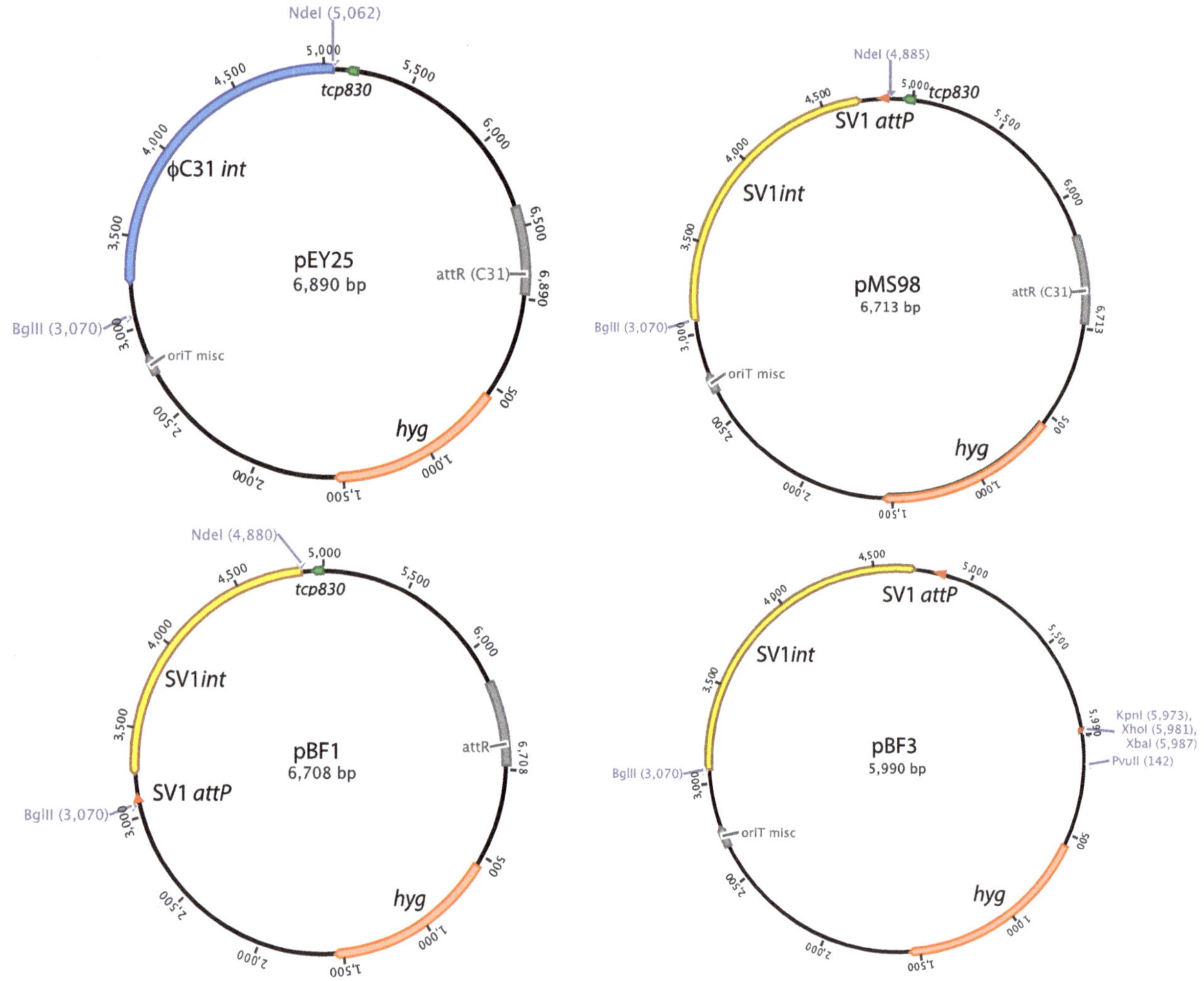

Figure 1 Plasmids constructed during the course of this work. pMS98 and pBF1 are derivatives of the plasmid pEY25, in which the ϕC31 *int* gene (blue arrow) is replaced with the SV1 *g27* (encoding SV1 integrase; yellow arrow) and *attP* (red arrow) in orientations in which the SV1 *g27* genes is being expressed from the *tcp830* promoter (pBF1) or is orientated opposite to the *tcp830* promoter (pMS98). pBF3 is a derivative of pMS98 in which the *tcp830* promoter and the ϕC31 *attR* sites have been removed. The hygromycin resistance gene is represented by the pale red arrow and the *tcp830* promoter by the green arrow. *oriT* is in grey.

pBF3 into *S. coelicolor* and *S. lividans* were reminiscent of those obtained with ϕBT1 and ϕC31 integration plasmids; the numbers of hygromycin resistant exconjugants for *S. coelicolor J1929, S. coelicolor M512,* and *S. lividans TK24* were greater than 1×10^5 (Table 2), and the integrations were stable after two rounds of sporulation without selection (see below). When *S. venezuelae* was used as the recipient, the hygromycin resistant exconjugants were, although numerous, very small and the integrations were not stable as after two rounds of sporulation, hygromycin resistance was lost. A BLAST search using the SV1 *attB* site from *S. coelicolor* revealed that highly similar sequences were indeed present in *S. lividans* and *S. avermilitis* but no homologous sequences were found in *S. albus* or *S. venezuelae*. Despite the presence of a putative *attB* site for SV1 in *S. avermitilis* the frequency of conjugation and integration of pBF3 was very low (Table 2).

The sequences of the SV1 *attB* and *attP* sites are distinct from the recombination sites for the other known phage integrases. We showed previously that integrating vectors derived using integrases from ϕC31 and ϕBT1 do not interfere with each other with respect to the frequency of integration or their stability [12]. We therefore tested whether the integration frequencies of ϕC31 or ϕBT1 derived integrating vectors were affected if the recipient already contained pBF3 integrated at the SV1 *attB* site. Conjugations were performed using *E. coli* donors containing either pSET152 (encoding ϕC31 *int/attP*) or pRT801 (encoding ϕBT1 *int/attP*), both plasmids conferring apramycin resistance, and *S. coelicolor* M512 containing pBF3 as recipient. Selection was for both hygromycin

Table 1 Conjugation frequnecies of various integrating plasmids into *S. coelicolor* J1929

E. coli ET12567 (pUZ8002) donor containing:	Origin of integrase and *attP*	Hygromycin resistant exconjugants/10^8 spores
pMS98	SV1	2.4×10^6
pBF1	SV1	1.1×10^7
pBF1*	SV1	9.9×10^6
pBF3	SV1	1.4×10^7
pEY25	ϕC31 (*int* only)	9×10^3

*The SV1 integrase was induced with 1 μg/ml anhydrotetracycline.

and apramycin. There was no great reduction in the conjugation frequency compared with the use of plasmid-free *S. coelicolor* M512 as a recipient (Table 2). SV1 vectors can therefore be used in combination without interference with ϕC31 and ϕBT1 derived vectors.

Hygromycin resistant colonies obtained after conjugation of *E. coli* containing pBF3 with *S. coelicolor J1929, S. coelicolor M512, S. lividans TK24 and S. venezuelae* were allowed to sporulate and were subcultured twice on medium without selection. Genomic DNA was isolated and analysed by Southern blotting (Figure 5). The expected 4.3 kbp band, indicative of integrated pBF3, was observed in the DNA from *S. coelicolor* and *S. lividans* exconjugants. The absence of the 4.3 kbp band from the *S. venezuelae* genomic DNA indicated that pBF3 did not persist in this strain and was lost. Stable hygromycin resistant *S. venezuelae* exconjugants were obtained with the ϕBT1 and ϕC31 derived vectors (pMS82 and ϕC31, respectively; Table 2). Paradoxically a lysogen of SV1 in *S. venezuelae* is perfectly stable and grows like the non-lysogen. We deduce that the interruption in SCO4383 caused by the integrating plasmid is toxic, but the toxicity is ameliorated by a prophage-encoded gene or by an unknown *cis* effect within the integrated prophage. Notably SV1 does not encode a homologue to SCO4383, or fragments of SCO4383 that could compensate for its truncation by integration of SV1 derived integrating vectors.

Conclusions

The activity of a novel phage integration system from bacteriophage SV1 has been demonstrated in *S. coelicolor* and *S. lividans* and the *attP* and *attB* sites identified. We believe that the new integrating vector pBF3 will be of use in the genetic manipulation of these and other *Streptomyces* strains. More generally the characterization of a new integrase and its substrates will provide biologists with new tools for DNA assembly in the genomes of a wide range of microorganisms and other model organisms.

Methods

Bacterial strains and culture

E. coli strain DH5α was used for plasmid construction. *E. coli* strain ET12567 (pUZ8002) is a methylation-defective strain (*dam-13: Tn9 dcm-6 hsdM*) and was used as the conjugation donor in plasmid conjugations from *E. coli* to *Streptomyces* [20].

Six *Streptomyces* strains were used as recipients for intergeneric conjugation: *Streptomyces coelicolor* J1929 (contains Δ*pglY* conferring sensitivity to ϕC31 and ϕBT1;

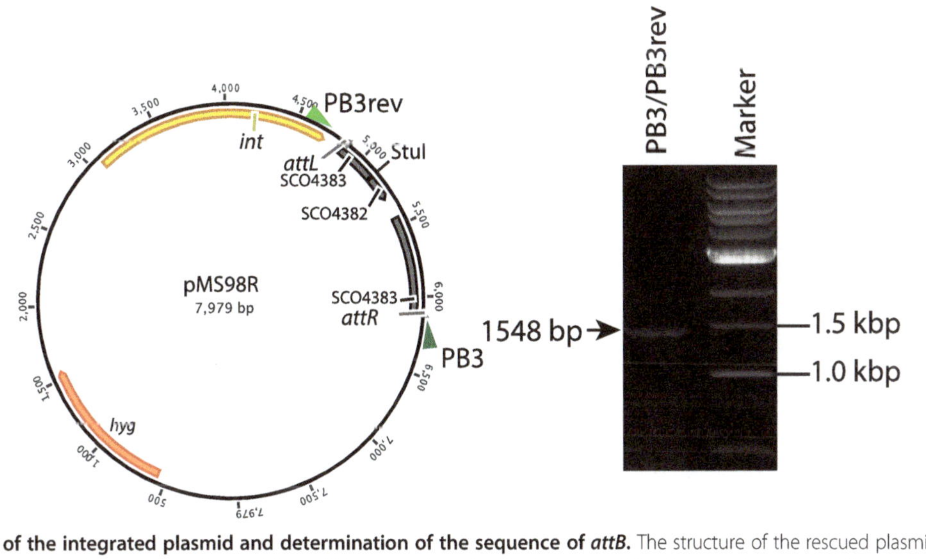

Figure 2 Rescue of the integrated plasmid and determination of the sequence of *attB*. The structure of the rescued plasmid, pMS98R, by digestion of genomic DNA from an S. coelicolor J1929 pMS98 exconjugant with StuI and self-ligation. The two primers PB3 and PB3rev were used to amplify the DNA reading out from the *attL* and *attR* sites produced on integration of pMS98 into the *attB* site. The PCR amplified DNA generated was separated by electrophoresis on a 0.8% agarose gel. The size of the band obtained is in agreement with the predicted 1548 bp fragment, after performing the manipulations *in silico* using the published *S. coelicolor* genome sequence [19].

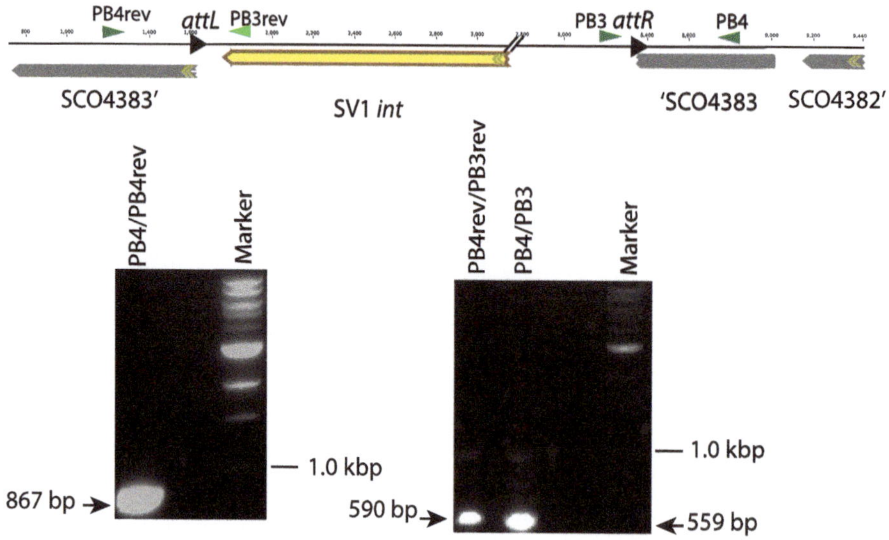

Figure 3 Validation of the position of the integrated plasmid, pMS98, in the *S. coelicolor* genome. Primers PB4 and PB4rev were designed to flank the SV1 *attB* site and were predicted to amplify a 867 bp fragment as shown in the agarose gel depicted on the left. Using pairs of primers PB4rev with PB3rev and PB3 with PB4, the *attL* and *attR* sites were amplified from *S. coelicolor* J1929:pMS98 genomic DNA. Fragments of the predicted sizes were obtained as depicted in the agarose gel (right).

[21]) *Streptomyces coelicolor* M512 (*ΔredD ΔactII-ORF4* SCP1$^-$ SCP2$^-$ Pgl$^+$) [22], *Streptomyces avermitilis MA-4680* [23], *Streptomyces venezuelae 10712* [24], *Streptomyces albus J1074* [25], *Streptomyces lividans TK24* (*str-6* SLP2$^-$, SLP3$^-$)[26].

The *E. coli* strains DH5α [27] and ET12567(pUZ8002) [20,26] were grown in Luria-Bertani broth (LB) or on LB agar at 37°C. *Streptomyces* strains were grown in Soya Mannitol (SM) agar at 30°C for routine maintenance [26]. Conjugations were performed on SM containing 10 mM $MgCl_2$ and Yeast extract malt extract medium was used for the preparation of genomic DNA [26]. Antibiotic concentrations for *E. coli* were 150 μg/ml hygromycin, 50 μg/ml apramycin, 50 μg/ml kanamycin, 25 μg/ml chloramphenicol and 100 μg/ml hygromycin, 50 μg/ml apramycin and 25 μg/ml nalidixic acid for selection with *Streptomyces*.

DNA manipulation

Plasmids preparations, *E. coli* transformations, DNA digestion by restriction enzymes, DNA fragment isolation and purification, and gel electrophoresis were carried

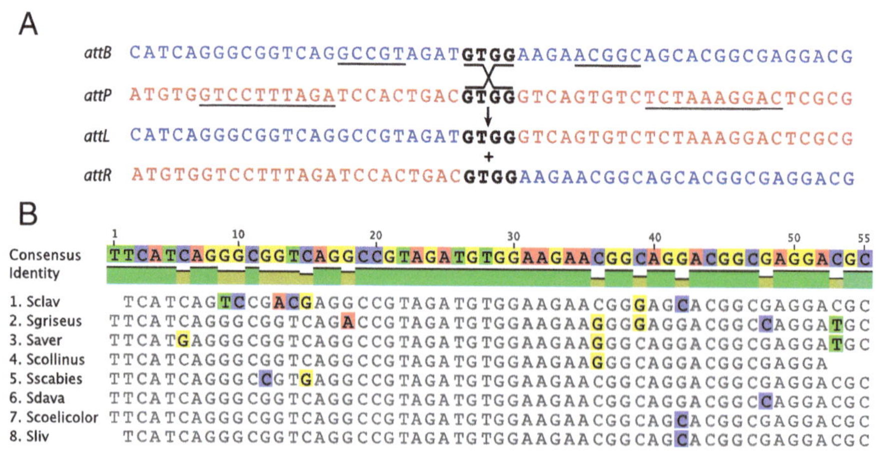

Figure 4 Sequences of the *attB, attP, attL* and *attR* sites used by SV1 integrase. Panel **A**. The crossover site occurs within the 4 bp sequence in black and the positions of the inverted repeats are underlined in the *attP* and *attB* sites. Panel **B**. The results of a BLAST search for the SV1 *attB* site in *Streptomyces* genomes deposited in the nucleotide sequence database. The default settings did not identify an SV1 *attB* site from *S. venezuelae* or *S. albus*.

Table 2 Conjugation frequencies per 10^8 spores of integrating plasmids into *Streptomyces* species

Streptomyces recipient:	*E. coli* donor, ET12567 (pUZ8002) containing plasmids:			
	pBF3 (SV1 *g27/attP*)	pMS82 (ϕBT1 *int/attP*)	pSET152 (ϕC31 *int/attP*)	pRT801 (ϕBT1 *int/attP*)
S. coelicolor J1929	2.5×10^6	3.6×10^6	6×10^5	ND[1]
S. coelicolor M512	1.7×10^7	6×10^7	6×10^6	ND
S. lividans	1.4×10^5	9.5×10^5	2.3×10^5	ND
S. venezuelae 10712	3.3×10^4	1.7×10^7	9.8×10^4	ND
S. avermitilis	40	3×10^5	1.8×10^3	ND
S. albus J1074	4.6×10^2	5×10^5	1×10^3	ND
S. coelicolor M512: pBF3	-	-	4×10^6	2.8×10^5

[1]Not Done.

out according to Sambrook *et al.* [27]. In-Fusion® cloning (Clontech®) was generally used for joining DNA fragments. DNA preparation from *Streptomyces* was performed following the *Streptomyces* manual [26].

Southern blotting was performed, according to the manufacturer's instructions, on Hybond-N nylon membrane (Amersham) using a fragment of DNA derived from the hygromycin resistant gene as the probe. The AlkPhos Direct Labeling and Detection System with CDP-Star kit (Amersham) was used for detection. 1 μg of *Nru*I (New England Biolabs) digested genomic DNA was loaded onto a 0.8% agarose gel in TBE buffer and electrophoreses overnight prior to capillary blotting.

Polymerase Chain Reaction (PCR) was carried out using Phusion® High-Fidelity DNA Polymerase (New England Biolabs) according to the manufacturer's instructions.

Plasmid constructions

pEY25 is a derivative of pAV11, an integration vector that encodes the ϕBT1 *int/attP* locus and the anhydrotetracycline inducible promoter, *tcp830*. To generate pEY25 the ϕBT1 *int* gene was deleted and the ϕC31 *int* gene was placed under the control of the *tcp830* promoter. pMS98 was constructed by PCR amplification of SV1 *g27/attP* locus using primers MS409 (5′ GCTTCATATGAAACGAGACCTACCAAG) and MS410 (5′CGTTAGATCTTCGCGCTCCGATGTGGTC) and In-Fusion® cloning into pEY25 cut with NdeI and BglII to replace the ϕC31 *int* gene. pBF1 was constructed in the same way but using primers PBF1for (5′ AAGGAGATATACATATGAAACGAGACCTACCAAGC- 3′) and PBF1rev (5′ CCATGAGCCAAGATCTTCGCGCTCCGATGTGGTCC- 3′). pBF3 was constructed as follows to remove unnecessary elements of pBF1: pBF1 was first cut with *Avr*II, and *Acc*65I, and the ends filled in with DNA Polymerase I, Large (Klenow) Fragment (New England Biolabs) to generate blunt ends for ligation. This blunt ended fragment was then self-ligated using Quick ligase enzyme (New England Biolabs) to produce pBF2. To remove the *tcp830* promoter, pBF2 was digested with *Nde*I, and *Ase*I and the 5985 bp fragment was self-ligated to form pBF3.

Inverse PCR

Inverse PCR was performed to identify the integration site of SV1 within *Streptomyces coelicolor* J1929. This procedure is designed for amplifying anonymous flanking genomic DNA regions. Genomic DNA was prepared from a strain containing the integrated plasmid, pMS98, digested with an enzyme that does not cut within the plasmid (StuI) and then ligation of DNA under dilute

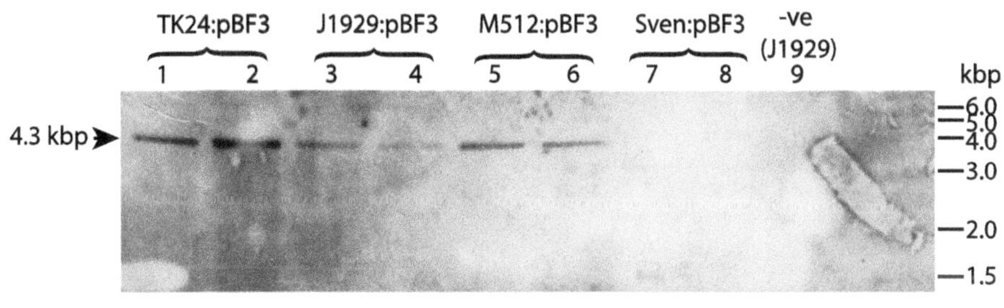

Figure 5 Southern blot to demonstrate the presence of the predicted hygromycin gene fragment in *S. coelicolor* and *S. lividans* exconjugants. Two independent exconjugants derived from an *E. coli* donor containing pBF3 and *S. coelicolor*, *S. lividans* and *S. venezuelae* as recipients were initially selected using the hygromycin resistance marker on pBF3 but then subsequently maintained without selection. Genomic DNA was then prepared from each line. The *hyg* gene was detected in *S. coelicolor*:pBF3 and *S. lividans*:pBF3 lines but not in the *S. venezuelae* lines indicating that pBF3 is not stably maintined in *S. venezuelae*.

DNA conditions to favour circularization. Finally, PCR amplification was performed using oligonucleotides PB3 for (5′ GTACGTCGGAGGTCTAGAGA) and PB3rev (5′ GCAGCTTCGAGTTTCATCCCG) that prime DNA synthesis from the known sequence within pMS98. To confirm the SV1 integration site, primers PB4 for (5′ CA CAGCCCCAACACCGTC) and PB4 rev (5′ -GTCGG TGAGGGAGACGATG) were designed to amplify the potential SV1 *attB* from the *S. coelicolor* J1929 DNA. These primers were also used with PB3 and PB3rev to amplify the potential *attR, attL* from the exconjugants *S. coelicolor* J1929 pMS98 DNA.

Competing interests
The authors declare that they have no competing interests.

Authors' contributions
BF designed and performed the experiments and helped to write the paper. EY and GT designed and performed experiments. MCMS designed experiments, raised funding and helped to write the paper. All authors read and approved the final manuscript.

Acknowledgements
We are grateful to Dr Maureen Bibb for providing SV1, *S. venezuelae*, the *S. venezuelae* SV1 lysogen, to Professor Mark Buttner for *S. coelicolor* strain J1929, to Professor Mervyn Bibb for *S. coelicolor* M512 and *E. coli* strain ET12567(pUZ8002), to Professor Keith Chater for *S. albus* J1074 and to the late Professor Simon Baumberg for *S. avermitilis*. We are also grateful to Dr Maureen Bibb for discussions on the manuscript. BF acknowledges receipt of a PhD studentship from the Egyptian Ministry of Higher Education and additional funding for his internship at the University of York. This work was funded by the Biotechnology and Biological Science Research Council, UK grant references BB/K003356 and BB/H001212.

Author details
[1]Department of Biology, University of York, York YO10 5DD, UK. [2]School of Medical Sciences, Institute of Medical Sciences, University of Aberdeen, Aberdeen AB25 2ZD, UK.

References
1. Chater KF: ***Streptomyces* inside-out: a new perspective on the bacteria that provide us with antibiotics.** *Philos Trans R Soc Lond B Biol Sci* 2006, **361**(1469):761–768.
2. Baltz RH: **Strain improvement in actinomycetes in the postgenomic era.** *J Ind Microbiol Biotech* 2011, **38**(6):657–666.
3. Baltz RH: ***Streptomyces* temperate bacteriophage integration systems for stable genetic engineering of actinomycetes (and other organisms).** *J Ind Microbiol Biotech* 2012, **39**(5):661–672.
4. Boccard F, Smokvina T, Pernodet JL, Friedmann A, Guerineau M: **Structural analysis of loci involved in pSAM2 site-specific integration in *Streptomyces*.** *Plasmid* 1989, **21**(1):59–70.
5. Smokvina T, Mazodier P, Boccard F, Thompson CJ, Guerineau M: **Construction of a series of pSAM2-based integrative vectors for use in actinomycetes.** *Gene* 1990, **94**(1):53–59.
6. Bierman M, Logan R, O'Brien K, Seno ET, Rao RN, Schoner BE: **Plasmid cloning vectors for the conjugal transfer of DNA from *Escherichia coli* to *Streptomyces* spp.** *Gene* 1992, **116**(1):43–49.
7. Kuhstoss S, Rao RN: **Analysis of the integration function of the streptomycete bacteriophage ϕC31.** *J Mol Biol* 1991, **222**(4):897–908.
8. Kuhstoss S, Richardson MA, Rao RN: **Plasmid cloning vectors that integrate site-specifically in *Streptomyces* spp.** *Gene* 1991, **97**(1):143–146.
9. Khaleel T, Younger E, McEwan AR, Varghese AS, Smith MCM: **A phage protein that binds ϕC31 integrase to switch its directionality.** *Mol Microbiol* 2011, **80**(6):1450–1463.
10. Lewis JA, Hatfull GF: **Control of directionality in integrase-mediated recombination: examination of recombination directionality factors (RDFs) including Xis and Cox proteins.** *Nucl Acid Res* 2001, **29**(11):2205–2216.
11. Brown WR, Lee NC, Xu Z, Smith MCM: **Serine recombinases as tools for genome engineering.** *Methods* 2011, **53**(4):372–379.
12. Gregory MA, Till R, Smith MCM: **Integration site for *Streptomyces* phage ϕBT1 and the development of novel site-specific integrating vectors.** *J Bacteriol* 2003, **185**(17):5320–5323.
13. Morita K, Yamamoto T, Fusada N, Komatsu M, Ikeda H, Hirano N, Takahashi H: **The site-specific recombination system of actinophage TG1.** *FEMS Microbiol Lett* 2009, **297**(2):234–240.
14. Stuttard C: **Generalized transduction in *Streptomyces* species.** In *Genetics and Molecular Biology of Industrial Microorganisms.* Edited by Hershberger CL, Queener SW, Hegeman G. Washington DC: Amercian Society of Microbiology; 1989:157–162.
15. Smith MCM, Hendrix RW, Dedrick R, Mitchell K, Ko CC, Russell D, Bell E, Gregory M, Bibb MJ, Pethick F, Jacobs-Sera D, Herron P, Buttner MJ, Hatfull G: **Evolutionary relationships among actinophages and a putative adaptation for growth in *Streptomyces* spp.** *J Bacteriol* 2013, **195**(21):4924–4935.
16. Hanssen A-M, Sollid JUE: **SCCmec in staphylococci: genes on the move.** *FEMS Immunol Med Microbiol* 2005, **46:**8–20.
17. Xu Z, Thomas L, Davies B, Chalmers R, Smith MCM, Brown WRA: **Accuracy and efficiency define Bxb1 integrase as the best of fifteen candidate serine recombinases for the integration of DNA into the human genome.** *BMC Biotech* 2013, **13:**87.
18. Santos CN, Koffas M, Stephanopoulos G: **Optimization of a heterologous pathway for the production of flavonoids from glucose.** *Metab Eng* 2011, **13**(4):392–400.
19. Bentley SD, Chater KF, Cerdeno-Tarraga AM, Challis GL, Thomson NR, James KD, Harris DE, Quail MA, Kieser H, Harper D, Bateman A, Brown S, Chandra G, Chen CW, Collins M, Cronin A, Fraser A, Goble A, Hidalgo J, Hornsby T, Howarth S, Huang CH, Kieser T, Larke L, Murphy L, Oliver K, O'Neil S, Rabbinowitsch E, Rajandream MA, Rutherford K: **Complete genome sequence of the model actinomycete *Streptomyces coelicolor* A3(2).** *Nature* 2002, **417**(6885):141–147.
20. MacNeil DJ: **Characterization of a unique methyl-specific restriction system in *Streptomyces avermitilis*.** *J Bacteriol* 1988, **170**(12):5607–5612.
21. Bedford DJ, Laity C, Buttner MJ: **Two genes involved in the phase-variable ϕC31 resistance mechanism of *Streptomyces coelicolor* A3(2).** *J Bacteriol* 1995, **177**(16):4681–4689.
22. Floriano B, Bibb M: ***afsR* is a pleiotropic but conditionally required regulatory gene for antibiotic production in *Streptomyces coelicolor* A3 (2).** *Mol Microbiol* 1996, **21**(2):385–396.
23. Ikeda H, Ishikawa J, Hanamoto A, Shinose M, Kikuchi H, Shiba T, Sakaki Y, Hattori M, Omura S: **Complete genome sequence and comparative analysis of the industrial microorganism *Streptomyces avermitilis*.** *Nat Biotechnol* 2003, **21**(5):526–531.
24. Pullan ST, Chandra G, Bibb MJ, Merrick M: **Genome-wide analysis of the role of GlnR in *Streptomyces venezuelae* provides new insights into global nitrogen regulation in actinomycetes.** *BMC Genomics* 2011, **12:**175.
25. Chater KF, Wilde LC: ***Streptomyces albus* G mutants defective in the SalGI restriction-modification system.** *J Gen Microbiol* 1980, **116**(2):323–334.
26. Kieser T, Bibb MJ, Buttner MJ, Chater KF, Hopwood DA: *Practical Streptomyces Genetics.* Norwich: The John Innes Foundation; 2000.
27. Sambrook J, Russell DW: *Molecular Cloning: A Laboratory manual.* New York: Cold Spring Harbor Laboratory Press; 2001.

16

Codon optimization and factorial screening for enhanced soluble expression of human ciliary neurotrophic factor in *Escherichia coli*

Jaakko M Itkonen[1], Arto Urtti[1,2*], Louise E Bird[3,4] and Sanjay Sarkhel[1*]

Abstract

Background: Neurotrophic factors influence survival, differentiation, proliferation and death of neuronal cells within the central nervous system. Human ciliary neurotrophic factor (hCNTF) has neuroprotective properties and is also known to influence energy balance. Consequently, hCNTF has potential therapeutic applications in neurodegenerative, obesity and diabetes related disorders. Clinical and biological applications of hCNTF necessitate a recombinant expression system to produce large amounts of functional protein in soluble form. Earlier attempts to express hCNTF in Escherichia coli (*E. coli*) were limited by low amounts and the need to refold from inclusion bodies.

Results: In this report, we describe a strategy to effectively identify constructs and conditions for soluble expression of hCNTF in *E. coli*. Small-scale expression screening with soluble fusion tags identified many conditions that yielded soluble expression. Codon optimized 6-His-hCNTF construct showed soluble expression in all the conditions tested. Large-scale culture of the 6-His-hCNTF construct yielded high (10 – 20 fold) soluble expression (8 – 9 fold) as compared to earlier published reports. Functional activity of recombinant 6-His-hCNTF produced was confirmed by its binding to hCNTF receptor (hCNTFRα) with an $EC_{50} = 36$ nM.

Conclusion: Our results highlight the combination of codon optimization and screening soluble fusion tags as a successful strategy for high yielding soluble expression of hCNTF in *E. coli*. Codon optimization of the hCNTF sequence seems to be sufficient for soluble expression of hCNTF. The combined approach of codon optimization and soluble fusion tag screen can be an effective strategy for soluble expression of pharmaceutical proteins in *E. coli*.

Keywords: Neurotrophic factors, Human CNTF, Codon optimization, Recombinant soluble expression, *E. coli*

Background

Neurotrophic factors play central role within the nervous system. They regulate cell survival, proliferation, differentiation, migration, dendrite and axonal growth, synaptic plasticity and interactions of neuronal and glial cells [1-4]. The neuropoietic cytokine, ciliary neurotrophic factor (CNTF), first identified in chick ciliary neurons [5,6], belongs to the interleukin-6 and leukemia inhibitory factor (LIF) family of four helix bundle cytokines [7,8]. Human CNTF (hCNTF), a polypeptide of approximate 23 kDa, is involved in the maintenance of ciliary and motor neurons. hCNTF exerts its signaling effect by binding to a heterotrimeric complex of hCNTF receptor (hCNTFRα), gp130 and LIF receptor (LIFRβ) [9-11]. Apart from neuroprotection [12,13], hCNTF is also known to influence energy balance [14-16]. Consequently, hCNTF has potential therapeutic applications not only in neurodegenerative diseases such as Amytropic Lateral Sclerosis (ALS) and Huntington's disease (HD), but also in obesity and related type II diabetes. hCNTF secreting implant has completed phase II studies as an effective therapy for dry age related macular degeneration (AMD) in the eye [17].

The therapeutic potential of CNTF necessitates a recombinant production system for large amounts of protein for clinical and biological applications. Bacterial expression systems are known to be rapid and economical for obtaining recombinant proteins [18-22]. However, attempts to express hCNTF in *Escherichia coli* (*E. coli*) have either yielded low soluble levels (approx. 13%) or insoluble

* Correspondence: arto.urtti@helsinki.fi; sanjay.sarkhel@helsinki.fi
[1]Centre for Drug Research, Division of Pharmaceutical Biosciences, Faculty of Pharmacy, University of Helsinki, P.O. Box 56, Viikinkaari 5E, Helsinki 00014, Finland
[2]School of Pharmacy, University of Eastern Finland, Kuopio, Finland
Full list of author information is available at the end of the article

inclusion bodies [8,23-26][a]. Soluble recombinant expression of pharmaceutical proteins is of utmost importance not only for direct applications but to be able to employ advanced molecular engineering methods to alter their physicochemical properties. Bacterial expression systems offer several advantages; they are relatively inexpensive to culture, can be easily modified genetically and have fast turnaround time. *E. coli*, a preferred bacterial host, is considered to be a 'laboratory workhorse' for recombinant protein expression. However, the advantages are easily offset by the challenges poised by recombinant production of eukaryotic proteins [27,28]. Bacterial hosts such as *E. coli* are often limited in amounts of tRNA for the codons that are used rather less frequently. Instances where codon bias of the gene is significantly different from that of the expression host *E. coli*, paucity of the tRNAs severely limits the translation process. This results in non-optimal translation of the RNA including stalling, termination, possible frame-shifting and low levels of protein expression [29,30]. Apart from codon usage being a determining factor in protein expression levels, structural features of mRNA that interfere with the initiation of translation is also believed to adversely affect protein expression [31,32]. In contrast, a recent study found that there was very less influence of mRNA structure on protein expression and that depletion of charged tRNAs was the main limiting factor for protein expression [33]. Codon optimization of the target gene and/or use of tRNA enhanced strains have become an attractive starting point for heterologous protein expression in *E. coli.* Codon optimization has been successfully utilized to express human pigment epithelium derived factor in *E. coli* [34] and thirty human short chain dehydrogenase/reductive genes showed vast improvement in expression with optimized codon and use of *E. coli* strains containing rare codon tRNAs [35]. The other relevant issue with heterologous expression in *E. coli* is the formation of unordered aggregates (inclusion bodies) due to improper folding of the polypeptide chain. High rates of protein expression or unfavorable conditions for protein folding cause inclusion bodies formation. Successful strategies to mitigate improper folding include decreasing the cultivation temperature [36], co-expression of folding modulators [37] and reducing the rate of gene expression [38]. An alternative strategy to improve the solubility has been to express the protein of interest fused to a solubility enhancing polypeptide 'tag' [39-42]. The soluble 'tag' also serves to provide a reliable translation initiation [43]. Although useful, the fusion 'tag' strategy is limited by the lack of a rational design for choice and a given 'tag' might be effective with only certain targets [40,44-46]. In general, production of soluble proteins in *E. coli* remains largely a trial-and-error process.

In this study, we report an effective strategy towards screening and optimization for soluble expression of hCNTF in *E. coli.* We address the twin issues of codon bias and possible improper folding by optimizing the hCNTF gene sequence for use in *E. coli* and adopting a factorial screening strategy. Small-scale screening methods have been successful in structural genomics initiatives [47-49] in providing reproducible qualitative and quantitative leads for large-scale expression. hCNTF gene was codon optimized and cloned into a suite of nine vectors harboring different fusion (soluble/affinity) tags. The constructs were screened for expression in two different *E. coli* strains and culture medias at different temperatures.

Results and discussion

The objective of our work was to achieve soluble, high yielding production of hCNTF in *E. coli.* Earlier attempts of hCNTF production in *E. coli* resulted in low amounts of soluble protein or required purification from insoluble inclusion bodies [8,23-26]. McDonald and co-workers have reported the presence of only 13% of soluble hCNTF in *E. coli* (BL21) cell extracts [23]. In another report, the insoluble fraction from bacterial cell lysate was found to contain 80% of recombinant hCNTF [24]. Masiakowski et al. have also reported purifying hCNTF from inclusion bodies even though the translation of recombinant hCNTF was 20 – 40% of total protein [25]. Purification of hCNTF from inclusion bodies requires an additional refolding step and such processes are usually not desirable since they are cumbersome and in many cases might result in non-native state of the target protein. Non-native state of proteins often leads to aggregation that can cause fatal immunogenic reactions from protein therapeutics [50]. Thus, for proteins of pharmaceutical relevance, such as hCNTF, it is highly desirable to have high yielding, soluble and cost effective production of protein. An effective strategy to express soluble proteins requires cost effective screening system where several factors could be tested in parallel. Structural genomics initiatives have effectively utilized such approaches to produce recombinant proteins in *E. coli* [47-49].

In our present work, we adopted the strategy of fusing the codon optimized sequence of hCNTF to nine tags (eight soluble tags and the 6–His tag) and carrying out a factorial screening for protein expression in different conditions (temperature/culture media/expression strains). The overall scheme is depicted in Figure 1. The codon optimized hCNTF sequence is shown in Figure 2. A proprietary algorithm (OptimumGene™, Genscript USA) from the vendor was used to increase the codon usage bias in *E. coli* by improving the codon adaptation index (CAI) to 0.87. GC content and unfavorable peaks were optimized to prolong the half-life of the mRNA. Stem-loop structures that have an impact on ribosomal binding and stability of mRNA, were disrupted. In addition, the codon optimization process screened and successfully modified

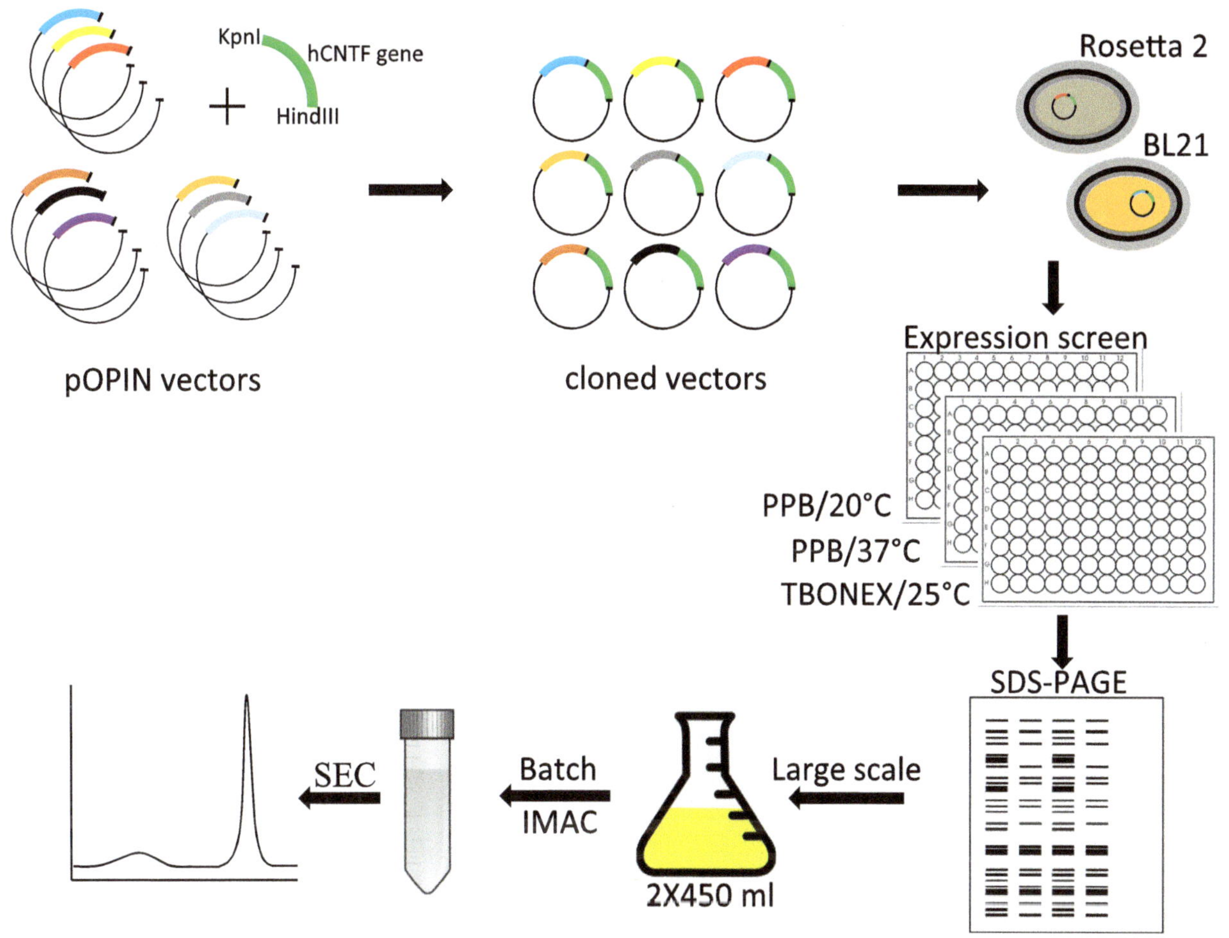

Figure 1 Scheme depicting the overall strategy of parallel cloning, small-scale expression screening and large-scale hCNTF production.

negative *cis*-acting sites. The construction of the pOPIN family of vectors have been described elsewhere [51-53]. Each vector has a N-terminal fusion 'tag' (Table 1) followed by Rhinovirus 3C protease cleavage site between the 'tag' and the inserted target sequence to facilitate 'tag' removal. The set of vectors have common cloning sites (KpnI and HindIII) that facilitate parallel cloning and construction of expression vectors. A cloning efficiency of 100% was achieved by picking three clones for each vector construction. The constructs were transformed into *E. coli* strains, BL21(DE3)pLysS and Rosetta 2(DE3)pLysS (a derivative of BL21 supplying tRNAs for 7 rare codons; AGA, AGG, AUA, CUA, GGA, CCC, and CGG) and small-scale expressions were carried out with culture medias Power Prime Broth (PPB) at 20°C/37°C and Overnight Expression Terrifc Broth (TBONEX) at 25°C.

The soluble fusion tags used in expression screening are listed in Table 1. The choice of soluble tags has been dictated by literature precedence on the success rate of soluble expression of a diverse set of proteins. No single soluble tag can increase the expression and solubility of all target proteins. However, some general trends on effectiveness of a limited set of tags (amongst those used in present study) have been reported in some studies. SUMO and NusA were found to be equally efficient in terms of total and soluble protein expression in studies on three model proteins [41,45]. The success of a particular soluble fusion tag is dependent on the target protein of interest and hence the need to screen a diverse set. Both the culture medias used, PPB and TBONEX, are rich and support high cell densities. TBONEX, being an auto-induction media, also offers the advantage of a simplified expression protocol by eliminating the monitoring and induction steps. Induction free protocol is particularly useful for high-throughput protein expression approaches. Culture growth in PPB was screened at temperatures 20°C and the normal 37°C. Lower culture temperature is known to facilitate proper folding and soluble protein expression [36].

The soluble fractions were treated with Ni-NTA magnetic beads to capture recombinant 6-His-(fusion tag)-hCNTF proteins. The eluted soluble, Ni-NTA affinity purified fractions were analyzed on a SDS-PAGE gel (Figure 3).

```
             10                  30                            60
wt  ATGGCGTTTACCGAACATAGCCCGCTGACCCCGCATCGCCGCGATCTGTGCAGCCGCAGC
co  ATGGCGTTTACCGAACATTCCCCGCTGACCCCGCATCGTCGTGACCTGTGTAGTCGCTCA
aa   M  A  F  T  E  H  S  P  L  T  P  H  R  R  D  L  C  S  R  S

             70                  90                           120
wt  ATTTGGCTGGCGCGCAAAATTCGCAGCGATCTGACCGCGCTGACCGAAAGCTATGTGAAA
co  ATCTGGCTGGCTCGCAAAATCCGCTCCGATCTGACCGCCCTGACGGAATCATATGTTAAA
aa   I  W  L  A  R  K  I  R  S  D  L  T  A  L  T  E  S  Y  V  K

            130                 150                           180
wt  CATCAGGGCCTGAACAAAAACATTAACCTGGATAGCGCGGATGGCATGCCGGTGGCGAGC
co  CATCAGGGCCTGAACAAAAATATTAACCTGGATTCGGCAGACGGTATGCCGGTCGCTAGC
aa   H  Q  G  L  N  K  N  I  N  L  D  S  A  D  G  M  P  V  A  S

            190                 210                           240
wt  ACCGATCAGTGGAGCGAACTGACCGAAGCGGAACGCCTGCAGGAAAACCTGCAGGCGTAT
co  ACCGACCAGTGGTCTGAACTGACGGAAGCGGAACGTCTGCAGGAAAATCTGCAAGCCTAC
aa   T  D  Q  W  S  E  L  T  E  A  E  R  L  Q  E  N  L  Q  A  Y

            250                 270                           300
wt  CGCACCTTTCATGTGCTGCTGGCGCGCCTGCTGGAAGATCAGCAGGTGCATTTTACCCCG
co  CGTACCTTTCATGTCCTGCTGGCACGCCTGCTGGAAGATCAGCAAGTGCACTTTACCCCG
aa   R  T  F  H  V  L  L  A  R  L  L  E  D  Q  Q  V  H  F  T  P

            310                 330                           360
wt  ACCGAAGGCGATTTTCATCAGGCGATTCATACCCTGCTGCTGCAGGTGGCGGCGTTTGCG
co  ACGGAAGGCGACTTCCATCAGGCTATCCACACGCTGCTGCTGCAGGTGGCAGCATTCGCG
aa   T  E  G  D  F  H  Q  A  I  H  T  L  L  L  Q  V  A  A  F  A

            370                 390                           420
wt  TATCAGATTGAAGAACTGATGATTCTGCTGGAATATAAAATTCCGCGCAACGAAGCGGAT
co  TATCAAATTGAAGAACTGATGATCCTGCTGGAATACAAAATTCCGCGCAATGAAGCGGAT
aa   Y  Q  I  E  E  L  M  I  L  L  E  Y  K  I  P  R  N  E  A  D

            430                 450                           480
wt  GGCATGCCGATTAACGTGGGCGATGGCGGCCTGTTTGAAAAAAAACTGTGGGGCCTGAAA
co  GGCATGCCGATCAACGTTGGTGACGGCGGTCTGTTTGAGAAAAAACTGTGGGGTCTGAAA
aa   G  M  P  I  N  V  G  D  G  G  L  F  E  K  K  L  W  G  L  K

            490                 510                           540
wt  GTGCTGCAGGAACTGAGCCAGTGGACCGTGCGCAGCATTCATGATCTGCGCTTTATTAGC
co  GTGCTGCAGGAACTGAGTCAATGGACCGTTCGTTCCATTCACGATCTGCGCTTCATCTCT
aa   V  L  Q  E  L  S  Q  W  T  V  R  S  I  H  D  L  R  F  I  S

            550                 570                           600
wt  AGCCATCAGACCGGCATTCCGGCGCGCGGCAGCCATTATATTGCGAACAACAAAAAAATG
co  TCACATCAGACGGGTATTCCGGCTCGTGGCTCGCACTACATCGCAAACAACAAGAAGATG
aa   S  H  Q  T  G  I  P  A  R  G  S  H  Y  I  A  N  N  K  K  M
```

Figure 2 Comparison of wild type (wt) and codon optimized (co) hCNTF sequence.

Table 1 ***Expression vectors and their fusion tags***

Lane	Vector	Fusion tag	BL21			Rosetta 2		
			20°C PPB	37°C PPB	25°C TBONEX	20°C PPB	37°C PPB	25°C TBONEX
1	pOPIN**F**	6-His-	1	1	1	1	1	1
2	pOPIN**S3C**	6-His-SUMO-	0.7	1.5	2.1	1.3	2.7	1.1
3	pOPIN**TRX**	6-His-Thioredoxin-	1.2	1.7	4.9	1.1	0.7	0.1
4	pOPIN**MSYB**	6-His-MsyB-	0.5	0.8	2.9	1.8	3.6	2.0
5	pOPIN**J**	6-His-GST-	0.3	1.3	0.4	0.1	0.5	0.1
6	pOPIN**HALO7**	6-His-Halo-	1.5	1.6	1.7	1.6	3.0	2.0
7	pOPIN**M**	6-His-MBP-	0.8	1.4	2.5	0.9	1.1	0.7
8	pOPIN**TF**	6-His-Trigger Factor-	0.5	1.1	1.7	0.3	0.6	0.5
9	pOPIN**NusA**	6-His-NusA-	0.6	1.2	1.1	0.7	0.8	0.1

Empirical scores of soluble fraction yield in small-scale expression relative to 6-His-construct. Highlighted text in bold within the vector name symbolize the fusion tag.

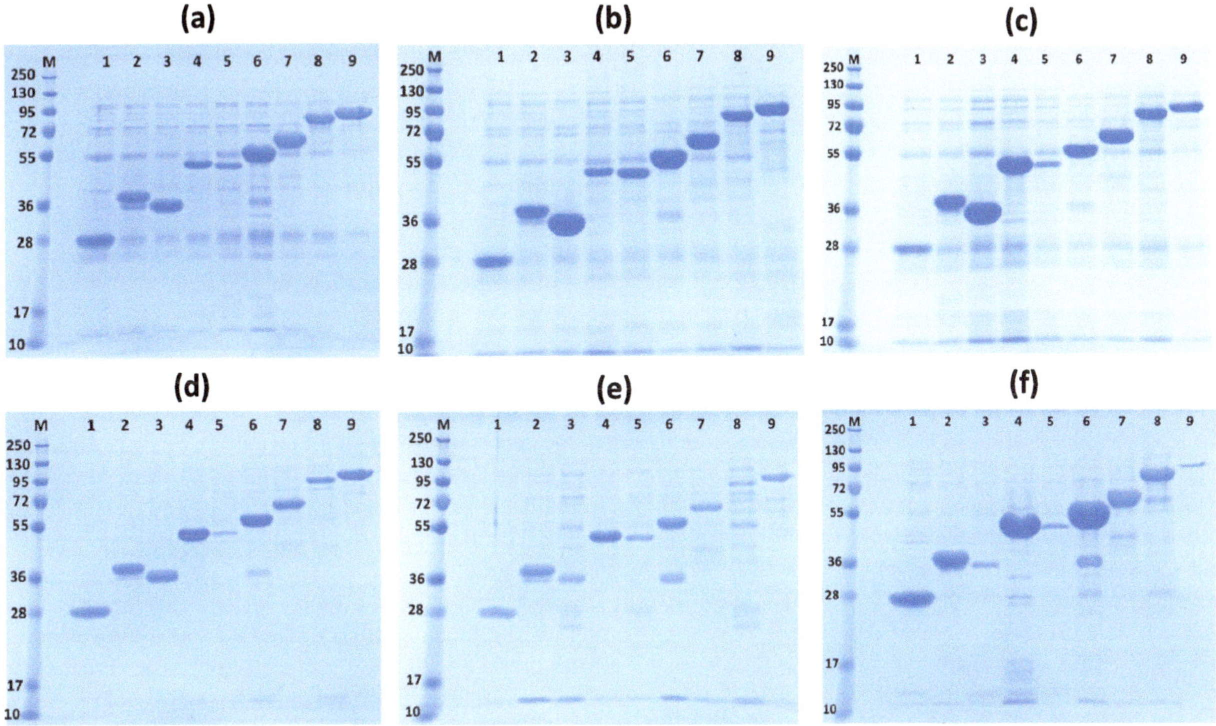

Figure 3 SDS-PAGE analysis of small-scale expression screen. **(a)** BL21/PPB/20°C; **(b)** BL21/PPB/37°C; **(c)** BL21/TBONEX/25°C; **(d)** Rosetta 2/PPB/20°C; **(e)** Rosetta 2/PPB/37°C; **(f)** Rosetta/TBONEX/25°C. Lanes M, 1–9 denote protein marker and Ni-NTA purified soluble hCNTF fusion constructs from expression vectors as listed in Table 1, respectively. The gel was stained with Coomassie Blue dye.

Most conditions (constructs/temperature/media) screened for expression showed the presence of soluble hCNTF as fusions (Figures 3 and 4). On an average, the percentage of hCNTF in the soluble fraction was close to 80% or more in majority of the conditions screened. Relative amounts of soluble hCNTF as fusions with soluble tags were empirically estimated (comparing intensity of bands) with respect to the 6-His-construct in each experimental condition (Table 1). The intensity of the bands depends on protein concentration and also on the size and content

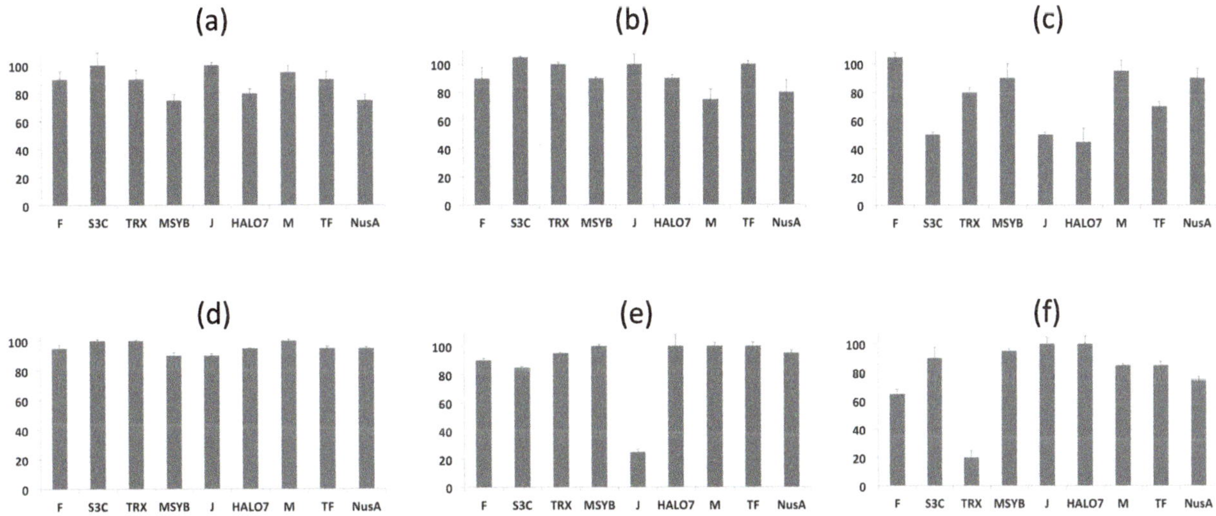

Figure 4 Histograms depicting percentage soluble fraction of hCNTF in small-scale expression screen. **(a)** BL21/PPB/20°C; **(b)** BL21/PPB/37°C; **(c)** BL21/TBONEX/25°C; **(d)** Rosetta 2/PPB/20°C; **(e)** Rosetta 2/PPB/37°C; **(f)** Rosetta/TBONEX/25°C. The estimates are based on measurements from three technical repeats.

of residue types of the construct. The empirical scores are not normalized against molecular size and residue content and as such the scores should not be used for direct comparisons but for easy visual summarization of relative trends. The efficacies of the fusion tags (scores averaged over the six expression conditions) were ranked as compared to the 6-His- tag. The trend observed was in the order of MsyB/Halo (1.9-fold); SUMO/Trx (1.6-fold); MBP/TF/NusA (approx. 1-fold) and GST (0.5 fold). Significant proteolysis seems to have taken place for some of the larger fusions (Figure 3; Halo, MBP, TF). This could possibly be due to improper folding of the larger fusion constructs. In a recent publication [53], Bird has reported similar observations with the larger fusion constructs. The most significant observation of our small scale screening experiment was that under all conditions, 6-His-hCNTF constructs showed appreciable amounts of expression. It is highly desirable to obtain soluble expression of proteins without a solubility enhancing fusion partner as the downstream purification process can avoid additional steps of tag cleavage and protease removal. Also, some proteins might exhibit markedly low solubility after the fusion tag has been cleaved off; limiting the utility of soluble fusion tags in such cases. The 6-His-tag is merely an affinity tag for purification and in all probability does not improve the solubility of recombinant hCNTF as such, unlike the other fusion tags. In fact, in a study on 20 human proteins, the 6-His-tag had a negative impact on protein solubility when present at either the N- or C-terminus [54]. In this context, our results for soluble expression of 6-His-hCNTF assume significant importance. Assuming that the 6-His-tag does not influence the solubility of recombinant hCNTF, or at least in a positive way, the soluble expression of hCNTF in the conditions tested can be attributed to codon optimization of the hCNTF sequence. Our experimental design was aimed at maximizing the chances of obtaining soluble protein in the backdrop of literature precedence [23-26]. However, we do note that in earlier published reports [23-26] of recombinant expression in *E. coli* with the wild type hCNTF; conditions of expression as varied as in our experimental set up (soluble fusion tags, lower temperatures and expression strain rich in tRNAs for rare codons) were not explored. Results from the earlier studies hint at the fact that most probably both translation (low yields of 5–10 mg/L) and folding (inclusion body formation) were the limiting factors in expression. It might thus be of interest to see the outcome of recombinant expression of the wild type sequence in a strain rich in tRNAs for rare codons, such as Rosetta 2 (DE3) or as a fusion construct with soluble tags in a strain, such as BL21(DE3). Even if the results of such experiments were positive, the codon optimized sequence used in the present study clearly results in a vast improvement since it allows high yielding soluble expression in BL21 at 37°C (approx. 90% or more in soluble fraction) without the aid of soluble fusion tags in contrast to the wild type sequence under similar conditions in previous studies [23-26]. With easy and cheap availability of synthetic genes, the optimized hCNTF codon sequence presents an ideal option for recombinant production in BL21 (a common laboratory strain) without the need for a soluble fusion tag.

Results from small-scale expression show varying efficacies of the fusion constructs in different conditions of growth as determined empirically (Table 1). However, since the overall yield after the cleavage of the fusion tag and purification might be very different from that of the fusion construct and also the need for additional tag cleavage and removal steps; 6-His-hCNTF was chosen for large scale expression with Rosetta 2(pLysS) and the auto inducing media, TBONEX at 25°C as a representative example. The analysis of the overall purification scheme is shown in Figure 5.

The amount of hCNTF protein obtained after final purification was 112 mg/L of culture. This is a vast improvement (greater than 10 – 20 fold) in overall yield from earlier reports of 5 – 10 mg/L [23,24]. Functional activity of recombinant hCNTF was confirmed by binding to hCNTFRα (Figure 6). The binding showed an EC_{50} value of 36 nM in accordance with an earlier published report [55]. A very recent report [56] published during the course of this manuscript preparation, has highlighted an attractive alternate strategy to express high yielding soluble hCNTF in *E. coli*. The study reports the usage of a single protein production (SPP) system that co-expresses *E. coli* toxin mRNA interferase, MazF and hCNTF. The strategy targets down-regulating the expression of endogenous bacterial proteins by selectively cleaving the ACA sites in host mRNAs by MazF. The hCNTF gene sequence was optimized for codon usage in *E. coli* and removing the ACA sites. So, whereas in the recently published study, a codon optimization strategy was directed at more specific use with the expression system, in our present study, a more sophisticated proprietary codon optimization algorithm OptimumGene™ (GenScript, NJ, USA) was used. hCNTF codons in our present study were not only optimized for codon usage in *E. coli* but parameters such as GC content, CpG dinucleotides content, mRNA secondary structure, cryptic splicing sites, premature PolyA sites, internal chi sites and ribosomal binding sites, negative CpG islands, RNA instability motif, repeat sequences were also optimized. The SPP system seems to be a promising approach, however, its robustness as a general strategy needs to be validated for a number of proteins. It appears that the SPP system might be useful in obtaining high yields of recombinant proteins indirectly by suppressing endogenous protein expression. However, protein sequences that might not fold properly once translated might still need soluble

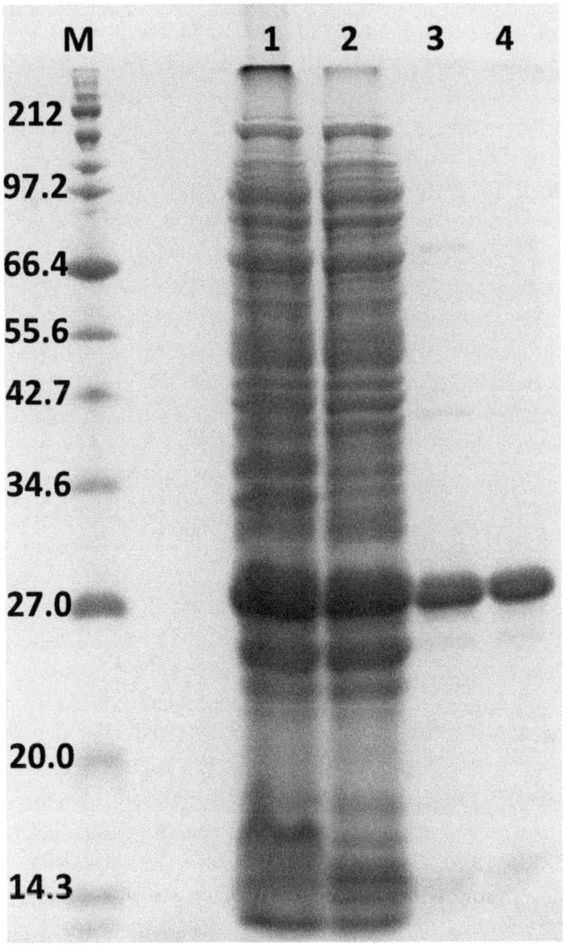

Figure 5 SDS-PAGE analysis of recombinant 6-His-hCNTF purification. M: protein marker; 1: crude cell lysate; 2: soluble fraction; 3: Ni-IDA purified hCNTF; 4: size exclusion chromatography purified hCNTF. The gel was stained with Coomassie Blue dye.

tags for proper folding. From such a perspective, our approach to screen with a suite of soluble tags might have a more versatile utility. It would be interesting to integrate our approach (a sophisticated codon optimization and screening with soluble tags in different strains of *E. coli*) with the SPP system (down-regulating endogenous bacterial proteins) for enhanced expression of difficult eukaryotic targets.

Conclusions

We report here high yielding soluble expression of functional hCNTF in *E. coli*. Our combined approach of codon optimization and screening for soluble expression with nine fusion tags under varying culture conditions identified several constructs producing soluble hCNTF expression (8 – 9 fold higher soluble fraction) and the overall yield of purified recombinant hCNTF achieved was significantly higher (greater than 10 – 20 fold) as compared to previous published reports. Soluble expression of the 6-His-hCNTF in all conditions tested points to the fact that codon optimization is sufficient for soluble expression of hCNTF in *E. coli*. However, for maximal yields, some fusion constructs might be more useful than 6-His-hCNTF. The strategy of combining codon optimization and fusion tag screening might be useful for soluble expression of pharmaceutical proteins in *E. coli*.

Methods

Codon optimization

The codons of the hCNTF gene were optimized for optimal usage in *E. coli* using a commercial proprietary algorithm, OptimumGene™ (GenScript, NJ, USA). Variety of parameters were optimized, including codon usage bias, GC content, CpG dinucleotides content, mRNA secondary structure, cryptic splicing sites, premature PolyA sites, internal chi sites and ribosomal binding sites, negative

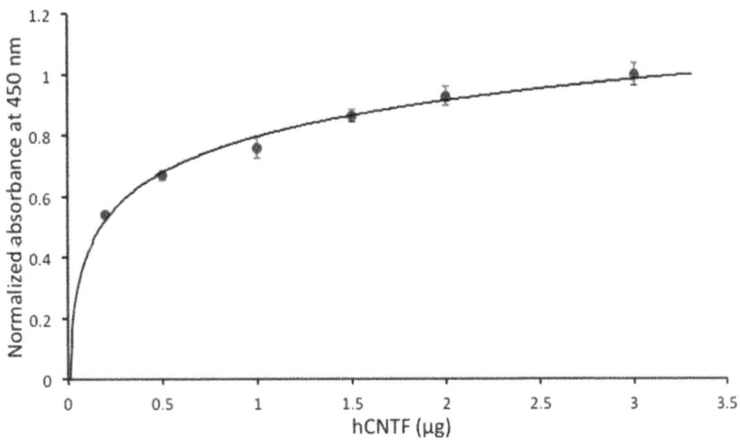

Figure 6 Binding of biotinylated hCNTF to hCNTFRα.

CpG islands, RNA instability motif, repeat sequences. The synthetic gene was purchased from GenScript, US.

Construction of expression vectors

The construction of pOPIN-fusion vector backbones has been described elsewhere [51-53]. The complete coding sequence of hCNTF was synthesized by GenScript, NJ, USA and provided as an insert within pUC57 plasmid. hCNTF gene was amplified by polymerase chain reaction (PCR) using KOD Hot Start (Novagen) with primers that had 5'-extensions of AAGTTCTGTTTCAGGGCCCG and ATGGTCTAGAAAGCTTTA on the forward and reverse primers, respectively (Additional file 1: Figure S1). The set of nine vectors harboring various fusion tags is listed in Table 1. The purified PCR product was inserted into digested pOPIN vectors between KpnI and HindIII restriction sites using Gibson Assembly™ cloning kit (New England Biolabs, MA, USA) and the reaction mixture transformed into NEB 5α (New England Biolabs, MA, USA) competent cells. Three colonies were screened for each construct and the presence of insert verified by PCR using a vector specific forward primer and hCNTF gene specific reverse primer (Additional file 1: Figure S2). pOPIN expression vectors 6-His-(fusion tag)-hCNTFs obtained were (6-His)-hCNTF, 6-His-(S3C)-hCNTF, 6-His-(Trx)-hCNTF, 6-His-(MSYB)-hCNTF, 6-His-(J)-hCNTF, 6-His-(HALO)-hCNTF, 6-His-(M)-hCNTF, 6-His-(TF)-hCNTF, 6-His-(NusA)-hCNTF.

Small-scale expression, affinity purification for soluble protein characterization

Expression strains BL21 (DE3)pLysS (Novagen) and Rosetta 2(DE3)pLysS (Novagen) were transformed with expression plasmids. Single colonies were used to inoculate 0.7 ml Power Prime Broth, PPB (AthenaEs) supplemented with ampicillin (100 μg/ml) and chloramphenicol (34 μg/ml) taken in a 2 ml 96 well plate. The plates were sealed with gas permeable adhesive seals and the seed cultures grown overnight at 37°C with shaking at 250 rpm. For the small-scale expression screen, 150 μl of BL21 (DE3)pLysS and 250 μl of Rosetta 2(DE3)pLysS seed cultures were used to inoculate 3 ml of PPB and Overnight Express™ Instant TB media, TBONEX (Novagen) taken in a 10 ml 24-deepwell plate and supplemented with ampicillin (100 μg/ml) and chloramphenicol (34 μg/ml). Cultures in PPB were grown at 37°C with shaking at 250 rpm. For expression at 37°C, 1 mM IPTG was added when the OD_{600} reached between 0.5 - 0.6. For expression at 20°C, the cultures were cooled for 15 minutes on a bench top before adding 1 mM IPTG and grown at 20°C overnight (18 h) with shaking at 250 rpm. TBONEX auto-inducing cultures were grown at 25°C for 24 h with shaking at 225 rpm. The cultures were harvested by centrifugation and the supernatant removed from the pellets. The pellets were stored at −80°C. The frozen pellets were resuspended in 210 μL of 50 mM NaH_2PO_4, 300 mM NaCl, pH 8.0; 10 mM imidazole, 1 mM dithiothreitol (DTT), protease inhibitor cocktail, PIC (Sigma). The pellets were freeze-thawed three times; resuspending each time in between the freeze thaw cycle. 1 μL of Lysonase (EMD Millipore) was added to each pellet suspension and incubated at room temperature with shaking at 270 rpm for 45 minutes. The lysates were then centrifuged at 21,100 g at 4°C for 40 minutes. 20 μL of suspended nickel nitriloacetic (Ni-NTA) magnetic beads (GenScript, NJ, USA) was dispensed into each well of a 96-well flat bottom plate. The beads were conditioned with 50 mM NaH_2PO_4, 300 mM NaCl, pH 8.0; 10 mM imidazole, 1% Tween 20. Supernatants from cleared lysates were transferred to the wells containing the magnetic beads and incubated at room temperature for 50 minutes with shaking at 270 rpm. The plate was then placed on a magnetic rack (New England Biolabs, MA, USA) for 5 minutes and the supernatant was removed. The magnetic beads were washed four times with 50 mM NaH_2PO_4, 300 mM NaCl, pH 8.0; 20 mM imidazole, 0.05% Tween 20 and 1 mM DTT. Proteins bound to the magnetic beads were eluted by adding 50 μL of 50 mM NaH_2PO_4, 300 mM NaCl, pH 8.0; 250 mM imidazole, 0.05% Tween 20 and 1 mM DTT and incubating with shaking (270 rpm) at room temperature for 10 minutes. The plate was placed on the magnetic rack for 7 minutes and 45 μL of the supernatant was removed. Protein samples were heated at 95°C with SDS-PAGE sample buffer and analyzed on a SDS-PAGE gel. The 6-His-fusion constructs of hCNTF were verified by specific detection of the 6-His tag using Pierce 6XHis protein tag stain reagent set (Thermo Fisher Scientific Inc., USA).

Estimation of soluble fraction of total recombinant protein expressed

Cell pellets from 3 ml expression culture were thawed and resuspended in 250 μL of 50 mM NaH_2PO_4, 300 mM NaCl pH 8.0; 0.2% Tween 20. The pellets were freeze-thawed three cycles and 50 μL of the suspended pellets were each treated with 1 μL of Lysonase. The pellets were incubated at room temperature for 45 minutes with shaking at 270 rpm. The mixture was divided into two fractions of 25 μL each. For total protein samples, 25 μL of the cell suspension was diluted with an equal volume of water followed by 10 μL of 6X SDS-PAGE sample buffer and boiled for 15 minutes at 95°C. The samples were centrifuged at high speed for 15 minutes and the supernatant run as the 'total protein' sample on a SDS-PAGE gel. For the soluble fraction, the 25 μL cell suspension was diluted with an equal volume of water and centrifuged at high speed for 15 minutes. The supernatant was removed and mixed with 10 μL of 6X denaturing SDS-PAGE sample buffer and boiled for 15 minutes at 95°C. The samples were analyzed as 'soluble fraction' on a SDS-PAGE gel. GelQuant.NET software (V 1.7.8) from biochemlabsolutions.com was used to quantitate band intensities to estimate the ratio of soluble to total protein. The values were averaged over three measurements. The intensity estimates for comparisons between constructs should be restricted to samples run within the same gel.

Large-scale expression and purification of hCNTF

6-His-hCNTF construct was chosen for large scale expression in Rosetta 2 (DE3)pLysS with the auto-inducing media TBONEX. 2 X 450 ml of cultures were inoculated with overnight seed cultures (30:1) and grown at 25°C with shaking at 270 rpm for 30 h. The cells were harvested by centrifugation at 10,000 g for 30 minutes at 4°C. and frozen at −80°C. The cell pellet was resuspended in lysis buffer (50 mM NaH_2PO_4, 300 mM NaCl, pH 8.0; 10 mM imidazole, 1 mM DTT and PIC) and freeze thawed three cycles, resuspending thoroughly after thawing each time. Lysonase was added to the suspension and incubated at room temperature for 60 minutes with gentle agitation. The lysate was centrifuged at 14,500 g at 4°C for 50 minutes and the cleared supernatant incubated with Ni-IDA resin (Macherey Nagel) at 4°C for 120 minutes with gentle agitation. The supernatant was removed by centrifugation at 500 g for 3 minutes and the Ni-IDA resin washed with 50 mM NaH_2PO_4, 300 mM NaCl, pH 8.0; 10 mM imidazole, 1 mM DTT and PIC. In subsequent washes, the imidazole concentration was increased. Bound hCNTF was eluted out of the resin (Additional file 1: Figure S3) by adding 250 mM imidazole to the buffer and the eluted fractions pooled, concentrated and buffer exchanged in 100 mM NaH_2PO_4, 300 mM NaCl, pH 8.0; 1 mM DTT using Amicon® Ultra-4 centrifugal filter units with a 3 kDa molecular weight cut off (Merck Millipore). The Ni-IDA eluted protein was further purified by size exclusion chromatography (SEC) by loading onto a HiLoad Superdex 200 prep grade filtration column (Additional file 1: Figure S4). The protein fractions were pooled, concentrated and buffer exchanged in 100 mM NaH_2PO_4, 50 mM NaCl, pH 8.0; 2 mM DTT using centrifugal filter units.

Binding assay of biotinylated hCNTF with hCNTFRα

hCNTF was labeled with biotin for binding studies with hCNTFRα. Briefly, hCNTF was reacted with a 20-fold molar excess of NHS-PEG4-biotin in a 100 mM carbonate buffer solution for 2 h at room temperature with gentle agitation. The reaction mixture was buffer exchanged with 100 mM NaH_2PO_4, 50 mM NaCl, pH 8.0 and concentrated using a spin filter having 3 kDa molecular weight cut off.

For the binding study, 1 μg of hCNTFRα (Sino Biologicals Inc.) in PBS was immobilized on the surface of a Nunc MaxiSorp flat-bottomed 96-well plate. The wells were washed with 50 mM Tris–HCl, 150 mM NaCl, 0.05% Tween 20, pH 7.0 (TBS-T) and blocked with a 2% BSA in TBS-T solution. After washings, biotinylated CNTF was added in varying amounts (0.2 – 3.0 μg) to the wells and incubated at room temperature for 2 h with gentle agitation. The wells were then washed with TBS-T followed by the addition of 50 ng of HRP (horseradish peroxidase)-conjugated streptavidin (Thermo Scientific) to each well and incubated at room temperature for 70 minutes with gentle agitation. The wells were washed and 200 μL of TMB (3,3', 5,5'-tetramethylbenzidine; Thermo Scientific) was added to each well. The reaction was stopped with 2 M H_2SO_4 and the optical density measured at 450 nm using a spectrophotometer (Varioskan Flash; Thermo Scientific).

Availability of supporting data

The data set supporting the results of this article are included within the article and in Additional file 1.

Additional file 1: Figures S1 - S4. Description: PCR amplified hCNTF, validation of hCNTF clones in expression vectors, affinity batch purification fractions, size exclusion chromatography profile.

Endnote

[a]During the course of this manuscript preparation, an alternate expression system expressing soluble hCNTF in *E. coli* was published. This has been discussed in the 'Results and Discussion' section.

Additional file

Additional file 1: Figure S1. PCR-amplified hCNTF insert. **Table S1** PCR primers for the amplification of the synthesized hCNTF. **Figure S2** PCR validation of cloned hCNTF in pOPIN vectors. **Figure S3** Representative eluted fractions from Ni-IDA batch purification of 2 × 450 ml cultures **(A and B)**. **Figure S4** Size exclusion chromatography (SEC).

Abbreviations

hCNTF: Human ciliary neurotrophic factor; E. coli: Escherichia coli; hCNTFRα: Human ciliary neurotrophic factor receptor alpha; LIFβ: LIF receptor; ALS: Amytropic lateral sclerosis; HD: Huntington's disease; AMD: Age related macular degeneration; tRNAs: Transfer RNAs; CAI: Codon adaptation index; TBONEX: Overnight expression terrific broth; PCR: Polymerase chain reaction; 6-His: 6-histidine; SUMO: Small ubiquitin related modifier; Trx: Thioredoxin; MsyB: *E. coli* acidic protein; GST: Glutathione S-transferase; Halo: Mutated dehalogenase; MBP: Maltose binding protein; TF: Trigger factor; NusA: N-utilization substance A; DTT: Dithiothreitol; Ni-NTA: Nickel nitriloacetic; PPB: Power prime broth; IPTG: Isopropyl β-D-1-thiogalactopyranoside; SDS-PAGE: Sodium dodecyl sulphate polyacrylamide gel electrophoresis; PIC: Protease inhibitor cocktail; Ni-IDA: Nickel iminodiacetic acid; SEC: Size exclusion chromatography; PBS: Phosphate buffered saline; NHS-PEG4: N-hydroxysuccinimide ester – polyethylene glycol; TBS-T: Tris buffered saline-tween; HRP: Horseradish peroxidase; TMB: 3,3', 5,5'-tetramethylbenzidine; SPP: Single protein production.

Competing interests

The authors declare that they have no competing interests.

Authors' contributions

JI carried out molecular cloning, protein expression studies, binding assay and prepared figures. AU provided financial and administrative support and participated in design of study and writing of the manuscript. LB prepared the pOPIN vectors and participated in experimental design. SS conceived of the study, its experimental design, wrote technical protocols, validated binding assay, did coordination, supervision and prepared the manuscript. All authors read and approved the final manuscript.

Acknowledgements

This study was supported by grants from Sigrid Juselius Foundation and Academy of Finland. The authors wish to thank the European Commission for funding under the FP7 programme (grant agreement number NMP4-SL-2010-246180). The OPPF-UK is funded by the Medical Research Council and the Biotechnology and Biological Sciences Research Council (grant no. MRC/K018779/1). Wellcome Trust Centre for Human Genetics is supported by the Wellcome Trust (grant no. 075491). The authors wish to thank Kornelia Mikula, Department of Biochemistry, University of Helsinki for her help with size exclusion chromatography.

Author details

[1]Centre for Drug Research, Division of Pharmaceutical Biosciences, Faculty of Pharmacy, University of Helsinki, P.O. Box 56, Viikinkaari 5E, Helsinki 00014, Finland. [2]School of Pharmacy, University of Eastern Finland, Kuopio, Finland. [3]Division of Structural Biology, Henry Wellcome Building for Genomic Medicine, University of Oxford, Roosevelt Drive, Oxford OX3 7BN, UK. [4]OPPF-UK, Research Complex at Harwell, R92 Rutherford Appleton Laboratory, Harwell Oxford, Didcot, Oxford OX11 OFA, UK.

References

1. Campenot RB: **Local control of neurite development by nerve growth factor.** *PNAS* 1977, **74:**4516–4519.
2. Chao MV: **Neurotrophins and their receptors: A convergence point for many signalling pathways.** *Nat Rev Neurosci* 2003, **4:**299–309.
3. Vergara C, Ramirez B: **CNTF, a pleiotropic cytokine: emphasis on its myotrophic role.** *Brain Res Brain Res Rev* 2004, **47:**161–173.
4. Weissmiller AM, Wu C: **Current advances in using neurotrophic factors to treat neurodegenerative disorders.** *Transl Neurodegener* 2012, **1:**14.
5. Adler R, Landa KB, Manthorpe M, Varon S: **Cholinergic neuronotrophic factors: intraocular distribution of trophic activity for ciliary neurons.** *Science* 1979, **204:**1434–1436.
6. Barbin G, Manthorpe M, Varon S: **Purification of the chick eye ciliary neuronotrophic factor.** *J Neurochem* 1984, **43:**1468–1478.
7. Bazan JF: **Neuropoietic cytokines in the hematopoietic fold.** *Neuron* 1991, **7:**197–208.
8. McDonald NQ, Panayotatos N, Hendrickson WA: **Crystal structure of dimeric human ciliary neurotrophic factor determined by MAD phasing.** *EMBO J* 1995, **14:**2689–2699.
9. Stahl N, Yancopoulos GD: **The alphas, betas, and kinases of cytokine receptor complexes.** *Cell* 1993, **74:**587–590.
10. Dy M, Vasquez A, Bertoglio J, The'se J: **General aspects of cytokine properties and function.** In *The Cytokine Network and Immune Functions.* Hong Kong: Oxford Univ. Press; 1999:1–13.
11. Sleeman MW, Anderson KD, Lambert PD, Yancopoulos GD, Wiegand SJ: **The ciliary neurotrophic factor and its receptor, CNTFRα.** *Pharm Acta Helv* 2000, **74:**265–272.
12. Emerich DF, Winn SR, Hantraye PM, Peschanski M, Chen EY, Chu Y, McDermott P, Baetge EE, Kordower JH: **Protective effect of encapsulated cells producing neurotrophic factor CNTF in a monkey model of Huntington's disease.** *Nature* 1997, **386:**395–399.
13. Dutta R, McDonough J, Chang A, Swamy L, Siu A, Kidd GJ, Rudick R, Mirnics K, Trapp BD: **Activation of the ciliary neurotrophic factor (CNTF) signalling pathway in cortical neurons of multiple sclerosis patients.** *Brain* 2007, **130:**2566–2576.
14. Lambert PD, Anderson KD, Sleeman MW, Wong V, Tan J, Hijarunguru A, Corcoran TL, Murray JD, Thabet KE, Yancopoulos GD, Wiegand SJ: **Ciliary neurotrophic factor activates leptin-like pathways and reduces body fat, without cachexia or rebound weight gain, even in leptin-resistant obesity.** *Proc Natl Acad Sci U S A* 2001, **98:**4652–4657.
15. Ettinger MP, Littlejohn TW, Schwartz SL, Weiss SR, McIlwain HH, Heymsfield SB, Bray GA, Roberts WG, Heyman ER, Stambler N, Heshka S, Vicary C, Guler H-P: **Recombinant variant of ciliary neurotrophic factor for weight loss in obese adults: a randomized, dose-ranging study.** *JAMA* 2003, **289:**1826–1832.
16. Matthews VB, Febbraio MA: **CNTF: a target therapeutic for obesity-related metabolic disease?** *J Mol Med* 2008, **86:**353–361.
17. Wen R, Tao W, Li Y, Sieving PA: **CNTF and retina.** *Prog Retin Eye Res* 2012, **31:**136–151.
18. Baneyx F: **Recombinant protein expression in Escherichia coli.** *Curr Opin Biotechnol* 1999, **10:**411–421.
19. Balbás P: **Understanding the art of producing protein and nonprotein molecules in Escherichia coli.** *Mol Biotechnol* 2001, **19:**251–267.
20. Peti W, Page R: **Strategies to maximize heterologous protein expression in Escherichia coli with minimal cost.** *Protein Expr Purif* 2007, **51:**1–10.
21. Huang C-J, Lin H, Yang X: **Industrial production of recombinant therapeutics in Escherichia coli & and its recent advancements.** *J Ind Microbiol Biotechnol* 2012, **39:**383–399.
22. Chen R: **Bacterial expression systems for recombinant protein production: E. coli and beyond.** *Biotechnol Adv* 2012, **30:**1102–1107.
23. McDonald JR, Ko C, Mismer D, Smith DJ, Collins F: **Expression and characterization of recombinant human ciliary neurotrophic factor from Escherichia coli.** *Biochim Biophys Acta* 1991, **1090:**70–80.
24. Negro A, Corona G, Bigon E, Martini I, Grandi C, Skaper SD, Callegaro L: **Synthesis, purification, and characterization of human ciliary neuronotrophic factor from E. coli.** *J Neurosci Res* 2004, **29:**251–260.
25. Masiakowski P, Liu H, Radziejewski C, Lottspeich F, Oberthuer W, Wong V, Lindsay RM, Furth ME, Panayotatos N: **Recombinant Human and Rat Ciliary Neurotrophic Factors.** *J Neurochem* 1991, **57:**1003–1012.
26. Wagener E-M, Aurich M, Aparicio-Siegmund S, Floss DM, Garbers C, Breusing K, Rabe B, Schwanbeck R, Grötzinger J, Rose-John S, Scheller J: **The amino acid exchange R28E in ciliary neurotrophic factor (CNTF) abrogates IL-6R- but retains CNTFR-dependent signaling via gp130/LIFR.** *J Biol Chem* 2014, in press.
27. Baneyx F, Mujacic M: **Recombinant protein folding and misfolding in Escherichia coli.** *Nat Biotechnol* 2004, **22:**1399–1408.
28. Khow O, Suntrarachun S: **Strategies for production of active eukaryotic proteins in bacterial expression system.** *Asian Pac J Trop Biomed* 2012, **2:**159–162.
29. Kane JF: **Effects of rare codon clusters on high-level expression of heterologous proteins in Escherichia coli.** *Curr Opin Biotechnol* 1995, **6:**494–500.
30. Calderone TL, Stevens RD, Oas TG: **High-level Misincorporation of Lysine for Arginine at AGA Codons in a Fusion Protein Expressed in Escherichia coli.** *J Mol Biol* 1996, **262:**407–412.
31. Welch M, Villalobos A, Gustafsson C, Minshull J: **You're one in a googol: optimizing genes for protein expression.** *J R Soc Interface* 2009, **6**(Suppl 4):S467–S476.
32. Kudla G, Murray AW, Tollervey D, Plotkin JB: **Coding-Sequence Determinants of Gene Expression in Escherichia coli.** *Science* 2009, **324:**255–258.
33. Welch M, Govindarajan S, Ness JE, Villalobos A, Gurney A, Minshull J, Gustafsson C: **Design parameters to control synthetic gene expression in Escherichia coli.** *PLoS ONE* 2009, **4:**e7002.
34. Gvritishvili AG, Leung KW, Tombran-Tink J: **Codon preference optimization increases heterologous PEDF expression.** *PLoS ONE* 2010, **5:**e15056.
35. Burgess-Brown NA, Sharma S, Sobott F, Loenarz C, Oppermann U, Gileadi O: **Codon optimization can improve expression of human genes in Escherichia coli: A multi-gene study.** *Protein Expr Purif* 2008, **59:**94–102.
36. Vasina JA, Baneyx F: **Expression of aggregation-prone recombinant proteins at low temperatures: a comparative study of the Escherichia coli cspA and tac promoter systems.** *Protein Expr Purif* 1997, **9:**211–218.
37. Wall JG, Plückthun A: **Effects of overexpressing folding modulators on the in vivo folding of heterologous proteins in Escherichia coli.** *Curr Opin Biotechnol* 1995, **6:**507–516.
38. Galloway CA, Sowden MP, Smith HC: **Increasing the yield of soluble recombinant protein expressed in E. coli by induction during late log phase.** *BioTechniques* 2003, **34:**524–526. 528, 530.

39. Davis GD, Elisee C, Newham DM, Harrison RG: **New fusion protein systems designed to give soluble expression in Escherichia coli.** *Biotechnol Bioeng* 1999, **65**:382–388.
40. Kapust RB, Waugh DS: **Escherichia coli maltose-binding protein is uncommonly effective at promoting the solubility of polypeptides to which it is fused.** *Protein Sci* 1999, **8**:1668–1674.
41. Butt T, Edavettal S, Hall J, Mattern M: **SUMO fusion technology for difficult-to-express proteins.** *Protein Expr Purif* 2005, **43**:1–9.
42. Esposito D, Chatterjee DK: **Enhancement of soluble protein expression through the use of fusion tags.** *Curr Opin Biotechnol* 2006, **17**:353–358.
43. Waugh DS: **Making the most of affinity tags.** *Trends Biotechnol* 2005, **23**:316–320.
44. Kataeva I, Chang J, Xu H, Luan C-H, Zhou J, Uversky VN, Lin D, Horanyi P, Liu ZJ, Ljungdahl LG, Rose J, Luo M, Wang B-C: **Improving solubility of Shewanella oneidensis MR-1 and Clostridium thermocellum JW-20 proteins expressed into Esherichia coli.** *J Proteome Res* 2005, **4**:1942–1951.
45. Marblestone J, Edavettal S, Lim Y, Lim P, Zuo X, Butt T: **Comparison of SUMO fusion technology with traditional gene fusion systems: Enhanced expression and solubility with SUMO.** *Protein Sci* 2006, **15**:182–189.
46. Bogomolovas J, Simon B, Sattler M, Stier G: **Screening of fusion partners for high yield expression and purification of bioactive viscotoxins.** *Protein Expr Purif* 2009, **64**:16–23.
47. Knaust RKC, Nordlund P: **Screening for Soluble Expression of Recombinant Proteins in a 96-Well Format.** *Anal Biochem* 2001, **297**:79–85.
48. Berrow NS, Büssow K, Coutard B, Diprose J, Ekberg M, Folkers GE, Levy N, Lieu V, Owens RJ, Peleg Y, Pinaglia C, Quevillon-Cheruel S, Salim L, Scheich C, Vincentelli R, Busso D: **Recombinant protein expression and solubility screening in Escherichia coli: a comparative study.** *Acta Crystallogr Sect D: Biol Crystallogr* 2006, **62**:1218–1226.
49. Saez NJ, Vincentelli R: **High-Throughput Expression Screening and Purification of Recombinant Proteins in E. coli.** In *Structural Genomics. Volume 1091.* Edited by Chen YW. Totowa, NJ: Humana Press; 2014:33–53.
50. Ratanji KD, Derrick JP, Dearman RJ, Kimber I: **Immunogenicity of therapeutic proteins: Influence of aggregation.** *J Immunotoxicol* 2013, **11**:99–109.
51. Berrow NS, Alderton D, Sainsbury S, Nettleship J, Assenberg R, Rahman N, Stuart DI, Owens RJ: **A versatile ligation-independent cloning method suitable for high-throughput expression screening applications.** *Nucleic Acids Res* 2007, **35**:e45.
52. Berrow NS, Alderton D, Owens RJ: **The precise engineering of expression vectors using high-throughput In-Fusion PCR cloning.** *Methods Mol Biol* 2009, **498**:75–90.
53. Bird LE: **High throughput construction and small scale expression screening of multi-tag vectors in Escherichia coli.** *Methods* 2011, **55**:29–37.
54. Woestenenk EA, Hammarström M, van den Berg S, Härd T, Berglund H: **His tag effect on solubility of human proteins produced in Escherichia coli: a comparison between four expression vectors.** *J Struct Funct Genomics* 2004, **5**:217–229.
55. Saggio I, Paonessa G, Gloaguen I, Graziani R, De Serio A, Laufer R: **Nonradioactive receptor binding assay for ciliary neurotrophic factor.** *Anal Biochem* 1994, **221**:387–391.
56. Wang K, Zhou F, Zhu L, Zhu X, Zhang K, Zhu L: **High level soluble expression, purification, and characterization of human ciliary neuronotrophic factor in Escherichia coli by single protein production system.** *Protein Expr Purif* 2014, **96**:8–13.

A versatile and highly efficient method for scarless genome editing in *Escherichia coli* and *Salmonella enterica*

Juhan Kim[1,2], Anthony M Webb[1], Jamie P Kershner[1,2], Stephen Blaskowski[1] and Shelley D Copley[1,2*]

Abstract

Background: Recently developed methods for genome editing in bacteria take advantage of the introduction of double-strand breaks by I-SceI in a mutation cassette to select for cells in which homologous recombination has healed the break and introduced a desired mutation. This elegantly designed method did not work well in our hands for most genes.

Results: We corrected a mutation in the gene encoding I-SceI that compromised the function of a previously used Red helper plasmid. Further, we found that transcription extending into the mutation cassette interferes with cleavage by I-SceI. Addition of two transcription terminators upstream of the cleavage site dramatically increases the efficiency of genome editing. We also developed an improved method for modification of essential genes. Inclusion of a segment of the essential gene consisting of synonymous codons restores an open reading frame when the mutation cassette is integrated into the genome and decreases the frequency of recombination events that fail to incorporate the desired mutation. The optimized protocol takes only 5 days and has been 100% successful for over 100 genomic modifications in our hands.

Conclusions: The method we describe here is reliable and versatile, enabling various types of genome editing in *Escherichia coli* and *Salmonella enterica* by straightforward modifications of the mutation cassette. We provide detailed descriptions of the methods as well as designs for insertions, deletions, and introduction of point mutations.

Keywords: Genome editing, I-SceI, *Escherichia coli*, *Salmonella enterica*

Background

A facile and completely general method for modification of any genomic sequence of interest in bacteria would be of enormous value for a wide range of applications, including construction of strains for metabolic engineering, studies of the effects of mutations acquired during adaptive evolution, and construction of strains for molecular biology studies that require chromosomal modification of genes such as allele replacement or epitope tagging.

The most frequently used techniques for genetic modification of *Escherichia coli* and *Salmonella enterica* are the two-step recombination method [1,2] and the 'gene gorging' method [3]. Both methods use the bacteriophage λ recombination system to promote gene replacement [4] in strains with a low intrinsic recombination rate. For this purpose, Datsenko and Wanner [1] developed the Red helper plasmid, pKD46, and the Blattner lab developed two derivatives, pACBSR [3] and pK-HT [2]. Some variations on these commonly used methods use a λ Red recombinase encoded in the genomic DNA [5-7] rather than a Red helper plasmid.

The two-step recombination method uses a selection marker to identify cells in which a mutation cassette composed of two homologous recombination fragments surrounding a selection marker has been integrated into the target site. Subsequently, the selection marker can be removed by FLP recombinase [1] or by cleavage with the I-SceI meganuclease followed by homologous recombination using the host's RecA system [2]. Use of FLP recombinase leaves a short FRT sequence as a scar. Use of the host's RecA system after double-strand breakage

* Correspondence: shelley.copley@colorado.edu
[1]Department of Molecular, Cellular, and Developmental Biology, University of Colorado Boulder, Boulder, CO 80309, USA
[2]Cooperative Institute for Research in Environmental Sciences, University of Colorado Boulder, Boulder, CO 80309, USA

by I-SceI leaves no scar at the end. In contrast, the gene gorging method relies upon screening to identify colonies in which a desired genetic change has been introduced by recombination of a mutation cassette into the genome. It also leaves no scar, but can be inefficient without the use of a special recombination strain [8] or a specially designed synthetic DNA with chemically modified bases [9]. These techniques have been used extensively and modified to enable deletions, insertions, and epitope tagging [5,10-18].

In our hands, genome editing by both the two-step recombination method and gene gorging was highly inefficient in most cases. We focused our attention on the two-step recombination method because of the advantage of employing selection to identify cells in which the mutation cassette has been integrated into the genome. We identified and corrected three mutations that limit the efficacy of a previously used helper plasmid and streamlined the procedure so that it can be carried out with higher efficiency and fewer experimental manipulations. However, we still observed a wide range of efficiencies depending on the identity of the target gene.

We discovered that the intransigence of certain genes was due to active transcription that continued into the inserted mutation cassette and prevented cleavage at the I-SceI site. Difficulties in modifying actively transcribed genes present an important limitation to the use of genome editing procedures. Inclusion of two successive terminator sequences in front of the I-SceI cleavage site obviated this problem and resulted in highly reliable chromosomal manipulation.

Modification of essential genes is also problematic using the two-step procedure because the first step results in interruption of the target gene by the mutation cassette. A solution to this dilemma is to include in the mutation cassette an extra fragment of the target gene that restores an open reading frame when the mutation cassette is integrated into the genome. However, since homologous recombination can occur anywhere within this extended extra fragment, many recombination events will fail to integrate the small region encoding the desired mutation into the genome. We addressed this problem by using a fragment of the target gene composed of synonymous codons to prevent homologous recombination within this extra fragment.

Our optimized procedure is summarized in Figure 1. The new helper plasmid pSLTS encodes I-SceI under the control of the anhydrotetracycline-inducible *tetA* promoter and the Lambda Red recombinase genes under the control of the arabinose-inducible *araB* promoter, and carries a Temperature-Sensitive origin of replication. (Descriptions of all plasmids used in this work are provided in Table 1). The mutation cassette consists of a fragment containing 100 bp homology regions (HR1 and HR2) that target the cassette to the desired chromosomal location. HR3 provides the region in which homologous recombination can occur after I-SceI cleavage of the genomic DNA. Two HR3 sequences flank a sequence encoding a double transcriptional terminator (depicted as the double hairpin, 'TT', in Figure 1) in front of the I-SceI recognition site and a selection marker. The first step in the procedure is introduction of the helper plasmid pSLTS into the target strain. Overexpression of the λ Red recombinase is then induced by L-arabinose (Figure 1, Step 1). The mutation cassette is then introduced into cells carrying the helper plasmid by electroporation (Figure 1, Step 2). Integration of the mutation cassette into the target location is accomplished by the λ Red recombinase (Figure 1, Step 3). Cells in which the mutation cassette has been integrated into the genome are selected on plates containing the antibiotic for which the selection cassette confers resistance. Subsequently, these cells are spread onto plates containing ampicillin (to maintain the helper plasmid) and anhydrotetracycline (to induce expression of I-SceI) (Figure 1, Step 4). Most of the cells do not survive the double-strand break introduced by I-SceI. However, in a minority of cells, homologous recombination between the HR3 sequences enables survival and accomplishes the desired chromosomal modification (Figure 1, Step 5). Here we describe the development and validation of this method with applications in both *E. coli* and *Salmonella*.

Results and discussion

Construction of an optimized helper plasmid

The ideal helper plasmid for two-step scarless chromosomal modification would have three characteristics: 1) genes encoding the λ-Red recombinase to increase the frequency of successful integration of the mutation cassette by recombination; 2) a gene encoding I-SceI; and 3) a temperature-sensitive origin of replication to facilitate removal of the helper plasmid after recombination of the mutation cassette into the genome. Further, the λ-Red recombinase genes and the gene encoding I-SceI should be inducible under different conditions. Three plasmids with these characteristics, pKDTS [2], pWRG99 [14] and pRedI [18], have been reported.

We used pKDTS as a helper plasmid to introduce a sequence encoding a 3 × FLAG tag at the 3'-end of *rpoD* in *E. coli*. We re-constructed this plasmid as previously described by inserting the 1407 bp NcoI fragment encoding P_{tet}-*I-sceI* from pK-HT [2] into the NcoI site of pKD46 [2]. We found that pKDTS effectively promoted recombination of the mutation cassette into the genome, as judged by growth of colonies on medium containing the antibiotic for which a resistance gene was carried on the mutation cassette. However, cleavage at the I-SceI site in the mutation cassette by I-SceI was extremely

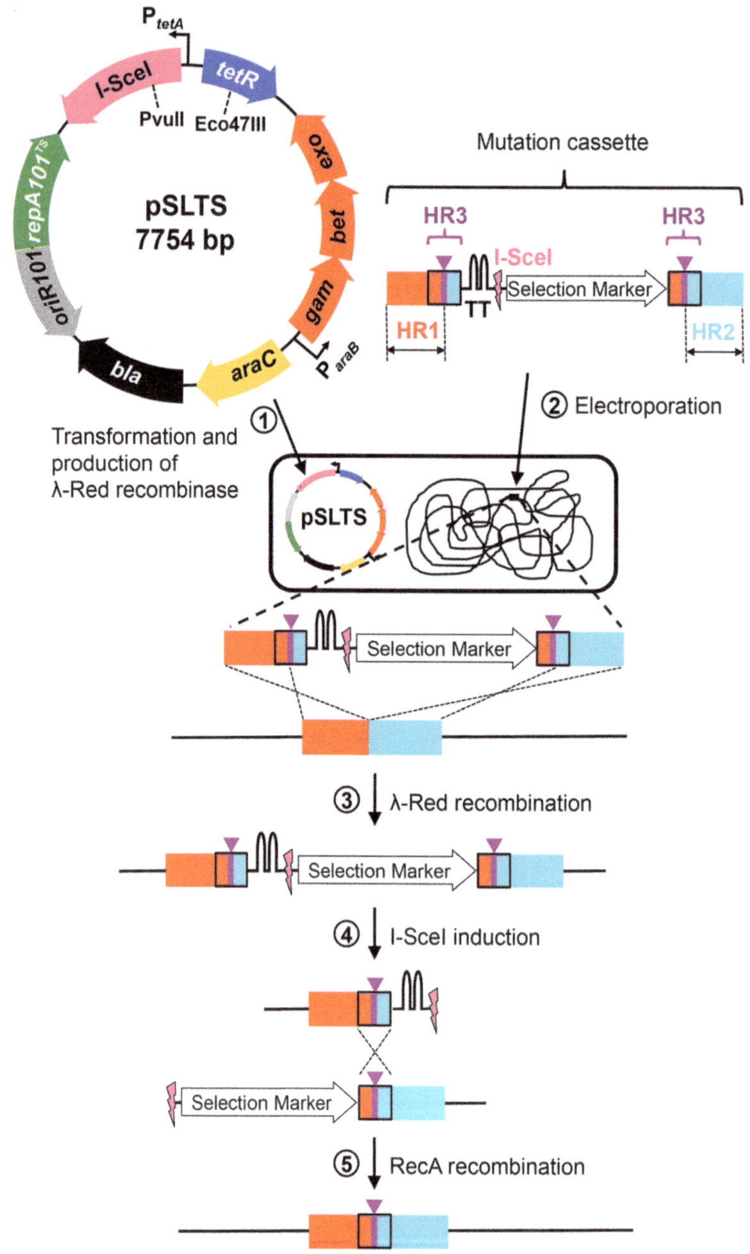

Figure 1 The general strategy for scarless genome editing. HR, homology region. HR1 and HR2 are identical to the target site in the genome. The purple bar and wedge in HR3 indicate the mutation that is to be introduced.

inefficient. After induction of I-SceI expression by anhydrotetracycline, only a small proportion of cells should survive; these cells have healed the double strand break introduced by I-SceI by a homologous recombination that removes the selection cassette. In this experiment, when 36 colonies were spread individually onto plates in the absence and presence of anhydrotetracycline, no difference in the number of colonies was observed (see Figure 2), suggesting that the intended cleavage of the genome by I-SceI was ineffective.

We discovered three mutations in the region of pKDTS derived from pK-HT. One of these, which results in replacement of 5 amino acid residues at the N-terminus of I-SceI with a single amino acid (MH*MKNIK*→MH*Q*), had been detected previously [2]. Two other mutations resulted in missense changes of Asn11 to Ser and Lys33 to Arg in TetR (AAC→AGC and AAG→AGG, respectively). All three mutations were present in the pK-HT plasmid from which the fragment containing genes for I-SceI and TetR was taken. Any or all of these mutations

Table 1 Plasmids used in this work

Plasmids	Characteristics	Reference
pSLTS*	ori SC101(Ts) Amp^r; P_{araB} for λ-Red; P_{tetR} for I-SceI	This work
pHA_{1887}	1887 bp fragment of pUC19 that contains the replication origin and the *bla* gene conferring ampicillin resistance	
pT2SC*	Amp^r Cm^r; a template plasmid for amplification of a selection cassette that includes two transcription terminator elements, an I-SceI site and a chloramphenicol resistance gene ligated into pHA_{1887}.	This work
pT2SK*	Amp^r Kan^r; a template plasmid for amplification of a selection cassette that includes two transcription terminator elements, an I-SceI site and a kanamycin resistance gene ligated into pHA_{1887}.	This work
pT2ST*	Amp^r $Trim^r$; a template plasmid for amplification of a selection cassette that includes two transcription terminator elements, an I-SceI site and a trimethoprim resistance gene ligated into pHA_{1887}.	This work
pT2SCb*	Amp^r Cm^r; a template plasmid for amplification of a selection cassette that includes two transcription terminator elements, an I-SceI site and a modified chloramphenicol resistance gene that provides a bacterial consensus ribosomal binding site at the 3'-end ligated into pHA_{1887}.	This work
pASC	Amp^r Cm^r; a template plasmid for amplification of a selection cassette that includes an I-SceI site and a chloramphenicol resistance gene ligated into pHA_{1887}.	This work
pASK	Amp^r Kan^r; a template plasmid for amplification of a selection cassette that includes an I-SceI site and a kanamycin resistance gene ligated into pHA_{1887}.	This work
pAST	Amp^r $Trim^r$; a template plasmid for amplification of a selection cassette that includes an I-SceI site and a trimethoprim resistance ligated into pHA_{1887}.	This work
pTSC	Amp^r Cm^r; a template plasmid for amplification of a selection cassette that includes a transcription terminator, an I-SceI site and a chloramphenicol resistance gene ligated into pHA_{1887}.	This work
pTSK	Amp^r Kan^r; a template plasmid for amplification of a selection cassette that includes a transcription terminator, an I-SceI site and a kanamycin resistance gene ligated into pHA_{1887}.	This work
pTST	Amp^r $Trim^r$; a template plasmid for amplification of a selection cassette that includes a transcription terminator, an I-SceI site and a trimethoprim resistance gene ligated into pHA_{1887}.	This work
pKD46	ori SC101(Ts) Amp^r; P_{araB} for λ-Red	[1]
pK-HT	ori SC101(Ts^-) Amp^r; P_{rha} for λ-Red; P_{tetR} for I-SceI	[2]

*These plasmids have been deposited to the addgene repository and are freely available for academic use.

might have been responsible for the failure of the I-SceI cleavage step. After correction of these three mutations, the new helper plasmid, pSLTS, was successfully used to introduce a sequence encoding a 3 × FLAG tag at the 3'-end of *rpoD* into *E. coli* (see Table 2 and Figure 2). The frequency of viable colonies after induction of I-SceI expression was 3.6 (±1.5) × 10^{-5}, attesting to the efficiency of the I-SceI cleavage step. Among eight colonies that lost the selection cassette after induction of I-SceI expression, all had the correct insertion (Table 2). pKDTS2, a helper plasmid in which only the mutation in I-SceI had been corrected, also successfully promoted recombination of the mutation cassette into the genome and showed the expected diminution in the number of colonies after the induction of I-SceI expression (Figure 2). However, only four of eight colonies that had lost the

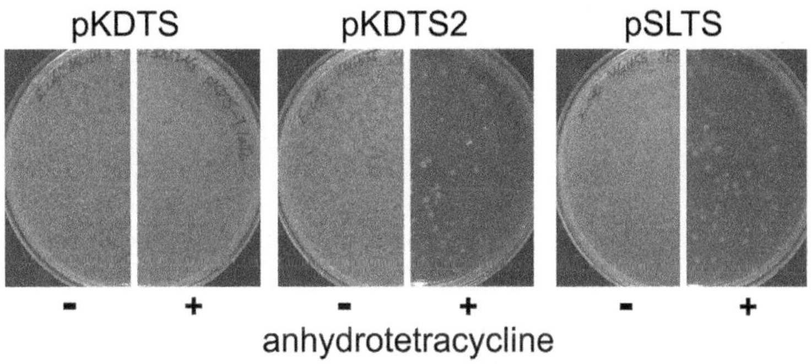

Figure 2 Comparison of the performance of helper plasmids in enabling I-SceI cleavage of a target site. Individual colonies of *E. coli* K-12 MG1655 cells containing the helper plasmid and into which a mutation cassette designed to introduce a C-terminal 3xFLAG tag on RpoD had been integrated were spread on agar plates containing LB Miller medium and 100 μg/mL ampicillin. Plates on the right half of each set contained 100 ng/ml anhydrotetracycline to induce expression of I-SceI. Successful cleavage of the mutation cassette results in a diminution of the number of viable colonies.

Table 2 Comparison of the performance of helper plasmids in enabling introduction of a sequence encoding a C-terminal 3XFLAG tag on RpoD in *E. coli* K-12 MG1655

Helper plasmid	pKDTS2	pSLTS
Fraction of colonies that lost the selection marker after double-strand cleavage of the mutation cassette by I-SceI	14/20	17/20
Fraction of colonies that had lost the selection marker for which a fragment of the correct size was amplified by colony PCR using primers flanking the gene of interest	4/8	8/8
Fraction of colonies for which the correct PCR product was obtained that contained the desired mutation	4/4	8/8

antibiotic resistance marker had the correct insertion (Table 2). The remaining four colonies were wild type.

Scarless deletion of *thrB* in *E. coli* K-12 MG1655 and *argC* in *S. enterica* SL1344 was easily accomplished using pSLTS (Additional file 1: Table S1). Again, all of the colonies that had lost the selection cassette after induction of I-SceI expression had the intended deletion. The general design of constructs for both insertion and deletion of sequences is diagrammed in Additional file 1: Figure S1.

Colony purification after introduction of the mutation cassette improves the efficiency of recovery of mutant strains

The recovery of colonies containing the desired mutation can be improved by re-streaking colonies obtained after introduction of the mutation cassette onto a second set of plates containing LB and ampicillin (LBA) and the antibiotic for which the mutation cassette confers resistance. Without this colony purification step, we routinely found some wild type colonies among those that grew on agar plates containing LB, ampicillin and anhydrotetracycline (LBAaTc). For example, in the experiment in which we introduced a sequence encoding a C-terminal 3 × FLAG tag on RpoD in *E. coli* K-12 MG1655 using pSLTS as a helper plasmid, 50% of the colonies obtained on LBAaTc agar plates were wild type when colony purification was not performed after integration of the mutation cassette. In contrast, no wild type cells were obtained when colony purification was performed (Table 2). When we attempted the same genetic modification in *S. enterica*, all of the 100 colonies tested after induction of I-SceI were wild type when colony purification was not done, while all of the 8 colonies tested had the desired modification when colony purification was performed. Apparently some wild type cells were able to survive in the vicinity of colonies that had integrated the mutation cassette and were able to detoxify the antibiotic in their immediate environment.

Manipulation of highly transcribed genes requires an additional element in the selection cassette

The two-step recombination method using pSLTS worked well in many cases, including the insertion of a sequence encoding a 3 × FLAG tag at the 3'-end of *rpoD* discussed above. However, we consistently failed to generate successful mutants when we attempted to introduce point mutations into *rpe* and *aceE* in *E. coli*, to delete an intergenic region between *thrL* and *thrA* and a small region in *pta* in *E. coli*, and to introduce a sequence encoding a 3XFLAG-tag at the C-terminus of GapA in *E. coli* and at the C-termini of GapA and FbaA in *S. enterica*. In each case, the mutation cassette was successfully integrated into the genome in the first step. However, spreading cells onto plates containing anhydrotetracycline to induce expression of I-SceI did not result in the expected decrease in the number of colonies due to introduction of a double-strand break (Figure 3). This result suggested that I-SceI failed to cleave at its target site in the inserted mutation cassette. We hypothesized that active transcription initiated at the promoter upstream of the target gene might extend into the mutation cassette and interfere with binding of I-SceI to its target site. To test this hypothesis, we introduced one or two terminator sequences upstream of the I-SceI site. We used the two most-efficient artificial terminators in the iGEM parts collection (http://parts.igem.org/Terminators/Catalog). When only one terminator element (BBa_B1006, 99% efficiency) was present before the cleavage site, only a slight decrease in the number of colonies was observed after induction of I-SceI, suggesting that cleavage at the I-SceI site was still inefficient. Insertion of two terminator elements (BBa_B1006 and BBa_B1002, 99 and 98% efficiencies, respectively) resulted in a dramatic increase in cleavage efficiency, evidenced by the substantial decrease in the number of colonies obtained when cells were spread on plates containing anhydrotetracycline (Figure 3). In both *E. coli* and *S. enterica*, it was possible to identify a correct mutant by screening fewer than a dozen colonies when the selection cassette contained a double terminator element, whereas it was necessary to screen forty colonies to find a correct mutant when the selection cassette contained a single terminator element (Table 3).

Introduction of mutations in essential genes

Introduction of mutations in essential genes is difficult because the typical strategy for initial integration of a mutation cassette into the genome disrupts the target gene. We have designed a modification of this strategy that enables efficient manipulation of essential genes.

frr and *ppa* are essential for growth of *E. coli* K-12 BW25113 in rich media such as LB [19]. *frr* encodes a ribosome recycling factor and *ppa* encodes an inorganic

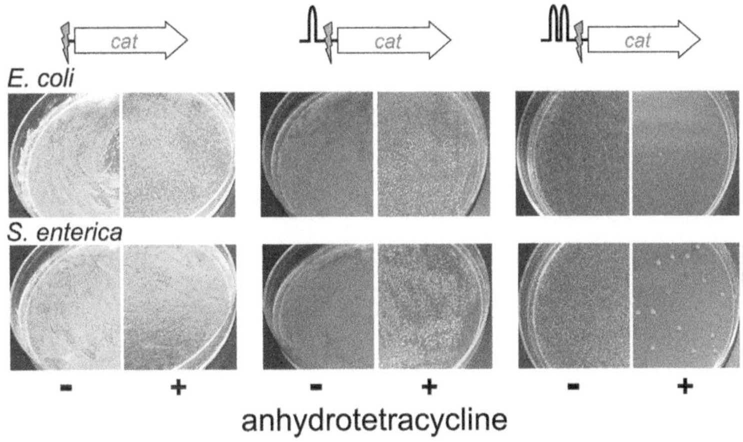

Figure 3 Terminator elements up-stream of the I-SceI cleavage site enhance the efficiency of I-SceI cleavage. One or two terminator elements were introduced ahead of the I-SceI cleavage site in mutation cassettes designed to introduce a sequence encoding a C-terminal 3 × FLAG tag on GapA in *E. coli* K-12 MG1655 and *S. enterica* Typhimurium SL1344. The efficiency of double-strand break formation by I-SceI was determined by spreading cells on agar plates containing LB Miller medium and 100 µg/mL ampicillin in the presence or absence of 100 ng/ml anhydrotetracycline.

pyrophosphatase. These genes were previously used to demonstrate a method for introduction of an amber stop codon that could subsequently be used in conjunction with a suppressor tRNA to enable condition-specific disruption of translation [2]. We targeted these genes to facilitate comparison of our approach and the previously published approach.

We designed a mutation cassette in which HR3 contains a silent mutation of a codon for serine (AGC→TCT or TCT→AGC) in either the 5'-half or the 3'- half of *frr* and *ppa.* To maintain the function of the encoded proteins during the procedure, we added an additional fragment that encodes part of the target gene so that an intact open reading frame was reconstituted either under control of its own promoter (Figure 4a) or under control of the constitutive promoter upstream of the selection marker (*cat*) preceding the target gene fragment (Figure 4b) after integration of the mutation cassette into the genome. (The latter strategy would be optimal for editing of the first gene in an operon because it would avoid polar effects on downstream genes). Each codon of this additional fragment was replaced with a synonymous codon to produce a sequence that encodes the same protein but cannot recombine with the target gene in the genome. An alternative design (Figure 5 and further discussion below) restores a complete open reading frame for the wild type sequence of the essential gene, which resolves to a modified gene after double-strand cleavage and homologous recombination. In addition to maintaining the function of the target gene throughout the procedure, this strategy helps confine the sequences capable of recombination to a small window, thereby decreasing the frequency of recombination events that regenerate the wild type sequence (Additional file 1: Figure S2). Such recombination events may account for the variable frequency of integration of the amber stop codon into target genes in a previous report [2]. Because unproductive recombination events increase the amount of screening required to identify colonies with the desired mutation, the synonymous codon strategy significantly improves the efficiency of the procedure, which is of particular importance for endeavors requiring a large number of modifications.

Table 3 Effect of terminator elements ahead of the I-SceI cleavage site on introduction of sequence encoding a C-terminal 3XFLAG tag on GapA in *E. coli* K-12 MG1655 and *S. enterica* Typhimurium SL1344

	E. coli K-12 MG1655			*S. enterica* Typhimurium SL1344		
Number of terminators in front of the I-SceI recognition site	0	1	2	0	1	2
Fraction of colonies that lost the selection marker after double-strand cleavage of the mutation cassette by I-SceI	0/100	2/40	5/11	0/100	1/40	4/7
Fraction of colonies that lost the selection marker for which a fragment of the correct size was amplified by colony PCR using primers flanking the gene of interest	NA*	1/2	1/2	NA*	1/1	2/2
Fraction of colonies for which the correct PCR product was obtained that contained the desired mutation	NA*	1/1	1/1	NA*	1/1	2/2

*NA, not applicable.

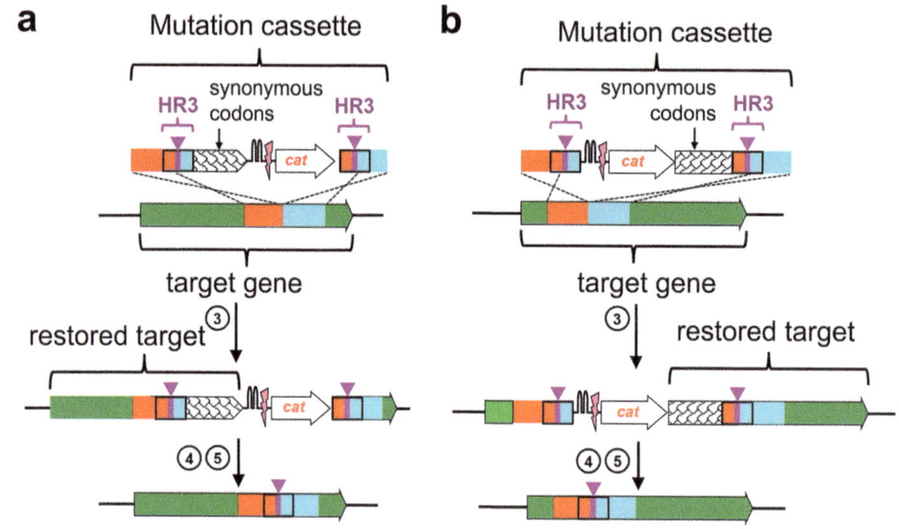

Figure 4 A gene fragment with synonymous codons maintains function of an essential gene during editing. An intact open reading frame can be reconstituted either under control of its own promoter **(a)** or under control of the constitutive promoter upstream of the selection marker (*cat*) preceding the target gene fragment **(b)**.

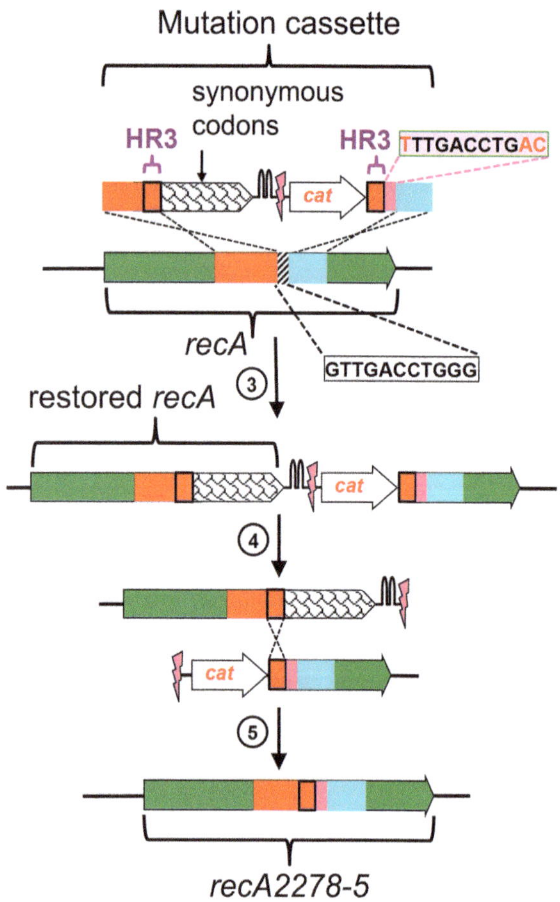

Figure 5 Introduction of multiple point mutations into *recA* in *E. coli*. The hatched area indicates the region to be replaced by the pink fragment of the mutation cassette.

We were able to introduce silent mutations into both the 5' and 3' halves of *frr* and the 3' half of *ppa* using this approach (Table 4). The mutation cassette was successfully integrated in 100% of the colonies obtained after selection on agar plates containing the antibiotic for which the selection marker conferred resistance. We typically selected four colonies from these plates and spread them onto plates containing ampicillin with or without anhydrotetracycline to induce expression of I-SceI. Effective cleavage by I-SceI was evidenced by a dramatic reduction in the number of colonies on the plates containing anhydrotetracycline. Subsequently, we patched up to five colonies from each plate containing anhydrotetracycline onto plates in the absence and presence of the antibiotic for which the mutation cassette confers resistance. All of the colonies that lost the antibiotic resistance marker contained the desired mutation (Table 4).

We were initially unable to introduce a silent mutation in the 5' half of *ppa* due to a failure to integrate the mutation cassette, suggesting that our construct did not enable production of sufficient Ppa to support growth. This failure might have been due to difficulties in transcription due to the dependence of *ppa* transcription on the constitutive promoter upstream of *cat*, although this was not an issue in the comparable experiment with *frr*. A more likely explanation is that translation of the synonymously coded fragment was sup-optimal. To explore this possibility, we introduced silent mutations in the *cat* gene upstream of the synonymous coding region of *ppa* in the mutation cassette to generate a ribosome binding site more closely resembling the bacterial consensus ribosome binding site [20]. This strategy was successful (see Figure 6), suggesting that a strong ribosome binding

Table 4 Efficiency of introduction of silent mutations into essential genes in *E. coli* K-12 BW25113 using pSLTS as a helper plasmid

Target gene	*frr*		*ppa*	
length of gene (bp) affected amino acid residue	558 Ser33		531	
Position of mutation (bp)	97-99	445-447 Ser149	109-111 Ser37	343-345 Ser115
Mutation	TCT→AGC	AGC→TCT	AGC→TCT	AGC→TCT
Efficiency of integration of the mutation cassette	100% (4/4)	100% (4/4)	67% (2/3)	100% (4/4)
Fraction of colonies that lost the selection marker after double-strand cleavage of the mutation cassette by I-SceI	2/7	3/6	4/10	9/20
Fraction of colonies that lost the selection marker for which a fragment of the correct size was amplified by colony PCR using primers flanking the gene of interest	2/2	3/3	2/2	4/4
Fraction of colonies for which the correct PCR product was obtained that contained the desired mutation	2/2	3/3	2/2	4/4

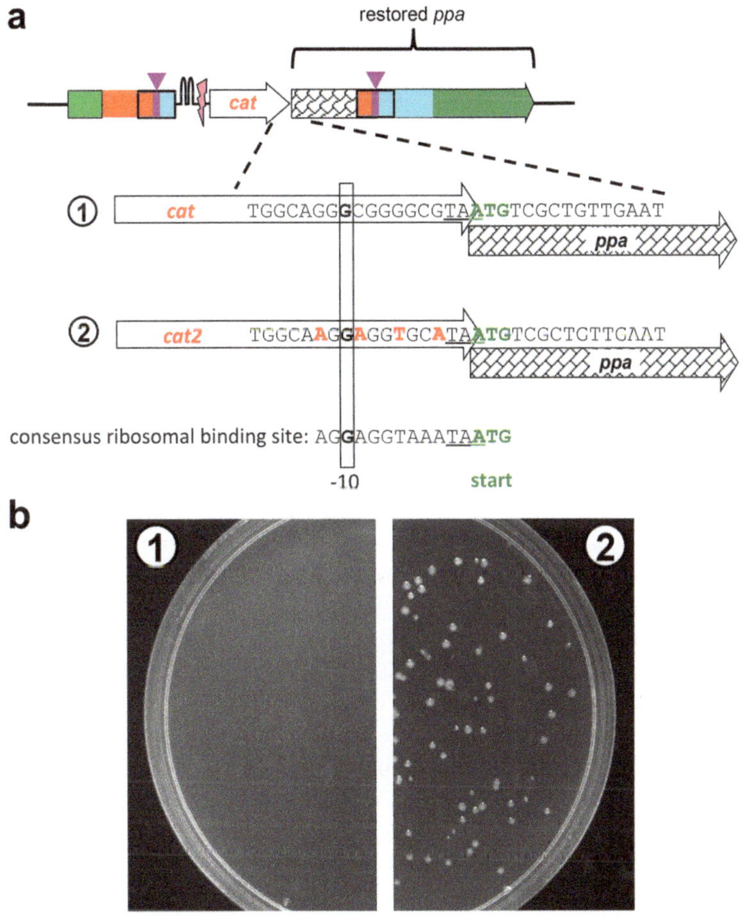

Figure 6 Introduction of a ribosome binding site improves production of the protein encoded by the downstream gene. (a) Four silent mutations at the 3′-end of *cat* were introduced to generate a more effective ribosomal binding site upstream of *ppa*. Sequences at the junction between *cat* and *ppa* are shown for the original (1) and modified (2) constructs. **(b)** Growth of cells on plates containing ampicillin (to force retention of the helper plasmid) and chloramphenicol (to confirm integration of the mutation cassette into the genome) after introduction of the mutation cassette by electroporation. The lack of colonies on the left indicates that construct 1 does not enable production of the essential protein Ppa. However, construct 2 allows growth, indicating that sufficient Ppa is produced.

site upstream of an essential gene that includes synonymous codons may be required to produce sufficient protein for cell survival in some cases.

λ-Red recombinase can be used to increase the efficiency of double-strand break repair

The synonymous codon strategy described above can be used to regenerate an open reading frame for either the mutant target gene (Figure 4) or the wild type target gene (Figure 5) after integration of the mutation cassette into the genome. The latter strategy is useful when it is desirable to maintain the wild type function of the gene until the final step. This was the case when we endeavored to introduce three mutations found in the *recA*2278-5 allele [21] into *recA* in *E. coli* because active RecA is needed to repair the double-strand break introduced by I-SceI [2] in the second step of the genome editing procedure. Even though wild type RecA should have been available, we found that the frequency of loss of the selection marker after induction of I-SceI was low; only two out of 25 colonies patched to LBA agar plates lost resistance to the antibiotic (Additional file 1: Table S2). The reconstitution of the *recA* gene using synonymous codons may have resulted in a lower level of RecA. To counter this difficulty, we over-expressed the Red recombinase before inducing expression of I-SceI. It has previously been reported that Exo and Beta of the Red recombinase increase the efficiency of double-strand break repair by 100-fold in cells containing wild type RecA [22]. Indeed, overproduction of Red recombinase dramatically improved the efficiency of the mutagenesis; all of the 25 colonies that were tested after the I-SceI induction step had lost the selection marker (Additional file 1: Table S2), indicating that homologous recombination had occurred.

Conclusions

Although impressive progress has been made in the past decade toward developing methods for introduction of specific genetic changes into the genomes of *E. coli* and *S. enterica*, the ease of performance and the efficiencies of even the best of these methods vary widely. We have identified the major impediments to the success of strategies that require creation of double-strand breaks by I-SceI followed by homologous recombination to accomplish genetic changes. In addition, we found that transcription initiated upstream of the mutation cassette apparently interferes with the ability of I-SceI to cleave at its recognition site. Introduction of two terminators upstream of the I-SceI site alleviates this problem.

Genetic modification of essential genes is particularly challenging because the function of the gene must be maintained throughout the procedure. We have shown that use of a mutation cassette in which synonymous codons restore the coding sequence of the gene after integration into the genome has the double benefit of maintaining function and limiting the region within which homologous recombination can take place, thereby increasing the proportion of crossovers that incorporate the desired change into the genome.

The method described here has enabled highly efficient modification of every gene we have targeted, including a number for which previously published strategies have been unsuccessful. The streamlined approach we have developed substantially decreases the time and effort required to generate genetically modified strains; we routinely generate desired mutants within one week. In addition, the approach is also highly versatile, allowing introduction of one or several point mutations, insertions and deletions into even difficult targets, using a common set of materials and procedures.

The method described in this paper compares favorably to the most recently developed methods for genome editing in *E. coli* and *S. enterica.* The CRISPR/Cas system developed for genome editing in eukaryotes has recently been applied to bacteria. This method is less convenient than the method described here because it requires construction of an editing fragment and a specific targeting RNA for each mutation to be introduced, whereas our two-step recombination method requires only a mutation cassette. Only two examples of successful gene editing have been reported in *E. coli* [7], so the general applicability of the system in bacteria is not yet proven.

A distinctly different method for accomplishing insertions, deletions and inversions in bacterial genomes using mobile group II introns and the Cre-*lox* system has been recently described [23]. This method is most useful for large-scale genomic changes, rather than smaller-scale changes, such as introduction of a point mutation or a sequence encoding a tag. Furthermore, this method requires initial disruption of a target gene, and cannot be used in its present form for essential genes.

Finally, a new two-plasmid method for scarless genome editing in *E. coli* has been reported by Yang et al. [24]. This method combines gene gorging with the use of a selection marker whose loss is an indication of introduction of the desired mutation into the genome. The first plasmid carries the genes for the Red recombinase and the I-SceI endonuclease, as well as a selection marker flanked by I-SceI sites and homology regions to target a particular genomic site. After integration of the selection marker, a second plasmid carrying a mutation cassette, again flanked by I-SceI sites, is introduced. After induction of I-SceI, cleavage at its recognition sites in the plasmid and the genome generates fragments that can recombine to integrate the mutation fragment into the genome. This procedure was shown to be very efficient in three cases. However, it is not well suited for editing of essential genes. The method was used to edit an essential gene (*metK*, which is essential for biosynthesis of S-adenosylmethionine). However, the procedure required a clever trick; initial replacement of *metK* with a gene encoding a transporter allowed uptake of S-adenosylmethionine and rendered MetK activity dispensable. This approach will not be applicable to most cases. Further, the Yang et al. method requires more time and experimental manipulation than the method described here.

Methods

Materials

E. coli K-12 BW25113 and K-12 MG1655 were obtained from the Nara Institute of Science and Technology and the American Type Culture Collection (ATCC), respectively. *S. enterica* Typhimurium SL1344 was obtained from the Detweiler laboratory at the University of Colorado Boulder. Chemically competent *E. coli* Mach1™ cells (Life Technologies, Inc., Grand Island, NY) were used for all cloning procedures. Bacteria were grown in Miller's modified Luria Broth (LB) or on Miller's Luria agar medium (Research Products International Corp.) unless otherwise specified. The following antibiotics were added to media as required for selection: ampicillin, 100 μg/mL; streptomycin, 50 μg/mL; chloramphenicol, 20 μg/mL; kanamycin, 50 μg/mL when the *kan* resistance gene was present on a high-copy-number plasmid and for selection after integration of the mutation cassette into the target DNA, and 20 μg/mL for subsequent experiments after the *kan* resistance gene had been integrated into the genomic DNA; trimethoprim, 50 μg/mL when the resistance gene was present on a high copy-number plasmid or 25 μg/mL when the resistance gene was integrated into the genomic DNA. Q5 High-Fidelity DNA polymerase (New England Biolabs) and Phusion High-Fidelity DNA polymerase (New England Biolabs) were used for polymerase chain reactions. Primer sequences and descriptions of synthetic DNA fragments are provided in the Additional file 1.

Construction of Red helper plasmids

pKD46 and pK-HT were obtained from the Blattner lab (University of Wisconsin-Madison). pKD46 encodes the λ Red recombinase genes under control of the *araB* promoter and a temperature-sensitive origin of replication. We re-constructed the previously described pKDTS by inserting the 1407 bp NcoI fragment from pK-HT, which encodes I-SceI under control of the anhydrotetracycline-inducible *tetA* promoter, into the NcoI site of pKD46 [2].

We constructed pSLTS by correcting three mutations found in pKDTS. A 555 bp fragment containing the three mutations and encompassing 405 bp of the 5'-end of the *tetR* gene, 81 bp of the *tetA* promoter, and 69 bp of the 5'-end of the I-SceI gene was removed from pKDTS by digesting the plasmid with AfeI (Eco47III) and PvuII. The remaining 7187 bp fragment was purified by gel extraction. The 5'-phosphate was removed using Antarctic phosphatase (New England BioLabs). A 610 bp synthetic DNA fragment (gBlock1) (Integrated DNA Technologies, Coralville, IA) was used to replace the removed fragment. gBlock1 contains 405 bp of the 5'-end *tetR* fragment with two mutations corrected based upon the sequence of the *tetR* gene of transposon Tn10 (GenBank Accession number, J01830.1), 81 bp of the *tetA* promoter, and 81 bp of the corrected 5'-end of the gene encoding I-SceI, as well as 21 and 22 bp complementary to the 7187 bp pKDTS fragment at the 5'- and 3'-ends, respectively. gBlock1 was assembled with the 7187 bp pKDTS fragment to make pSLTS by the one-step enzymatic DNA assembly method [25] using the Gibson Assembly Master Mix (New England BioLabs). The same procedure was used to make pKDTS2 from the 7187 bp fragment of pKDTS and gBlock2, which encodes intact I-SceI but still contains two mutations in *tetR*. Correct construction of pSLTS, pKDTS2, and pKDTS was confirmed by sequencing the modified parts of the plasmids using sequencing primers pKDTS-F and pKDTS-R.

Construction of selection cassette template plasmids

We constructed a set of template plasmids from which selection cassettes containing the I-SceI site and a positive selection marker can be amplified for subsequent incorporation into mutation cassettes. These plasmids were constructed by ligating a selection cassette containing an I-SceI recognition site followed by an antibiotic resistance gene into pHA_{1887} (see further below). The resulting plasmids, pASC, pASK and pAST, contain selection cassettes that confer resistance to chloramphenicol, kanamycin and trimethoprim, respectively. A second generation of template plasmids was constructed by adding sequences encoding one or two artificial transcription terminators upstream of the I-SceI site. These plasmids were assembled from pHA_{1877}, a sequence containing one or two transcriptional terminators (STS or STTS), and a selection cassette (amplified from a pAS series plasmid using primers MarkerF and MarkerR), using the one-step enzymatic DNA assembly method [25]. The names of these plasmids indicate the number of terminators and the antibiotic for which they confer resistance. pTS series plasmids contain one terminator element, an I-SceI recognition site and a gene conferring resistance to chloramphenicol (pTSC), kanamycin (pTSK), or trimethoprim (pTST). pT2S series plasmids contain two terminator elements, an I-SceI recognition site and a gene conferring resistance to chloramphenicol (pT2SC and pT2SCb), kanamycin (pT2SK) or trimethoprim (pT2ST). Correct insertion of selection cassettes into pHA_{1887} was confirmed in each case by sequencing inserted elements using sequencing primers pHA.seq.F and pHA.seq.R.

Selection cassettes to be used for construction of mutation cassettes were amplified from selection cassette template plasmids using primers MF and MR for pTS and pT2S series plasmids (except for pT2SCb, for which MR2 was used instead of MR) and primers MarkerF and MR for pAS series plasmids. After treatment of the amplification mixture with DpnI to remove the template plasmid, amplified selection cassettes were gel-purified.

The following paragraphs describe the origin and assembly of the various pieces of the selection cassette template fragments.

pHA_{1887} is an 1887 bp fragment amplified from pUC19 using primers pHAFor and pHARev. pHA_{1887} contains the replication origin that confers high copy number and the *bla* gene that confers ampicillin resistance. PCR-amplified pHA_{1887} was treated with DpnI to remove the template plasmid and gel-purified.

Genes encoding antibiotic resistance were amplified using primers that introduced extra sequences for use in either PCR amplification of selection cassettes or in the genome editing procedure itself. The *cat* gene was amplified from pACYC184 using primers ISceIcatF and catR. ISceIcatF contains the I-SceI site recognition site followed by 18 bp of the 5' end of *cat*. catR is complementary to the 3' end of the *cat* gene. The *kan* gene was amplified from pACYC177 with primers ISceIkanF and kancat16R. ISceIkanF contains the I-SceI recognition site followed by 18 bp of the 5'-end of *kan*. kancat16R contains the last 16 bp of *cat* followed by the 3'-end of *kan*. (The 16 bp *cat* sequence was included so that a common primer (MR) could be used for subsequent amplification of selection cassettes containing either *kan* or *cat*). Amplified fragments containing either *cat* or *kan* were treated with DpnI to remove the template plasmid, gel-purified and ligated into the pHA_{1887} fragment using the Rapid DNA ligation kit (Thermo Fisher Scientific, Inc.) to make pASC and pASK, respectively.

gISceIdfrA is a synthetic DNA fragment designed to contain the I-SceI site followed by the 630 bp *dfrA* sequence (from the EZ-Tn5™ < DHFR-1 > (Epicentre)) and the 16 bp common primer binding sequence. Twenty bp sequences complementary to pHA_{1887} were included on each side. gISceIdfrA was assembled with pHA_{1887} to make pAST using the one-step enzymatic DNA assembly method [25].

gISceIcat2 is a synthetic DNA fragment designed to contain the I-SceI recognition site and a modified *cat* gene (*cat2*) in which the 3'-end sequence was modified to generate a sequence resembling the consensus bacterial ribosomal binding site [20].

The sequences of the terminator fragments STS and STTS (see Additional file 1: list of synthetic DNA fragments) were designed based upon iGEM parts. The sequence of STS, which contains one terminator flanked by two spacer elements, was patched together from spacer BBa_K259002, terminator BBa_B1006, and spacer BBa_B0040 parts. The sequence of STTS, which contains two terminator elements flanked by two spacer elements, was designed to minimize secondary structure in the spacer regions using UNAFold (Integrated DNA Technologies) and was patched together from part of spacer BBa_K259002, part of spacer BBa_B0040, terminator BBa_B1002, part of spacer BBa_B0040, terminator BBa_B1006, and part of spacer BBa_K259002. STS and STTS were prepared as minigenes (Integrated DNA Technologies) and were amplified by PCR using primers STSF and STSR. The amplification mixture was treated with DpnI to remove the template plasmid and the STS or STTS fragments were gel-purified.

Construction of mutation cassette template plasmids

We constructed mutation cassette template plasmids to facilitate subsequent production of the mutation cassette by PCR amplification (see Additional file 1: Figure S3). Mutation cassettes contain a selection marker preceded by two transcriptional terminators and flanked by sequences homologous to the genomic target (see Figure 1). While 50 bp homology regions (HR1 and HR2) on each side of the selection marker are generally sufficient for homologous recombination into the genome, we typically obtained more colonies using 100 bp homology regions. Because the efficiency of recombination varies among genomic targets, we routinely use the longer 100 bp homology regions to ensure successful integration. The mutation cassette contains two identical regions (HR3) that span the desired genomic modification. The length of HR3 was 30–50 bp, sufficient for RecA-mediated recombination (see Step 5 in Figure 1), but short enough to minimize the frequency of recombination events that fail to incorporate the desired genetic change (see Additional file 1: Figure S2a).

Mutation template plasmids were assembled by mixing 5'- and 3'- mutation fragments (125 fmol each) with pHA_{1877}, a linear fragment of pUC19 obtained by PCR (see Additional file 1: Information) (25 fmol), and a selection cassette encoding an antibiotic resistance gene (75 fmol). The volume of the mixture was adjusted to 10 μL with DNase-free water and then 10 μL of Gibson Assembly Master Mix (New England BioLabs) was added. After incubation for 1 hour at 50° C, 2 μL of the reaction mixture was introduced into chemically competent *E. coli* Mach1^{TM} cells following the manufacturer's protocol. The transformed cells were spread on LB with ampicillin and the antibiotic to which the selection cassette conferred resistance. Typically, four colonies were selected for confirmation of the correct assembly of the mutation fragments and the correct sequence of the mutation cassette using primers pHA.seq.F and pHA.seq.R (see Additional file 1: Table S3).

Most mutation fragments used in the experiments described in this paper were obtained as synthetic double-stranded DNAs (gBlocks) from Integrated DNA Technologies, Inc., but mutation fragments can also be amplified from plasmid or genomic DNA. Generally, the 5'- mutation fragment contained approximately 200 bp of the target gene flanked by a sequence that overlaps the 3'-end of pHA_{1887} and a sequence that overlaps the 5'-end of a selection cassette. The 3'-mutation fragment also contains approximately 200 bp of the target gene flanked by a sequence that overlaps the 3'-end of the selection cassette and a sequence that overlaps the 5'-end of pHA_{1887}. Both fragments contain a 30–50 bp region designated HR3 at the 3'-end of the 5'-mutation fragment and at the 5'-end of the 3'-mutation fragment. The desired mutation can be included in HR3 (Figure 1 and Figure 4), or immediately following HR3 (Figure 5).

Mutation cassettes were amplified from mutation template plasmids by PCR using appropriate primers (see Additional file 1: Figure S3). The template plasmid was removed by digestion with DpnI and the mutation cassette was purified by electrophoresis on a 1% TAE-agarose gel.

Preparation of electrocompetent cells carrying a Red helper plasmid

A Red helper plasmid was introduced into *E. coli* K-12 BW25113 or *E. coli* K-12 MG1655 using the method of Chung et al. [26]. The plasmid (100 ng) was mixed with competent cells prepared from a 1 mL LB culture. A Red helper plasmid was introduced into electrocompetent *S. enterica* Typhimurium SL1344 cells by electroporation. (*S. enterica* Typhimurium SL1344 cells were rendered electrocompetent as described in Molecular Cloning [27]). The transformants were spread on LB Miller agar medium containing ampicillin (100 μg/mL). After overnight growth at 30°C, a single colony was used to inoculate 5 mL of LB Miller medium containing ampicillin (100 μg/mL) (LBA) for *E. coli* K-12 strains or ampicillin (100 μg/mL) and streptomycin (50 μg/mL) (LBAS) for *S. enterica* Typhimurium SL1344. After overnight incubation with shaking at 30°C, 1 mL of the culture was inoculated into 100 mL of LBA or LBAS in a 500 mL Erlenmyer flask. The culture was incubated with shaking for an hour at 30°C. L-Arabinose was added to a final concentration of 1 mM for *E. coli* K-12 BW25113 or 2 mM for *E. coli* K-12 MG1655 and *S. enterica* Typhimurium SL1344 to induce expression of the λ-Red recombinase. Incubation at 30°C with shaking was continued for 2–3 hours. When the OD_{600} of the culture reached 0.7-0.9, the cells were harvested by centrifugation at 4500 × g and washed twice with ice-cold 10% glycerol. The cells were resuspended in 10% glycerol and stored at −70°C before use.

Genome editing procedure

A mutation cassette (50 to 100 ng) was mixed with 50 μL of electrocompetent cells carrying a Red helper plasmid on ice. After electroporation, 450 μL SOC [27] was added to the cells and the mixture was transferred to a 15 mL culture tube. Cells were incubated for 3 hours in a shaking incubator at 30°C and then spread onto plates containing LB plus ampicillin (LBA) for *E. coli* or LB plus ampicillin and streptocymcin (LBAS) for *S. enterica* Typhimurium SL1344 and an additional antibiotic (chloramphenicol (20 μg/mL), kanamycin (50 μg/mL), or trimethoprim (50 μg/mL)) as needed to select for cells in which the mutation cassette had been integrated into the genome. After overnight incubation at 30°C, four colonies were picked and streaked onto fresh selection plates. After incubation overnight at 30°C, one colony from each plate was suspended in sterile phosphate-buffered saline (PBS [27]). Because the frequency of double strand break repair is about 1 in 10^5, it is only necessary to plate a small proportion of the cells in a colony in order to obtain enough colonies for the subsequent step. Aliquots of the cell suspensions were spread on plates containing LBA agar and LBA agar plus anhydrotetracycline (100 ng/mL) (LBAaTc). For *S. enterica*, the plates also contained streptomycin (50 μg/mL). In general, plating 50 to 100 μL of a cell suspension in which a single colony had been resuspended in 500 μL PBS resulted in several to tens of colonies on plates containing anhydrotetracycline. To confirm that double strand break repair had resulted in excision of the selection cassette, five to ten colonies from the LBAaTc plate were patched onto LB agar and onto LB agar supplemented with the antibiotic for which the mutation cassette provided resistance. (Streptomycin was also added for *S. enterica* Typhimurium SL1344). Ampicillin can be included in these plates to force retention of the helper plasmid if further rounds of mutations are planned. Colonies that grew on the LB agar plates but not on plates supplemented with the antibiotic were streaked onto fresh LB agar plates. After overnight incubation, a few individual colonies were used to prepare freezer stocks. The introduction of the desired mutation was verified by amplifying the target region by PCR and sequencing.

When necessary to increase the efficiency of double-strand break repair, expression of the Red recombinase was induced before expression of I-SceI. A single colony in which the mutation cassette had been integrated was resuspended in 50 μL of sterile DNase-free water. An aliquot (20 μL) was boiled at 100°C prior to amplification of the mutation cassette by PCR to confirm correct integration of the mutation cassette into the genome. The remaining aliquot (30 μL) was mixed with 120 μL of sterile PBS. Fifty μL of the suspended cells were spread on plates containing LBA or LBAaTc. The remaining 50 μL were inoculated into 1 mL of LBA in a 15 mL culture tube. The culture was incubated in a shaking incubator for an hour at 30°C. L-Arabinose was added to a final concentration of 1 mM for *E. coli* K-12 BW25113 or 2 mM for *E. coli* K-12 MG1655 and *S. enterica* to induce production of the Red recombinase. After 2 hours of further incubation at 30°C in a shaking incubator, 50 μL aliquots were spread on plates containing LBA or LBAaTc.

Additional file

Additional file 1: Figure S1. Diagrams for design of mutation cassettes for insertion **(a)** and deletion **(b)**. **Figure S2.** Synonymous codon fragment increases the efficiency of genome editing of the essential target. **(a)** Homologous recombination of the mutation cassette into the target gene in step 3 can fail to introduce mutations in HR3 (purple band and wedge) of the 3'-end of the mutation fragment when the second cross-over occurs in either the green or red region before HR3. The consequence is regeneration of the wild type sequence during the second homologous recombination step (step 4). **(b)** The use of a synonymous codon fragment restricts the positions at which homologous recombination can take place in step 3, increasing the likelihood that the sequence that includes the mutations is properly integrated into the genome and that the desired mutation will be inserted into the target gene after double-strand cleavage (step 4) and homologous recombination (step 5). **Table S1.** Deletion of genes in *S. enterica* and *E. coli* using pSLTS. **Table S2.** λ Red recombinase contributes to double-strand break repair during chromosomal modification of *recA* in *E. coli* K-12 BW25113. **Table S3.** PCR and sequencing primers. **Table S4.** Components used to generate mutation cassettes. Supplemental List of synthetic DNA fragments.

Competing interests

The University of Colorado Boulder has filed a patent application related to this publication.

Authors' contributions

JK designed and carried out experiments and wrote the paper. AMW and SB carried out experiments. JPK designed the method for assembly of the mutation template plasmid. SDC supervised the experiments and wrote the paper. All authors read and approved the final manuscript.

Acknowledgements

This work was supported by the National Institutes of Health [5RO1 GM083285-06 to S.D.C.].

References

1. Datsenko KA, Wanner BL: **One-step inactivation of chromosomal genes in *Escherichia coli* K-12 using PCR products.** *Proc Natl Acad Sci U S A* 2000, **97**(12):6640–6645.
2. Herring CD, Blattner FR: **Conditional lethal amber mutations in essential *Escherichia coli* genes.** *J Bacteriol* 2004, **186**(9):2673–2681.
3. Herring CD, Glasner JD, Blattner FR: **Gene replacement without selection: regulated suppression of amber mutations in *Escherichia coli*.** *Gene* 2003, **311**:153–163.
4. Murphy KC: **Use of bacteriophage lambda recombination functions to promote gene replacement in *Escherichia coli*.** *J Bacteriol* 1998, **180** (8):2063–2071.
5. Tischer BK, von Einem J, Kaufer B, Osterrieder N: **Two-step red-mediated recombination for versatile high-efficiency markerless DNA manipulation in *Escherichia coli*.** *Biotechniques* 2006, **40**(2):191–197.
6. Wang HH, Isaacs FJ, Carr PA, Sun ZZ, Xu G, Forest CR, Church GM: **Programming cells by multiplex genome engineering and accelerated evolution.** *Nature* 2009, **460**(7257):894–898.
7. Jiang W, Bikard D, Cox D, Zhang F, Marraffini LA: **RNA-guided editing of bacterial genomes using CRISPR-Cas systems.** *Nat Biotechnol* 2013, **31**(3):233–239.
8. Mosberg JA, Gregg CJ, Lajoie MJ, Wang HH, Church GM: **Improving lambda red genome engineering in *Escherichia coli* via rational removal of endogenous nucleases.** *PLoS One* 2012, **7**(9):e44638.
9. Wang HH, Xu G, Vonner AJ, Church G: **Modified bases enable high-efficiency oligonucleotide-mediated allelic replacement via mismatch repair evasion.** *Nucleic Acids Res* 2011, **39**(16):7336–7347.
10. Wendland J: **PCR-based methods facilitate targeted gene manipulations and cloning procedures.** *Curr Genet* 2003, **44**(3):115–123.
11. Cho BK, Knight EM, Palsson BO: **PCR-based tandem epitope tagging system for *Escherichia coli* genome engineering.** *Biotechniques* 2006, **40**(1):67–72.
12. Cox MM, Layton SL, Jiang T, Cole K, Hargis BM, Berghman LR, Bottje WG, Kwon YM: **Scarless and site-directed mutagenesis in *Salmonella enteritidis* chromosome.** *BMC Biotechnol* 2007, **7**:59.
13. Stringer AM, Singh N, Yermakova A, Petrone BL, Amarasinghe JJ, Reyes-Diaz L, Mantis NJ, Wade JT: **FRUIT, a scar-free system for targeted chromosomal mutagenesis, epitope tagging, and promoter replacement in *Escherichia coli* and *Salmonella enterica*.** *PLoS One* 2012, **7**(9):e44841.
14. Blank K, Hensel M, Gerlach RG: **Rapid and highly efficient method for scarless mutagenesis within the Salmonella enterica chromosome.** *PLoS One* 2011, **6**(1):e15763.
15. Blount ZD, Barrick JE, Davidson CJ, Lenski RE: **Genomic analysis of a key innovation in an experimental *Escherichia coli* population.** *Nature* 2012, **489**(7417):513–518.
16. Stanek MT, Cooper TF, Lenski RE: **Identification and dynamics of a beneficial mutation in a long-term evolution experiment with *Escherichia coli*.** *BMC Evol Biol* 2009, **9**:302.
17. Charusanti P, Conrad TM, Knight EM, Venkataraman K, Fong NL, Xie B, Gao Y, Palsson BO: **Genetic basis of growth adaptation of *Escherichia coli* after deletion of pgi, a major metabolic gene.** *PLoS Genet* 2010, **6**(11):e1001186.
18. Yu BJ, Kang KH, Lee JH, Sung BH, Kim MS, Kim SC: **Rapid and efficient construction of markerless deletions in the *Escherichia coli* genome.** *Nucleic Acids Res* 2008, **36**(14):e84.
19. Baba T, Ara T, Hasegawa M, Takai Y, Okumura Y, Baba M, Datsenko KA, Tomita M, Wanner BL, Mori H: **Construction of *Escherichia coli* K-12 in-frame, single-gene knockout mutants: the Keio collection.** *Mol Syst Biol* 2006, **2**:2006 0008.
20. Nelson DL, Cox MM: *Lehninger Principles of Biochemistry.* 4th edition. W. H. Freeman and Company: New York, NY; 2005.
21. Alexseyev AA, Bakhlanova IV, Zaitsev EN, Lanzov VA: **Genetic characteristics of new *recA* mutants of *Escherichia coli* K-12.** *J Bacteriol* 1996, **178**(7):2018–2024.
22. Takahashi N, Kobayashi I: **Evidence for the double-strand break repair model of bacteriophage lambda recombination.** *Proc Natl Acad Sci U S A* 1990, **87**(7):2790–2794.
23. Enyeart PJ, Chirieleison SM, Dao MN, Perutka J, Quandt EM, Yao J, Whitt JT, Keatinge-Clay AT, Lambowitz AM, Ellington AD: **Generalized bacterial genome editing using mobile group II introns and Cre-lox.** *Mol Syst Biol* 2013, **9**:685.
24. Yang J, Sun B, Huang H, Jiang Y, Diao L, Chen B, Xu C, Wang X, Liu J, Jiang W, Yang S: **High-Efficiency Scarless Genetic Modification in *Escherichia coli* by Using Lambda Red Recombination and I-SceI Cleavage.** *Appl Environ Microbiol* 2014, **80**(13):3826–3834.
25. Gibson DG, Young L, Chuang RY, Venter JC, Hutchison CA 3rd, Smith HO: **Enzymatic assembly of DNA molecules up to several hundred kilobases.** *Nat Methods* 2009, **6**(5):343–345.
26. Chung CT, Niemela SL, Miller RH: **One-step preparation of competent *Escherichia coli*: transformation and storage of bacterial cells in the same solution.** *Proc Natl Acad Sci U S A* 1989, **86**(7):2172–2175.
27. Sambrook J, Fritsch EF, Maniatis T: *Molecular Cloning; a laboratory manual.* 2nd edition. Cold Spring Harbor, NY: Cold Spring Harbor Laboratory Press; 1989.

Developing a xylanase XYNZG from *Plectosphaerella cucumerina* for baking by heterologously expressed in *Kluyveromyces lactis*

Fei Xiang Zhan[1], Qin Hong Wang[2], Si Jing Jiang[1], Yu Ling Zhou[1*], Gui Min Zhang[1*] and Yan He Ma[2]

Abstract

Background: Xylanase can replace chemical additives to improve the volume and sensory properties of bread in the baking. Suitable baking xylanase with improved yield will promote the application of xylanase in baking industry. The xylanase XYNZG from the *Plectosphaerella cucumerina* has been previously characterized by heterologous expression in *Pichia pastoris*. However, *P. pastoris* is not a suitable host for xylanase to be used in the baking process since *P. pastoris* does not have GRAS (Generally Regarded As Safe) status and requires large methanol supplement during the fermentation in most conditions, which is not allowed to be used in the food industry. *Kluyveromyces lactis*, as another yeast expression host, has a GRAS status, which has been successfully used in food and feed applications. No previous work has been reported concerning the heterologous expression of xylanase gene *xynZG* in *K. lactis* with an aim for application in baking.

Results: The xylanase gene *xynZG* from the *P. cucumerina* was heterologously expressed in *K. lactis*. The recombinant protein XYNZG in *K. lactis* presented an approximately 19 kDa band on SDS-PAGE and zymograms analysis. Transformant with the highest halo on the plate containing the RBB-xylan (Remazol Brilliant Blue-xylan) was selected for the flask fermentation in different media. The results indicated that the highest activity of 115 U/ml at 72 h was obtained with the YLPU medium. The mass spectrometry analysis suggested that the hydrolytic products of xylan by XYNZG were mainly xylobiose and xylotriose. The results of baking trials indicated that the addition of XYNZG could reduce the kneading time of dough, increase the volume of bread, improve the texture, and have more positive effects on the sensory properties of bread.

Conclusions: Xylanase XYNZG is successfully expressed in *K. lactis*, which exhibits the highest activity among the published reports of the xylanase expression in *K. lactis*. The recombinant XYNZG can be used to improve the volume and sensory properties of bread. Therefore, the expression yield of recombinant XYNZG can be further improved through engineered strain containing high copy numbers of the XYNZG, and optimized fermentation condition, making bread-baking application possible.

Keywords: Xylanase, Heterologous expression, *Kluyveromyces lactis*, Baking

Background

Nowadays, more and more attentions are paid on food safety and nutrition, and providing non-contaminated and fiber-rich food is becoming an important public issue. Accordingly, developing improved and new methods to produce whole wheat bread and reduce the use of chemical additives are challenges for the baking industry [1]. Traditional chemical food additives have been used in the baking industry to enlarge loaf volume, lengthen shelf life, and improve the taste of breads, etc. However, some of these compounds may threaten the health of consumers. For example, potassium bromate, the most widely used food additive, is now known to be a human carcinogen and has been banned by most of countries [2]. Azodicarbonamide, a bleaching and improving agent, is only a permitted food additive in certain countries. It partially degrades under the heat process to form trace amounts of

* Correspondence: zyl609@sina.com.cn; zhangguimin@hubu.edu.cn
[1]Hubei Collaborative Innovation Center for Green Transformation of Bio-Resources, College of Life Science, Hubei University, Wuhan 430062, People's Republic of China
Full list of author information is available at the end of the article

semicarbazide, which shows carcinogenicity and has been proved to cause tumors [1]. Therefore, it is very urgent to find safe food additives to replace the previous harmful chemical additives. Recombinant enzymes, as safe substituents, were firstly applied in baking industry in the 1970s due to the ever-increasing demand for more natural products. In the past 40 years, many enzymes, including α-amylase, cellulase, hemicellulase, and xylanase, have been successfully applied in the baking industry [2,3].

Xylanase is used as baking additives to improve processing and product quality. It affects enhancements in dough and bread quality leading to improved dough flexibility, machinability and stability as well as a larger loaf volume and an improved crumb structure. However, xylanase has not been applied extensively in baking industry because of high cost and poor effects, especially in the developing country. To reduce the price of enzymes, it is highly desirable to adopt gene engineering to produce enzymes with better performance. Thus, hundreds of xylanase genes from bacteria, fungi and actinomycetes have been cloned and expressed heterolougously [4,5].

Now, many xylanases have been used to efficiently express in the yeast host, especially in *Pichia pastoris* within the past teens years. However, *P. pastoris* cannot be used in the food industry since it does not have GRAS status by the FDA (US Food and Drug Administration) and requires large methanol supplement during the fermentation in most conditions. Contrarily, *Kluyveromyces lactis*, as another yeast expression host, has a GRAS status by FDA, which permits its use in food and feed applications. Meanwhile, compared with other yeast expression systems, *K. lactis* has advantages of multicopy gene integration, easy genetic manipulation, the availability of a fully sequenced genome [6], and it can easily grow to a high density on inexpensive lactose-based media [7]. Therefore, *K. lactis* has been used to efficiently express many proteins in the past few years, and the best example for its use is commercial production of the milk clotting enzyme, bovine chymosin [7]. Thus, several researchers tried to express different xylanase in this system. XynAs from the extreme thermophile *Thermotoga* sp. strain FjSS3B.1 [8], and the *Dictyoglomus thermophilum* strain Rt46B.1 [9], respectively, and Xyn11A from *Bacillus halodurans* strain C-125 [10] were expressed in *K. lactis* using episomal vector. However, they would have the risk of instability due to lacking of selective pressure. Only XynB from the *T. maritima* MSB8 was expressed stable in *K. lactis* based on the integration vector [11]. However, these xylanases were not investigated about application in baking industry. Additionally, thermophilic and halophilic xylanases are not favorable in the dough baking since high temperature and salt are not necessary.

Xylanase gene *xynZG* was cloned from *P. cucumerina* and had been successfully expressed in *P. pastoris* in our previous report [12]. Recombinant xylanase XYNZG has high specific activity and the optimal pH and temperature of 6 and 40°C, respectively [12]. These features may indicate that XYNZG has the potential use in baking. Thus, we try to express xylanase gene *xynZG* in *K. lactis* and study the effect of recombinant XYNZG on dough and bread in bread-making. Additionally, xylo-oligosaccharides (XOs) of the hydrolytic products were recently reported to possess a remarkable potential for stimulating the growth of intestinal bifidobacteria and promoting the intestinal health [13]. Because of this reason, we investigated the hydrolytic products of XYNZG in this study.

Results

Construction of plasmid pKLAC2-*xynZG* and expression in *K. lactis*

A 600 bp DNA fragment corresponding to *xynZG* was cloned into the pKLAC2 vector, fusing with the MF-alpha leader sequence for secretory expression in *K. lactis* GG799. The recombinant plasmid was named as pKLAC2-*xynZG* (Figure 1). For bio-safety and integration consideration, the ampicillin resistance gene on recombinant plasmid pKLAC2-*xynZG* was removed by *Sac*II digestion before transformation, so that the engineering strain producing xylanase does not contain any known antibiotic resistance genes. This is of vital importance for food enzymes [14]. The linearized plasmid was transformed into *K. lactis* and the transformants were cultured on the YPD plates containing 1% RBB xylan (Figure 2). One transformant with the largest halo GKX21 was chosen for shake-flask fermentation. The supernatant of the fermentation was performed on a 12% SDS-PAGE gel and the resolved proteins were visualized by staining with Coomassie brilliant blue. A band of proximately 19 kDa was observed, which accorded with the deduced molecular weight of XYNZG from *P. pastoris*. In addition, to detect the activity in situ and identify the target protein, the zymogram analysis was performed and the result revealed that there was only one about 19 kDa active band detected by Amido Black staining (Figure 3), consistent with our previous report [12].

The fermentation of XYNZG in different media

The recombinants GKX21 were inoculated into YPL, YLP, YLU and YLPU medium, respectively, and underwent further shaken flask fermentation for 96 h. The activities and biomass were all measured during 96 h (Figure 4a) and the results indicated the activity of XYNZG reached highest level at 72 h with 115 U/ml in YLPU. In order to illustrate the relationship between the activities and the biomass, Figure 4b showed the biomass and enzyme activities in different media. No obvious biomass difference existed in the four media whereas the enzyme activity differed significantly. Thus, the medium YLPU was the optimal medium for *xynZG* expression in *K. lactis*.

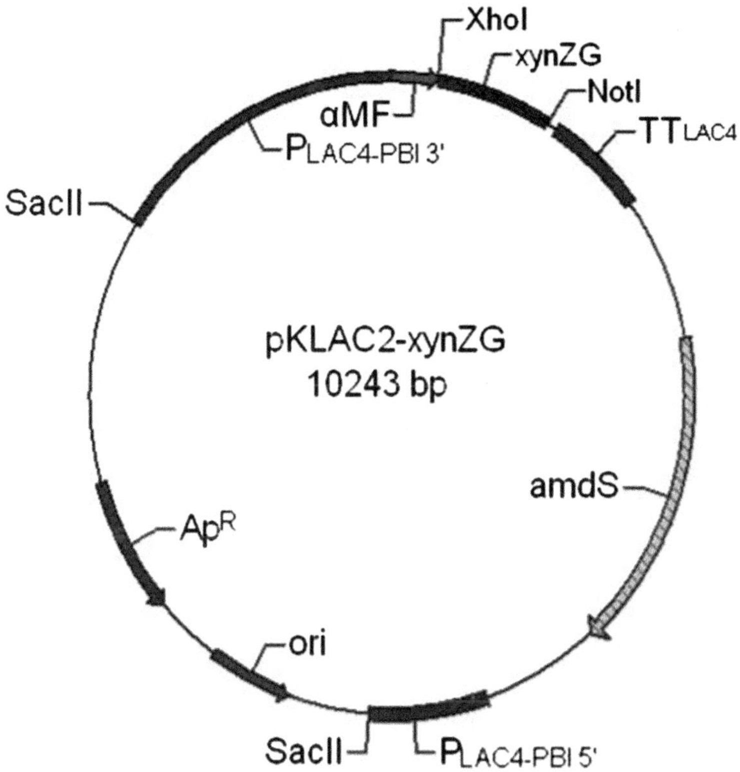

Figure 1 The map for the expression plasmid pKLAC2-*xynZG* in *K. lactis*. xynZG (the target gene encoding XYNZG), $P_{LAC4-PBI}$ (LAC4 Promoter), α-MF (α-mating factor secretion leader sequence), Ap^R (Ampicillin resistance gene), amdS (encoding acetamidase to break down acetamide for use as a nitrogen source), ori (origin of *E. coli*), TT_{LAC4} (LAC4 transcription terminator).

Mass spectrometry analysis of xylan hydrolysate

The hydrolyzate was analyzed using electrospray ion source mass spectrometry in the positive ion reflective mode. As shown in Figure 5, there were two *m/z* peaks of 305.08, and 437.12, which corresponded to xylobiose ($[M+Na]^+$) and xylotriose ($[M+Na]^+$), respectively. Therefore, the results showed that xylobiose and xylotriose were the main hydrolysis products released from beechwood xylan by XYNZG and the yield of xylobiose was greater than that of xylotriose.

Effects of XYNZG on the bread making

XYNZG has a broad working temperature range, which has an apparent optimal temperature of 40°C and retains approximately 75% and 55% of its maximum activity at 35°C and 25°C, respectively [12]. The mesophilic property of XYNZG makes it exhibit different characteristics from other fungi family11 xylanase, thus it may be highly suited for use in the baking industry as it is generally optimally active at the room temperatures most frequently used for dough preparation and proofing [2].

Bread quality assessment by sensory properties is largely based on personal judgment and subjective qualitative evaluation, whereas the results cannot be absolute but reflect the influences of consumer preferences [15]. The effects of XYNZG at different concentration on the attributes of dough or bread were shown in Table 1. Compared to the control lacking of xylanase, the dough handling time reduced from 6.5 min to 6.0 min and the height-diameter ratio of dough increased with the increasing concentration of XYNZG. Meanwhile, the addition of xylanase remarkably modified organoleptic properties of fresh whole wheat bread, so that the total sensory properties score reached 93 when 1200 ppm of 50 U XYNZG was supplemented, while the control only got a score of 85. However, the dough handling properties became a little sticky when the concentration of XYNZG reached 2000 ppm which was not suitable for manufacturing conditions which required the dough to be more 'machine-friendly' to avoid sticking to machinery parts. The result was consistent with the previous report that extra addition of the xylanase was not necessary in the baking process [16]. Thus, the proper concentration of XYNZG was 1200 ppm.

In order to avoid the personal sensory difference, the volume and texture of bread were also evaluated by the Volume Measuring Device and the Texture Analyzer, respectively. As shown in Figure 6, the addition of XYNZG to the baking dough led to the increase of volume compared to the negative control lacking xylanase. At the

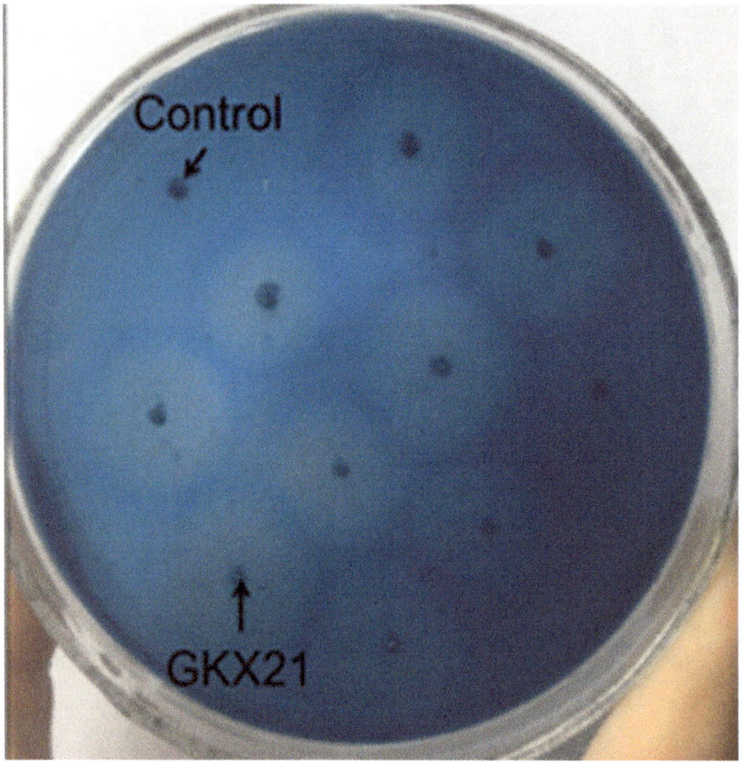

Figure 2 The transformants of activity screening. The transformants were cultured on the YPD plates containing 1% RBB xylan. The plates were photographed after 12 h incubation at 28°C. Negative control is *K. lactis* transformant harboring pKLAC2.

optimal dosage of 1200 ppm, XYNZG increased the specific volume of 19.1%. The effects of XYNZG on the bread texture profile were also shown in Table 2. With the increase of XYNZG, the initial crumb hardness and chewiness were both improved. The hardness was increased by 17.2% at the dosage of 1200 ppm, while chewiness changed slightly. In addition, XYNZG did not have a significant effect on bread springiness.

Discussion

Enzymatic baking serves as a good alternative to improve bread quality and strengthen food safety. Therefore, developing a cost-efficient bio-baking process of bread by enzyme represents a major future trend of baking industry. Many bacteria, fungi, and actinomycetes can produce xylanase while the yield is not very high to apply in industry. The low-cost of xylanase will be a key issue for accelerating bio-baking application in the dough treatment process. So it is important to get the cost efficient and high specific activity xylanase with an improved expression level in an ideal host. To reduce the price of enzymes, finding new xylanase gene and heterologous expression are highly desirable to produce enzymes with better performance. In our previous study, a xylanase gene *xynZG* was cloned from *P. cucumerina* and heterologously expressed in *P. pastoris* with high specific activity of 362 U/ml [12]. Recombinant xylanase XYNZG has optimal pH and temperature of 6 and 40°C, respectively [12]. In spite of the high yield and simple purification process, *P. pastoris* is not permitted for use in the food industry because it is not a GRAS and requires methanol supplement during its fermentation in most condition. Compared to *P. pastoris*, *K. lactis* as a heterologous expression host, which has been used to efficiently express many proteins in the past few years, and has a GRAS status that permits their use in food and feed applications. Thus, in this study, we successfully expressed the xylanase gene *xynZG* in *K. lactis* despite of expression yield in *K. lactis* was lower than that in *P. pastoris.*

So far, several researchers expressed four different xylanase in this system. However, the activity of XYNZG in this study can reach 115 U/ml based on the integration vector, which is much higher than 49 U/ml of XynB from the hyperthermophilic bacterium *Thermotoga maritima* MSB8, which was uniquely expressed stable in *K. lactis* based on the integration vector [11]. The other three xylanase expression in *K. lactis* based on episomal vector would have the risk of instability without selective pressure in spite that their activity are a little lower than XYNZG, 98 U/ml of XynA from the extreme thermophile *Thermotoga* sp. strain FjSS3B [8], 95 U/ml of XynA from

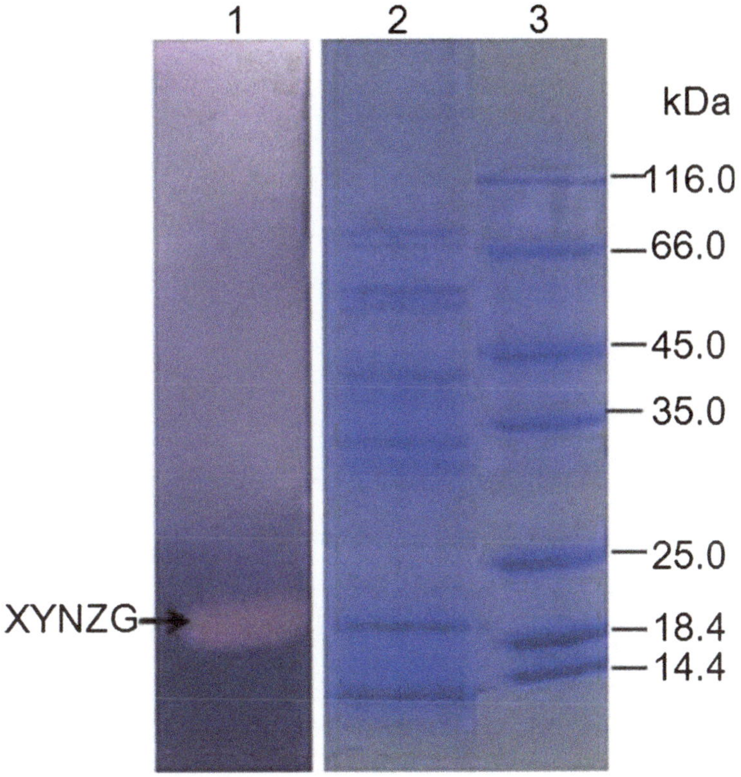

Figure 3 SDS-PAGE and zymogram analysis of XYNZG. Lane 1, zymogram analysis of the fermentation supernatant. Lane 2, the fermentation supernatant. Lane 3, the protein marker.

D. thermophilum Rt46B.1 [9] and 98 U/ml of Xyn11A from *B. halodurans* strain C-125 [10]. Additionally, neither of them was investigated or mentioned about application in baking industry. Their features are not suitable for baking use, either. What's more, the expression yield of recombinant XYNZG can be further improved through engineered strain containing high copy numbers of the XYNZG, and optimized fermentation condition, making bread-baking application possible.

In addition,xylanase was used to produce xylo-oligosaccharides (XO) in recent years. Owing to the special properties of XOs, they possess a remarkable potential for practical utilization in many fields, including pharmaceuticals, feed formulations and agricultural applications [17]. Among these XOs, xylobiose and xylotriose (DP = 2 and 3) are considered to be the main xylooligosaccharides for food applications since xylobiose and xylotriose can stimulate the growth of intestinal bifidobacteria. The vitro assays showed that both xylobiose and xylotriose can be utilized by *Bifidobacterium spp.* and *B. adolescentis* [13,17]. Contrarily, some harmful microorganisms, such as *Staphylococcus, E. coli* and many *Clostridium* spp. cannot utilize xylobiose and xylotriose [18]. Moreover, the sweetness of xylobiose is equivalent to 40% of sucrose while it does not cause increase of blood sugar [17]. In this study, the main products of hydrolytes produced by XYNZG are xylobiose and a small part of xylotriose. The results of baking trial also indicate that addition of XYNZG can increase the bread volume and improve the crumb structure, which is similar to the enzymes in GH11, a previously used industrial xylanase family. Therefore, addition of XYNZG can not only increase the bread volume and the crumb structure, but also improve the flavor and the health value of bread.

Conclusions

In general, the xylanase gene xynZG can be stably and highly expressed in *K. lactis* as a functional enzyme, and XYNZG can efficiently increase bread volume and improve the flavor and function of bread. In this study, the xylanase gene xynZG of *P. cucumerina* was first successfully expressed in *K. lactis* and the expression activity was the highest activity reported so far. This is also the first report of using *K. lactis* expressed xylanase used in baking industry and the baking experiments showed that addition of this enzyme can efficiently improve the volume and sensory properties of bread.

Methods

Strains, plasmids and media

The strain *K. lactis* GG799 (New England Biolabs, USA) was used as the host for the expression of recombinant

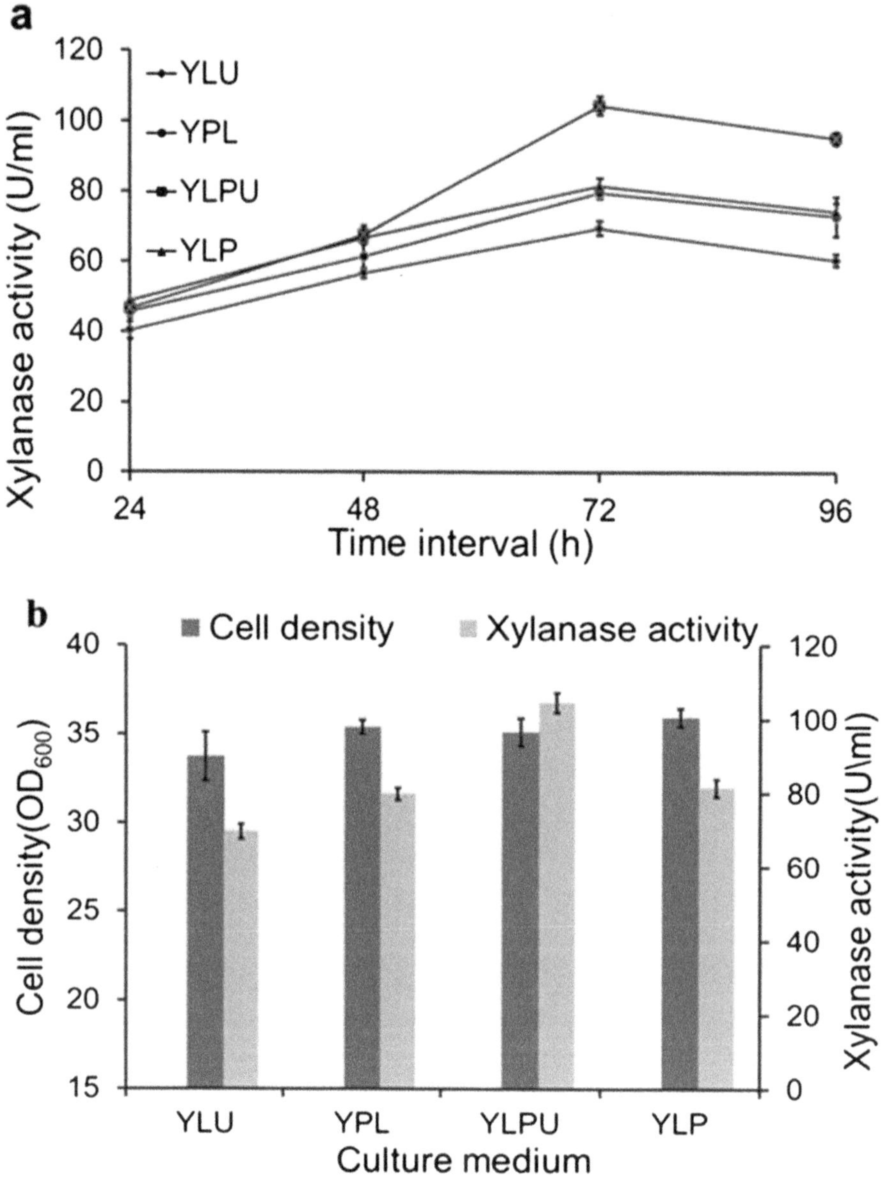

Figure 4 The fermentation of xylanase XYNZG in different media. a: The activity changes of XYNZG in YPL, YLP, YLU and YLPU medium during 96 h fermentation. The samples were taken and measured every 24 h. **b**: The comparison of biomass and activities at 72 h culture. The experiments were performed in triplicate.

protein and *E. coli* DH5α was used as the host for cloning and plasmid amplification. Plasmid pKLAC2 (New England Biolabs, USA) was employed as the expression vector of *K. lactis*. LB medium was used for the storage and culture of *E. coli* DH5α, while YPD (1% yeast extract, 2% peptone, 2% glucose) was used for *K. lactis*. Transformants were selected on YCB medium (1.17% yeast carbon base, 0.03 M sodium phosphate buffer, pH 7, New England Biolabs) with 5 mM acetamide and were grown on YPD plates containing 1% RBB-xylan for activity screening. Media YPL (1% yeast extract, 2% peptone, 2% lactose), YLP (1% yeast extract, 2% lactose, 1.5% peptone), YLPU (1.2% yeast extract, 2.6% lactose,1.2% peptone, 0.3% urea) and YLU (1.2% yeast extract, 2.6% lactose, 0.5% urea) were used for the recombinant xylanase fermentation of *K. lactis* [11].

Construction of the expression vector pKLAC2-*xynZG*

The *xynZG* (GenBank accession number DQ157736) was cloned previously in our laboratory [12], and the encoding product XYNZG was recorded in UniProt accession Q49UB8. In this study, the *xynZG* was amplified by PCR without the signal peptide sequence, using primers xynF (CCG*CTCGAG*AAAAGAATGGCGCCTGCGACTG

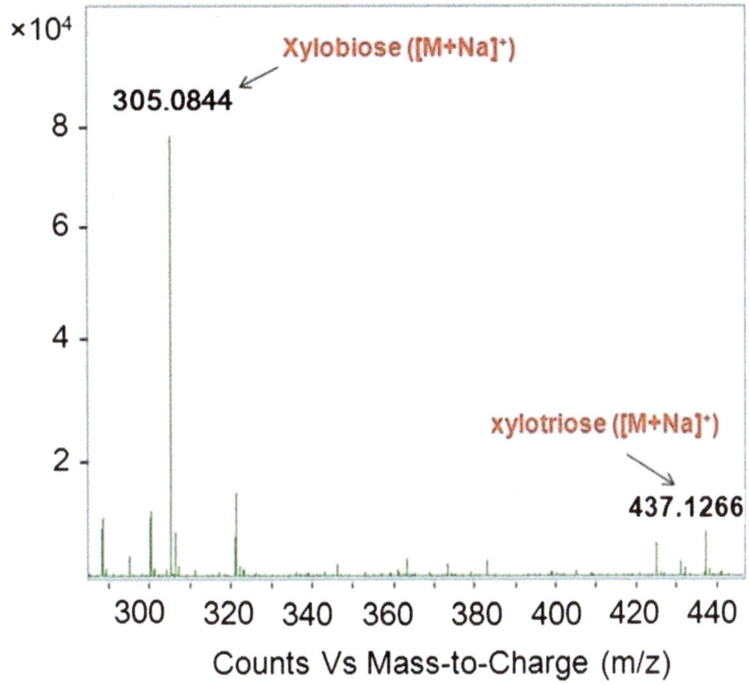

Figure 5 Mass spectrometry analysis of xylan hydrolysate.

Table 1 Effects of xylanase XYNZG on dough and bread properties

Trial number	Maximal score	1	2	3	4
50 U xylanase (ppm)		0	800	1200	2000
Kneading dough time (min)		6.5	6.0	6.0	6.0
Height-diameter ratio of round bread		6.28/10.7	6.40/11.0	6.46/11	6.33/11.52
Dough handling properties		Inadhesion	Inadhesion	Inadhesion	A little stickiness
Volume of hand bag	35	25	30	32	33
Color of crust	10	8	8	8	8
Color of crumb	10	8	8	8	8
Skin and shape	5	3	4	4	4
Smoothness	10	8	8	8	8
Texture structure	25	22	21	20	19
Springiness	10	7	9	9	9
Taste	5	4	4	4	4
Total score	100	85	92	93	93

The experiment was performed by one step fermentation. Water, yeast, sodium chloride, modifier and oil bread were mixed according to the fixed ratio, except for XYNZG. The dough was proofed at 38°C for 150 min, and baked at 225°C for 20 min. Static (texture profile analysis, firmness and relaxation test), image analysis, sensory analysis and color measurements (colorimeter and photoshop system) were used for bread quality evaluation.

ATACCCC) and xynR (ATTT*GCGGCCGC*TTAACCAG AGTCCGAAACAGTGATCCTA) with restriction sites *Xho*I and *Not*I (*underlined*), respectively. It was cloned into pKLAC2 vector according to manufacturer's instructions. The expression vector pKLAC2-*xynZG* was linearized by *Sac*II digestion before transformation, and the transformation of *K. lactis* GG799 was performed by lithium chloride as described by Miklenic et al. [19]. The transformants were screened on the YCB agar plates with 5 mM acetamide, and further selected on YPD plates containing 1% RBB-xylan.

Expression of the recombinant protein in *K. lactis*

The recombinant strain with the highest halo was named GKX21, and inoculated into a flask containing 50 ml YPD medium, and incubated at 28°C, 200 rpm for 48 h, and then the 50 ml YPD medium was transferred to 50 ml YPL medium for another 3 days. Samples were taken every 24 h for SDS-PAGE and enzyme activity assays.

SDS -PAGE and zymogram analysis of recombinant XYNZG

The fermentation supernatant was loaded on a 12% polyacrylamide gel for SDS-PAGE, followed by staining with Coomassie Brilliant Blue G-250. The zymogram analysis was performed by the published method of Zhang et al. [20] with birchwood xylan replaced by beechwood xylan (Sigma).

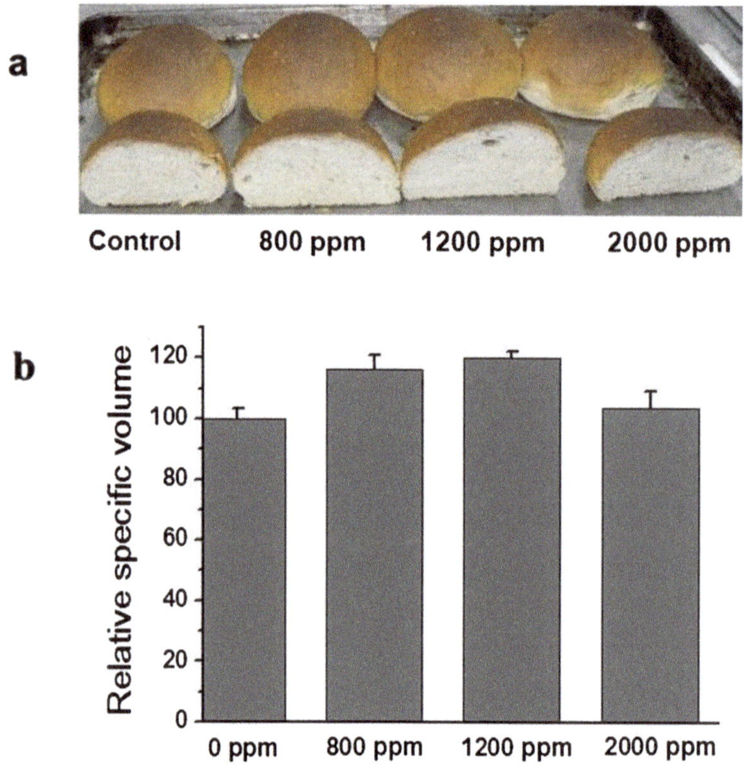

Figure 6 Pictorial showing the effects of various concentrations of XYNZG on the volume of wheat bread. a: Photograph of bread and cut bread. **b**: The relative specific volume of bread. Data represent mean±standard deviation of three replicates. The activity of XYNZG was 50 U/ml.

Determination of xylanase activity

Xylanase were assayed by measuring the reducing groups released from beechwood xylan by the dinitrosalicylic acid method (DNS) [12]. Reaction mixture containing 100 μl of diluted enzyme solution and 2.4 ml of 10 mg/ml suspension of xylan in 0.05 M sodium phosphate buffer (pH 6.0) was incubated at 40°C for 10 min. The reducing sugar was determined by the DNS procedure at the absorbance of 540 nm, using xylose as a standard. One unit of enzyme activity was defined as the amount of enzyme capable of releasing 1 μmol of reducing sugar from xylan per minute under the assay condition.

Table 2 Effects of xylanase XYNZG on the texture profiles of wheat bread

50 U Xylanase dosage (ppm)	Hardness	Springness	Chewiness
0	334.2±12.0	0.86±0.04	197.6±5.7
800	357.0±2.5	0.86±0.02	206.9±3.5
1200	391.7±3.6	0.87±0.07	211.1±8.0
2000	371.7±15.8	0.86±0.01	215.9±9.3

The measurements were performed after baking with additional 2-hour standby at room temperature. The texture profiles of wheat bread were measured using a Texture Analyzer. Data represent mean±standard deviation of three replicates.

The fermentation of xylanase XYNZG in different media

The medium components for XYNZG fermentation were optimized according to the published method with a slight modification [11]. The transformant GKX21 was precultured in 50 ml YPD medium and shaken at 200 rpm and 28°C for 48 h. Then the preculture was transferred into 50 ml media, including YPL, YLP, YLPU and YLU at a concentration of 1%, respectively, for another 96 h culture. The fermentation samples were taken every 24 h to measure the xylanase activities and the biomass.

Mass spectrometry analysis of xylan hydrolysate

The 5.0% (w/v) beechwood xylan in sodium phosphate buffer (pH 6.0) was hydrolyzed by XYNZG at 40°C for 12 h. The hydrolyzate was centrifuged at 12, 000 rpm for 20 min, and then the supernatant was filtered with 0.22 μm filter membrane (Merck Millipore Ltd). The sample was analyzed with electrospray ion source mass spectrometry in the positive ion reflective mode (Agilent, USA).

Baking trials

Recombinant XYNZG was tested for its effectiveness in baking applications. The experiment was performed by Applied Technology Center of SUNSON Industrial

Group Company Limted, Ningxia, China. The formulation of the bread dough was as follows (per liter): 52 ml of water, 100 g of wheat flour (Ningxia Saibei Company, China), 1.5 g of dry yeast, 20 g of sugar, 1 g of edible salt, 8 g of butter oil, and bread improver (5 ppm Novozymes fungal amylase, 10 ppm DSM glucose oxidase, 10 ppm Novozymes FBG lipase, and 100 ppm Vitamin vitamin C). The various concentrations of XYNZG (50 U, 400 ppm-2000 ppm) were mixed for 6.0 or 6.5 min in a mixer. The dough was then proofed at 38°C for 150 min and baked at 225°C for 20 min.

Determination of bread characteristics

Bread quality assessment by sensory score is largely based on the personal judgment and the subjective qualitative evaluation. Although the assessment system is not unified, it still reflects the influence of consumer preferences. Sensory properties of control and experimental breads were performed according to the 100-point evaluation system [21] to evaluate different attributes of bread such as visual, textural, biting, and organoleptic characteristics.

In order to avoid the subjective judgment difference, bread quality was also determined on the basis of weight, volume, specific volume, and crumb firmness by instruments. All measurements were performed after baking with additional 2 hours standby at room temperature. The loaf volume was measured using the Volume Measuring Device (BVM-L370LC, Sweden). Crumb texture was measured using the Texture Analyzer (TA.XT-plus Texture Analyzer, USA). Bread was equally sliced into 2 cm pieces, and only three central pieces were used to measure the force required to compress a slice of 1.0 mm/s [15].

Competing interests

The authors declare that they have no competing interests.

Authors' contributions

FX Zhan carried out the xylanase gene expression in *K. lactis* and measure the enzyme activity; QH Wang participated in fermentation and the writing of the manuscript; SJ Jiang carried out zymogram analysis. YL Zhou designed the whole study and carried out mass spectrometry analysis of xylan hydrolysate. GM Zhang designed the whole study and drafted the manuscript. YH Ma participated in its design and coordination and helped to draft the manuscript. All authors read and approved the final manuscript.

Acknowledgements

We thank Carolyn Katie for critical reading of the manuscript. This study was funded by the Ministry of Science and Technology of China (863 program 2012AA022203C), the National Natural Science Foundation of China (31240008), the Natural Science Foundation of Hubei Province (2011CDA00302), State Key Laboratory of Microbial Metabolism (Shanghai Jiao Tong University, MMLKF13-070), and the Key Deployment Program of Chinese Academy of Sciences (KSZD-EW-Z-015)

Author details

[1]Hubei Collaborative Innovation Center for Green Transformation of Bio-Resources, College of Life Science, Hubei University, Wuhan 430062, People's Republic of China. [2]Tianjin Institutes of Industrial Biotechnology, Chinese Academy of Science, Tianjin 300308, China.

References

1. Ye J, Wang XH, Sang YX, Liu Q: **Assessment of the determination of azodicarbonamide and its decomposition product semicarbazide: investigation of variation in flour and flour products.** *J Agric Food Chem* 2011, **59**(17):9313–9318.
2. Collins T, Hoyoux A, Dutron A, Georis J, Genot B, Dauvrin T, Arnaut F, Gerday C, Feller G: **Use of glycoside hydrolase family 8 xylanases in baking.** *J Cereal Sci* 2006, **43**(1):79–84.
3. Courtin C, Delcour JA: **Arabinoxylans and endoxylanases in wheat flour bread-making.** *J Cereal Sci* 2002, **35**(3):225–243.
4. Shrivastava S, Shukla P, Deepalakshmi PD, Mukhopadhyay K: **Characterization, cloning and functional expression of novel xylanase from *Thermomyces lanuginosus* SS-8 isolated from self-heating plant wreckage material.** *World J Microbiol Biotechnol* 2013, **29**(12):2407–2415.
5. Zhang G, Mao L, Zhao Y, Xue Y, Ma Y: **Characterization of a thermostable xylanase from an alkaliphilic *Bacillus* sp.** *Biotechnol Lett* 2010, **32**(12):1915–1920.
6. Dujon B, Sherman D, Fischer G, Durrens P, Casaregola S, Lafontaine I, De Montigny J, Marck C, Neuvéglise C, Talla E: **Genome evolution in yeasts.** *Nature* 2004, **430**(6995):35–44.
7. van Ooyen AJ, Dekker P, Huang M, Olsthoorn MM, Jacobs DI, Colussi PA, Taron CH: **Heterologous protein production in the yeast *Kluyveromyces lactis*.** *FEMS Yeast Res* 2006, **6**(3):381–392.
8. Walsh DJ, Gibbs MD, Bergquist PL: **Expression and secretion of a xylanase from the extreme thermophile, *Thermotoga* strain FjSS3B. 1, in *Kluyveromyces lactis*.** *Extremophiles* 1998, **2**(1):9–14.
9. Walsh DJ, Bergquist PL: **Expression and secretion of a thermostable bacterial xylanase in *Kluyveromyces lactis*.** *Appl Environ Microbiol* 1997, **63**(8):3297–3300.
10. Wamalwa BM, Zhao G, Sakka M, Shiundu PM, Kimura T, Sakka K: **High-level heterologous expression of *Bacillus halodurans* putative xylanase xyn11a (BH0899) in *Kluyveromyces lactis*.** *Biosci Biotechnol Biochem* 2007, 71(3):688–693.
11. Yin T, Miao L, Guan F, Wang G, Peng Q, Li B, Guan G, Li Y: **Optimized medium improves expression and secretion of extremely thermostable bacterial xylanase, XynB, in *Kluyveromyces lactis*.** *J Microbiol Biotechnol* 2010, **20**(11):1471.
12. Zhang GM, Huang J, Huang GR, Ma LX, Zhang XE: **Molecular cloning and heterologous expression of a new xylanase gene from *Plectosphaerella cucumerina*.** *Appl Microbiol Biotechnol* 2007, **74**(2):339–346.
13. Chapla D, Pandit P, Shah A: **Production of xylooligosaccharides from corncob xylan by fungal xylanase and their utilization by probiotics.** *Bioresour Technol* 2012, **115**:215–221.
14. Olempska-Beer ZS, Merker RI, Ditto MD, DiNovi MJ: **Food-processing enzymes from recombinant microorganisms—a review.** *Regul Toxicol Pharmacol* 2006, **45**(2):144–158.
15. Shah AR, Shah RK, Madamwar D: **Improvement of the quality of whole wheat bread by supplementation of xylanase from *Aspergillus foetidus*.** *Bioresour Technol* 2006, **97**(16):2047–2053.
16. Zheng H, Guo B, Chen X-L, Fan S-J, Zhang Y-Z: **Improvement of the quality of wheat bread by addition of glycoside hydrolase family 10 xylanases.** *Appl Microbiol Biotechnol* 2011, **90**(2):509–515.
17. JLA MJV'z, Domı'nguez H, Parajo' JC: **Xylooligo-saccharides: manufacture and applications.** *Trends Food Sci Technol* 2000, **11**:387–393.
18. Vazquez M, Alonso J, Domınguez H, Parajo J: **Xylooligosaccharides: manufacture and applications.** *Trends Food Sci Technol* 2000, **11**(11):387–393.
19. Miklenic M, Stafa A, Bajic A, Zunar B, Lisnic B, Svetec I: **Genetic Transformation of the Yeast Dekkera/Brettanomyces bruxellensis with Non-Homologous DNA.** *J Microbiol Biotechnol* 2013, **23**(5):674–680.
20. Zhang GM, Hu Y, Zhuang YH, Ma LX, Zhang XE: **Molecular cloning and heterologous expression of an alkaline xylanase from *Bacillus pumilus* HBP8 in *Pichia pastoris*.** *Biocatalysis and Biotransformation* 2006, **24**(5):371–379.
21. Angioloni A, Collar C: **Bread crumb quality assessment: a plural physical approach.** *Eur Food Res Technol* 2009, **229**(1):21–30.

Generation in yeast of recombinant virus-like particles of porcine circovirus type 2 capsid protein and their use for a serologic assay and development of monoclonal antibodies

Juozas Nainys[1], Rita Lasickiene[1], Rasa Petraityte-Burneikiene[1], Jonas Dabrisius[1], Raimundas Lelesius[2], Vilimas Sereika[2], Aurelija Zvirbliene[1], Kestutis Sasnauskas[1] and Alma Gedvilaite[1*]

Abstract

Background: Porcine circovirus type 2 (PCV2) is considered to be an important emerging pathogen associated with a number of different syndromes and diseases in pigs known as PCV2-associated diseases. It has been responsible for significant mortality among pigs and remains a serious economic problem to the swine industry worldwide leading to significant negative impacts on profitability of pork production.

Results: In this study we have demonstrated that PCV2 capsid (Cap) protein based virus-like particles (VLPs) were efficiently produced in yeast *S. cerevisiae* and induced production of monoclonal antibodies (MAbs) reactive with virus-infected cells. Moreover, PCV2 Cap VLPs served as a highly specific recombinant antigen for the development of an indirect IgG PCV2 Cap VLP-based ELISA for the detection of virus-specific IgG antibodies in swine sera. Four hundred-nine serum samples collected from pigs in Lithuania were tested for PCV2-specific IgG to determine the sensitivity and specificity of the newly developed ELISA in parallel using a commercial SERELISA test as a gold standard. From 409 tested serum samples, 297 samples were positive by both assays. Thirty-nine sera from 112 serum samples were determined as negative by SERELISA but were found to be positive both in the newly developed indirect IgG PCV2 Cap VLP-based ELISA and the PCR test.

Conclusions: We have demonstrated that *S. cerevisiae* expression system is an alternative to insect/baculovirus expression system for production of homogenous in size and shape PCV2 Cap protein-based VLPs similar to native virions. Yeast expression system tolerated native virus genes encoding PCV2 Cap protein variants as well as the codon-optimized gene. Moreover, yeast-derived PCV2 Cap VLPs were capable to induce the generation of PCV2-specific MAbs that did not show any cross-reactivity with PCV1-infected cells. The high sensitivity and specificity of the indirect IgG PCV2 Cap VLP-based ELISA clearly suggested that this assay is potentially useful diagnostic tool for screening PCV2–suspected samples.

Keywords: Virus-like particles, Porcine circovirus 2, Monoclonal antibodies

* Correspondence: agedv@ibt.lt
[1]Institute of Biotechnology, Vilnius University, Graiciuno 8, LT-02241 Vilnius, Lithuania
Full list of author information is available at the end of the article

Background

Circoviruses belong to the Circoviridae family and infect a variety of animal and plant species. They are named after their circular single-stranded DNA (ssDNA) genome and represent the smallest DNA viruses infecting mammals. Porcine circovirus 2 (PCV2) is found all over the world in the domestic pigs and the wild boars which seldom display the clinical condition. PCV2 serological prevalence in the wild boars is almost as high as in domestic pigs [1]. PCV2 is considered to be an important emerging pathogen associated with a number of different syndromes and diseases in pigs known as PCV2-associated diseases (PCVAD). Overall, PCVAD includes clinical syndromes and diseases such as postweaning multisystemic wasting syndrome (PMWS), porcine dermatitis and nephropathy syndrome (PDNS), porcine reproductive disorders, respiratory disease complex, granulomatous enteritis, congenital tremor, necrotizing tracheitis and lymphadenitis and exudative epidermitis [2-6]. PCV2 has been responsible for significant mortality among pigs and remains a serious economic problem to the swine industry worldwide leading to significant negative impacts on profitability of pork production [2]. The ubiquitous nature of the virus and the retrospective evidence of this infection long before disease appearance provided the reason for debate about the real capabilities of this virus in its association with the disease. How the virus causes disease is still not fully understood. Infection with PCV2 in combination with still unknown co factors can cause immunopathological disorders such as PMWS, or induce protective immunity. It might be that PCV2-induced diseases are the result of PCV2-caused immune disorder. It was demonstrated that pigs with PMWS display a marked depletion of T- and B-lymphocytes, and their lymphoid organs are infiltrated by histiocytes and multinucleated giant cells and might cause deficiency of total and neutralizing antibody response [7,8]. The interaction of PCV2 with the innate immune system have been mostly studied *in vitro* looking for the effects of the virus on dendritic cells (DCs), macrophages and peripheral blood mononuclear cells [7]. It was shown that PCV2 infection of plasmacytoid DCs impairs the induction of interferon-α and tumor necrosis factor-α, which silences responsiveness to pathogen-associated recognition signals by compromising immune defense development against other pathogens. As a result this may lead to a host susceptibility to secondary infections which actually are responsible for developing of PCVAD [9].

PCV2 is a non-enveloped virus with circular single stranded DNA molecule of approximately 1,767 bases packed in icosahedral 12–23 nm diameter virion particle. Virus genome sequences with nucleotide diversity cut-off of 3.5% were divided into four genotypic groups [1]. The two major groups designated as PCV2a, PCV2b are found worldwide and have been demonstrated to be virulent enough to trigger PCVAD [1,10,11]. The two more genotype types - PCV2c reported from Denmark in the 1980s and PCV2d found in China are not distributed so widely [12,13].

The PCV2 genome contains two major open reading frames (ORFs). The replicase (Rep and Rep') proteins responsible for initiation of viral replication are encoded by ORF1 located on the viral plus strand. A sole structural protein of the viral coat encoded by ORF2 on complementary strand is the unique structural capsid (Cap) protein [14]. Cap protein is the major immunogenic protein containing specific epitopes of PVC2 [15]. The Cap protein reacts strongly with the serum of PCV2-infected pigs and therefore, it is a major candidate antigen for the development of serological diagnostic methods.

There were multiple attempts to produce recombinant Cap protein in different expression systems including *E. coli*, *Lactococcus lactis*, *Pichia pastoris*, baculovirus or pseudorabies virus expression systems [15-21]. In most cases the expression of ORF2 gene of PCV2 was achieved by optimizing codon composition, removing N terminal part of the protein rich of arginine residues or by adding different tags useful for purification. These approaches have been shown to be successful to yield truncated or full length monomeric Cap protein. Expression of PCV2 Cap in insect cells resulted in recombinant protein self-assembled into virus-like particles (VLPs) that were structurally and antigenically indistinguishable from regular PCV2 capsids [15]. Recombinant VLPs due to their repetitive surface epitopes, the virus-like structure and the capability to induce humoral and/or CTL responses are more suitable for the development of immunogenic vaccines and sensitive diagnostic tests as compared to monomeric viral antigens [22]. At present, few baculovirus-derived PCV2a subunit vaccines are available [23]. PCV2 Cap VPLs produced in baculovirus expression systems were shown to be a suitable antigen for detection of PCV2 infection by the indirect IgG PCV2 ELISA [24,25]. However, the possibilities to employ cheaper antigens for diagnostic purposes are considered mostly by using monomerus PCV2 Cap protein [18,25-27].

In the current study we demonstrate that PCV2b-derived Cap VLPs can be efficiently produced in yeast *Sacharomyces cerevisiae* and are excellent antigens for generation of monoclonal antibodies (MAbs) reactive with virus-infected cells. Moreover, we demonstrate that yeast-derived recombinant PCV2 Cap VLPs represent a promising antigen for a highly sensitive and specific immunoassay for serologic diagnosis of PCV2 infection.

Results

Production of PCV2 VLPs in yeast

PCV2-positive samples collected between 2009 and 2013 from domestic pigs from 9 regions of Lithuania were used for the amplification of full-length PCV2 genomic DNA with overlapping primers. DNA sequence alignment of the 45 cloned and sequenced PCV2 genomes circulating in Lithuania revealed six slightly different genome variants (GenBank: KJ128269, KJ128270, KJ128271, KJ128272, KJ128273, KJ128274). All six PCV2 genome variants belonged to genotype PCV2b according to the previously proposed criteria [10,11].

Three open reading frame 2 (ORF2) variants amplified by PCR from native PCV2 genomes: variant 622 (GenBank: KJ128269), variant 521 (GenBank: KJ128273) and variant 113 (GenBank: KJ128274) as well as ORF2 of the most representative PCV2-Cap gene variant 622 which was codon-optimized by *S. cerevisiae* codon usage (GenBank: KJ128275) were cloned into pFX7 expression vector [28] and used for the expression in yeast *S. cerevisiae.*

The expression of four cloned genes was detected in SDS-PAGE and confirmed by Western blot using pig serum (Figure 1A,B). All recombinant PCV2-Cap proteins were found in the soluble fraction of yeast cell lysate which allowed their efficient purification (Figure 1A, B, lanes 2, 4, 6, 8). All four variants of PCV2-Cap proteins generated in yeast were purified by density gradient centrifugation and examined using electron microscope (EM). Despite the slight differences in expression levels of PCV2-Cap gene variants, all four purified Cap proteins were assembling into regular homogenous in size VLPs about 17–22 nm in diameter similar to native PCV2 virions (Figure 1C). PCV-Cap VLPs encoded by the most representative PCV2-Cap gene variant 622 were used for generation of MAbs and development of an indirect ELISA for detection of PCV2-specific IgG.

Generation and characterization of monoclonal antibodies against PCV2b-Cap VLPs

To generate the MAbs, mice were immunized with recombinant PCV2 Cap 622 VLPs and tested for the presence of Cap-specific serum antibodies. Spleen cells of a mouse showing the highest titer of Cap-specific antibodies were used to generate hybridomas. After fusion, all hybrid clones were screened for the reactivity with PCV2 Cap VLPs and specific ones were subjected to isotyping and recloning. After these procedures, three stable hybridoma cell lines producing PCV2 Cap-specific MAbs of IgG isotype were obtained. Two of them (clones #14G10 and #7C4) produced MAbs of IgG2a subtype and one clone (#24C7) of IgG1 subtype. The specificity of all three MAbs was proven by ELISA using all three variants of yeast-expressed Cap proteins: PCV2 Cap521 protein, PCV2 Cap113 protein, and PCV2 Cap622 protein. No reactivity of the MAbs with Hantaan (Fojnica) and Tioman nucleocapsid proteins [29,30] used as a negative control was observed, which confirmed the specificity of these MAbs to the PCV2 Cap protein (data not shown).

Western blot analysis of MAb specificity using the lysate of yeast expressing PCV2 Cap protein revealed that only one MAb (clone #14G10) recognized SDS-denaturated PCV2 Cap protein (data not shown). Meanwhile, the MAbs #7C4 and #24C7 did not react with PCV2 Cap protein in Western blot thus indicating that they are directed against conformational epitopes (data not shown).

To evaluate the diagnostic potential of the generated MAbs, their reactivity with PCV2-infected (virus strain T-657) swine testicle cells (PT-1) and PCV1-infected porcine kidney (PK) cells was tested by an indirect immunofluorescence assay (IFA) using commercially available slides (VMRD, Inc, Pullman, WA, USA). The MAbs #7C4 and #14G10 were reactive with PCV2-infected swine testicular cells but did not recognize PCV1-infected PK cells thus confirming their specificity to PCV2 Cap protein (Figure 2). Meanwhile, the MAb #24C7 was not reactive in both IFA tests.

Development of an indirect ELISA for detection of PCV2 Cap-specific IgG

As a first step in the development of PCV2-Cap VLP-based ELISA, checkerboard titrations were performed to determine the optimal concentration of the Cap antigen and the test serum. To optimize plate coating, the recombinant PCV2-Cap 622 protein was immobilized on ELISA plates at five different concentrations: 2; 1; 0.5; 0.25; 0.125 μg/mL. The optimal antigen concentration for plate coating as determined by a checkerboard titration was 0.5 μg/mL. To determine the optimal serum sample dilution, serum samples identified by the commercial SERELISA test (Synbiotics, Lyon, France) as high positive, low positive and negative were diluted in serial two fold dilutions ranging from 1:100 to 1:3200. The OD450 values obtained in an indirect ELISA were plotted against the dilutions (Figure 3). High OD readings with positive control serum and low background signals with negative sera were obtained at serum dilution 1:800. The absorbance values declined at serum dilution 1:1600. As the commercial SERELISA test uses serum dilution 1:1000 and the differences between OD values at serum dilutions 1:800 and 1:1000 were insignificant in our newly developed ELISA, for convenience the appropriate dilution (1:1000) was selected for the indirect IgG PCV2 Cap-based ELISA. Control antigen – yeast expressed Hamster polyomavirus (HaPyV) VP1 protein VLPs [28] was included in the test to determine the specificity of the assay. The positive samples revealed significantly higher OD values (0.9–2.5) with PCV2 Cap protein as compared to the control HaPyV-VP1 antigen (0.05–0.1) (Figure 4, Dotted lines).

The cut-off value was calculated as the mean OD450 value of 42 serum specimens negative by SERELISA plus

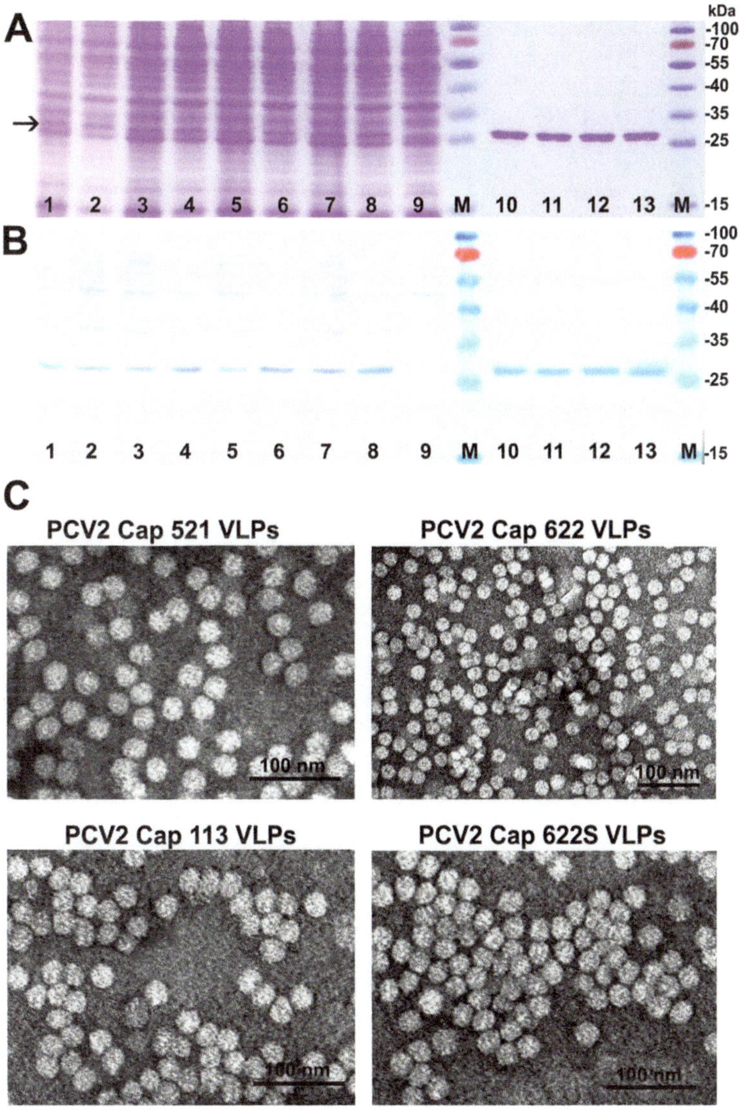

Figure 1 Analysis of recombinant PCV2 Cap protein-derived VLPs produced in yeast. (A) Coomassie blue-stained SDS-PAGE; **(B)** Western blot with pig serum. **(A,B)**: In lanes: 1 - whole crude lysate of yeast transformed with pFX7-PCV2Cap-521 plasmid, 2 - the soluble fraction of the lysate of yeast transformed with pFX7-PCV2Cap-521 plasmid; 3 - whole crude lysate of yeast transformed with pFX7- pFX7-PCV2Cap-113 plasmid; 4 - the soluble fraction of the lysate of yeast transformed with pFX7-PCV2Cap-113 plasmid; 5 - whole crude lysate of yeast transformed with pFX7-PCV2Cap-622 plasmid; 6 - the soluble fraction of the lysate of yeast transformed with pFX7-PCV2Cap-622 plasmid; 7 - whole crude lysate of yeast transformed with pFX7-PCV2Cap-622S plasmid; 8 - the soluble fraction of the lysate of yeast transformed with pFX7-PCV2Cap-622S plasmid; 9 - whole crude lysate of yeast transformed with pFX7 plasmid, M- pre-stained protein weight marker (Thermo Fisher Scientific Baltics); 10 - Cap521 protein VLPs purified by density gradient centrifugation; 11 - Cap113 protein VLPs purified by density gradient centrifugation; 12 – Cap622 protein VLPs purified by density gradient centrifugation; 13 - Cap622S protein VLPs (Cap622 protein VLPs generated using synthetic codon-optimized gene) purified by density gradient centrifugation. **(C)**: Electron microscopy pictures of VLPs formed by PCV2-Cap521, protein PCV2-Cap113, protein PCV2-Cap622 and protein PCV2-Cap622S protein stained with 2% aqueous uranyl acetate solution and examined by Morgagni 268 electron microscope.

3 standard deviations (mean +3SD). This calculation gives 99% confidence. The mean OD450 value and SD were 0.146 and 0.053, respectively. Therefore, the positive threshold determined for the newly developed indirect IgG PCV2 Cap VLP-based ELISA was 0.3. A serum sample was considered positive when its OD450 value was greater than the positive threshold (OD450 > 0.3).

Four hundred-nine serum samples were tested to determine the sensitivity and specificity of the newly developed ELISA. The sensitivity and specificity were calculated for the indirect IgG PCV2-Cap VLP-based ELISA using SERELISA test as a gold standard. All serum samples were tested in parallel with a commercial SERELISA kit. From 409 serum samples tested, 297

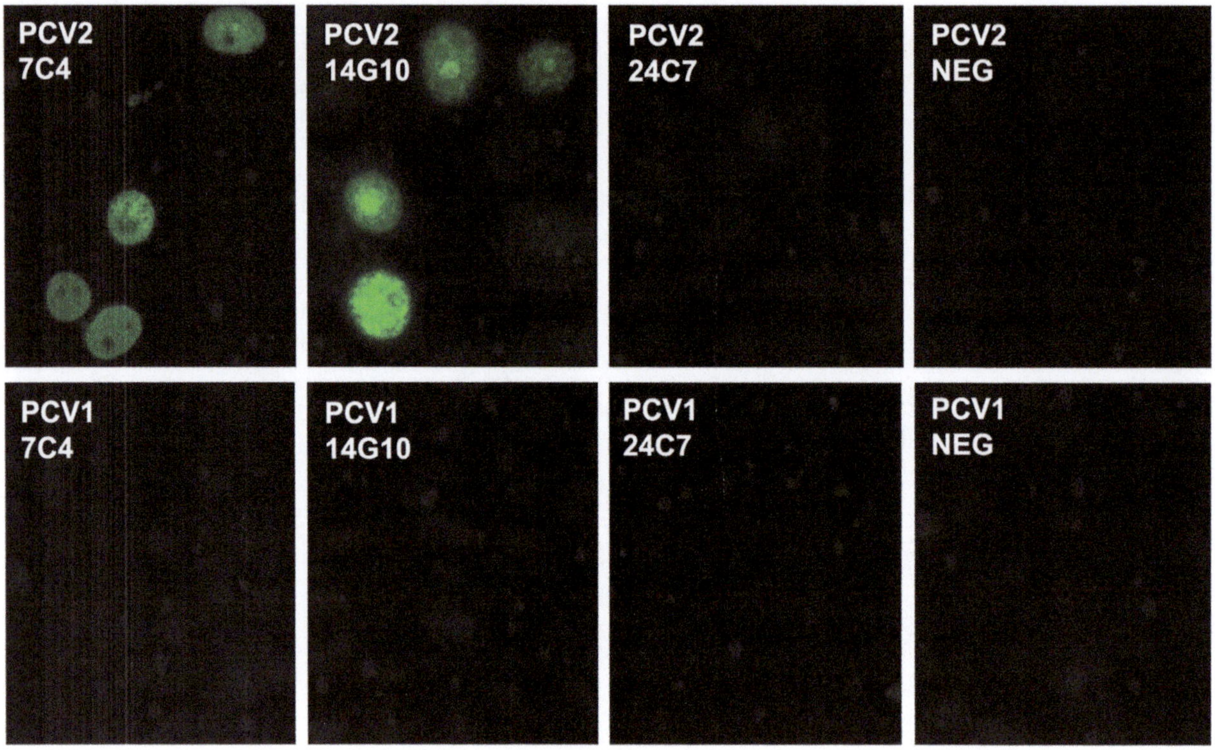

Figure 2 Indirect immunofluorescence staining of PCV2 infected swine testicle (PT-1) cells and PCV1-infected PK cells with MAbs #7C4, #14G10 and #24C7. NEG - MAb #4 F11 directed against parainfluenza type 3 N protein [41].

samples were positive by both assays. Thus, the calculated sensitivity of the new ELISA test was 100% (297/297).

One hundred-twelve serum samples negative by SERELISA were tested to determine the specificity of the indirect IgG PCV2-Cap VLP-based ELISA. Seventy-three sera were below and 39 serum samples were above the cut-off by indirect IgG PCV2-Cap VLP-based ELISA (Figure 4, Table 1). Indirect IgG PCV2-Cap VLP-based ELISA identified these samples as positive (Figure 4, samples labeled with opened circles), although in the commercial SERELISA test these samples were found to be negative. Discrepant serum samples were re-tested by

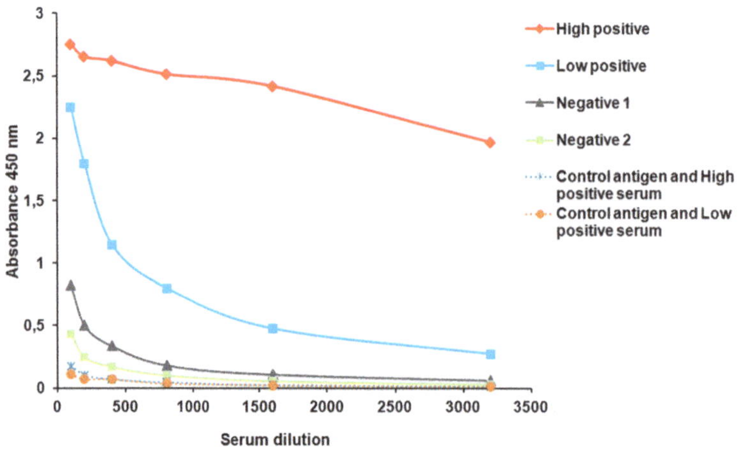

Figure 3 Titration curves of swine serum samples. Either positive or negative for PCV2 antibodies swine serum samples were titrated by the newly developed indirect IgG PCV2 Cap VLP-based ELISA using recombinant PCV2 Cap VLPs at a concentration of 0.5 μg/ml (solid lines) and control antigen (HaPyV-VP1) at a concentration of 2 μg/ml (dotted line). Sera were tested in serial duplicate twofold dilutions ranging from 1:100 to 1:3200. The results are expressed as the OD_{450} and correspond to the average of the values obtained in at least two different analyses.

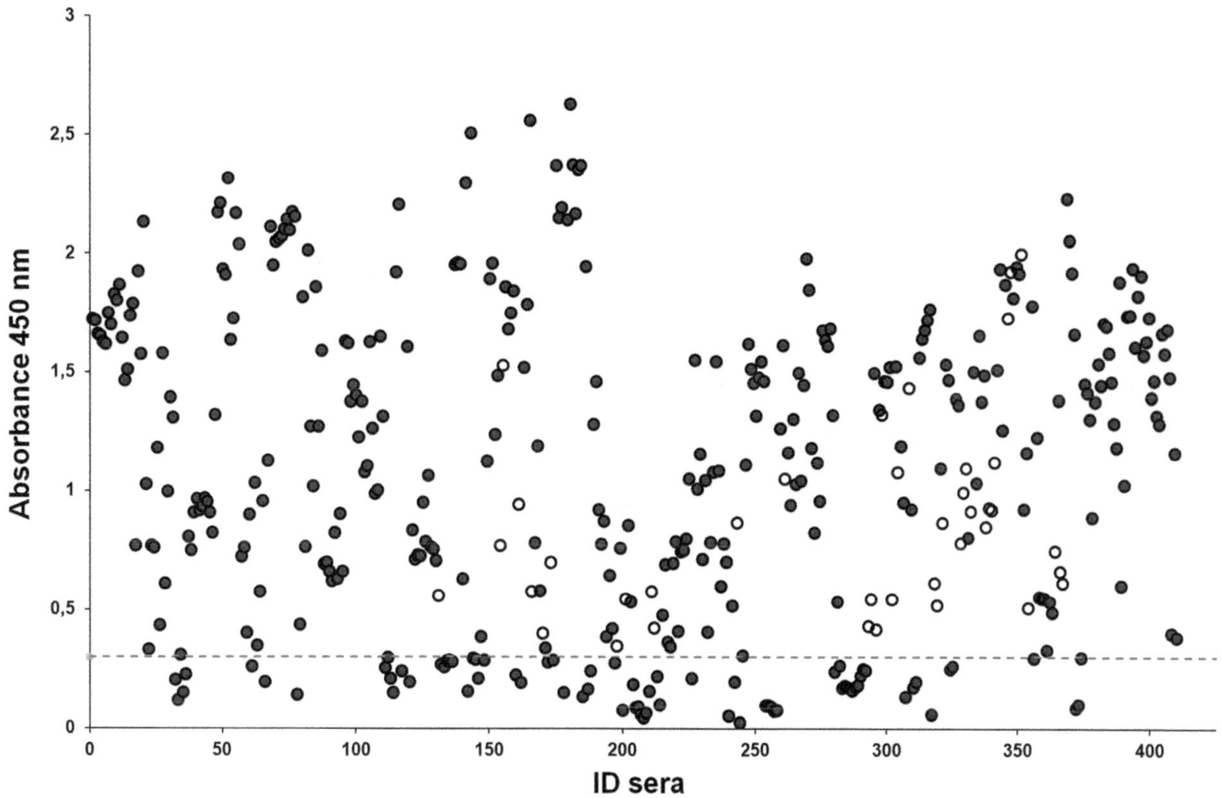

Figure 4 Optical density results of serum samples in the newly developed indirect IgG PCV2 Cap VLP-based ELISA. Four hundred and nine serum samples previously characterized by the commercial SERELISA test were analyzed by the newly developed indirect PCV2-Cap VLP IgG ELISA. The absorbance at 450 nm obtained for the serum samples are shown. The dotted line indicates the assay cut-off, which was established to be 0.3. for the indirect PCV2 Cap VLP IgG ELISA. Opened circles represent discrepant results, filled circles indicate matching results with commercial SERELISA test. ID sera, serum sample number.

PCR. Overall, all samples of the suspected PCV2 cases were PCV2 DNA positive. Thus, samples positive by the newly developed indirect IgG PCV2-Cap VLP-based ELISA were confirmed by PCR but missed by commercial SERELISA IgG test. All serum samples were tested in duplicate and the correlation coefficient (r-value) between two measurements was calculated. The r value was shown to be 0.972, which indicates very high reproducibility between tests.

These results indicate that the newly developed indirect IgG PCV2-Cap VLP-based ELISA had higher sensitivity as compared to the commercial SERELISA test, which failed to detect some PCV2-positive samples as positives.

Table 1 Comparative evaluation of the indirect IgG PCV2 Cap VLP-based ELISA with commercial SERELISA kit for detection of anti-PCV2 antibodies

Indirect IgG ELISA	Commercial SERELISA test		
	Positive	Negative	Total
Positive	297	39	**336**
Negative	0	73	**73**
Total	**297**	**112**	

The values represent the numbers of sera tested.

Discussion

Virus-like particles present viral antigens in a more authentic conformation than monomeric structural proteins. Therefore, recombinant VLPs are more preferable for vaccine or diagnostic purposes [22]. VLPs mimic the structure of authentic virus particles they are derived from and are easily recognized by the immune system and able to stimulate both B-cell and T-cell immune responses. Multiple expression systems such as *E. coli, L. lactis, P. pastoris* allowed generation of monomeric PCV2 Cap protein but not VLPs [16-20]. Different modifications of the gene encoding PCV2 protein were also necessary for generation of recombinant PCV2 Cap protein including removal of the N terminal part with nuclear localization signal and fusing it with different tags or yeast signal peptide as well as the optimization of codon usage according to the host [16-21]. Even for the investigation of the structure of PCV2 Cap-derived VLPs, both the full length and an N-terminally truncated monomeric Cap protein was purified from *E.coli* and reassembled into VLPs *in vitro* [31]. On the other hand, baculovirus/insect cell expression system supported PCV2-Cap protein self-assembly into VLPs [14,15]. Yeast *Saccharomyces cerevisiae* might provide an alternative expression system for PCV2 Cap protein production as there are several reports which confirm that yeast can support VLP auto-assembly of viral proteins [32-34]. In a Bucarey's study, the failure of PCV2 Cap protein production in *S. cerevisiae* using native PCV2-Cap coding sequence was reported [35]. It was demonstrated that only codon optimized gene allowed generation of PCV2 Cap VLPs in yeast. Moreover, in Bucarey's study generated VLPs were not homogenous in size and shape [35]. This is opposite to our results presented in the current study (Figure 1). We demonstrate that PCV2b-Cap VLPs were efficiently produced in yeast *S.cerevisiae* using native virus ORF2 versions of Cap-encoding genes as well as the codon-optimized gene. We purified three slightly different PCV2 Cap proteins variants encoded by native Cap-encoding sequences and confirmed that they all formed homogenous in size and shape VLPs similar to native virions or VLPs produced in baculovirus expression system [14].

Yeast-produced PCV2 Cap VLPs were highly immunogenic and induced PCV2-specific immune response in mice. We generated three MAbs with different capabilities to recognize both the recombinant PCV2 Cap protein and virus-infected cells. The MAb #14G10 demonstrated exclusive and very specific reactivity with linear PCV2 Cap protein epitope in Western blot, ELISA and IFA implying the localization of this epitope on the surface of VLPs. The MAb #24C7 was non-reactive with virus-infected cells and did not recognize SDS-denaturated PCV2 Cap protein, which suggests that its epitope is accessible only in native PCV2 Cap VLPs. The MAb #7C4 also recognized conformational PCV2 epitope but it was accessible on virus-infected cells as demonstrated in IFA (Figure 2). All three MAbs were highly specific and did not show any cross-reactivity with PCV1 Cap protein in all assays despite a significant homology between PCV2 and PCV1 (a non-pathogenic virus) in terms of amino-acid sequences. Therefore, the newly generated MAb #7C4 and especially the MAb #14G10 could be used as highly specific reagents for PCV2 detection by IFA or monoclonal antibody-based blocking ELISA.

The Cap protein is a major candidate antigen for the development of serologic diagnostic methods as it is the major immunogenic protein containing specific epitopes of PVC2 [15]. In the previous studies recombinant PCV2 Cap protein expressed in either prokaryotes or eukaryotes was used in an indirect ELISA as coating antigen. It was shown that PCV2 Cap VLPs or Cap protein expressed in insect cells as well as Cap protein expressed in *E. coli* showed high immunoreactivity with serum antibodies [18,24-26]. The diagnostic sensitivity, specificity, and accuracy of ELISAs performed with PCV2 Cap VLPs and PCV2 were similar (>90%) to IFA [24]. However the sensitivity of ELISA's developed using *E. coli*-expressed recombinant proteins were lower. This is likely because natural virus particles are perfect in structure whereas recombinant *E. coli*-expressed proteins are not always refolded completely. Sun and colleagues developed protein-based ELISA for PCV2 specific diagnosis using *E. coli*-expressed antigenic domain of PCV2 Cap protein [36]. The developed assay failed to detect some PCV2-positive serum samples identified by IFA. The specificity was lower also in a monoclonal antibody-based blocking ELISA developed by Huang and colleagues [37]. Its sensitivity and specificity were determined to be 98.8% and 88.5%, respectively. Higher sensitivity was achieved in a quantitative immunofluorescence assay (QIFA) using the recombinant PCV2 nucleocapsid protein expressed in Vero cells by a lentivirus system, but the specificity was lower when compared to ELISA [38]. In the current study, the newly developed indirect IgG PCV2-Cap VLP-based ELISA proved to be a highly sensitive and specific tool for the serologic diagnosis of PCV2 infection and was even more sensitive than the commercial SERELISA test. It might be that not only VLP structure of yeast-expressed Cap protein influenced the higher sensitivity of the newly developed indirect IgG PCV2-Cap VLP-based ELISA. Based on the GenBank database and several studies published in different countries a shift on genotype prevalence from PCV2a to b have been observed [1]. Our analysis of PCV2 genotype spread in Lithuania pig farms in the samples collected between 2009 and 2013 confirmed that all isolated virus genomes belonged to the PCV2b genotype. Most of commercial diagnostic tests are based on PCV2a genotype-derived Cap proteins. Our preliminary data showed that some PCV2-negative serum samples identified by a commercial SERELISA test were negative in an indirect IgG Cap VLP-based ELISA when PCV2a genotype-derived Cap VPLs were used but were positive when PCV2b genotype-derived Cap VLPs were used as a coating antigen. However to prove the hypothesis more serological assays should be performed.

Conclusions

Recombinant VLPs are more preferable for vaccine development or diagnostic purposes as they present viral antigens in a proper conformation. In this study we have demonstrated that *S.cerevisiae* expression system is an alternative to insect/baculovirus expression system for production of homogenous in size and shape PCV2 Cap protein-based VLPs similar to native virions. Yeast cultures are easy to maintain, which makes them faster and less expensive to use than other eukaryotic expression systems. We showed that yeast expression system tolerated native virus genes encoding PCV2 Cap protein variants as well as the codon-optimized gene. Yeast-derived PCV2 Cap VLPs were capable to induce the generation of PCV2-specific MAbs that did not show any cross-reactivity with PCV1-infected cells. Moreover, the employment of PCV2 Cap VLPs as an antigen in the indirect IgG PCV2 Cap VLP-based ELISA demonstrated that this newly developed PCV2 diagnostic test was more sensitive and specific than the commercial SERELISA test and is potentially useful as a diagnostic assay in screening PCV2-suspected swine samples.

Methods

Generation of yeast expression plasmids

All DNA manipulations were carried out according to standard procedures [39]. Enzymes and kits for DNA manipulations were purchased from UAB "Thermo Fisher Scientific Baltics" Vilnius, Lithuania). Recombinants were screened in *E. coli* DH5α cells.

PCV2 genome sequences were amplified using Phusion High-Fidelity DNA Polymerase (Thermo Fisher Scientific Baltics) from PCV2-positive samples of domestic pigs collected between 2008 and 2013 from 9 regions of Lithuania, sequenced with overlapping primers and aligned. PCV2 genome sequences with nucleotide variations were submitted to GenBank (GenBank: KJ128269, KJ128270, KJ128271, KJ128272, KJ128273, KJ128274). DNA sequences encoding three different variants of PCV2 capsid protein sequences - 622 (GenBank: KJ128269), 521 (GenBank: KJ128273) and 113 (GenBank: KJ128274) - were PCR amplified from DNA isolated from porcine serum or tissue samples collected in Lithuanian pig farms. Collection of samples was carried out in accordance with institutional guidelines and permission of the national authorities (Permission No 0241). The PCV2 Cap-encoding DNA fragments were amplified using Phusion High-Fidelity DNA Polymerase (Thermo Fisher Scientific Baltics) to introduce BcuI sites at the 5′ end 3′ ends of the PCR fragment using primers CV-CP-Dir 5′CGACTAGTATGACGTATCCA AGGAGGCGTTAC CG and CV-CP-rev 5′CGACTAG TTAGGG GTTAAGTGGGGGGTCTTTAAG. Cloned gene variants were sequenced. Additionally, the gene encoding PCV2b-Cap protein 622 which appeared most common in Lithuanian pigs samples was codon optimized for expression in yeast (GenBank KJ128275). The gene of PCV2b Cap 622 with yeast-optimized codons (622S) was synthesized by GenScript. All four PCV2 Cap protein-encoding genes were cut with BcuI and subcloned into the XbaI site of the yeast expression vector pFX7 for target protein expression [28].

In addition, the gene encoding PCV1 Cap protein was synthesized by GenScript and subcloned into the XbaI site of the yeast expression vector pFX7.

Detection of PCV2 virus DNA by PCR

Two-hundred microliters of porcine serum samples were used to isolate viral genomic DNA with GeneJET™ Viral RNA/DNA Purification Kit (Thermo Fisher Scientific Baltics) according to the manufacturer's recommendations. Three pairs of primers (Rep-Bdir 5′ gccacctgggtgtggtaaaag caaatg and PCV2-Xdir 5′tctagatgataactttgtaacaaaggc; Rep-Bdir 5′ gccacctgggtgtggtaaaagcaaatg and PCV-cap-Csq 5′ ccgtgtaaccatgtatgtacaattcag; Rep-Brev 5′gtggccccacaat gacgtgtacattag and PCV2-revsq cggtaacgcctccttggatac gtc) were used to amplify PCV2 viral DNA fragments from isolated DNA.

Yeast strains, growth media and cultivation conditions

The constructed plasmids pFX7-PCV2Cap-521, pFX7-PCV2Cap-113, pFX7-PCV2Cap-622, pFX7-PCV2Cap-622S and pFX7 were used for the transformation of *S. cerevisiae* strain AH22-214 (*a, leu2-3,112, his4-519*). Yeast transformants harboring plasmids with PCV2-cap encoding genes were grown in YEPD medium (yeast extract 1%, peptone 2%, and glucose 2%) supplemented with 5 mM formaldehyde overnight at 30°C and recombinant protein expression was induced after transferring yeast cells into induction medium YEPG (yeast extract 1%, peptone 2%, and galactose 3%) supplemented with 5 mM formaldehyde and culturing for additional 18 h. Yeast biomass harboring recombinant proteins was harvested by centrifugation and was stored at −20°C until use. Yeast transformed with vector pFX7 without any insert was used as a negative control.

SDS-PAGE and western blotting

Proteins were analyzed by electrophoresis on 12% sodium dodecylsulfate-polyacrylamide gels (SDS-PAGE) followed by Coomassie brilliant blue staining. Briefly, 20–50 mg of yeast cell pellets were collected by centrifugation, washed with distillate water and suspended in 20–50 μL of DB 150 buffer (150 mM NaCl, 1 mM $CaCl_2$, 0.001% Trition X-100, 0.25 M L-Arginine 10% glycerol in 10 mM Tris/HCl-buffer, pH 6.5). An equal volume of glass beads was added and the cells were lysed by vortexing for 7 min with cooling on ice for 1 min between each vortexing step. The samples of whole yeast lysates, the supernatant obtained after centrifugation of whole yeast lysate and the purified proteins were analyzed by SDS-PAGE. For this purpose, the samples were mixed with the SDS-PAGE sample buffer (Thermo Fisher Scientific Baltics), boiled for 5 min, applied to a SDS-PAGE and run in SDS-Tris-glycine buffer. Protein bands in SDS-PAGE were visualized by staining with Coomassie brilliant blue (Sigma–Aldrich, St. Louis, MO, USA) or electro-transferred to Immobilon P membrane (Millipore, Bedford, MA, USA). The membranes were blocked with 5% milk in TTBS (0.1% Tween 20, 0.1 M Tris, 0.3 M NaCl, pH 7.4) and incubated overnight in the relevant antibody solutions at room temperature (RT). Porcine serum and the MAbs were diluted 1: 600 and 1:1,000, respectively, in TBS with 0.1% Tween 20. After incubation with the diluted antibodies, the membranes were incubated for 2 hours with HRP-labeled anti-mouse IgG conjugate (Bio-Rad, Hercules, CA, USA) diluted 1:5000 or rabbit anti-pig IgG (Fab'2) conjugate (LifeSpan Biosciences, Seattle, USA) diluted 1:10000. The peroxidase-mediated staining was performed by adding 4-chloro-1-naphtol and H_2O_2 (Fluka, Buchs, Switzerland).

Purification and characterization of PCV2-Cap VLPs

After centrifugation of yeast lysate, the supernatant containing PCV2-Cap proteins was collected and loaded onto a 20–69% sucrose gradient. Subsequently, a centrifugation at 100,000 × g (Beckman Optima LE-80 K Ultracentrifuge, Brea, CA, USA) overnight at 4°C followed. Thereafter, fractions of 3 mL were collected and samples were subjected to SDS-PAGE followed by Coomassie brilliant blue staining. Fractions showing a protein band with an apparent molecular mass of approximately 27 kDa corresponding to PCV2 Cap protein were pooled and diluted in DB150 buffer. The mixture was subjected to the ultracentrifugation on CsCl gradient with densities from 1.23 to 1.46 g/mL overnight at 100,000 × g (Beckman). Collected fractions were analyzed as described above. After pooling of the Cap-containing fractions, recombinant PCV2-Cap VLPs were diluted and precipitated by ultracentrifugation for 4 h at 100,000 × g (Beckman). The pellets were dissolved in PBS and dialyzed against PBS overnight. The dialyzed PCV2-Cap VLPs were stored in PBS in 50% glycerol at −20°C until use.

The samples of purified recombinant PCV2-Cap proteins were placed on 300-mesh carbon coated palladium grids, negatively stained with 2% aqueous uranyl acetate solution and examined by Morgagni 268 electron microscope (FEI Inc., Hillsboro, OR, USA).

Immunization of mice and generation of monoclonal antibodies

All procedures involving laboratory mice were performed at the breeding colony of the Center for Innovative Medicine (Vilnius, Lithuania) under controlled laboratory conditions in strict accordance with Lithuanian and European legislation.

To generate MAbs against the recombinant PCV2-Cap protein, three BALB/c mice were injected subcutaneously 3 times with 50 μg of recombinant PCV2-Cap VLPs at 21-day intervals. For a primary immunization, the antigen was emulsified in Complete Freund's adjuvant (Sigma-Aldrich Co), the subsequent immunizations were performed with PCV2 Cap VLPs dissolved in PBS without adjuvant. After the 3rd immunization, serum specimens were collected to determine the titers of specific antibodies. Finally, the mouse with the highest anti-PCV2-Cap titer was boosted with the same dose of PCV2-Cap VLPs in PBS and sacrificed for hybridoma development.

Hybridomas were generated essentially as described previously [40]. Four days after the booster immunization, spleen cells of the immunized mouse were fused with mouse myeloma Sp2/0 cells using PEG 1500 as a fusion agent (PEG/DMSO solution, HybriMax, Sigma-Aldrich Co). Hybrid cells were selected in growth medium supplemented with hypoxanthine, aminopterin and thymidine (50 × HAT media supplement, Sigma-Aldrich Co). Viable clones were screened by an indirect ELISA using 96-well microtiter plates coated with PCV2-Cap VLPs. Positive clones were stabilized by limiting dilution cloning on macrophage feeder layer. Hybridoma cells were maintained in complete Dulbecco's modified Eagle's medium (DMEM) containing 15% fetal calf serum (FCS, Biochrom, Cambridge, UK) and antibiotics. Heavy chain types of MAbs were determined by ELISA using Monoclonal Antibody Isotyping Kit (ISO-2, Sigma-Aldrich Co).

Indirect immunofluorescence assay

Indirect immunofluorescence assay (IFA) was performed using commercial IFA slides for detection of PCV2-infected (virus strain T-657) swine testicle cells (PT-1) and PCV1-infected PK cells using commercially available slides (VMRD, Inc, Pullman, WA, USA) as recommended by the manufacturer. Briefly, 50 μL of cell culture supernatant of hybridomas #14G10, #24C7 and #7C4 was added to the slides in a dilution of 1:50 in PBS with 1% BSA or undiluted. The negative control MAb #4 F11 directed against parainfluenza type 3 N protein [41] was used at the same dilution. Detection was done by incubation with a fluorescein isothiocyanate (FITC)-conjugated anti-mouse immunoglobulin (H + L) goat IgG (DakoCytomation) diluted 1:20 in PBS. Cells were counter-stained with Evans Blue and IFA slides were analyzed using Olympus IX-70 fluorescence microscope (Olympus, Tokyo, Japan) with 40× objective at an excitation of 495 nm and Image-pro Plus Version 7.0.

Serum samples

The porcine sera used in the study consisted of a collection of 409 samples obtained from 4–7 month old pigs from different regions in Lithuania. Samples were stored at −70°C prior to testing.

Indirect ELISA

The specificity of mouse antisera, MAbs and swine sera was tested by an indirect ELISA. For testing the mouse antisera and the MAbs microtiter plates (Nunc MaxiSorp, Roskilde, Denmark) were coated with the respective antigens by adding 100 μL of the antigen solution (2 μg/mL) in coating buffer (50 mM sodium carbonate, pH 9.5) and incubation overnight at +4°C. The plates were blocked for 30 min at RT with 1% bovine serum albumin (BSA) in PBS and then incubated with either undiluted hybridoma supernatant or serially diluted MAbs for 1 h at RT.

For the indirect IgG PCV2-Cap VLP ELISA microtiter plates (Nerbe plus, Winsen/Luhe, Germany) were coated with 100 μL per well of recombinant PCV2 Cap VLPs (variant 622, GenBank: AHN49776) diluted in coating buffer to the concentration of 0.5 μg/mL and incubated overnight at +4°C. This concentration was determined by titration to reach optimal conditions for sensitivity and specificity. Control antigen - hamster polyomavirus (HaPyV) VP1 protein VLPs [28] was diluted to the concentration of 2 μg/mL. The coated plates were blocked with 150 μL/well of Roti-block (Carl Roth, Karlsruhe, Germany) for 1 h at room temperature. Plates were rinsed three times with washing buffer (PBS containing 0.1% Tween-20). The optimal serum dilution was determined by checkerboard titration (1:100, 1:200, 1:400, 1:800, 1:1600, 1:3200) in serum diluent (PBS containing 1% BSA and 0.05% Tween-20).

After 1 h of incubation the plates were rinsed five times with washing buffer (PBS containing 0.1% Tween-20) and incubated for 1 h at RT with HRP-labeled anti-mouse IgG (Bio-Rad, diluted 1:5000 in PBS-T) or rabbit anti-pig IgG (Fab'2; LifeSpan Biosciences, Seattle, USA) diluted 1:15 000 in PBS, containing 1% BSA and 0.05% Tween-20. The plates were washed as described above. Binding of specific antibodies was detected by the addition of 100 μL/well of TMB ready-to-use substrate (Sigma) or TMB substrate (Clinical Science Products, Mansfield, USA). After 10 min of incubation the reaction was stopped by adding 100 μL/well of 10% sulphuric acid and the optical density (OD) was measured at 450 nm (reference filter 620 nm) in a microtiter plate reader (Tecan, Groedig, Austria).

Characterization of serum samples by the commercial test

Serum samples were assayed for the presence of anti-PCV2 antibodies using a commercial SERELISA PCV2 Ab Mono Blocking Kit (Synbiotics, Lyon, France) as recommended by the manufacturer. This ELISA is an epitope-blocking assay in which labeled MAb against PCV2 Cap protein is blocked from binding to the PCV2-coated microtiter plate if the serum sample contains antibody against PCV2. The test was performed according to the recommendations of the manufacturer. Data management and calculations were performed using software provided by the manufacturer. Any sample presenting a S/N ratio ≤ to 0.4 and a titer >350 is considered positive for the presence of PCV2 Cap –specific antibodies in pig serum. Any sample presenting a S/N ratio >0.4 and the titer ≤350 is considered as "low antibody in serum" or non-reactive.

Abbreviations

PCV2: Porcine circovirus type 2; Cap: Capsid protein; VLPs: Virus-like particles; MAbs: Monoclonal antibodies; PCVAD: PCV2-associated diseases; PMWS: Postweaning multisystemic wasting syndrome; PDNS: Nephropathy syndrome; DCs: Dendritic cells; HaPyV: Hamster polyomavirus; IFA: Immunofluorescence assay; QIFA: Quantitative immunofluorescence assay; DMEM: Dulbecco's modified Eagle's medium.

Competing interests

The authors declare that they have no competing interests.

Authors' contributions

JN was involved in construction of plasmids, purification and characterization of PCV2 VLPs, data collection and analysis; RiL was involved in generation and characterization of monoclonal antibodies; RPB developed an indirect ELISA for PCV2 diagnostics, was involved in data analysis and interpretation and the part of manuscript drafting; JD was involved in cloning and sequencing of PCV2 genomes, participated in the sequence alignment, analysis of PCV2 genomes and purification of PCV2 and PCV1 VLPs; RaL participated in sample collection and characterization using molecular diagnostic method; VS was involved in sample collection, data analysis and interpretation; AZ carried out the immunoassay, participated in characterization of monoclonal antibodies and edited the manuscript; KS conceived of the study, participated in its design and was involved in data analysis and interpretation; AG conceived of the study, was involved in all aspects of the experimental design and coordination, data analysis and interpretation, drafted and edited the manuscript. All authors read and approved the final version of the manuscript.

Acknowledgement

This work was funded by European Social Fund, grant No VP1-3.1-ŠMM-10-V-02-017.

Author details

[1]Institute of Biotechnology, Vilnius University, Graiciuno 8, LT-02241 Vilnius, Lithuania. [2]Institute of Microbiology and Virology, Veterinary Faculty of Veterinary Academy, Lithuanian University of Health Sciences, Tilzes 18, LT-47181 Kaunas, Lithuania.

References

1. Segales J, Kekarainen T, Cortey M: **The natural history of porcine circovirus type 2: from an inoffensive virus to a devastating swine disease?** *Vet Microbiol* 2013, **165**:13–20.
2. Segalés J, Allan GM, Domingo M: **Porcine circovirus diseases.** *Anim Health Res Rev* 2005, **6**(2):119–142.
3. Chae C: **A review of porcine circovirus 2-associated syndromes and diseases.** *Vet J* 2005, **169**:326–336.
4. de Boisseson C, Béven V, Bigarré L, Thiéry R, Rose N, Eveno E, Madec F, Jestin A: **Molecular characterization of Porcine circovirus type 2 isolates from post-weaning multisystemic wasting syndrome-affected and non-affected pigs.** *J Gen Virol* 2004, **85**:293–304.
5. Rosell C, Segalés J, Ramos-Vara JA, Folch JM, Rodríguez-Arrioja GM, Duran CO, Balasch M, Plana-Durán J, Domingo M: **Identification of porcine circovirus in tissues of pigs with porcine dermatitis and nephropathy syndrome.** *Vet Rec* 2000, **146**:40–43.
6. West KH, Bystrom JM, Wojnarowicz C, Shantz N, Jacobson M, Allan GM, Haines DM, Clark EG, Krakowka S, McNeilly F, Konoby C, Martin K, Ellis JA: **Myocarditis and abortion associated with intrauterine infection of sows with porcine circovirus 2.** *J Vet Diagn Invest* 1999, **11**(6):530–532.
7. Kekarainen T, McCullough K, Fort M, Fossum C, Segalés J, Allan GM: **Immune responses and vaccine-induced immunity against Porcine circovirus type 2.** *Vet Immunol Immunopathol* 2010, **136**:185–193.
8. Fort M, Olvera A, Sibila M, Segalés J, Mateu E: **Detection of neutralizing antibodies in postweaning multisystemic wasting syndrome (PMWS)-affected and non-PMWS-affected pigs.** *Vet Microbiol* 2007, **125**:244–255.
9. Vincent IE, Balmelli C, Meehan B, Allan G, Summerfield A, McCullough KC: **Silencing of natural interferon producing cell activation by porcine circovirus type 2 DNA.** *Immunology* 2007, **120**(1):47–56.

10. Cortey M, Olvera A, Grau-Roma L, Segales J: **Further comments on porcine circovirus type 2 (PCV2) genotype definition and nomenclature.** *Vet Microbiol* 2011, **149:**522–523.
11. Segales J, Olvera A, Grau-Roma L, Charreyre C, Nauwynck H, Larsen L, Dupont K, McCullough K, Ellis J, Krakowka S, Mankertz A, Fredholm M, Fossum C, Timmusk S, Stockhofe-Zurwieden N, Beattie V, Armstrong D, Grassland B, Baekbo P, Allan G: **PCV-2 genotype definition and nomenclature.** *Vet Rec* 2008, **162:**867–868.
12. Dupont K, Nielsen EO, Baekbo P, Larsen LE: **Genomic analysis of PCV2 isolates from Danish archives and a current PMWS casecontrol study supports a shift in genotypes with time.** *Vet Microbiol* 2008, **128:**56–64.
13. Guo LJ, Lu YH, Wei YW, Huang LP, Liu CM: **Porcine circovirus type 2 (PCV2): genetic variation and newly emerging genotypes in China.** *Virol J* 2010, **7:**273.
14. Nawagitgul P, Morozov I, Bolin SR, Harms PA, Sorden SD, Paul PS: **Open reading frame 2 of porcine circovirus type 2 encodes a major capsid protein.** *J Gen Virol* 2000, **81:**2281–2287.
15. Fan H, Ju C, Tong T, Huang H, Lv J, Chen H: **Immunogenicity of empty capsids of porcine circovirus type 2 produced in insect cells.** *Vet Res Commun* 2007, **31:**487–496.
16. Zhou JY, Shang SB, Gong H, Chen QX, Wu JX, Shen HG, Chen TF, Guo JQ: ***In vitro* expression, monoclonal antibody and bioactivity for capsid protein of porcine circovirus type II without nuclear localization signal.** *J Biotechnol* 2005, **118**(2):201–211.
17. Trundova M, Celer V: **Expression of porcine circovirus 2 ORF2 gene requires codon optimized *E. coli* cells.** *Virus Genes* 2007, **34:**199–204.
18. Marcekova Z, Psikal I, Kosinova E, Benada O, Sebo P, Bumba L: **Heterologous expression of full-length capsid protein of porcine circovirus 2 in *Escherichia coli* and its potential use for detection of antibodies.** *J Virol Methods* 2009, **162:**133–141.
19. Wang K, Huang L, Kong J, Zhang X: **Expression of the capsid protein of porcine circovirus type 2 in *Lactococcus lactis* for oral vaccination.** *J Virol Methods* 2008, **150:**1–6.
20. Tu Y, Wang Y, Wang G, Wu J, Liu Y, Wang S, Jiang C, Cai X: **High-level expression and immunogenicity of a porcine circovirus type 2 capsid protein through codon optimization in *Pichia pastoris.*** *Appl Microbiol Biotechnol* 2012, doi:10.1007/s00253-012-4540-z
21. Ju C, Fan H, Tan Y, Liu Z, Xi X, Cao S, Wu B, Chen H: **Immunogenicity of a recombinant pseudorabies virus expressing ORF1-ORF2 fusion protein of porcine circovirus type 2.** *Vet Microbiol* 2005, **109:**179–190.
22. Pumpens P, Grens E: **Artificial genes for chimeric virus-like particles.** In *Artificial DNA: Methods and Applications.* Edited by Khudyakov YE, Fields HA. Boca Raton: CRC Press LLC; 2002:249–327.
23. Beach NM, Meng X-J: **Efficacy and future prospects of commercially available and experimental vaccines against porcine circovirus type 2 (PCV2).** *Virus Res* 2012, **164:**33–42.
24. Nawagitgul P, Harms PA, Morozov I, Thacker BJ, Sorden SD, Lekcharoensuk C, Paul PS: **Modified indirect porcine circovirus (PCV) type 2-based and recombinant capsid protein (ORF2)-based enzyme-linked immunosorbent assays for detection of antibodies to PCV.** *Clin Diagn Lab Immunol* 2002, **9**(1):33–40.
25. Liu C, Ihara T, Nunoya T, Ueda S: **Development of an ELISA based on the baculovirus-expressed capsid protein of porcine circovirus type 2 as antigen.** *J Vet Med Sci* 2004, **66:**237–242.
26. Wu PC, Chien MS, Tseng YY, Lin J, Lin WL, Yang CY, Huang C: **Expression of the porcine circovirus type 2 capsid protein subunits and application to an indirect ELISA.** *J Biotechnol* 2008, **133:**58–64.
27. Shang SB, Li YF, Guo JQ, Wang ZT, Chen QX, Shen HG, Zhou JY: **Development and validation of a recombinant capsid protein-based ELISA for detection of antibody to porcine circovirus type 2.** *Res Vet Sci* 2008, **84:**150–157.
28. Sasnauskas K, Buzaite O, Vogel F, Jandrig B, Razanskas R, Staniulis J, Scherneck S, Krüger DH, Ulrich R: **Yeast cells allow the high-level expression and formation of polyomavirus-like particles.** *Biol Chem* 1999, **380:**381–386.
29. Razanskiene A, Schmidt J, Geldmacher A, Ritzi A, Niedrig M, Lundkvist A, Kruger DH, Meisel H, Sasnauskas K, Ulrich R: **High yields of stable and highly pure nucleocapsid proteins of different hantaviruses can be generated in the yeast *Saccharomyces cerevisiae.*** *J Biotechnol* 2004, **111:**319–333.
30. Petraityte R, Tamosiunas PL, Juozapaitis M, Zvirbliene A, Sasnauskas K, Shiell B, Russell G, Bingham J, Michalski WP: **Generation of Tioman virus nucleocapsid-like particles in yeast *Saccharomyces cerevisiae.*** *Virus Res* 2009, **145**(1):92–96.
31. Khayat R, Brunn N, Speir JA, Hardham JM, Ankenbauer RG, Schneemann A, Johnson JE: **The 2.3-angstrom structure of porcine circovirus 2.** *J Virol* 2011, **85**(15):7856–7862.
32. Sasnauskas K, Bulavaite A, Hale A, Jin L, Gedvilaite A, Dargeviciute A, Bartkeviciute D, Zvirbliene A, Staniulis J, Brown DWG, Ulrich R: **Generation of recombinant virus-like particles of human and non-human polyomaviruses in yeast *Saccharomyces cerevisiae.*** *Intervirology* 2002, **45:**471–482.
33. Samuel D, Sasnauskas K, Beard L-JS, Zvirbliene A, Gedvilaite A, Cohen B: **High level expression of recombinant mumps nucleoprotein in *Saccharomyces cerevisiae* and its evaluation in mumps IgM serology.** *J Med Virol* 2002, **66**(1):123–130.
34. Galao RP, Scheller N, Alves-Rodrigues I, Breinig T, Meyerhans A, Díez J: ***Saccharomyces cerevisiae*: a versatile eukaryotic system in virology.** *Microb Cell Factories* 2007, **6:**32. doi:10.1186/1475-2859-6-32.
35. Bucarey SA, Noriega J, Reyes P, Tapia C, Sáenz L, Zuñiga A, Tobar JA: **The optimized capsid gene of porcine circovirus type 2 expressed in yeast forms virus-like particles and elicits antibody responses in mice fed with recombinant yeast extracts.** *Vaccine* 2009, **27:**5781–5790.
36. Sun S-Q, Guo H-C, Sun D-H, Yin S-H, Shang Y-J, Cai X-P, Liu X-T: **Development and validation of an ELISA using a protein encoded by ORF2 antigenic domain of porcine circovirus type 2.** *Virol J* 2010, **7:**274. doi:10.1186/1743-422X-7-274.
37. Huang L, Lu Y, Wei Y, Guo L, Liu C: **Development of a blocking ELISA for detection of serum neutralizing antibodies against porcine circovirus type 2.** *J Virol Methods* 2011, **171:**26–33. doi:10.1016/j.jviromet.2010.09.023.
38. Kim Y-H, Kweon C-H, Kang S-W, Oh YI, Song JY, Lee KK, Park SC: **Vero cells expressing porcine circovirus type 2-capsid protein and their diagnostic application.** *J Virol Methods* 2013, **194**(1–2):26–32. doi:10.1016/j.jviromet.2013.07.030.
39. Sambrook J, Russell DW: *Molecular Cloning, A Laboratory Manual.* Cold Spring Harbour: Cold Spring Harbor Press; 2001.
40. Köhler G, Milstein C: **Continuous cultures of fused cells secreting antibody of predefined specificity.** *Nature* 1975, **256:**495–497.
41. Zvirbliene A, Sezaite I, Pleckaityte M, Kucinskaite-Kodze I, Juozapaitis M, Sasnauskas K: **Mapping of an antigenic site on the nucleocapsid protein of human parainfluenza virus type 3.** *Viral Immunol* 2009, **22:**181–188.

Saccharification of rice straw by cellulase from a local *Trichoderma harzianum* SNRS3 for biobutanol production

Nooshin Rahnama[1†], Hooi Ling Foo[1,2†], Nor Aini Abdul Rahman[1,4†], Arbakariya Ariff[1,4†] and Umi Kalsom Md Shah[1,3,4*]

Abstract

Background: Rice straw has shown to be a promising agricultural by-product in the bioconversion of biomass to value-added products. Hydrolysis of cellulose, a main constituent of lignocellulosic biomass, is a requirement for fermentable sugar production and its subsequent bioconversion to biofuels such as biobutanol. The high cost of commercial enzymes is a major impediment to the industrial application of cellulases. Therefore, the use of local microbial enzymes has been suggested. *Trichoderma harzianum* strains are potential CMCase and β-glucosidase producers. However, few researches have been reported on cellulase production by *T. harzianum* and the subsequent use of the crude cellulase for cellulose enzymatic hydrolysis. For cellulose hydrolysis to be efficiently performed, the presence of the whole set of cellulase components including exoglucanase, endoglucanase, and β-glucosidase at a considerable concentration is required. Biomass recalcitrance is also a bottleneck in the bioconversion of agricultural residues to value-added products. An effective pretreatment could be of central significance in the bioconversion of biomass to biofuels.

Results: Rice straw pretreated using various concentrations of NaOH was subjected to enzymatic hydrolysis. The saccharification of rice straw pretreated with 2% (w/v) NaOH using crude cellulase from local *T. harzianum* SNRS3 resulted in the production of 29.87 g/L reducing sugar and a yield of 0.6 g/g substrate. The use of rice straw hydrolysate as carbon source for biobutanol fermentation by *Clostridium acetobutylicum* ATCC 824 resulted in an ABE yield, ABE productivity, and biobutanol yield of 0.27 g/g glucose, 0.04 g/L/h and 0.16 g/g glucose, respectively. As a potential β-glucosidase producer, *T. harzianum* SNRS3 used in this study was able to produce β-glucosidase at the activity of 173.71 U/g substrate. However, for cellulose hydrolysis to be efficient, Filter Paper Activity at a considerable concentration is also required to initiate the hydrolytic reaction. According to the results of our study, FPase is a major component of cellulose hydrolytic enzyme complex system and the reducing sugar rate-limiting enzyme.

Conclusion: Our study revealed that rice straw hydrolysate served as a potential substrate for biobutanol production and FPase is a rate-limiting enzyme in saccharification.

Keywords: Rice straw, Saccharification, Biobutanol, *Trichoderma harzianum* SNRS3

* Correspondence: umikalsom@upm.edu.my
†Equal contributors
[1]Department of Bioprocess Technology, Faculty of Biotechnology and Biomolecular Sciences, Universiti Putra Malaysia, 43400 UPM Serdang, Selangor, Malaysia
[3]Institute of Tropical Forestry and Forest Products, Universiti Putra Malaysia, 43400 UPM Serdang, Selangor, Malaysia
Full list of author information is available at the end of the article

Background

Lignocellulosic substrates are the most abundant, low cost renewable resources worldwide [1]. Cellulose, hemicelluloses and lignin are the three main components which make up lignocellulosic biomass. Via enzymatic hydrolysis, polysaccharides in the lignocellulosic substrates can be hydrolyzed to monosaccharides [2]. Saccharification is therefore a crucial step for sugar production [1]. Acid hydrolysis and enzymatic hydrolysis are the two commonly used methods for hydrolysis of cellulosic materials. Enzymatic hydrolysis is advantageous over acid hydrolysis. It has low environmental impact and the reaction is carried out under mild conditions [3]. Recently, due to the constant increase in the oil price, the significance of biofuel production from lignocellulosic biomass as an alternative energy source has been intensified and efficient conversion of lignocelluloses to biofuel is gaining interest [4].

Rice is a major crop produced in large quantities in the world [5] and rice straw is the most abundant agricultural residue worldwide. Approximately 700-800 million tons of rice straw is produced annually, most of which is found in Asia. Hence, rice straw is a suitable feedstock for biofuel production especially in Asia [2]. Various agricultural wastes have been used as the substrate for biobutanol production. However, not much research has focused on biobutanol production from rice straw as the feedstock for fermentation and limited research has been reported [6-9].

As fermentable sugar production from agricultural waste is a prerequisite for bioenergy production, enormous research attempt has focused on bioconversion of cellulose into fermentable sugar [10]. Yet, more research is required to improve the process. Hence, this work focused on fermentable sugar production from alkali pretreated rice straw by using cellulase produced from a local *T. harzianum* SNRS3 under solid state cultivation. Rice straw hydrolysate was subsequently utilized as substrate for biobutanol production.

Results and discussion

Effect of alkali pretreatment of rice straw on rice straw chemical composition

Alkali pretreatment of rice straw using 2% (w/v) NaOH showed a rise in cellulose content of rice straw, whereas the content of lignin and hemicelluloses was decreased. Table 1 demonstrates cellulose, hemicellulose, lignin, and ash content of rice straw both in untreated and alkali-pretreated rice straw. As shown in the table, as a result of pretreatment, rice straw was delignified and a decrease in lignin content from 9.22% to 3.78% was observed. There was a decrease in hemicelluloses content from 26.03% to 17.63% after rice straw was pretreated as well. However, cellulose content of rice straw was promoted from 39.74% in untreated rice straw to 70.9% after rice straw was subjected to alkali pretreatment.

In a study, [2% (w/v) NaOH] was used to pretreat rice straw and a decrease in lignin and hemicelluloses content of rice straw was reported from 15.8% and 25.8% to 9.8% and 18%, respectively, and an increase in cellulose content of alkali-pretreated rice straw from 36.8% to 52.2% was noted as well [1]. As reported in another research, chemical composition of rice straw was different before and after pretreatment by [2% (w/v) NaOH], and a decrease in lignin and hemicelluloses content of rice straw after alkali pretreatment was observed. Whereas, an increase in cellulose content occurred when rice straw was pretreated. As a result of alkali pretreatment of rice straw with [2% (w/v) NaOH], cellulose content was increased from 38.3% to 59.3% in alkali-pretreated rice straw and lignin and hemicelluloses content was decreased from 14.9% and 28% in untreated rice straw to 9.5% and 10.9% in pretreated rice straw, respectively [11].

Effect of different concentrations of NaOH for pretreatment of rice straw on reducing sugar production

Rice straw pretreated with various concentrations of NaOH [1%, 2%, 3%, and 4% (w/v)] was used as saccharification substrate and enzymatic hydrolysis of rice straw was performed. Figure 1 illustrates the reducing sugar production profile from rice straw pretreated by using different concentrations of NaOH. The reducing sugar production was increased significantly ($p < 0.05$) when NaOH-pretreated rice straw was used as substrate in enzymatic hydrolysis. As for the effect of incubation time, there was significant difference ($p < 0.05$) for the production of reducing sugar from 0-72 h from NaOH-pretreated rice straw used in enzymatic hydrolysis, except for 1% (w/v) NaOH, where the production of reducing sugar was only increased significantly from 0-48 h. The detail of the statistical analysis is presented in Table 2.

According to the results obtained in this study, the enzymatic hydrolysis of rice straw pretreated by using 2% (w/v) NaOH resulted in the highest production of reducing sugars at 5.82 g/L at 72 h of incubation, as compared to 0.99 g/L reducing sugars produced from untreated rice straw.

Table 1 Chemical composition of rice straw before and after alkali pretreatment

Pretreatment	*Cellulose (%)	*Hemicelluloses (%)	*Lignin (%)	*Ash (%)
Untreated	39.74 ± 3.69	26.03 ± 0.30	9.22 ± 3.01	12.48 ± 0.38
2% (w/v) NaOH	70.9 ± 0.81	17.63 ± 3.07	3.78 ± 0.95	1.55 ± 0.05

*Results are based on the mass of dry matter.

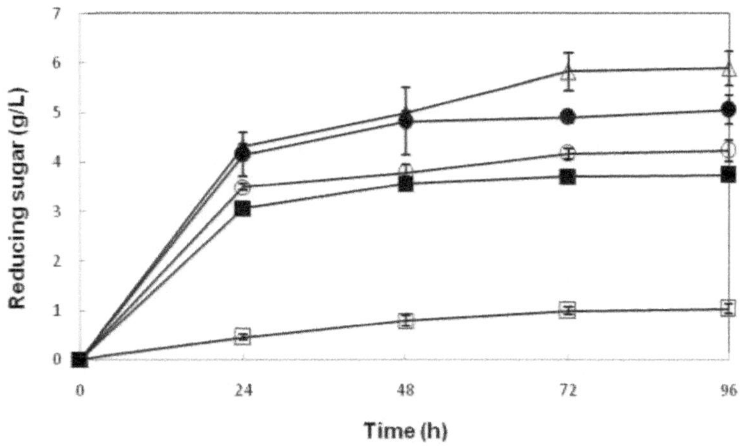

Figure 1 Enzymatic hydrolysis of rice straw pretreated with different concentrations of NaOH. Values are means of 3 replicates ± SD. Symbols represent: □: Untreated; •: 1%; Δ: 2%; ○: 3%; ■: 4%.

In a study, the effect of different concentrations of NaOH on enzymatic hydrolysis of bagasse using cellulases of *Bacillus subtilis* was investigated. As reported, the enzymatic hydrolysis of the substrate pretreated with 2% (w/v) NaOH resulted in higher production of reducing sugars, as compared to enzymatic hydrolysis of the substrate pretreated with [1%, 3%, and 4% (w/v)] NaOH. Low yield of reducing sugar obtained at [3% and 4% (w/v)] NaOH concentration could be due to the loss of carbohydrates that might have been solubilized while the substrate was being pretreated. In addition, the low saccharification could be attributed to the inaccessible insoluble cellulose. This might contribute to low sugar yield when saccharification substrates were pretreated at higher concentrations of NaOH [12].

As reported in our previous research [13], after rice straw was pretreated by using alkali, not only chemical composition of rice straw was changed, but also changes in the cellulose morphological structure occurred. The results of X-ray diffraction analysis revealed an increase in relative crystallinity of cellulose in alkali-pretreated rice straw, as compared to untreated rice straw. Hydrolyzation and peeling of amorphous regions during pretreatment could result in the rise in relative crystallinity of cellulose in alkali-pretreated rice straw. However, alkali pretreatment decreased the absolute crystallinity of cellulose. As the SEM images showed, the pretreatment process disrupted the hemicellulose and lignin, which might have resulted in the changes in the structure of cellulose. According to the results of FTIR analysis, alkali pretreatment caused lignin removal and the changes in cellulose structure.

Table 2 Reducing sugar production (g/L) by crude cellulase from *T. harzianum* SNRS3 from rice straw pretreated with different concentrations of NaOH

		NaOH Concentration (w/v)			
Time (h)	Untreated	1%	2%	3%	4%
0	0.00^{Aa}	0.00^{Aa}	0.00^{Aa}	0.00^{Aa}	0.00^{Aa}
24	0.44^{Ba}	4.15^{Bb}	4.31^{Bb}	3.49^{Bc}	3.05^{Bd}
48	0.78^{Ca}	4.82^{Cb}	4.98^{Cb}	3.78^{Cc}	3.55^{Cc}
72	0.99^{Da}	4.90^{Cb}	5.82^{Dc}	4.17^{Dd}	3.70^{De}
96	1.03^{Da}	$5.06^{Cb,c}$	5.89^{Db}	4.23^{Dc}	3.73^{Dd}

Note: Means expressed with different superscript capital letters within the same column was significantly different at $p < 0.05$.
Means expressed with different superscript small letters within the same row was significantly different at $p < 0.05$.

Effect of enzyme loadings on reducing sugar production

Enzyme loading is an important parameter in enzymatic hydrolysis, affecting reducing sugar yield. Therefore, the effect of enzyme concentration on enzymatic hydrolysis of rice straw pretreated with 2% (w/v) NaOH was investigated. Various enzyme loadings were used for hydrolysis of 5% (w/v) rice straw pretreated with 2% (w/v) NaOH as saccharification substrate. The reducing sugar production profile was studied over a period of 96 h. Figure 2 shows that the reducing sugar production was increased significantly ($p < 0.05$) according to the cellulases concentration used in enzymatic hydrolysis, except for the cellulase concentration of FPase: 125 U/g; CMCase: 2226.2 U/g; β-glucosidase: 3474.2 U/g and FPase: 93.75 U/g CMCase: 1669.65 U/g; β-glucosidase: 2605.65 U/g. As for the effect of incubation time there was significant difference ($p < 0.05$) for the production of reducing sugar from 0-72 h for all cellulase concentrations used in enzymatic hydrolysis. The concentration of crude cellulase at the activity of FPase: 93.75 U/g CMCase: 1669.65 U/g; β-glucosidase: 2605.65 U/g that resulted in the production of 29.87 g/L reducing sugar was selected for the subsequent study since the reducing sugar production was

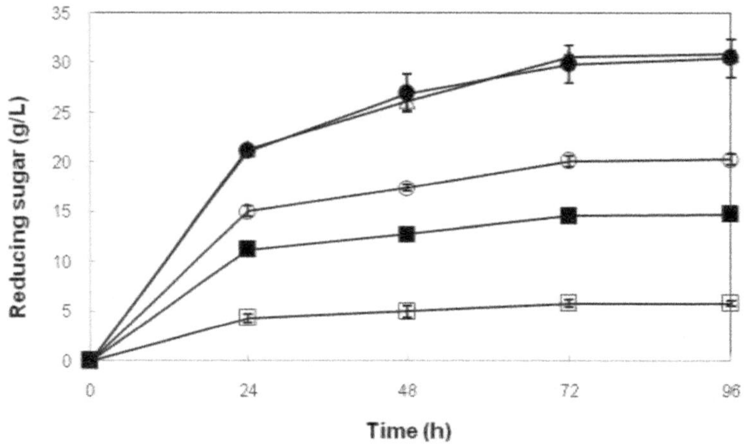

Figure 2 Effect of enzyme concentration on hydrolysis of rice straw by cellulases from *T. harzianum* SNRS3. Values are means of 3 replicates ± SD. Symbols represent enzyme activity (U/g substrate): □: FPase 6.25, CMCase 111.31, β-glucosidase 173.71; ■: FPase 31.25, CMCase 566.55, β-glucosidase 868.55; ○: FPase 62.5, CMCase 1113.1, β-glucosidase 1737.1; ●: FPase 93.75, CMCase 1669.65, β-glucosidase 2605.65; Δ: FPase 125, CMCase 2226.2, β-glucosidase 3474.2.

not significantly ($p > 0.05$) different from the cellulase at the activity of FPase: 125 U/g; CMCase: 2226.2 U/g; β-glucosidase: 3474.2 U/g. The detail of the statistical analysis is presented in Table 3.

The cost of cellulase enzymes contributes to the total cost of saccharification process. Hence, it is suggested to minimize enzyme dosage as much as possible [14]. An increase in enzyme concentration resulted in significant rise of hydrolysis rate of rice straw, wheat straw, and bagasse. However, further increase in enzyme concentration did not result in significant rise of hydrolysis rate. This might be due to improper mixing and suspension of the slurry. In fact, the rise in enzyme concentration should increase hydrolysis rate. However, it makes the process uneconomical [12].

Table 3 Reducing sugar production (g/L) from rice straw using various concentrations of crude cellulase from *T. harzianum* SNRS3

	Enzyme Concentration (U/g substrate)				
Time (h)	*1	*2	*3	*4	*5
0	0.00^{Aa}	0.00^{Aa}	0.00^{Aa}	0.00^{Aa}	0.00^{Aa}
24	4.31^{Ba}	11.19^{Bb}	15.05^{Bc}	21.14^{Bd}	21.34^{Bd}
48	4.98^{Ca}	12.78^{Cb}	17.40^{Cc}	26.93^{Cd}	26.20^{Cd}
72	5.82^{Da}	14.57^{Db}	20.14^{Dc}	29.87^{Dd}	30.55^{Dd}
96	5.84^{Da}	14.77^{Db}	20.31^{Dc}	30.47^{Dd}	30.87^{Dd}

Note: *1: FPase 6.25 U/g, CMCase 111.31 U/g, β-glucosidase 173.71 U/g.
*2: FPase 31.25 U/g, CMCase 566.55 U/g, β-glucosidase 868.55 U/g.
*3: FPase 62.5 U/g, CMCase 1113.1 U/g, β-glucosidase 1737.1 U/g.
*4: FPase 93.75 U/g, CMCase 1669.65 U/g, β-glucosidase 2605.65 U/g.
*5: FPase 125 U/g, CMCase 2226.2 U/g, β-glucosidase 3474.2 U/g.
Means expressed with different superscript capital letters within the same column was significantly different at $p < 0.05$.
Means expressed with different superscript small letters within the same row was significantly different at $p < 0.05$.

In this study, a similar trend was noted for fermentable sugar production and conversion yield percentage. A comparison between fermentable sugar production and hydrolysis yield (%) is illustrated in Table 4.

The increase in enzyme loading favoured enzymatic hydrolysis of corncob residue [12], wheat straw, rice straw, and bagasse [12]. A similar trend was also noted for fermentable sugar production and conversion yield percentage with the increase in concentration of the enzyme used [15]. As reported in other studies, a rise in reducing sugar yield occurred as the enzyme concentration increased [12,15]. Therefore, the results of this research are in good agreement with other studies. The optimal ratio between enzyme and substrate is an important factor affecting efficiency of enzymatic hydrolysis. In fact, more active sites of the enzyme would be involved in conversion of the substrate into reducing sugars via enzymatic hydrolysis when enzyme concentration increases [5].

Table 4 Fermentable sugars production and hydrolysis yield (%) obtained by using various concentrations of crude cellulases from *T. harzianum* SNRS3

Cellulase of *T. harzianum* SNRS3(U/g)			Fermentable sugar (g/L)	Conversion/ Hydrolysis yield (%)
FPase	CMCase	β-glucosidase		
6.25	111.31	173.71	5.82 ± 0.37^{a}	11.83
31.25	566.55	868.55	14.57 ± 0.47^{b}	29.63
62.50	1113.10	1737.10	20.14 ± 0.53^{c}	40.96
93.75	1669.65	2605.65	29.87 ± 1.87^{d}	60.75
125	2226.20	3474.20	30.55 ± 0.23^{d}	62.13

Note: Means expressed with different superscript letters within the same column was significantly different at ($p < 0.05$).

Effect of substrate concentration on reducing sugar production

Efficiency of enzymatic hydrolysis is determined by the optimal ratio between enzyme and substrate. Hence, the effect of concentration of the substrate on reducing sugar production was investigated by using different concentrations of substrate [1%, 3%, 5%, and 7% (w/v)] and the optimal unit of enzyme activity (FPase: 93.75 U/g, CMCase: 1669.65 U/g, β-glucosidase: 2605.65 U/g) based on the results obtained in the previous experiment. The reducing sugar profile was studied over a period of 96 h. Figure 3 shows the reducing sugar production was increased significantly ($p < 0.05$) according to the substrate concentration used in enzymatic hydrolysis, except for the substrate concentration at 5% and 7% (w/v). As for the effect of incubation time there was significant difference ($p < 0.05$) for the production of reducing sugar from 0-72 h for all substrate concentrations used in enzymatic hydrolysis. 5% (w/v) substrate that resulted in the production of 29.87 g/L reducing sugars was selected as substrate concentration for reducing sugar production in subsequent study since the reducing sugar production was not significantly different ($p > 0.05$) from 7% (w/v) substrate concentration. The detail of the statistical analysis is presented in Table 5.

Enzymatic hydrolysis of 10% (w/v) rice straw resulted in the production of higher reducing sugar, as compared to [4% and 6% (w/v)] substrates. As a result, 18.5 g/L reducing sugar was obtained [12]. Effect of substrate concentration on reducing sugar production was investigated by using various lignocellulosic substrates such as corncob, sunflower stalks, wheat straw, rice straw, and bagasse and it was shown that the increase in substrate concentration favoured reducing sugar production [12,15,16]. In this study, an opposite trend was noted between reducing sugar concentration and hydrolysis yield (%). With the rise in substrate concentration, sugar production increased. However, hydrolysis yield percentage decreased due to end product inhibition and mixing problems. Similar results have been reported by other researchers [12,15,16]. Enzymatic hydrolysis of various lignocellulosic materials using a variety of enzyme sources has been reported by other researchers. In a study, a comparison was made between the use of celluclast and a combination of celluclast and *Periconia* sp. bcc 2871 as the source of enzyme for hydrolysis of rice straw and a reducing sugar production of 84 and 132 mg/g substrate was reported, respectively [17]. Cellulase from *T. reesei* ZM4F3 was used for enzymatic hydrolysis of rice straw and as a result, 733 mg/g substrate reducing sugar was produced [17]. Use of cellulase by *Trametes hirsuta* for enzymatic hydrolysis of rice straw resulted in the production of 685 mg/g substrate reducing sugars and a hydrolysis yield of 88% was achieved [1]. Rice straw was enzymatically hydrolyzed by cellulases from *T. reesei* A1 and *Penicillium* sp. B1. As a result, a reducing sugar concentration of 374 and 316 mg/g substrate and a hydrolysis yield of 70.50% and 57.90% were obtained, respectively [18]. As reported in a research, enzymatic hydrolysis of rice hull by a combination of celluclast 1.5 L and Novozyme 188 resulted in the production of 154 mg/g substrate reducing sugar and 32% hydrolysis yield was achieved [19]. Bagasse was hydrolyzed by cellulases from *Penicillium janthinellum* NCIM 1171 and 846 mg/g substrate reducing sugar equivalent to 94.60% hydrolysis yield was obtained [20]. Enzymatic hydrolysis of maize straw by cellulase from *T. reesei* ZU-02 produced 814 mg/g substrate reducing sugar and 83.5% conversion yield was obtained [14]. In this study, as a result of enzymatic hydrolysis of alkali-pretreated rice straw using cellulase

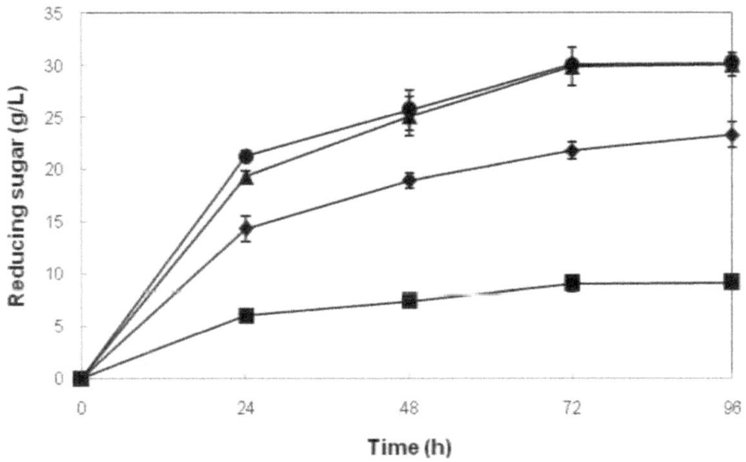

Figure 3 Enzymatic hydrolysis of rice straw using different concentrations of the substrate (w/v). Values are means of 3 replicates ± SD. Symbols represent different concentrations of the substrate (w/v): ■: 1%; ♦: 3%; ▲: 5%; ●: 7%.

Table 5 Reducing sugar production (g/L) by crude cellulase from *T. harzianum* SNRS3 using various concentrations of rice straw as substrate

Time (h)	Substrate Concentration (w/v)			
	1%	3%	5%	7%
0	0.00^{Aa}	0.00^{Aa}	0.00^{Aa}	0.00^{Aa}
24	6.02^{Ba}	14.33^{Bb}	19.30^{Bc}	21.85^{Bd}
48	7.39^{Ca}	18.94^{Cb}	25.10^{Cc}	25.70^{Cc}
72	9.03^{Da}	21.79^{Db}	29.87^{Dc}	30.07^{Dc}
96	9.15^{Da}	23.30^{Eb}	30.70^{Dc}	30.23^{Dc}

Note: Means expressed with different superscript capital letters within the same column was significantly different at $p < 0.05$.
Means expressed with different superscript small letters within the same row was significantly different at $p < 0.05$.

from *T. harzianum* SNRS3, a reducing sugar yield of 600 mg/g substrate and a hydrolysis yield of 60.75% were achieved, as compared to reducing sugar production of 800 mg/g substrate and 81.35% conversion yield obtained when celluclast was used.

As a result of enzymatic hydrolysis of alkali-pretreated rice straw using cellulase from *T. harzianum* SNRS3, a reducing sugar yield of 600 mg/g substrate and a hydrolysis yield of 60.75% were achieved.

FPase as the rate-limiting enzyme

Cellulose enzymatic hydrolysis needs all the three componets of cellulase enzyme complex system to act synergistically. However, to study the significant role of FPase in cellulose enzymatic hydrolysis, four combinations of cellulase enzyme were used in this study. The reducing sugar profile was studied over a period of 96 h. Table 6 shows that the production of reducing sugar was increased significantly ($p < 0.05$) according to the concentration of FPase. As for the effect of incubation time, the reducing sugar production was significantly increased ($p < 0.05$) from 0-72 h for all the enzyme concentrations used in enzymatic hydrolysis of rice straw. The detail of the statistical analysis is presented in Table 7.

As shown in Table 6, at 72 h of enzymatic hydrolysis and at concentration of FPase, 9.9 U/mL and 9.86 U/mL for celluclast and the crude cellulase, respectively, regardless of the significant difference in the concentration of β-glucosidase present in celluclast (5.55 U/mL) and that in the crude cellulase (265.2 U/mL), 36.30 g/L and 26.33 g/L reducing sugars were obtained by using celluclast and the crude cellulase of *T. harzianum* SNRS3, respectively. Moreover, glucose concentration was detected at 11.0 g/L and 10.02 g/L for celluclast and the crude cellulase, respectively. However, when β-glucosidase activity was fixed at 22.2 U/mL and 22.1 U/mL for celluclast and crude cellulase while FPase concentration for celluclast and the crude cellulase was different at the activity of 39.6 U/mL and 0.83 U/mL for celluclast and the crude cellulase, respectively, the reducing sugar analysis revealed a much higher production of reducing sugars by celluclast, as compared to the crude cellulase. As a result, 57.27 g/L and 5.94 g/L reducing sugars were obtained by using celluclast and the crude cellulase, respectively. The results of HPLC analysis revealed that glucose at the concentration of 21.15 g/L and 1.16 g/L was obtained when celluclast and the crude cellulase were used, respectively.

Cellulases consist of three main enzymes, endoglucanase or CMCase, exoglucanase or cellobiohydrolase and β-glucosidase, which act synergistically in the cellulose hydrolysis process [21-26]. Endoglucanase attacks intramolecular β-1,4-glucosidic bonds in cellulose polymeric chain, leaving new chain ends for exoglucanase to cleave. Exoglucanase or cellobiohydrolase is the second hydrolytic enzyme attacking cellulose chain in the ends producing cellobiose or glucose, while β-glucosidase converts cellobiose to glucose. Therefore, endoglucanase and exoglucanase are mainly responsible for depolymerization of cellulose polymer chains and this enzymatic depolymerization step is

Table 6 Reducing sugar production using celluclast and crude cellulase from *T. harzianum* SNRS3 at different concentrations of FPase (U/mL) and β-glucosidase (U/ mL)

Enzyme source	FPase (U/mL)	β-glucosidase (U/mL)	Reducing sugar (g/L)	Glucose (g/L)
Celluclast (5 FPU)	9.90 ± 0.07	5.55 ± 0.88	36.30 ± 0.26	11.00 ± 0.45
Crude cellulase from *T. harzianum* SNRS3	9.86 ± 0.05	265.20 ± 7.22	26.33 ± 0.75	10.02 ± 0.2
Celluclast (20 FPU)	39.60 ± 0.08	22.20 ± 0.84	57.27 ± 0.63	21.15 ± 0.89
Crude cellulase from *T. harzianum* SNRS3	0.83 ± 0.07	22.10 ± 0.33	5.94 ± 0.31	1.16 ± 0.1

Note: Values are means of 3 replicates ± SD.

Table 7 Reducing sugar production from rice straw using celluclast and crude cellulase from *T. harzianum* SNRS3 at different concentrations of FPase (U/mL) and β-glucosidase (U/ mL)

Time (h)	Enzyme Concentration (U/mL)			
	*1	*2	*3	*4
0	0.00^{Aa}	0.00^{Aa}	0.00^{Aa}	0.00^{Aa}
24	23.55^{Ba}	16.15^{Bb}	39.56^{Bc}	4.47^{Bd}
48	29.19^{Ca}	22.28^{Cb}	51.10^{Cc}	5.05^{Cd}
72	36.30^{Da}	26.33^{Db}	57.27^{Dc}	5.94^{Dd}
96	37.36^{Da}	27.56^{Db}	58.89^{Dc}	6.23^{Dd}

Note: *1: Celluclast: FPase 9.90 U/mL, β-glucosidase 5.55 U/mL.
*2: Crude cellulase: FPase 9.86 U/mL, β-glucosidase 265.20 U/mL.
*3: Celluclast: FPase 39.60 U/mL, β-glucosidase 22.20 U/mL.
*4: Crude cellulase: FPase 0.83 U/mL, β-glucosidase 22.10 U/mL.
Means expressed with different superscript capital letters within the same column was significantly different at $p < 0.05$.
Means expressed with different superscript small letters within the same row was significantly different at $p < 0.05$.

the rate-limiting step in cellulose hydrolysis process [27]. The effect of different ratios of cellulase and β-glucosidase on glucose production was studied and it was shown that the ratio of cellulase to β-glucosidase 5:1 gave higher glucose concentration, as compared to the ratio of 1: 5, where the lowest glucose concentration was produced. In fact, the latter ratio only resulted in the excess of β- glucosidase in the reaction mixture [3]. Therefore, the results of this study are in accordance with those reported in other research [3,27] that laid emphasis on the need for the presence of FPase at a substantial concentration in the cellulase enzyme complex for an efficient hydrolysis. From the results of this study, it is inferred that for an efficient cellulose enzymatic hydrolysis, the presence of the whole set of cellulose hydrolytic enzymes is required. FPase is a major component of cellulase enzymes complex, lack of which causes inefficient cellulose hydrolysis. This leads to a low production of reducing sugars and the presence of β-glucosidase at a high concentration would not help in such a condition since this latter enzyme is only accumulated as the relative activity of β-glucosidase to FPase is much higher.

Production of ABE by *C. acetobutylicum* ATCC 824 using rice straw hydrolysate

Rice straw hydrolysate was used as the substrate for ABE production. Based on the results of HPLC, rice straw hydrolysate contained glucose, xylose, and arabinose as individual sugars. Concentration of the individual sugars is shown in Figure 4.

In order to investigate the application of fermentable sugars produced from pretreated rice straw using the crude cellulase, rice straw hydrolysate obtained, was employed as fermentation substrate in ABE fermentation by *C. acetobutylicum* ATCC 824. In order to compare the results of ABE fermentation obtained in this study, glucose (10 g/L) was used as a model substrate. As can be seen from Figure 5, fermentation of (10 g/L) glucose as a model substrate resulted in the production of a total ABE concentration of 2.79 g/L (0.91 g/L acetone, 1.81 g/L butanol, and 0.07 g/L ethanol). While, as illustrated in Figure 6, the results of rice straw hydrolysate fermentation by *C. acetobutylicum* ATCC 824 indicates a maximum total ABE production of 2.73 g/L (0.82 g/L acetone, 1.62 g/L butanol, and 0.29 g/L ethanol).

In a research, fermentation of 10 g/L, and 20 g/L glucose by *C. acetobutylicum* ATCC 824, resulted in the production of 1.76 g/L total ABE (0.41 g/L acetone, 1.18 g/L butanol, and 0.17 g/L ethanol), and 5.76 g/L total ABE (1.36 g/L acetone, 4.08 g/L butanol, and 0.32 g/L ethanol), when 10 g/L, and 20 g/L glucose were used, respectively [28]. In another study, domestic organic waste containing (14 g/L glucose) as carbon source was used in fermentation by *C. acetobutylicum* DSM 1731 and 1.5 g/L total ABE was produced [29].

In fermentation of rice straw hydrolysate, at 48 h of fermentation, glucose as the primary and preferred carbon source was totally consumed. It is worth noticing

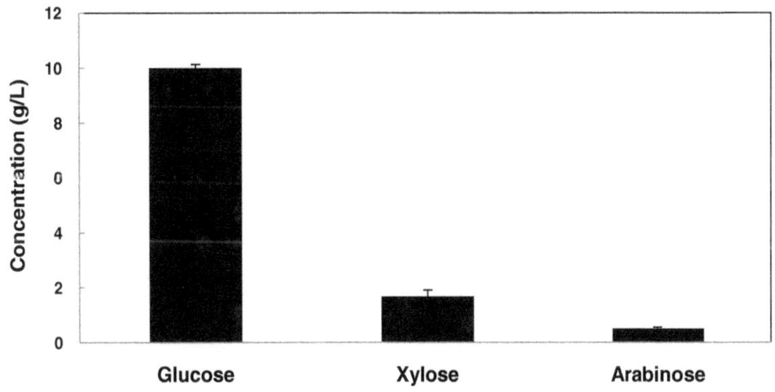

Figure 4 Composition of rice straw hydrolysate. Values are means of 3 replicates ± SD.

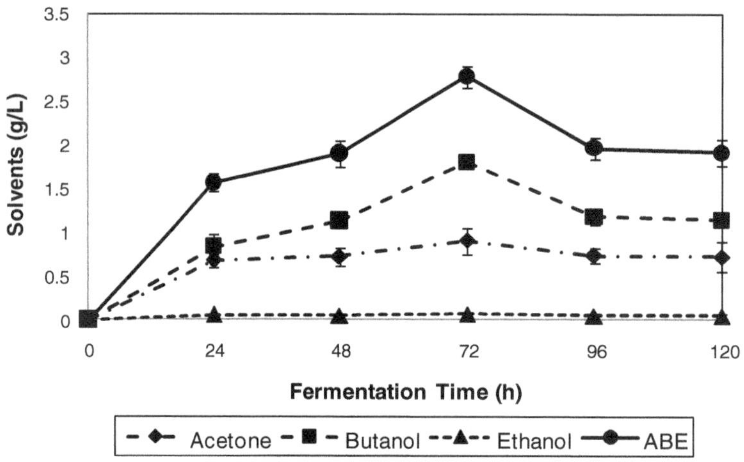

Figure 5 Production of ABE from glucose by *C. acetobutylicum* ATCC 824. Values are means of 3 replicates ± SD. Symbols represent: Acetone, Butanol, Ethanol, ABE.

that Clostridia are able to consume pentoses as well as hexoses [30]. However, arabinose was not consumed by the strain, whereas xylose was consumed. Hence, fermentable sugars produced via enzymatic hydrolysis of rice straw showed to be potential sugars as substrate for ABE production and can be applied for production of biofuels such as biobutanol.

The results of kinetic studies are indicated in Table 8. As can be seen from the table, kinetic studies indicated a cell yield and biomass productivity of 0.19 g cell/g glucose, and 0.04 g/L/h, respectively. Fermentation of rice straw hydrolysate resulted in an ABE yield and productivity of 0.27 g/g glucose, and 0.04 g/L/h, respectively. A butanol yield of 0.16 g/g glucose was achieved.

Conclusions

The present study confirmed that the use of NaOH as an alkali reagent for pretreatment of rice straw could be effective. The subsequent enzymatic hydrolysis of alkali-pretreated rice straw and the use of rice straw hydrolysate as the substrate for biobutanol production was shown to be successful. The use of 2% (w/v) NaOH was shown to be the most effective concentration of NaOH for pretreatment of rice straw. Enzymatic hydrolysis of rice straw pretreated with 2% (w/v) NaOH resulted in a reducing sugar yield of 0.6 g/g substrate and 60.75% saccharification was obtained. Reducing sugar production increased with the increase in the concentration of the enzyme and a similar trend was

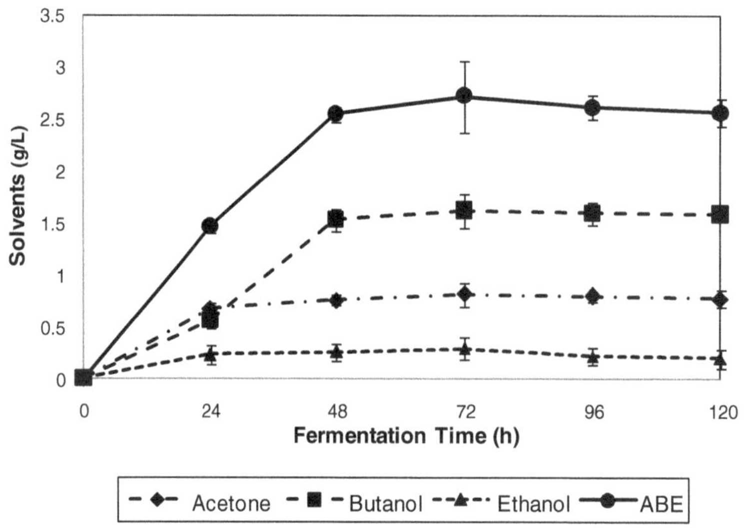

Figure 6 Production of ABE from rice straw hydrolysate by *C. acetobutylicum* ATCC 824. Values are means of 3 replicates ± SD. Symbols represent: Acetone, Butanol, Ethanol, ABE.

Table 8 Performance of ABE fermentation by *C. acetobutylicum* ATCC 824 on rice straw hydrolysate

Parameter	ABE fermentation by *C.acetobutylicum* ATCC 824
Initial glucose concentration (g/L)	10
Final glucose concentration (g/L)	0
Maximum acetone concentration(g/L)	0.82 ± 0.10
Maximum butanol concentration(g/L)	1.62 ± 0.16
Maximum ethanol concentration(g/L)	0.29 ± 0.12
Total solvents concentration(g/L)	2.73 ± 0.34
Fermentation time(h)	72
Solvents yield(g ABE/g glucose)	0.27
Butanol yield (g butanol/g glucose)	0.16
Yield of cells (g cells/g glucose)	0.19
Solvents productivity (g/L/h)	0.04
Productivity of biomass (g/L/h)	0.04

observed between fermentable sugar production and hydrolysis yield (%).With the rise in substrate concentration, fermentable sugar production increased as well. However, an opposite trend was observed between substrate concentration and hydrolysis yield (%) due to stirring difficulties, poor mixing, and end product inhibition. The use of 5% (w/v) substrate resulted in the highest concentration of reducing sugars. Rice straw hydrolysate was successfully utilized as substrate for biobutanol production resulting in an ABE yield and productivity of 0.27 g/g glucose and 0.04 g/L/h, respectively. A butanol yield of 0.16 g/g glucose was achieved. The study also confirmed that, as an enzyme complex, in order for cellulase to be efficient, the presence of all the components at substantial concentration is highly required and that the presence of β-glucosidase at a high concentration would not help in case FPase is not present at high enough concentration to initiate the cellulolytic reaction. In fact, in such a condition the addition of β-glucosidase is not effective since this enzyme is only accumulated due to the lack of cellobiose, the substrate for this enzyme.

Methods

Preparation and pretreatment of the substrate

Rice straw was collected from a paddy field in Sekinchan, Selangor, Malaysia. By using an electric grinder (Model CW-1, Hsiang Tai, Taiwan, 220 V, 50 Hz, 10 A), rice straw was ground to 2 mm length in size, kept in a cold room at 4°C prior to use. Pretreatment method was described in detail in our previous paper [13]. Rice straw pretreatment was performed by using different concentrations of NaOH [1%, 2%, 3%, and 4% (w/v)]. Effect of NaOH at various concentrations on substrate pretreatment and the subsequent sugar production was investigated.

Microorganism and inoculum preparation

A local isolate of *T. harzianum* SNRS3 was used in this study. The fungus was isolated from rice straw obtained from a rice field in Sekinchan, Selangor, Malaysia (unpublished data). Inoculum preparation was described in our previous study [13].

Solid state fermentation of rice straw by *T. harzianum* SNRS3 for cellulase production

Solid state cultivation was conducted and the crude enzyme mixture was extracted as described in the previous research [13]. The crude enzyme was kept at 4°C prior to use for enzymatic hydrolysis.

Enzymatic hydrolysis

Effect of pretreatment of rice straw using NaOH at various concentrations on reducing sugar production

To investigate the most effective concentration of NaOH for pretreatment of rice straw as the substrate for saccharification, NaOH at different concentrations [1%, 2%, 3%, and 4% (w/v)] was used. Untreated substrate was used as control. Pretreated substrate was placed in 50 mL Erlenmeyer flasks followed by being autoclaved at 121°C, for 15 min. Crude cellulase enzyme was added to each flask at the concentration of FPase 6.25 U/g substrate, CMCase 111.31 U/g substrate and β-glucosidase 173.71 U/g substrate. Enzymatic hydrolysis of alkali-pretreated rice straw was carried out in a reaction mixture containing [5% (w/v)] substrate in 20 mL 50 mM sodium citrate buffer (pH 5). The reaction mixture was incubated in a shaker incubator at 50°C and rotated at 150 rpm for 96 h. Samples (1 mL) were withdrawn at 0, 24, 48, 72 and 96 h. Samples were centrifuged for 10 min at 10,000 rpm. The supernatant was used for reducing sugar analysis and kept at -20°C prior to analysis. Concentration of NaOH yielding the highest concentration of reducing sugars was applied for pretreatment of rice straw in subsequent study.

Effect of enzyme loadings on reducing sugar production

Effect of enzyme loadings on enzymatic hydrolysis was investigated using rice straw pretreated with 2% (w/v) NaOH as the substrate. In order to study the effect of enzyme concentration on sugar production, preparation of the flasks and the substrate was done as described above. [5% (w/v)] substrate pretreated with 2% (w/v) NaOH was placed in each flask autoclaved at 121°C, for 15 min. It was then followed by the addition of various concentrations of the enzyme, each in a separate flask. Enzymatic saccharification was performed under the same conditions as described above and the samples were taken at regular interval. Samples were centrifuged and kept at -20°C prior to use for sugar analysis. Total activity of the enzyme was calculated based on the following equation:

$$\text{Total activity of the enzyme}(U/g) = \text{Enzyme activity } (U/mL) x \text{ Volume of the enzyme}(mL) x\ 10^{a} \quad (1)$$

[a]coefficient of the conversion of unit of enzyme activity (U/ mL) to (U/g substrate).

Enzyme activity was calculated based on the following Equations:

$$FPase\ (U/mL) = \frac{final\ abs - c}{m}\ x\ \frac{df}{sample\ volume}\ x\ \frac{1}{time}\ x\ \frac{1000\ \mu g}{1\ mg}\ x\ \frac{1\ \mu mole}{180.16\ \mu g} \quad (2)$$

$$CMCase\ (U/mL) = \frac{final\ abs - c}{m}\ x\ \frac{df}{sample\ volume}\ x\ \frac{1}{time}\ x\ \frac{1000\ \mu g}{1\ mg}\ x\ \frac{1\ \mu mole}{180.16\ \mu g} \quad (3)$$

$$\beta\text{-}glucosidase\ (U/mL) = \frac{final\ abs - c}{m}\ x\ \frac{df}{sample\ volume}\ x\ \frac{1}{time}\ x\ \frac{1000\ \mu g}{1\ mg}\ x\ \frac{1\ \mu mole}{139.1\ \mu g} \quad (4)$$

Where m is the slope from the standard curve, abs is the absorbance, c is the intercept, and df is the dilution factor. Time is the reaction time and is expressed in minute. To convert U/mL to U/g, the value for the activity of the enzyme (U/mL) was divided by the amount of rice straw in each flask (3 g) times volume of sodium citrate buffer used to extract the enzyme (30/mL).

Effect of substrate concentration on reducing sugar production

To investigate the effect of substrate concentration on enzymatic hydrolysis [1%, 3%, 5%, and 7% (w/v)] substrate, pretreated with 2% (w/v) NaOH was placed in different flasks. Crude cellulase concentration yielding the highest reducing sugar concentration from the previous study determined the enzyme concentration to be added to the substrate in each flask. Therefore, crude cellulase at the concentration of FPase 93.75 U/g substrate, CMCase 1669.65 U/g substrate and β-glucosidase 2605.65 U/g substrate was added to the substrate in flasks, each containing a different concentration of the substrate. Sampling was done at regular interval and the samples were centrifuged at 10,000 rpm for 10 min followed by being kept at -20°C. All the experiments were done in triplicates.

FPase as the rate-limiting enzyme

To study the significant role of FPase in cellulose enzymatic hydrolysis, FPase activity was fixed both for celluclast (5 FPU) and the crude cellulase from *T. harzianum* SNRS3. FPase activity was fixed at 9.90 U/mL and 9.86 U/mL for celluclast and the crude cellulase, respectively. However, β-glucosidase activity for celluclast and the crude cellulase enzyme was significantly different at 5.55 U/mL and 265.2 U/mL, respectively. Enzymatic hydrolysis of 5% (w/v) rice straw was carried out in 20 mL of 50 mM sodium citrate buffer solution, using a shaker incubator at 50°C, 150 rpm for 96 h. The initial pH was adjusted to 5.0 which is the optimum pH for celluclast [31]. Alternatively, in another experiment, β-glucosidase activity was fixed both for celluclast (20 FPU) and the crude cellulase enzyme. β-glucosidase activity was fixed at 22.20 U/mL and 22.10 U/mL for celluclast and the crude cellulase enzyme, respectively. However, the concentration of FPase in celluclast and the crude cellulase enzyme was significantly different at 39.6 U/mL and 0.83 U/mL, respectively. The enzymatic hydrolysis was performed under the afore-mentioned conditions. Samples were kept at -20°C prior to the analysis of reducing sugar concentration.

Substrate preparation for Acetone-Butanol-Ethanol (ABE) fermentation

As substrate for ABE fermentation, rice straw pretreated with 2% (w/v) NaOH was used. Following pretreatment, the pretreated substrate was subjected to enzymatic hydrolysis as described earlier and the enzyme and substrate concentration were used according to the results of the previous experiments. Rice straw hydrolysate was used as the substrate in ABE fermentation for the growth of the bacteria and subsequent biobutanol production.

Inoculum maintenance and preparation for ABE fermentation

C. acetobutylicum ATCC 824 from American Type Culture Collection was used as the ABE producing microorganism. The stock culture was grown on Reinforced Clostridial Medium (RCM) agar under anaerobic conditions using anaerobe container system Gaspak^{TM} EZ at 37°C. The single colony was grown on RCM broth at 37°C for 48 h. The medium was sparged with N_2 gas to provide anaerobic conditions for the growth of the bacteria. Bacterial cells were maintained as spore suspension on 60% (v/v) sterile glycerol at the ratio 2: 1 (culture: glycerol) and stored at -40°C prior to use.

Chemical composition and RCM preparation method is as follows:

RCM is composed of Meat extract 10 g/L, Peptone 5 g/L, Yeast extract 3 g/L, D-(+)-glucose 5 g/L, Starch 1 g/L, NaCl 5 g/L, Sodium acetate 3 g/L, L-Cysteine hydrochloride 0.5 g/L, Agar 0.5 g/L. Medium pH was adjusted to pH 6.0 using 1 M sodium hydroxide (NaOH) before being transferred into serum bottles. 100 mL of the prepared media was transferred into 125 mL serum bottles and sparged with nitrogen gas for 15 min and closed tightly using rubber septa before it was autoclaved at 121°C for 15 min. The media was prepared a day before inoculum preparation and stored at 4°C prior to use for inoculum preparation. Inoculum was heat shocked at 80°C for 2 min prior to being inoculated into freshly prepared RCM. Inoculum was then kept at 37°C for 48 h before being transferred into fermentation production medium.

ABE production medium

Tryptone-Yeast extract-Acetate (TYA) medium was used as ABE production medium. Chemical composition and preparation method is as follows:

Yeast extract 2 g/L, Tryptone 6 g/L, CH_3COONH_4 3 g/L, $FeSO_4.7\ H_2O$ 10 mg/L, KH_2PO_4 0.5 g/L, $MgSO_4$ 0.3 g/L, rice straw hydrolysate (10 g/L glucose, 1.66 g/L xylose, 0.5 g/L arabinose). For preparation of TYA medium, all the chemical components were dissolved in rice straw hydrolysate (carbon source) and transferred into 125 mL serum bottles. Medium pH was adjusted to 5.5. The medium was then sparged with nitrogen gas for 15 min for anaerobiosis before being autoclaved at 115°C for 10 min. 5% of freshly prepared inoculum was inoculated into fermentation production medium. Fermentation was carried out at 37°C, 120 rpm for 120 h. Sampling was done at 0, 24, 48, 72, 96, and 120 h. Samples were kept at -20°C prior to being analyzed for solvents and optical density.

Analytical procedure

Cellulose, hemicelluloses, lignin, and ash content of rice straw were measured using a standard method [32]. Reducing sugar analysis was performed by using Dinitrosalicylic acid (DNS) reagent and according to a standard method [33]. Enzyme activity was assayed according to the standard method [34]. (%) Saccharification was calculated according to the following Equation: (%) Saccharification = reducing sugars × 0.9 × 100)/ carbohydrates in substrate [1]. Bacterial cell concentration was determined based on correlation between dry cell weight (DCW) and optical density (OD) using spectrophotometer at 680 nm. Individual sugars concentration was measured by using High Performance Liquid Chromatography (HPLC). Solvents were analyzed by using Gas Chromatography (GC). Yield was defined as grams of ABE produced per gram of glucose utilized. ABE productivity was calculated as total ABE produced in g/L divided by the fermentation time and is expressed as g/L/h.

Yield of cells was defined as grams of cells produced per grams of glucose utilized. Productivity of biomass was calculated as maximum cell concentration in g/L divided by the fermentation time and is expressed as g/L/h.

Statistical analysis

The data was analyzed by one-way analysis of variance (ANOVA). t Tests (LSD) was used to compare the difference of means among treatment groups. Differences of $p < 0.05$ were considered significant.

Competing interests

The authors declare that they have no competing interests.

Authors' contributions

NR carried out all the experimental work which included enzymatic hydrolysis of rice straw as well as the subsequent use of rice straw hydrolysate for biobutanol production and also drafted the manuscript. All authors contributed to the design and interpretation of the experimental results, as well as editing and revising the manuscript. All authors have read and approved the final manuscript.

Acknowledgements

The authors would like to extend their gratitude for the financial support generously provided by Ministry of Science, Technology and Innovation (MOSTI), Malaysia, grant NO. 02-01-04-SF1036.

Author details

[1]Department of Bioprocess Technology, Faculty of Biotechnology and Biomolecular Sciences, Universiti Putra Malaysia, 43400 UPM Serdang, Selangor, Malaysia. [2]Institute of Bioscience, Universiti Putra Malaysia, 43400 UPM Serdang, Selangor, Malaysia. [3]Institute of Tropical Forestry and Forest Products, Universiti Putra Malaysia, 43400 UPM Serdang, Selangor, Malaysia. [4]Bioprocessing and Biomanufacturing Research Center, Faculty of Biotechnology and Biomolecular Sciences, Universiti Putra Malaysia, 43400 UPM, Serdang, Selangor, Malaysia.

References

1. Jeya M, Zhang YW, Kim IW, Lee JK: **Enhanced saccharification of alkali-treated rice straw by cellulase from *Trametes hirsuta* and statistical optimization of hydrolysis conditions by RSM.** *Bioresour Technol* 2009, **100:**5155–5161.
2. Hsu TC, Guo GL, Chen WH, Hwang WS: **Effect of acid pretreatment of rice straw on structural properties and enzymatic hydrolysis.** *Bioresour Technol* 2010, **101:**4907–4913.
3. Hamzah F, Idris A, Shuan TK: **Preliminary study on enzymatic hydrolysis of treated oil palm (Elaeis) empty fruit bunches fibre (EFB) by using combination of cellulase and β- 1-4 glucosidase.** *Biomass Bioenergy* 2011, **35:**1055–1059.
4. Bak JS, Ko JK, Han YH, Lee BC, Choi IG, Kim KH: **Improved enzymatic hydrolysis yield of rice straw using electron beam irradiation pretreatment.** *Bioresour Technol* 2009, **100:**1285–1290.
5. Chang KL, Thitikorn-amorn J, Hsieh JF, Ou BM, Chen SH, Ratanakhanokchai K, Huang PJ, Chen ST: **Enhanced enzymatic conversion with freeze pretreatment of rice straw.** *Biomass Bioenergy* 2011, **35:**90–95.
6. Amiri H, Karimi K, Zilouei H: **Organosolv pretreatment of rice straw for efficient acetone, butanol, and ethanol production.** *Bioresour Technol* 2014, **152:**450–456.
7. Chen WH, Chen YC, Lin JG: **Evaluation of biobutanol production from non-pretreated rice straw hydrolysate under non-sterile environmental conditions.** *Bioresour Technol* 2013, **135:**262–8.
8. Gottumukkala LD, Parameswaran B, Valappil SK, Mathiyazhakan K, Pandey A, Sukumaran RK: **Biobutanol production from rice straw by a non-acetone producing *Clostridium sporogenes* BE01.** *Bioresour Technol* 2013, **145:**182–7.

9. Chen SW, Ma X, Wang LS, Zhao XH: **Acetone-butanol fermentation of rice straw enzymatic hydrolysate.** *Indust Microbiol* 1998, **28:**30–34.
10. Kumar P, Barrett DM, Delwiche MJ, Stroeve P: **Methods for pretreatment of lignocellulosic biomass for efficient hydrolysis and biofuels production.** *Ind Eng Chem Res* 2009, **46:**3713–3729.
11. Zhang Q, Cai W: **Enzymatic hydrolysis of alkali-pretreated rice straw by *Trichoderma reesei* ZM4-F3.** *Biomass Bioenergy* 2008, **32:**1130–1135.
12. Akhtar MS, Saleem M, Akhtar MW: **Saccharification of lignocellulosic materials by the cellulases of *Bacillus subtilis*.** *Int J Agr Biol* 2001, **3:**199–202.
13. Rahnama N, Mamat S, Shah UKM, Ling FH, Rahman NAA, Ariff AB: **Effect of alkali pretreatment of rice straw on cellulase and xylanase production by local *Trichoderma harzianum* SNRS3 under solid state fermentation.** *BioResources* 2013, **8:**2881–2896.
14. Chen M, Zhao J, Xia L: **Enzymatic hydrolysis of maize straw polysaccharides for the production of reducing sugars.** *Carbohydr Polymers* 2008, **71**(3):411–415.
15. Ouyang J, Li Z, Li X, Ying H, Yong Q: **Enhanced enzymatic conversion and glucose production via two-step enzymatic hydrolysis of corncob residue from xylo-oligosaccharides producer's waste.** *BioResources* 2009, **4:**1586–1599.
16. Sharma SK, Kalra KL, Grewal HS: **Enzymatic saccharification of pretreated sunflower stalks.** *Biomass Bioenergy* 2002, **23:**237–243.
17. Harnpicharnchai P, Champreda V, Sornlake W, Lily Eurwilaichitr L: **A thermotolerant β-glucosidase isolated from an endophytic fungi, *Periconia* sp., with a possible use for biomass conversion to sugars.** *Protein Expr Purif* 2009, **67:**61–69.
18. Vlasenko EY, Ding H, Labavitch JM, Shoemaker SP: **Enzymatic hydrolysis of pretreated rice straw.** *Bioresour Technol* 1997, **59:**109–119.
19. Saha BC, Cotta MA: **Lime pretreatment, enzymatic saccharification and fermentation of rice hulls to ethanol.** *Biomass Bioenergy* 2008, **32:**971–977.
20. Adsul MG, Ghule JE, Shaikh H, Singh R, Bastawde KB, Gokhale DV, Varma AJ: **Enzymatic hydrolysis of delignified bagasse polysaccharides.** *Carbohydr Polymers* 2005, **62:**6–10.
21. Henrissat B: **Cellulases and their interaction with cellulose.** *Cellulose* 1994, **1994**(1):169–196.
22. Knowles J, Lehtovaara, Teeri T: **Cellulase families and their genes.** *Trends Biotechnol* 1987, **5:**255–261.
23. Lynd LR, Weimer PJ, Van Zyl WH, Pretorius IS: **Microbial cellulose utilization: fundamentals and biotechnology.** *Microbiol Mol Biol Rev* 2002, **66:**506–577.
24. Teeri TT: **Crystalline cellulose degradation: new insight into the function of cellobiohydrolases.** *Trends Biotechnol* 1997, **15:**160–167.
25. Wood TM, Garcia-Campayo V: **Enzymology of cellulose degradation.** *Biodegradation* 1990, **1:**147–161.
26. Zhang YHP, Lynd LR: **Toward an aggregated understanding of enzymatic hydrolysis of cellulose: noncomplexed cellulase systems.** *Biotechnol Bioeng* 2004, **88:**797–824.
27. Zhang YHP, Himmel ME, Mielenz JR: **Outlook for cellulase improvement: Screening and selection strategies.** *Biotech Adv* 2006, **24:**452–481.
28. Monot F, Martin JR, Petitdemange H, Gay R: **Acetone and butanol production by *Clostridium acetobutylicum* in a synthetic medium.** *Appl Environ Microbiol* 1982, **44:**1318–1324.
29. Classen PAM, Budde MAW, Lopez-Contreras AM: **Acetone, Butanol and Ethanol production from domestic organic waste by solventogenic Clostridia.** *J Mol Microbiol Biotechnol* 2000, **2:**39–44.
30. Qureshi N, Saha BC, Hector RE, Hughes SR, Cotta MA: **Butanol production from wheat straw by simultaneous saccharification and fermentation using *Clostridium beijerinckii*: Part I- Batch fermentation.** *Biomass Bioenergy* 2008, **32:**168–175.
31. Macarron R, Acebal C, Castillon MP, Dominguez JM, Mata I, Pettersson G, Tomme P, Claeyssens M: **Mode of action of endoglucanase III from *Trichoderma reesei*.** *Biochem J* 1993, **289:**867–873.
32. Goering HK, Van Soest PJ: **Forage fiber analysis (apparatus, reagents, procedures and some applications).** In *Agricultural Research Service: Agricultural Handbook NO. 379.* Washington DC: United States Department of Agriculture; 1970.
33. Miller GL: **Use of dinitrosalicylic acid for determination of reducing sugar.** *Anal Chem* 1959, **31:**426–428.
34. Wood TM, Bhat KM: **Measuring cellulase activities.** *Meth Enzymol* 1988, **160:**87–112.

A rapid culture independent methodology to quantitatively detect and identify common human bacterial pathogens associated with contaminated high purity water

Elizabeth Minogue[1], Nina L Tuite[1], Cindy J Smith[2], Kate Reddington[1] and Thomas Barry[1*]

Abstract

Background: Water and High Purity Water (HPW) distribution systems can be contaminated with human pathogenic microorganisms. This biocontamination may pose a risk to human health as HPW is commonly used in the industrial, pharmaceutical and clinical sectors. Currently, routine microbiological testing of HPW is performed using slow and labour intensive traditional microbiological based techniques. There is a need to develop a rapid culture independent methodology to quantitatively detect and identify biocontamination associated with HPW.

Results: A novel internally controlled 5-plex real-time PCR Nucleic Acid Diagnostics assay (NAD), was designed and optimised in accordance with Minimum Information for Publication of Quantitative Real-Time PCR Experiments guidelines, to rapidly detect, identify and quantify the human pathogenic bacteria *Stenotrophomonas maltophilia*, *Burkholderia* species, *Pseudomonas aeruginosa* and *Serratia marcescens* which are commonly associated with the biocontamination of water and water distribution systems. The specificity of the 5-plex assay was tested against genomic DNA isolated from a panel of 95 microorganisms with no cross reactivity observed. The analytical sensitivities of the *S. maltophilia*, *B. cepacia, P. aeruginosa* and the *S. marcescens* assays are 8.5, 5.7, 3.2 and 7.4 genome equivalents respectively.
Subsequently, an analysis of HPW supplied by a Millipore Elix 35 water purification unit performed using standard microbiological methods revealed high levels of naturally occurring microbiological contamination. Five litre water samples from this HPW delivery system were also filtered and genomic DNA was purified directly from these filters. These DNA samples were then tested using the developed multiplex real-time PCR NAD assay and despite the high background microbiological contamination observed, both *S. maltophilia* and *Burkholderia* species were quantitatively detected and identified. At both sampling points the levels of both *S. maltophilia* and *Burkholderia* species present was above the threshold of 10 cfu/100 ml recommended by both EU and US guidelines.

Conclusions: The novel culture independent methodology described in this study allows for rapid (<5 h), quantitative detection and identification of these four human pathogens from biocontaminated water and HPW distribution systems. We propose that the described NAD assay and associated methodology could be applied to routine testing of water and HPW distribution systems to assure microbiological safety and high water quality standards.

Keywords: Human microbial pathogens, Multiplex real-time PCR NAD assays, Biocontaminated HPW distribution systems, Biological water quality assurance

* Correspondence: thomas.barry@nuigalway.ie
[1]Nucleic Acid Diagnostics Research Laboratory (NADRL), Microbiology, School of Natural Sciences, National University of Ireland Galway, Galway, Ireland
Full list of author information is available at the end of the article

Background

Maintaining high water quality standards is crucial, as water is essential to human health, development and well-being. The efficient operation and biological control of water distribution systems is important as High Purity Water (HPW), municipal and recreational water contaminated with waterborne pathogens such as non-fastidious Gram-negative bacteria non *Enterobacteriaceae* (GNB-NE) can pose a serious risk to human health. Water, and in particular HPW, is a vital component of both the clinical and many industrial manufacturing sectors such as the pharmaceutical, food and beverage and semiconductor industries [1-3].

In healthcare facilities, such as neonatal units, intensive care units, dental practices and clinical diagnostics testing laboratories, water and HPW is routinely used for hand washing, bathing, cleaning semi-critical areas washing and rinsing patient medical devices and in the preparation and processing of medicines and other health products [4,5]. A recent traditional culture based microbiological study of water and water distribution systems in a number of hospital departments revealed high levels of contamination by GNB-NE [6]. In such settings, this microbial contamination with opportunistic pathogens could result in the transmission of Healthcare Associated Infections (HCAIs) in immunocompromised patient groups [7-10]. There have been many reported cases of HCAIs associated with microbial contamination of water and HPW delivery systems and exposure to these water-borne pathogens in hospitals can result in increased patient morbidity and mortality [11-13]. It has also been observed that the use of biocontaminated water and HPW for clinical culture based microbiological diagnostics can result in the outbreak of pseudoepidemics [14,15]. Such clinical derived pseudoepidemics can result in costly additional and unnecessary diagnostic procedures and therapeutic interventions for patients [16-19].

In industrial settings, microbial contamination of water and HPW distribution systems can result in the biodeterioration of manufacturing equipment and as previous studies have demonstrated contamination with microorganisms can severely compromise the quality of industrial raw materials and finished products [20-22].

Previous studies have demonstrated that the microbial load present in water distribution systems can be significantly underestimated using the traditional plate count microbial culture method as a result of the presence of physiologically active bacteria that are unable to form colonies on culture media [23,24]. These viable but non-culturable bacteria may therefore be undetectable by standard microbial cultivation [25]. It has also been observed that variables such as the sampling technique and culture medium used can be crucial in obtaining accurate and reproducible results when testing these water samples [26]. Such studies demonstrate the need for and importance of real-time monitoring for the onset of biocontamination of HPW purification units and HPW delivery systems by rapid microbial culture independent methods such that prompt remedial action may be undertaken.

Therefore, there is a need to develop rapid, sensitive and high throughput methodologies for the quantitative detection and identification of biocontamination associated with HPW to ensure appropriate standards and quality can be achieved. Nucleic Acid Diagnostics (NAD) technologies offer the potential to analyse in real-time the onset and progress of microbial biocontamination associated with HPW distribution systems. Previously, we have reported the design and development of an internally controlled multiplex real-time PCR NAD assay for both *P. aeruginosa* and *Burkholderia* species for use in the quantitative detection and identification of these GNB-NE in biocontaminated HPW under laboratory conditions [27]. For the present study, additional novel NAD targets for two more GNB-NE, *S. maltophilia and S. marcescens* have been incorporated resulting in a 5-plex real-time PCR NAD assay, inclusive of an Internal Amplification Control (IAC).

This study which was designed in accordance with Minimum Information for Publication of Quantitative Real-Time PCR Experiments (MIQE) guidelines [28] describes a methodology combining water filtration, DNA purification and NAD technologies to detect, identify and absolutely quantify these common human pathogenic contaminates of HPW. Furthermore, we have also demonstrated the developed methodology by testing a HPW delivery system supplied by a Millipore Elix 35 water purification unit distributing HPW based in a public building facility. The Millipore Elix 35 system uses both reverse osmosis and electrodeionization for production of HPW. Using the developed methodology, we determined that the HPW delivery system was contaminated with the human pathogens, *S. maltophilia* and *Burkholderia* species.

Methods

Bacterial strains, microbial culture media and growth conditions

Ten *Stenotrophomonas* species/strains, 10 *Burkholderia* species, 26 *Pseudomonas* species/strains and 10 *Serratia* species/strains were used in this study (Additional file 1: Table S1). All target species/strains were purchased from either the German Collection of Microorganisms and Cell Cultures (Deutsche Sammlung von Mikroorganismen und Zellkulturen GmbH, DSMZ. Braunschweig, Germany) or from BCCM Belgian Co-ordinated Collections of Microorganisms bacterial collection (LMG, Gent, Belgium) (Additional file 1: Table S1).

All *Stenotrophomonas*, *Burkholderia*, *Pseudomonas* and *Serratia* species were grown in tryptone soya broth (Oxoid, Cambridge, UK) or R2A broth (Lab M limited, Lancashire, United Kingdom) at either 30°C or 37°C for approximately 12 - 24 h or until sufficient growth was observed. For all other microorganisms tested in the study, total genomic DNA was provided from microbial stocks held within this laboratory (Additional file 1: Table S1).

DNA isolation and quantification

Total genomic DNA was isolated from overnight broth culture using the Quick-gDNA MiniPrep kit (Zymo Research, California, USA). Briefly, 1.5 ml of culture was centrifuged in a bench-top centrifuge at 18,000 g for two min. The supernatant was discarded and the pellet was resuspended in 400 μl Genomic Lysis Buffer by vortexing briefly, then incubated at room temperature for 5-10 min. Subsequently, steps 2-5 of the purification of total DNA from cell suspensions and proteinase K digested samples procedure were followed according to the manufacturer's guidelines (http://www.zymoresearch.com/downloads/dl/file/id/18/d3006i.pdf).

DNA integrity was assessed on a 1% w/v agarose gel. Concentrations for all total genomic DNA used in this study were determined using the Quant-iT™ dsDNA HS Assay Kit and the Qubit fluorometer (Invitrogen Corporation, California, USA). All DNA samples were stored at -20°C prior to use.

Diagnostics target identification and development

The diagnostics target genes used in this study were identified by *in silico* evaluation of a number of genes currently under investigation and evaluation in this laboratory for use in nucleic acid based diagnostics test systems. These include the *ssrA*, *lepA*, *rnpB*, 16S rRNA, and genes and the 16 – 23 S rRNA ITS. These gene targets have all previously been reported as useful species specific diagnostics targets for use in nucleic acid based diagnostics [29-35]. Nucleotide sequence information was retrieved from the National Center for Biotechnology Information (NCBI) website (http://www.ncbi.nlm.nih.gov), the Functional Gene Pipeline and Repository website (http://fungene.cme.msu.edu), the RNaseP database (http://www.mbio.ncsu.edu/rnasep/) and the tmRNA databse (http://www.ag.auburn.edu/mirror/tmRDB/). Alignments of these nucleotide sequences were carried out using the clustalW2 multiple sequence alignment programme (http://www.ebi.ac.uk/Tools/clustalw2/index.html).

Conventional and real-time PCR primers and Taqman hydrolysis probe design

All oligonucleotide primers and hydrolysis probes were designed manually according to recommended general guidelines [28,36]. Nucleotide sequencing primers were designed to have a melting temperature (Tm) of between 60-62°C. The oligonucleotide primers used for real-time PCR were designed to have a Tm of 58-61°C, with all probes designed to have a Tm 6-8°C higher (Table 1). To determine the specificity of each assay designed *in silico* all primers and probes were BLAST analysed. Subsequently, the performance characteristics of each primer set was analysed in the laboratory along with the specific probe. The most robust assay combination, as reflected by PCR efficiency, which had an LoD of < 10 GE was used in the final assay. Oligonucleotide primers and probes were supplied by *Eurofins* MWG Operon (Ebensburg, Germany), TIB MOLBIOL (Berlin, Germany) or Integrated DNA Technology (Leuven, Belgium).

Conventional PCR for sequencing

Conventional PCR was performed on the iCycler iQ thermal cycler (Bio-Rad Laboratories Inc., California, USA). PCR reactions were carried out in 0.2 ml PCR tubes, containing forward and reverse primer (0.2 μM final conc.), 2.5 μl 10X buffer (15 mM $MgCl_2$), 0.5 μl dNTP mix (10 mM: deoxynucleoside triphosphate set, 0.5 U Taq DNA polymerase Roche Diagnostics, Basel, Switzerland), 1 μl of template DNA and adjusted to a final volume of 25 μl with the addition of nuclease-free water (Applied Biosystems, California, USA). The thermal cycling parameters consisted of an initial denaturation step at 95°C for 4 min, followed by 35 cycles of denaturation at 95°C (30 s), amplification at 50°C (30 s), and extension at 72°C (30 s), followed by a final elongation at 72°C for 7 min. Sequencing primers are outlined in Table 1.

Nucleic acid diagnostics target sequencing

Nucleotide sequence data used for real-time PCR NAD assay design and development was obtained either from publicly available databases or were generated in this study. Sequencing primers (Table 1) were designed to amplify the *ssrA* gene of *Stenotrophomonas*, and *Serratia* species. Amplification of the *ssrA* gene of *P. aeruginosa* and the *rnpB* gene of *Burkholderia* species was performed as previously described [27]. Conventional PCR products were generated as described above, followed by purification using the High Pure PCR product purification kit (Roche Diagnostics). The purified PCR products were sequenced externally (Sequiserve, Vaterstetten, Germany).

Internal Amplification Control for multiplex real-time PCR NAD assay

The IAC used in this multiplex real-time diagnostics assay as previously described [27] was developed using a modified version of the composite primer approach

Table 1 Oligonucleotide primers and probes used in this study

Name	Function	Sequence (5'-3')	Target gene
CIAC F	Forward composite primer for IAC generation	TGAGGAAGACTTATTGGCTGATACCCAACTTGGAATG	Chimeric PCR product
CIAC R	Reverse composite primer for IAC generation	CGGTACATTGTGGTCTTTAAGTCTTCACCAGAATAAAATTG	Chimeric PCR product
Ser F	Sequencing forward primer	GTGGAGCTGGGGGGAT	*ssrA*
Ser R	Sequencing forward primer	GGGGCTGATTCTGGATTCG	*ssrA*
Psd F1	Sequencing forward primer	GGATTTGAACCCCCGTCC	*ssrA*
Psd R2	Sequencing reverse primer	AGGATTCGACGCCGGT	*ssrA*
BF1	Sequencing forward primer	GAGGAAAGTCCGGACTCC	*rnpB*
BF2	Sequencing forward primer	GGCAGGGTGATGGCTAA	*rnpB*
BR2	Sequencing reverse primer	GATAAGCCGGATTCTGTGC	*rnpB*
Sten F	Sequencing forward primer	GGGGMYTACGGWTTCGAC	*ssrA*
Sten R	Sequencing forward primer	GGGARTCGAACCRSGTCC	*ssrA*
IAC F	IAC forward primer	TGAGGAAGACTTATTGGCTG	Chimeric PCR product
IAC R	IAC reverse primer	CGGTACATTGTGGTCTTTAAG	Chimeric PCR product
IAC P	IAC specific hydrolysis probe	TYE665-TCCTTAGATGGTACCAGTGCCA- IBRQSp	Chimeric PCR product
S. mar F	*S. marcescens* forward primer	GCACTACATGCTTAGTCCAAT	*ssrA*
S. mar R	*S. marcescens* reverse primer	CAGCTTAATACCCTGCTTAGAG	*ssrA*
S. mar P	*S. marcescens* specific hydrolysis probe	ROX-CACGCTACTGACAGACTAGCCT-BHQ2	*ssrA*
BF1	*Burkholderia* forward primer	GAGGAAAGTCCGGACTCC	*rnpB*
BR1	*Burkholderia* reverse primer	TCTTACCGCACCGTTTCA	*rnpB*
Burk P	*Burkholderia* genus specific hydrolysis probe	FAM-ACACGCGGAACAGGGCAA-BHQ1	*rnpB*
PF2	*P. aeruginosa* forward primer	GACAGTCGTTTCGGGTTTAC	*ssrA*
PR1	*P. aeruginosa* reverse primer	CGACGACAACTACGCTCT	*ssrA*
P. aer P	*P. aeruginosa* specific hydrolysis probe	HEX-GCATCCCCTAGCGACTGCT-BHQ1	*ssrA*
S. malt F	*S. maltophilia* forward primer	TACATGCTTAGCTCACCGT	*ssrA*
S. malt R	*S. maltophilia* reverse primer	CGAAACTGCTTGTGTCCAT	*ssrA*
S. malt P	*S. maltophilia* specific hydrolysis probe	CYAN500-CTGTTGTGTTCTAGTGCCGGAC-BHQ1	*ssrA*

[37-39]. Briefly, the IAC amplifies and detects a chimeric PCR product containing nucleotide sequence of both *Candida albicans ALS1* gene and the *aqualysin 1* gene from *Thermus aquaticus.*

Real-time PCR

Real-time PCR NAD assays were initially tested in a monoplex format to determine the specificity and analytical sensitivity. Monoplex real-time PCR reactions were performed on the LightCycler 2.0 instrument using the LightCycler TaqMan Master kit (Roche Diagnostics). Each monoplex reaction contained 5X master mix (4 μl), forward and reverse primer (0.5 μM final conc.), probe (0.2 μM final conc.), 0.5 U uracil-DNA glycosylase (UDG,New England Biolabs), 5% v/v Dimethyl sulphoxide (DMSO), template DNA (2 μl) and the volume adjusted to 20 μl with the addition of nuclease-free water. The thermal cycling parameters consisted of 10 min at 37°C to incubate UDG followed by 10 min incubation at 95°C to deactivate UDG, activate the polymerase and denature the template DNA. Subsequently, 50 cycles of 95°C for 10 s and 60°C for 30 s were performed, with a final a cooling step at 40°C for 10 s. The temperature transition rate for all cycling steps was 20°C/s.

For multiplex real-time PCR the *S. maltophilia, Burkholderia* species and *P. aeruginosa* specific probes were labelled with Cyan 500, FAM, HEX and Black Hole Quencher 1 (BHQ1) respectively. The *S. marcescens* probe was labelled with ROX and Black Hole Quencher 2 (BHQ2). The IAC assay probe was labelled with TYE665 and Iowa Black® RQ-Sp (IBRQSp). The multiplex real-time PCR NAD assay was optimised to the required standards routinely used in this laboratory and the specificity and analytical sensitivity was determined.

Multiplex real-time PCR reactions were subsequently performed using the LightCycler 480 probes master kit (Roche Diagnostics) on the LightCycler 480 System I. Each multiplex reaction contained 2 X master mix (10 μl), forward and reverse primers (0.5 μM final conc.) CYAN500, FAM, HEX, ROX and TYE665 probes (0.2 µM final conc.), UDG (0.4 U), DMSO (5% v/v). Template DNA (5 µl) and IAC DNA (1×10^4 recombinant plasmid copies) were added to each reaction mix. Nuclease-free water was added to a final volume of 20 µl. A no template control reaction was included in each experiment. The cycling parameters used for multiplex reactions were in accordance with monoplex parameters (see previous paragraph). The temperature transition rate, referred to as the ramp rate on the LightCycler 480, was 4.4°C/s while heating and 2.2°C/s on cooling. As outlined by the manufacturer [40] a colour compensation file was generated prior to experimental analysis to avoid fluorescence leakage between channels.

Determining the specificity of the multiplex real-time diagnostic assay

To demonstrate the specificity of the designed multiplex real-time PCR NAD assay, it was tested against genomic DNA isolated from a panel of 95 microorganisms (Additional file 1: Table S1) at a concentration of 10^4 Genome Equivalents (GE) per reaction. Genome sizes were available from the NCBI website for all microorganisms of interest and were used to calculate the GE of each microorganism tested using previously published guidelines [41].

Analytical sensitivity of the multiplex real-time diagnostic assay determined using probit regression analysis

The analytical sensitivity of each assay in multiplex format was established by testing twelve replicates of eight bacterial genomic DNA concentrations, using probit regression analysis [42]. To determine the impact of genomic DNA from other microorganisms commonly associated with biocontaminated HPW distribution systems the LOD was also investigated in the presence of a background of pooled genomic DNA from 10 other bacteria commonly associated with water system contamination [24,43,44]. The final total concentration of the pooled bacterial genomic DNA was 10^4 GE per reaction. These bacteria were *Klebsiella pneumonia* DSM 9377, *Ralstonia sp.* DSM 13640, *Enterobacter aerogenes* NCTC 10006, *Proteus mirabilis* DSM 4479, *Staphylococcus aureus* DSM 21705, *Acinetobacter baumannii* DSM 30007, *Bacillus cereus* DSM 31 and 3 out of 4 of the following *S. maltophilia* DSM 50170, *B. cepacia* DSM 7288, *P. aeruginosa* DSM 19880 and *S. marcescens* DSM 30121 (genomic DNA from bacterial target subject to analytical sensitivity testing not included in background genomic DNA). The IAC, at a concentration of 10^4 plasmid equivalents per reaction, was also included in all samples during sensitivity testing.

Generation of external standard curves

Three overnight microbial cultures of either *S. maltophilia*, *Burkholderia* species (for standard curve generation *B. cepacia* DSM 7288 the type strain of the type species of the *Burkholderia* genus was used), *P. aeruginosa* or *S. marcescens* grown in R2A broth with an optical density at 600 nm (OD_{600}) of 0.1 were serially diluted tenfold (6 dilutions) in sterile phosphate buffered saline (PBS) (Oxoid). One hundred µl of each dilution were spread on R2A agar and incubated for 48 h at 30°C. These plates were used to count cfu. DNA was isolated from two hundred µl aliquots from each dilution (in triplicate) using the Quick-gDNA MiniPrep kit (Zymo Research) as per protocol outlined above. DNA was eluted in 50 µl sterile nuclease free water which was then used as template DNA in the multiplex real time PCR reaction. Standard curves were derived by plotting threshold cycle (C_T) values against overnight plate counts (CFU ($\log_{10}$)).

Sampling from HPW delivery system for microbiological analysis

An Elix 35 HPW purification unit (Merck Millipore, Massachusetts, USA) and HPW delivery system based in a public building facility was investigated for bacterial contamination. The last full service of the Elix 35 HPW purification unit was carried out approximately 36 months prior to first sampling. Sampling of HPW was performed at two point of use sources. Sampling point A was an outlet valve adjacent to the Elix 35 HPW purification unit and point B was a laboratory tap approximately 25 - 30 m from the Elix 35 HPW purification unit. During the course of this study the Elix 35 HPW purification unit had a new Progard TL1 System Cl2 Autoclean Pretreatment Pack and the Q-Gard TL Polishing Pack fitted. Samples of the HPW were subsequently taken from both points after this process was carried out. Before taking HPW samples, the outlets exteriors were cleaned with 70% ethanol, and HPW was subsequently allowed to flow for 2 min (~300 ml/min). Five litre samples were collected into sterile containers and filtered using 0.45 µm, 47 mm cellulose nitrate/acetate membrane filters (Millipore SAS, Molsheim Cedex, France) [45]. Each filter was placed in 1500 µl Genomic Lysis Buffer (Quick-gDNA MiniPrep kit, Zymo Research) and incubated on rotating platform at room temperature for 5 - 10 min followed by steps 2 - 5 of the total DNA isolation from cell suspension procedure according to the manufacturer's guidelines. The DNA was eluted in 50 µl sterile nuclease free water.

Microbiological analysis of HPW delivery system

(i) Conventional culture methods for the microbiological analysis of HPW delivery system.

Heterotrophic Plate Counts (HPCs) were performed by plating 100 µl of HPW onto R2A agar and incubating at 37°C for 48 h. Additionally, 100 µl samples of HPW were plated on four selective agars. These were *Pseudomonas* agar, L-arabinose ornithine irgasan medium [46],

B. cepacia medium, and *B. cepacia* specific agar (Fannin, Ireland). (Growth of *S. maltophilia* on *B. cepacia* medium was confirmed by culture of *S. maltophilia* strains (DSM 50170^{T} and DSM 50173) at 30°C for 24 h). Plates were incubated at 30°C for 48 h. Single colonies were isolated and these microbial cultures were used to inoculate the BBL Crystal Enteric/Nonfermenter ID system panels (Becton Dickinson, New Jersey, USA), which were incubated at 37°C for 20 h. Results were evaluated using BBL Crystal ID System Electronic Codebook.

(ii) Rapid NAD quantitative detection and identification of bacteria from a biocontaminated HPW delivery system.

Total genomic DNA isolated from filtered HPW samples was used as template DNA in the developed multiplex real-time PCR NAD assay to determine if the HPW samples were contaminated with *S. maltophilia*, *Burkholderia* species, *P. aeruginosa* or *S. marcescens*. In order to determine the degree of contamination of the HPW delivery system with these bacteria each positive result was absolutely quantified by comparison to one of the established external standard curves generated as described in section Generation of external standard curves.

Results

Diagnostics target identification and development

The NAD targets used in this study for the specific detection of the *Burkholderia* species and *P. aeruginosa* were previously described [27]. For the specific detection of the pathogens *S. maltophilia* and *S. marcescens in silico* analysis of a number of genes for use as a nucleic acid target was carried out (data not shown). The usefulness of these NAD target genes (*ssrA*, *lepA*, *rnpB*) has been previously demonstrated with the development of NAD assays to specifically detect microorganisms of interest [29-34,47]. The analysis of the *ssrA* gene revealed significant intragenic nucleotide sequence variation between closely related species within the nucleotide gene sequences of *S. maltophilia* and *S. marcescens* to allow for the design of species specific probes for both these microorganisms of interest. A BLAST analysis of each target probe did not show any cross reactivity between specific probes and any other non target species/strains (data not shown).

Specificity of the multiplex real-time diagnostic assay

The specificity of each real-time PCR NAD assay was confirmed both in monoplex and multiplex formats using the panel of microorganisms listed in Additional file 1: Table S1. The *S. maltophilia* and *S. marcescens* specific real-time PCR NAD assays each detected five positive control strains and did not detect any of the remaining 90 microorganisms tested in this study. The *P. aeruginosa* specific real-time PCR NAD assay detected the 14 *P. aeruginosa* strains in our collection. The remaining 81 microorganisms tested were not detected. Ten *Burkholderia* species were tested using the *Burkholderia* genus specific real-time PCR NAD assay. All 10 *Burkholderia* species were detected and none of the remaining 85 microorganisms tested were detected. The specificity of the IAC assay was also tested using the same panel of microorganisms and no cross reactivity was observed. For complete results of specificity testing see Table 2.

Analytical sensitivity testing of the multiplex real-time PCR NAD assay

The LOD of each diagnostics assay was evaluated in both monoplex and multiplex real-time PCR formats. Total genomic or plasmid DNA was quantified and serial dilutions were prepared from 10^5 to 0.1 genome/plasmid equivalents. The dilution series was run in triplicate for each monoplex real-time PCR NAD assay. In a monoplex format, all five diagnostics assays had an analytical sensitivity of less than 10 GE. The LOD of the real-time multiplex diagnostics assay was established using probit regression analysis with 12 replicates for each dilution. With 95% confidence, the analytical sensitivities of the *S. maltophilia*, *B. cepacia*, *P. aeruginosa* and the *S. marcescens* assays are 8.5 GE, 5.7 GE, 3.2 GE and 7.4 GE respectively. The LOD for each assay when investigated in the presence of a background of pooled genomic DNA from 10 other microorganisms at a concentration of 10^4 GE per reaction are 9.5 GE, 6.5 GE, 3.3 GE and 7.7 GE respectively. The IAC, at a concentration of 10^4 plasmid equivalents per reaction, was included in all samples during sensitivity testing and was detected at C_T values from 27.2 – 28.6. The amplification efficiency (E) of the multiplex real-time PCR reactions ranged from 94 – 102%.

Limit of quantification (LOQ)

DNA extracted from different concentrations of overnight culture was used as template DNA in the multiplex real-time PCR reaction. Standard curves were derived by plotting threshold cycle (C_T) values against overnight plate counts (CFU (log_{10})). The dynamic range of each assay was established and the limit of quantification for each assay was determined. The LOQ of the *S. maltophilia*, *B.*

Table 2 Microbial specificity testing of the developed multiplex assay

Assay	Diagnostics gene target	Inclusivity testing	Exclusivity testing
S. maltophilia	*ssrA*	5/5	0/90
Burkholderia genus	*rnpB*	10/10	0/85
P. aeruginosa	*ssrA*	14/14	0/81
S. marcescens	*ssrA*	5/5	0/90

cepacia, P. aeruginosa and the *S. marcescens* assays are 88.5 cfu, 59.7 cfu, 73.2 cfu and 87.4 cfu respectively.

Microbiological analysis of a HPW delivery system

(i) Conventional culture methods for the microbial analysis of a HPW delivery system.

HPW samples were cultured on R2A agar to determine HPCs. HPCs observed in the HPW delivery system ranged from $2.93 \times 10^6 - 3.67 \times 10^6$ / L (Additional file 1: Table S2). This demonstrates naturally occurring high bacterial contamination in the HPW delivery system. A biochemical analysis of microbial colonies isolated on R2A agar and selective agar was performed using the BBL crystal enteric NF system. This analysis confirmed the presence of both *S. maltophilia* and *Burkholderia* species in the HPW delivery system. However, neither *P. aeruginosa* nor *S. marcescens* were detected using this analysis. Other bacteria identified in the HPW delivery system using the BBL crystal enteric NF system were *Sphingomonas* sp., *Flavinonas* sp and *Vibrio* sp. The Turn Around Time (TAT) to result for the microbiological analysis of the HPW delivery system using conventional microbial culture methods was > 72 h.

(ii) Rapid NAD quantitative detection and identification of bacteria in a HPW delivery system.

The dynamic range for the external standard curves used in this study for the absolute quantification of pathogenic bacteria present in HPW delivery system was 10^2-10^7 cfu. For the quantitative detection and identification of bacteria in the HPW delivery system total genomic DNA was extracted from each filter and used as a template in the multiplex real-time PCR NAD assay developed in this study. The results were absolutely quantified by comparison to the external standards curves (see section Generation of external standard curves). From this analysis, it was determined that the HPW delivery system was contaminated with both *S. maltophilia* and *Burkholderia* species. The concentration of *S. maltophilia* detected in the HPW ranged from 5.75×10^1 - 7.65×10^1 cfu/100 ml at sampling point A and $1.30 \times 10^2 - 1.72 \times 10^2$ cfu/100 ml at sampling point B. The concentration of *Burkholderia* species observed in the HPW was higher with concentrations of 4.06×10^2-7.42×10^2 cfu/100 ml at sampling point A and $6.88 \times 10^2 - 1.17 \times 10^3$ cfu/100 ml at sampling point B. *P. aeruginosa* and *S. marcescens* were not detected. Both European and US guidelines for microbial analysis of HPW recommend that under normal conditions an appropriate action level is a microbial count of 10 cfu/100 ml [48,49]. The concentration of both *S. maltophilia* and *Burkholderia* species detected in the HPW distribution system tested was above the recommended action level. For complete results of this analysis see Table 3. The TAT to result for the microbiological analysis of the HPW delivery system using the multiplex real-time PCR NAD assay was < 5 h.

Discussion

The use of rapid quantitative multiplex real-time PCR NAD assays for the analysis of microbial contamination in water and HPW distribution systems can potentially overcome some of the limitations that are associated with conventional culture based microbiological diagnostics techniques traditionally used to monitor water quality. Conventional microbial analysis of water and HPW distribution systems by traditional culture techniques are labour-intensive and can take between 2 - 7 days, which can result in significant delays in identifying the contaminating microorganism [50]. Furthermore, conventional microbial culture methodologies traditionally used to analyse the biocontamination of water and HPW distribution systems assess the number of bacteria

Table 3 Microbiological analysis of a HPW delivery system using developed methodology

Sampling point	Sampling time	C_T value (SD) *S. maltophilia*	[a]Mean calculated cfu/100 ml (SD) *S. maltophilia*	C_T value (SD) *Burkholderia*	[a]Mean calculated cfu/100 ml (SD) *Burkholderia*	Above threshold EU/US guidelines action level of > 10 cfu/100 ml
A	1	29.02 (0.07)	7.7×10^1 (37)	27.14 (0.15)	7.4×10^2 (145)	Yes
A	2	29.25 (0.21)	6.5×10^1 (103)	27.81 (0.23)	6.2×10^2 (97)	Yes
A	3	29.40 (0.04)	5.8×10^1 (13)	28.47 (0.22)	4.1×10^2 (560)	Yes
B	1	27.90 (0.14)	1.7×10^2 (162)	27.64 (0.07)	6.9×10^2 (333)	Yes
B	2	28.30 (0.27)	1.3×10^2 (246)	27.16 (0.03)	9.4×10^2 (167)	Yes
B	3	28.10 (0.12)	1.5×10^2 (139)	26.83 (0.16)	1.2×10^3 (867)	Yes

C_T Threshold cycle.
[a]Mean calculated cfu/100 ml calculated by multiplex real time PCR.
A Valve adjacent to Elix 35 HPW purification system.
B Laboratory tap approximately 25 - 30 m from the Elix 35 HPW purification system.
[1]Before Pretreatment and Polishing Pack change.
[2]one day post Pretreatment and Polishing Pack change.
[3]one week post Pretreatment and Polishing Pack change.

that can form visible colonies on a solid medium under specific test growth conditions (e.g. incubation temperature, culture medium and incubation time). However, these approaches only reveal the presence of bacteria culturable under these conditions and may not grow all the contaminating microorganisms present [51]. Therefore, monitoring water quality by HPCs enumeration with conventional microbial culture methods may only represent a relatively small proportion of the total viable microbial population in a biocontaminated water or HPW distribution system [25].

It is possible that the use of NAD technologies such as real-time PCR for the analysis of microbial contamination in water and HPW distribution systems could result in the detection of DNA from dead microorganisms. This may result in an overestimation of the microbial contamination or the risk of false positive results. However, the presence of any DNA from pathogenic microorganisms in a water sample indicates that there is biological contamination within the system. Any contamination with pathogenic microorganisms within water distribution systems even if present at low concentrations should be considered a risk.

In a healthcare setting, if the water or HPW distribution system is suspected as a potential source of HCAIs, the ability to specifically detect and identify pathogenic microorganisms rapidly would allow the required infection control and prevention measures to be taken promptly. A delay in this analysis could result in an increased risk of patients contracting HCAIs and subsequently a significant increase in both the duration and cost of a healthcare stay. In this setting, where there is an increasing proportion of immunologically compromised patients there is added potential for infection by pathogenic microorganisms originating from biocontaminated water and HPW distribution systems. A recent commentary on waterborne pathogen detection [52] highlights that although hospital water and HPW distribution systems have been implicated in the spread of HCAIs there is a lack of evidence of direct transmission from water distribution systems. However, it concludes that a microbiological investigation of the hospital water distribution system, such as that performed by Cristina *et al.* [6], may be beneficial in identifying the position, frequency and extent of contamination. Such analysis could then be used to assess the degree of risk to patients. Therefore an ability to specifically detect and identify pathogenic microorganisms in the HPW distribution system rapidly may aid in determining the potential source and causative agent of the HCAIs, thereby allowing the required infection control and prevention measures to be taken promptly.

In this current study the internally controlled novel multiplex real-time PCR diagnostics assay developed uses NAD targets for the rapid specific quantitative detection and identification of four important human pathogenic bacteria associated with the contamination of HPW distribution systems [24,43,53]. The developed multiplex real-time PCR NAD assay was determined to be 100% specific when tested against a panel of 95 microorganisms with no cross reactivity observed (Table 2 and Additional file 1: Table S1). To our knowledge, this is the first report of an internally controlled 5-plex assay for the quantitative detection and identification of *S. maltophilia*, *Burkholderia* species, *P. aeruginosa* and *S. marcescens* for use in the microbial analysis of a HPW distribution system.

The analytical sensitivity of all four diagnostics assays described in this study is less than 10 GE allowing for the potential detection of low levels of pathogenic bacteria present in a biocontaminated HPW distribution system. The 5-plex PCR NAD assay described does not require microbial culture pre-enrichment of the HPW sample to increase analytical sensitivity prior to running the diagnostics assay. This ability to detect contaminating microorganisms without microbial culture pre-enrichment is a significant advantage, as the inclusion of a microbial culture pre-enrichment step eliminates the ability to absolutely quantify the contaminating microorganism in the HPW. This is important as routine testing of HPW needs to be both qualitative and quantitative [48]. Another major advantage of not including a pre-incubation step is that the assay TAT is significantly reduced. Using the developed multiplex real-time PCR NAD assay, the TAT to result inclusive of HPW sampling and filtration, DNA preparation, real-time PCR reaction, and data analysis is < 5 h in contrast to 2-7 days using traditional microbial culture diagnostics methodologies [49].

A HPW delivery system supplied by an Elix 35 HPW purification unit was evaluated for microbiological contamination using both traditional microbial culture methodologies and the developed NAD assay and associated methodology. This analysis was performed both before and after a pretreatment and polishing pack change of the Millipore Elix 35 HPW purification unit. The developed multiplex real-time PCR NAD assay determined that the HPW delivery system was contaminated with the opportunistic human pathogens *S. maltophilia* and *Burkholderia* species. The numbers of both of these GNB-NE present in the HPW distribution system was above the action level threshold of 10 cfu/100 ml recommended by both European and US guidelines [48,49]. The presence of both *S. maltophilia* and *Burkholderia* species in the HPW delivery system was also confirmed by traditional microbial culture methods using the BBL crystal enteric NF system (Becton Dickinson). Neither *P. aeruginosa* nor *S. marcescens* was detected in the HPW using the developed NAD test or by traditional microbial culture methods.

This current study, which was designed in accordance with MIQE guidelines [28], has focused on the development of a rapid methodology for the absolute quantification, detection and identification of four pathogenic bacteria from an operating HPW distribution system. Other studies have demonstrated that biocontamination of water and HPW distribution systems can occur with a number of other important human pathogenic microorganisms which were not investigated in this study [24,43,54]. We propose that with the inclusion of additional NAD tests, this methodology could be used to detect and identify a more comprehensive panel of pathogenic microorganisms (bacterial and fungal) for use in the analysis of microbial contaminants of water and HPW distribution systems. While the developed rapid culture independent methodology was demonstrated using a limited number of samples from a HPW distribution system the same methodology could be applied to other water and HPW distribution systems such as those in a clinical setting. We intend to evaluate a number of other water and HPW distribution systems over a longer period to further demonstrate the robustness of the developed methodology.

Conclusion

As water and HPW is routinely used for a wide range of applications in the clinical and industrial sectors, it is essential that water purification units and water distribution systems are subject to strict maintenance schedules and stringent microbiological analysis. The rapid culture independent methodology described here may be a valuable diagnostics tool and have many useful applications in both these sectors.

Additional file

Additional file 1: Table S1. Bacterial and fungal species and strains used in this study. **Table S2.** TVC from microbiological analysis of HPW delivery system by conventional culture methodologies.

Abbreviations

HPW: High purity water; NAD: Nucleic acid diagnostics; PCR: Polymerase chain reaction; TAT: Turnaround time; HCAIs: Healthcare Associated Infections; GE: Genome equivalents; CFU: Colony forming units; TVC: Total viable counts; LOD: Limit of detection; HPCs: Heterotrophic Plate Counts (HPCs).

Competing interests

The authors declare that they have no competing interests.

Authors' contributions

EM performed all laboratory experiments, analysed the data and wrote the manuscript. TB directed the study, contributed to writing of the manuscript and is the corresponding author. KR, NT and CS contributed to certain aspects of the direction of the study and they also contributed to the editing and revision of the manuscript. All authors read and approved the final manuscript.

Acknowledgements

We wish to acknowledge the College of Science and the Thomas Crawford Hayes Fund, National University of Ireland, Galway who have supported EM with a postgraduate scholarship. This publication, in part, has also emanated from research conducted with the financial support of Science Foundation Ireland under Grant SFI 13/TIDA/I2597. The authors also wish to acknowledge the European Union's Seventh Programme for research.

Author details

[1]Nucleic Acid Diagnostics Research Laboratory (NADRL), Microbiology, School of Natural Sciences, National University of Ireland Galway, Galway, Ireland. [2]Marine Microbial Ecology Laboratory, Microbiology, School of Natural Sciences, National University of Ireland Galway, Galway, Ireland.

References

1. Guo J-H, Skinner GW, Harcum WW, Barnum PE. Pharmaceutical applications of naturally occurring water-soluble polymers. Pharmaceutical Sci Tech Today. 1998;1(6):254–61.
2. Poretti M. Quality control of water as raw material in the food industry. Food Control. 1990;1(2):79–83.
3. Meltzer TH: High-purity water preparation for the semiconductor, pharmaceutical, and power industries: Tall Oaks Publishing
4. Walker JT, Marsh PD. A review of biofilms and their role in microbial contamination of dental unit water systems (DUWS). Int Biodeter Biodegr. 2004;54(2–3):87–98.
5. Walker J: Decontamination in Hospitals and Healthcare: Woodhead Publishing; 2014
6. Cristina MLBP, Spagnolo AMBP, Casini BBMS, Baggiani AMD, Giudice PDMD, Brusaferro SMD, et al. The impact of aerators on water contamination by emerging gram-negative opportunists in at-risk hospital departments. Infect Control Hosp Epidemiol. 2014;35(2):122–9.
7. Anaissie EJ, Penzak SR, Dignani MC. The hospital water supply as a source of nosocomial infections: a plea for action. Arch Intern Med. 2002;162(13):1483–92.
8. Bert F, Maubec E, Bruneau B, Berry P, Lambert-Zechovsky N. Multi-resistant Pseudomonas aeruginosa outbreak associated with contaminated tap water in a neurosurgery intensive care unit. J Hosp Infect. 1998;39(1):53–62.
9. Roberts LA, Collignon PJ, Cramp VB, Alexander S, McFarlane AE, Graham E, et al. An Australia-wide epidemic of Pseudomonas pickettii bacteraemia due to contaminated "sterile" water for injection. Med J Aust. 1990;152(12):652–5.
10. Moreira BM, Leobons MBGP, Pellegrino FLPC, Santos M, Teixeira LM, de Andrade ME, et al. Ralstonia pickettii and Burkholderia cepacia complex bloodstream infections related to infusion of contaminated water for injection. J Hosp Infect. 2005;60(1):51–5.
11. Loveday HP, Wilson JA, Kerr K, Pitchers R, Walker JT, Browne J. Association between healthcare water systems and Pseudomonas aeruginosa infections: a rapid systematic review. J Hosp Infect. 2014;86(1):7–15.
12. Walker JT, Jhutty A, Parks S, Willis C, Copley V, Turton JF, et al. Investigation of healthcare-acquired infections associated with Pseudomonas aeruginosa biofilms in taps in neonatal units in Northern Ireland. J Hosp Infect. 2014;86(1):16–23.
13. Verweij PE, Meis JFGM, Christmann V, Van Der Bor M, Melchers WJG, Hilderink BGM, et al. Nosocomial outbreak of colonization and infection with Stenotrophomonas maltophilia in preterm infants associated with contaminated tap water. Epidemiology Infect. 1998;120(03):251–6.
14. Lai KK, Brown BA, Westerling JA, Fontecchio SA, Zhang Y, Wallace RJ. Long-Term Laboratory Contamination by Mycobacterium abscessus Resulting in Two Pseudo-Outbreaks: Recognition with Use of Random Amplified Polymorphic DNA (RAPD) Polymerase Chain Reaction. Clin Infect Dis. 1998;27(1):169–75.
15. El Sahly HM, Septimus E, Soini H, Septimus J, Wallace RJ, Pan X, et al. Mycobacterium simiae pseudo-outbreak resulting from a contaminated hospital water supply in Houston. Texas Clin Infect Dis. 2002;35(7):802–7.
16. Verschraegen G, Claeys G, Meeus G, Delanghe M. Pseudomonas pickettii as a cause of pseudobacteremia. J Clin Microbiol. 1985;21(2):278–9.
17. Heard S, Lawrence S, Holmes B, Costas M. A pseudo-outbreak of Pseudomonas on a special care baby unit. J Hosp Infect. 1990;16(1):59–65.

18. Forman W, Axelrod P, St John K, Kostman J, Khater C, Woodwell J, et al. Investigation of a pseudo-outbreak of orthopedic infections caused by Pseudomonas aeruginosa. Infect Control Hosp Epidemiology Official J Society Hosp Epidemiologists America. 1994;15(10):652–7.
19. Farmer JJ, Davis BR, Hickman FW, Presley DB, Bodey GP, Negut M, et al. Detection of Serratia outbreaks in hospital. Lancet. 1976;2(7983):455–9.
20. Wiemer KE, Anderson A, Stewart B. The importance of water quality for media preparation. Hum Reprod. 1998;13 suppl 4:166–72.
21. Tothill I, Seal K. Biodeterioration of waterborne paint cellulose thickeners. Int Biodeter Biodegr. 1993;31(4):241–54.
22. Jimenez L, Ignar R, Smalls S, Grech P, Hamilton J, Bosko Y, et al. Molecular detection of bacterial indicators in cosmetic/pharmaceuticals and raw materials. J Ind Microbiol Biotechnol. 1999;22(2):93–5.
23. Byrd JJ, Xu HS, Colwell RR. Viable but nonculturable bacteria in drinking water. Appl Environ Microbiol. 1991;57(3):875–8.
24. Kulakov LA, McAlister MB, Ogden KL, Larkin MJ, O'Hanlon JF. Analysis of bacteria contaminating ultrapure water in industrial systems. Appl Environ Microbiol. 2002;68(4):1548–55.
25. Colwell RR, Brayton PR, Grimes DJ, Roszak DB, Huq SA, Palmer LM. Viable but Non-Culturable Vibrio cholerae and Related Pathogens in the Environment: Implications for Release of Genetically Engineered Microorganisms. Nat Biotech. 1985;3(9):817–20.
26. Mittelman M, Geesey GG: Biological fouling of industrial water systems: a problem solving approach: Water Micro Assoc; 1987
27. Minogue E, Reddington K, Dorai-Raj S, Tuite N, Clancy E, Barry T. Diagnostics method for the rapid quantitative detection and identification of low-level contamination of high-purity water with pathogenic bacteria. J Ind Microbiol Biotechnol. 2013;40(9):1005–13.
28. Bustin SA, Benes V, Garson JA, Hellemans J, Huggett J, Kubista M, et al. The MIQE guidelines: minimum information for publication of quantitative real-time PCR experiments. Clin Chem. 2009;55(4):611–22.
29. Gopalan V, Vioque A, Altman S. RNase P: variations and uses. J Bio Chem. 2002;277(9):6759–62.
30. Innings Å, Ullberg M, Johansson A, Rubin CJ, Noreus N, Isaksson M, et al. Multiplex real-time PCR targeting the RNase P RNA gene for detection and identification of Candida species in blood. J Clin Microbiol. 2007;45(3):874–80.
31. Clancy E, Glynn B, Reddington K, Smith T, Barry T. Culture confirmation of Listeria monocytogenes using tmRNA as a diagnostics target. J Microbiol Methods. 2012;88(3):427–9.
32. Reddington K, Zumla A, Bates M, van Soolingen D, Niemann S, Barry T, et al. SeekTB, a two-stage multiplex real-time-PCR-based method for differentiation of the mycobacterium tuberculosis complex. J Clin Microbiol. 2012;50(7):2203–6.
33. O'Grady J, Lacey K, Glynn B, Smith TJ, Barry T, Maher M. tmRNA – a novel high-copy-number RNA diagnostic target – its application for Staphylococcus aureus detection using real-time NASBA. FEMS Microbiol Lett. 2009;301(2):218–223.
34. Wernecke M, Mullen C, Sharma V, Morrison J, Barry T, Maher M, et al. Evaluation of a novel real-time PCR test based on the ssrA gene for the identification of group B streptococci in vaginal swabs. BMC Infectious Diseases. 2009;9(1):148.
35. Barry T, Colleran G, Glennon M, Dunican L, Gannon F. The 16s/23s ribosomal spacer region as a target for DNA probes to identify eubacteria. Genome Res. 1991;1(1):51–6.
36. Robertson JM, Walsh-Weller J. An introduction to PCR primer design and optimization of amplification reactions. Methods Mol Biol. 1998;98:121–54.
37. Hoorfar J, Ahrens P, Radstrom P. Automated 5′ Nuclease PCR Assay for Identification of Salmonella enterica. J Clin Microbiol. 2000;38(9):3429–35.
38. Hoorfar J, Malorny B, Abdulmawjood A, Cook N, Wagner M, Fach P. Practical considerations in design of internal amplification controls for diagnostic PCR assays. J Clin Microbiol. 2004;42(5):1863–8.
39. O' Grady J, Sedano-Balbás S, Maher M, Smith T, Barry T. Rapid real-time PCR detection of Listeria monocytogenes in enriched food samples based on the ssrA gene, a novel diagnostic target. Food Microbiol. 2008;25(1):75–84.
40. Anon: LightCycler® 480 User guide. In: Roche LightCycler® 480 Instrument Operator's Manual 2008: 248-256
41. Yun JJ, Heisler LE, Hwang IIL, Wilkins O, Lau SK, Hyrcza M, et al. Genomic DNA functions as a universal external standard in quantitative real-time PCR. Nucleic Acids Res. 2006;34(12):e85.
42. Motulsky H: Fitting Models to Biological Data Using Linear and Nonlinear Regression: A Practical Guide to Curve Fitting: Oxford University Press; 2004
43. Bohus V, Tóth EM, Székely AJ, Makk J, Baranyi K, Patek G, et al. Microbiological investigation of an industrial ultra pure supply water plant using cultivation-based and cultivation-independent methods. Water Res. 2010;44(20):6124–32.
44. Penna VT, Martins SA, Mazzola PG. Identification of bacteria in drinking and purified water during the monitoring of a typical water purification system. BMC Public Health. 2002;2:13.
45. Anon: British Pharmacopoeia Volume I & 2. In: Medicinal and Pharmaceutical Substances: Water for Injection Ph Eur monograph 1927. London, United Kingdom: British Pharmacopoeia Commission 2012
46. Gibson S, Friedman H. L-Arabinose-ornithine-Irgasan medium for differentiating Serratia species. J Clin Microbiol. 1978;7(3):279–81.
47. McGuinness S, McCabe E, O'Regan E, Dolan A, Duffy G, Burgess C, et al. Development and validation of a rapid real-time PCR based method for the specific detection of Salmonella on fresh meat. Meat Sci. 2009;83(3):555–62.
48. Anon: European Directorate for the Quality of Medicines. In: Purified water. vol. European Pharmacopoeia 6.3. Strasbourg, France: EDQM Council of Europe; 2009: pp 4344–4346.
49. Anon: U.S Pharmacopeia. In: Water for Pharmaceutical Purposes General Chapter 1231. Maryland, USA: United States Pharmacopeial Convention; 2011.
50. Sartory DP, Watkins J. Conventional culture for water quality assessment: is there a future? J Appl Microbiol. 1998;85(S1):225S–33.
51. Reasoner DJ, Geldreich EE. A new medium for the enumeration and subculture of bacteria from potable water. Appl Environ Microbiol. 1985;49(1):1–7.
52. Decker BKMD, Palmore TNMD. Commentary: waterborne pathogen detection: more than just "location, location, location". Infect Control Hosp Epidemiol. 2014;35(2):130–1.
53. Favero MS, Carson LA, Bond WW, Petersen NJ. Pseudomonas aeruginosa: growth in distilled water from hospitals. Science. 1971;173(3999):836–8.
54. Baba T, Matsumoto R, Yamaguchi N, Nasu M. Bacterial population dynamics in a reverse-osmosis water purification system determined by fluorescent staining and PCR-denaturing gradient Gel electrophoresis. Microbes Environ. 2009;24(2):163–7.

Soluble expression, purification, and characterization of active recombinant human tissue plasminogen activator by auto-induction in *E. coli*

Xiaobin Long[1,2†], Yeran Gou[1,2†], Miao Luo[2,3,4†], Shaocheng Zhang[2,3], Hongpeng Zhang[2,3], Lei Bai[1,2], Shuang Wu[2,3], Quan He[1], Ke Chen[2], Ailong Huang[2], Jianzhong Zhou[1*] and Deqiang Wang[2,3*]

Abstract

Background: Human tissue plasminogen activator (tPA) belongs to the serine protease family. It converts plasminogen into plasmin and is used clinically to treat thrombosis. Human tPA is composed of 527 amino acids residues and contains 17 disulfide bonds. *Escherichia coli* has been used only rarely for the efficient production of recombinant tPA. However, the functional expression of full-length tPA that contains multiple disulfide bonds on an industrial scale remains challenging. Here, we describe the soluble expression and characterization of full-length tPA by auto-induction in *E. coli*.

Results: We achieved optimal levels of gene expression, minimized negative effects related to the production of heterologous proteins, and optimized cytoplasmic yields. Three different *E. coli* strains, BL21 (DE3), Rosetta, and Origami 2, could express tPA using an auto-induction mechanism. In addition, similar yields of recombinant protein were produced at temperatures of 33, 35, and 37°C. The *E. coli* strain origami 2 could increase disulfide bond formation in cytoplasmic tPA and produce purified soluble recombinant protein (~0.9 mg/l medium). The full-length tPA was monomeric in solution, and fibrin plate assays confirmed that the recombinant tPA displayed serine protease activity.

Conclusions: This is the first report that describes the heterologous expression of correctly folded active full-length tPA. This could provide valuable information for using prokaryotic auto-induction expression systems to produce tPA at industrial and pharmaceutical levels without *in vitro* refolding during the production step.

Keywords: tPA, Recombinant protein, Soluble expression, Autoinduction, Thrombus

Background

Human tissue plasminogen activator (tPA), a serine protease that selectively cleaves plasminogen, is found in the fibrinolytic system of blood vessel endothelial cells [1]. The human protein comprises 527 amino acids residues, including 35 cysteine residues that participate in the formation of 17 disulfide bonds [1]. tPA contains five distinct structural domains: an N-terminal finger domain (F domain, residues 4–50), the epidermal growth factor-like domain (E domain, residues 50–87), two kringle domains (K1 domain, residues 87–176; and K2 domain, residues 176–256), and a serine protease catalytic domain (P domain, residues 276–527). Its binding to fibrin and the subsequent modulation of protease activity are regulated primarily by the F and K2 domains, respectively. tPA converts plasminogen into plasmin, and has been used clinically to treat thrombosis [2].

However, the functional preparation of tPA containing multiple disulfide bonds remains the bottleneck for its

* Correspondence: zjzcqu@163.com; wangd@ustc.edu
†Equal contributors
[1]Department of Cardiology, The First Affiliated Hospital of Chongqing Medical University, Chongqing 400016, China
[2]Key Laboratory of Molecular Biology on Infectious Disease (Ministry of Education), The Second Affiliated Hospital, Chongqing Medical University, Chongqing 400016, China
Full list of author information is available at the end of the article

production on an industrial scale. The expression and purification of human tPA from eukaryotic and prokaryotic sources, such as human uterus or Chinese hamster ovary cells, has been described in several reports [3-5]. In contrast, tPA that is purified from mammalian expression systems is expensive, and the resulting glycosylated protein is cleared rapidly from the blood [6]. Consequently, several recombinant hosts, such as *Saccharomyces cerevisiae* and insect systems, have been used for the industrial preparation of tPA. However, these have also been associated with several problems including hyperglycosylation, poor export, and improper folding [7-10].

Due to its low cost and simplicity, the *Escherichia coli* expression system is the preferred choice for the production of therapeutic proteins. Generally, the overexpression of eukaryotic proteins in *E coli* is tightly regulated by the inclusion of an inducible promoter [11]. Bacterial systems utilize the *lac* promoter, which is induced by Isopropyl β-D-1-Thiogalactopyranoside (IPTG). In addition, some background expression of host proteins might occur in the bacterial system, even when the *lac* operator sequence is present [12]. The expression of eukaryotic proteins, particularly those of human origin, in such heterologous systems has not been achieved, which is problematic for biophysical or structural biology experiments that require milligram quantities of highly purified protein.

There are several obstacles to the expression of tPA in prokaryotic systems, including disulfide bond formation, rare codon usage, and cytotoxicity [13]. Auto-induction, an alternative to IPTG induction, depends on glucose catabolite repression and lactose (substrate) induction to supply tightly control protein expression [14]. Because of the reduced need for sample processing (e.g., no OD_{600}-dependent induction window) and ease by which the culture size can be scaled-up, this system is a very attractive method for achieving high-throughput protein expression. In addition, auto-induction results in a high cell density prior to induction, which often results in several-fold higher yields of the target protein compared with than obtained using conventional IPTG induction [14].

To facilitate the structural and functional analysis of tPA, we sought to improve the expression and purification strategies to obtain milligram quantities of purified and untagged full-length tPA protein. tPA with a six-histidine tag (His-tag) was expressed using auto-induction in *E. Coli* (Figure 1). After Ni^{2+}-nitrilotriacetic acid (NTA) chromatography and elution with imidazole, the His-tag was cleaved using PreScission Protease (PSP), leaving only two additional residues (Gly and Pro) at the N-terminal. The RGDS, as the integrin binding motif, fused in C-terminal of the recombinant [15]. Gel filtration and fibrin plate experiments suggested that the recombinant tPA was an active monomer in solution.

Results

Auto-induction of soluble tPA

The aim of this study was to use bacterially expressed tPA in structural and biophysical experiments that require mg quantities of highly purified protein. To the best of our knowledge, high yields of full-length tPA using these bacterial expression systems have been reported previously only rarely. Therefore, we sought to improve the current expression and purification strategies to obtain milligram amounts of highly purified and untagged tPA protein. BL21 is the most widely used prokaryotic host for protein expression and has the advantage of being deficient in the *lon* and *ompT* proteases. Rosetta host strains are BL21 derivatives designed to enhance the expression of eukaryotic proteins containing rare codons in *E. coli* [16]. Additionally, Origami 2 host strains have mutations in both the thioredoxin reductase (*trx*B) and glutathione reductase (*gor*) genes to enhances disulfide bond formation in the cytoplasm [17]. The expression of tPA was examined in three cell lines using different IPTG concentrations and in the absence of IPTG (data not shown). The cells were then harvested after 4–20 h of induction. All the cell lines showed significant leaky expression in the presence or absence of IPTG (data not shown).

A reductive cytoplasmic environment and cytotoxicity are both possible obstacles for the active expression of tPA, a heterologous protein with multiple disulfide bonds, in wild-type *E coli*. Studier developed a reliable protocol for the *lac* operon/promoter-dependent auto-induction of genes in *E coli* [14]. The amount of human proteases expressed using auto-induction is far greater than that achieved using IPTG-based induction. In addition, supplying rare tRNAs (using the Rosetta 2 and Origami 2 strains) did not increase expression compared with BL21 (Figure 2). Origami 2 cells enhanced tPA disulfide bond formation in the cytoplasm; therefore,

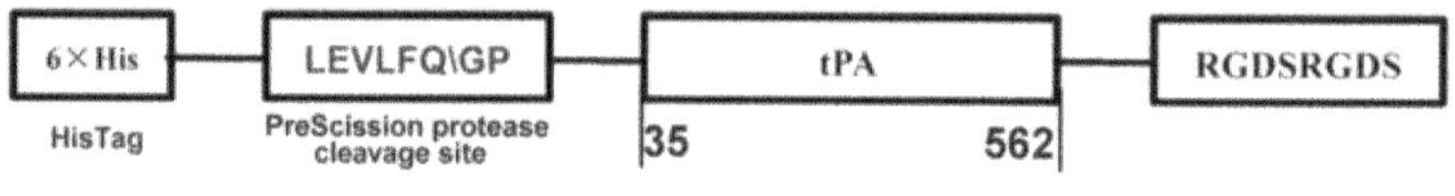

Figure 1 Recombinant full-length tPA protein. A scheme showing the functional domains of HisTag-PSP-tPA-RGDS protein (calculated MW 66.7 kDa) generated using auto-induction in *E. coli* Origami 2 cells. The protein was further processed using His-affinity purification and cleavage with PSP to generate a tag-free 528 amino acid full-length tPA with a calculated MW of 60.8 kDa.

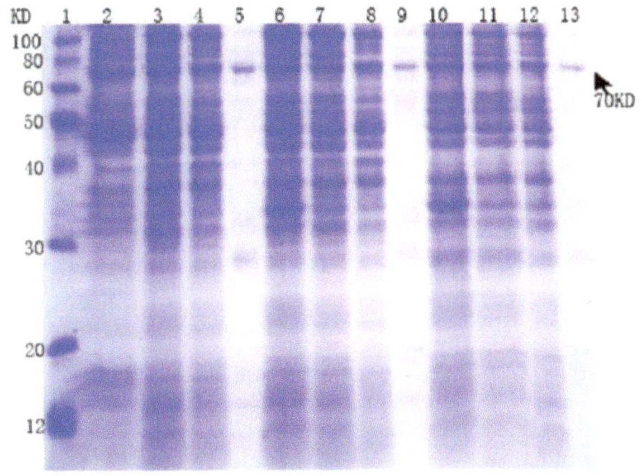

Figure 2 Soluble expression of tPA in different *E. coli* strains. After the auto-induction of tPA expression at 37°C for 24 h. The lysates from *E coli*, supernatant (soluble) fractions, flow-through proteins that did not bind to the column, and recombinant full-length tPA (His-tag-PSP-tPA-RGDS) were analyzed using SDS-PAGE and Coomassie blue staining. Lane 1, protein markers; lane 2, Origami 2 cell lysate; lane 3, Origami 2 supernatant; lane 4, flow-through; lane 5, elution with 200 mM imidazole from Ni^{2+}-NTA; lane 6, BL21 lysate; lane 7, BL21 supernatant; lane 8, flow-through; lane 9, elution with 200 mM imidazole from Ni^{2+}-NTA; lane 10, Rosetta™ 2 lysate; lane 11, Rosetta™ 2 supernatant; lane 12, flow-through; lane 13, elution with 200 mM imidazole from Ni^{2+}-NTA.

Origami 2 was the preferred choice for the expression of tPA.

Next, the effect of temperature was tested to optimize the expression of recombinant tPA. As shown in Figure 3, the recombinant protein was expressed at three different temperatures: 33, 35, and 37°C. Similar yields of recombinant protein were produced at each temperature. These results suggest that expressing the protein at a higher temperature increased the yield of recombinant tPA significantly.

The molecular mechanisms of protein folding have been studied extensively over previous decades both experimental and computationally. However, the refolding of inclusion bodies *in vitro* is a difficult and complex task, particularly for proteins that have multiple domains and complex disulfide bonds. The refolding of tPA, which contains five domains and 17 disulfide bonds, is an even greater challenge. It was presumed that a soluble form of the target protein would be folded in its native state and exhibit biological activity. To this end, different strains harboring the pET28-HisTag-tPA-RGDS plasmid were cultured and auto-induced at 37°C in auto-induction medium, which provides a cell density that is typically several-fold higher than obtained using conventional IPTG induction in LB medium [18]. The auto-induced cells were harvested and lysed by sonication. The cell lysates were separated by high-speed centrifugation, and the expression of tPA in the supernatant and pellets was determined using SDS-PAGE. The Origami 2 strains produced the protein of interest in a soluble form, although ~50% of the tPA was present in the insoluble protein fraction (Figures 2 and 3).

Purification of soluble tPA protein

To facilitate the purification of tPA, the N-terminal His-tag of pET28-HisTag-tPA-RGDS was captured using metal ion affinity chromatography. This purification strategy prevents the cumbersome refolding processes that lead to extremely poor efficiency, particularly for proteins with several disulfide bonds.

After the cultivation period, cells were harvested by centrifugation, resuspended in cold lysis buffer, and disrupted by sonication. The supernatants containing soluble recombinant tPA protein were applied to an Ni^{2+}-NTA affinity column, and were subsequently purified using ion exchange chromatography (DEAE, GE Healthcare, England). The target fractions were then combined and incubated with PSP to cleave the His-tag from tPA. Finally, the recombinant protein was subjected to size-exclusion chromatography. The final purified tPA preparation yielded only one major protein band, at ~65 kDa, after analysis by SDS-PAGE followed by Coomassie Brilliant Blue staining (Figure 4). Although no weak overexpressed band of tPA was observed in total protein lysates from IPTG induction cultures, we used the powerful auto-induction system to generate ~1.8 mg tPA protein (calculated using the BCA method, Thermo Scientific, USA) from an auto-induced *E coli* culture.

Characterization and activity of the recombinant tPA

SDS-PAGE analysis of the different fractions obtained during purification revealed the progressive enrichment of a ~70 kDa protein (Figure 4). This was the only protein band that was present in the electrophoretically homogeneous final enzyme preparation; it was the was the same size as a tPA monomer (Figure 4). Western blotting confirmed that this band (~70 kDa) was detected by an anti-tPA monoclonal antibody, suggesting that the purified band represents recombinant tPA (Figure 4).

Previous studies revealed that the P domain of tPA exists as a monomer under native conditions [10]. Therefore, we next aimed to further characterize the molecular weight of tPA in solution. In a size-exclusion chromatography experiment, tPA was eluted in a single peak at 72.1 ml, corresponding to the estimated monomeric mass of ~66.7 kDa (Figure 5). Because tPA has a calculated molecular mass of ~63.8 kDa, the recombinant protease exists mostly as a monomer under native conditions rather than in an intermolecular disulfide-bonded oligomeric form. Both the P domain and full-length tPA exist in a monomeric state in solution, implying that the monomer serves as its Physiological aggregation.

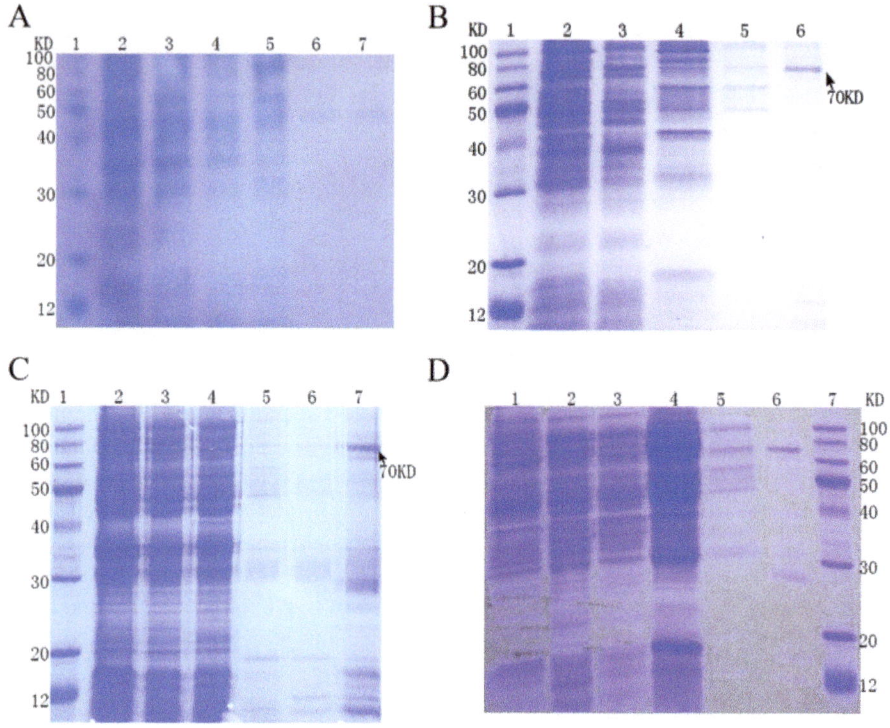

Figure 3 The auto-induction of Origami 2 (pET28-HisTag-tPA-RGDS) at various temperatures. A, Recombinant tPA after auto-induction at 25°C. Lane 1, protein marker; lane 2, lysate; lane 3, supernatant; lane 4, flow-through; lanes 5 and 6, washes with 20 and 40 mM imidazole, respectively; lane 7, elution with 200 mM imidazole. **B,** Recombinant tPA after auto-induction at 33°C. Lane 1, protein marker; lane 2, supernatant; lane 3, flow-through; lanes 4 and 5, washes with 20 and 40 mM imidazole, respectively; lane 6, elution with 200 mM imidazole. **C,** Recombinant tPA after auto-induction at 35°C. Lane 1, protein marker; lane 2, lysate; lane 3, supernatant; lane 4, flow-through; lanes 5 and 6, washes with 20 and 40 mM imidazole, respectively; lane 7, elution with 200 mM imidazole. **D,** Recombinant tPA after auto-induction at 37°C. Lane 1, lysate; lane 2, supernatant; lane 3, flow-through; lanes 4 and 5, washes with 20 and 40 mM imidazole, respectively; lane 6, elution with 200 mM imidazole; Lane 7, protein marker.

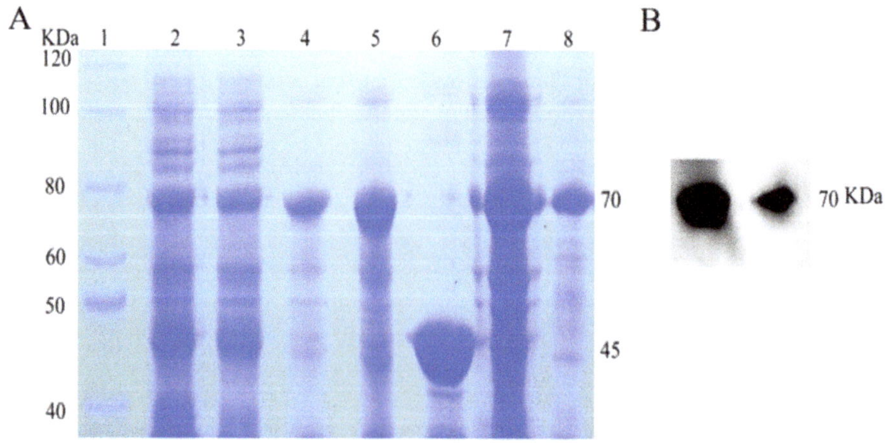

Figure 4 Purification of recombinant tPA. A, SDS-PAGE analysis of tPA purity. Lane 1, protein markers; lane 2, lysate; lane 3, supernatant; lane 4, elution with 200 mM imidazole; lane 5, recombinant tPA purified using ion exchange chromatography; lane 6, PSP; lane 7, PSP enzyme reaction mixtures; lane 8, tPA purification using size-exclusion chromatography. **B,** Western blotting of the enzyme reaction products (lane 7 of Figure 4A) and the purified fraction (lane 8 of Figure 4A) using anti-tPA antibodies.

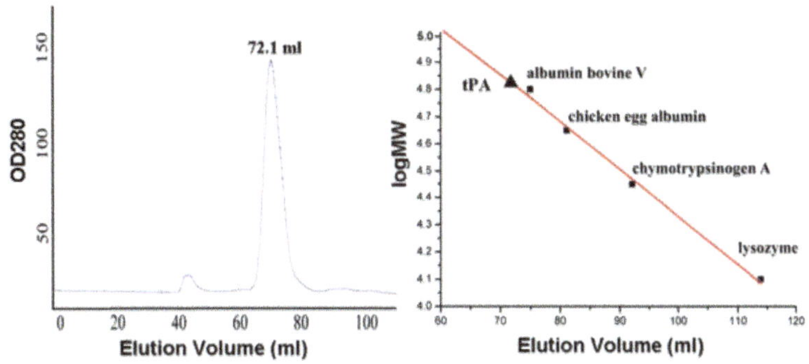

Figure 5 Size exclusion gel filtration chromatography of tPA using a Superdex 200 column. tPA (MW ~60.8 kDa) was eluted as a single peak at a volume of 72 ml, which corresponds to a molecular weight of 66.68 kDa on the calibrated column. Standard protein samples (black data points) were used for calibration: albumin bovine V (66.2 kDa), chicken egg albumin (44.2 kDa), chymotrypsinogen A (24.5 kDa), and lysozyme (14.4 kDa).

Disulfide bonds, which play multiple critical roles in protein stability and function, are often abundantly present in secreted proteins. The functional expression of human proteins with multiple disulfide bonds, such as tPA, in bacterial systems has proved challenging. Therefore, we used fibrin plate assays to assess whether the recombinant tPA was actively folded. Plasminogen was used as the tPA substrate in the medium on the plate. Active tPA can bind to plasminogen and cleave it into plasmin, which degrades fibrin and results a clear lysed zone on the fibrin/agar plate. RGDS, as the binding motif of integrin, will improve the specific affinity of the recombinant tPA to substrate [15]. Similar to tPA, Urokinase could cleave plasminogen into plasmin and was used as a positive control [19]. As is shown in Figure 6A and B, tPA and urokinase exhibited similar halo patterns, suggesting that the His-tag-PSP and RGDS, which were attached to the C- and N-terminus of tPA, respectively, did not affect either tPA receptor-ligand binding or the resulting signal transduction. In addition, the similar diameter of the halo patterns of HisTag-PSP-tPA-RGDS and tPA-RGDS suggests that they exhibit similar amidolytic activities *in vitro*. The activity of recombinant tPA was similar to that of urokinase.

Discussion

The production of recombinant proteins in their soluble active forms mainly uses eukaryotic and/or prokaryotic expression systems. Because of their low cost and the simplicity of bacterial cultivation, *E. coli* expression systems remain the preferred choice for the production of therapeutic proteins both on a laboratory and an industrial scale. The expression of soluble and active full-length tPA in *E. coli* is a major bottleneck for investigators. Under normal physiological conditions, the reductive environment of *E. coli* cytoplasm makes it challenging to prepare recombinant proteins with multiple disulfide bonds. However, native disulfide bond formation is critical for the proper folding of active proteins. tPA is an important thrombolytic agent that contains five distinct structural

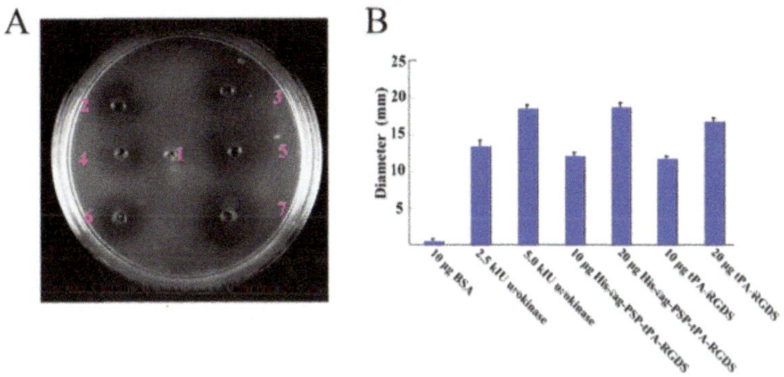

Figure 6 Lysis on a fibrin plate using urokinase and tPA. tPA-RGDS; HisTag-PSP-tPA-RGDS; urokinase standard (10×10^3 IU ml^{-1}. positive control); BSA (bovine serum albumin standard, negative control). The fibrinolytic activity of tPA was measured and compared with urokinase (positive control) and bovine serum albumin (BSA; negative control). **A,** dot 1, 10 μg BSA; dot 2, 2.5×10^3 IU urokinase; dot 3, 5.0×10^3 IU urokinase (positive control); dot 4, 10 μg His-tag-PSP-tPA-RGDS; dot 5, 20 μg His-tag-PSP-tPA-RGDS; dot 6, 10 μg tPA-RGDS; dot 7, 20 ug tPA-RGDS. **B,** the diameters of the lysis dots on the fibrin plate from at four independent evaluations. The bars indicate standard error of the mean.

domains with 17 disulfide bonds; therefore, it is typically unfolded or misfolded and so inactive when expressed in *E. coli*. Most recent studies have focused on improving the oxidation state of the cytoplasm, for example by selecting the periplasmic space or fusing granules, to express tPA in *E. coli* [20-22]. Nevertheless, because of the rare codon usage, cellular toxicity, and the narrow periplasmic space, the soluble expression of full length tPA has not yet been reported in a prokaryotic system. In the current study, we described a novel and efficient strategy to express and purify high yields of soluble, biologically active tPA.

Although we used the conventional *E. coli* expression system with multiple strategies, such as different host strains, temperatures, and/or IPTG concentrations, we did not observe any overexpressed tPA in the total protein lysates. As a protease, human tPA might be toxic to its host bacteria, which leads to low levels of expression [13]. In addition, the rare code of tPA might hamper the overexpression of large amounts of this protein. Therefore, we explored several novel methods to express soluble tPA. Studier previously developed a reliable protocol for the *lac* operon/promoter-dependent auto-induction of genes in *E. coli* [14]. Auto-induction media contains three different carbon sources: glucose, glycerol, and lactose. At the initial stage, gene expression is suppressed by glucose, and the cells grow to a very high density. Subsequently, the expression is induced in a lactose-dependent manner until the glucose is consumed. Compared with IPTG induction, auto-induction has several advantages including a greater cellular biomass, regulated expression, and reduced sample handling. All three *E. coli* lines (BL21, Rosetta 2, and Origami 2) yielded an obvious target protein band in the cell lysates. The supply of rare tRNAs in Rosetta 2 and Origami 2 did not enhance expression compared with that obtained in BL21 (Figure 2). However, Origami 2 cells enhanced the formation of disulfide bonds; therefore, they were used for the preparation of full-length tPA.

Protein solubility is an important indicator of its correct folding [23,24]. Interestingly, the auto-induction expression system provides advantages for protein expression including enhanced folding and protein solubility. In the current study, ~50% of the tPA was expressed in the soluble form, which represents the first report of soluble tPA expressed using a prokaryotic expression system. Because the recombinant protein is expressed in a soluble form, the purification of tPA could be achieved easily in three steps using affinity systems, ion exchange chromatography, and size-exclusion chromatography (Figure 4). The high proteolytic activity of the recombinant tPA confirmed that an active protease had been expressed successfully (Figure 6). The method used to purify tPA described here is relatively simple and, more importantly, allows the large-scale production of active protein. This suggests that the auto-induction system is an optimal expression system for achieving the high-yield production of soluble tPA in *E. coli.*

Approximately 33% of all known proteases identified to date are serine proteases with a wide variety of functions, including roles in blood clotting, protein digestion, cell signaling, inflammation, and protein processing [25]. As a serine protease, human tPA contains the so-called "classical" catalytic Ser195/His322/Asp371 triad, and cleaves the inactive proenzyme plasminogen to form plasmin, the active enzyme [1,26]. The aggregation state of serine proteases in solution, such as monomers, dimers, and trimers, is related to its biological activity. For some serine proteases, such as thrombin, alpha-synuclein, and Rv3671c from *Mycobacterium tuberculosis*, the monomeric form is active in solution [27-29]. In contrast, some serine proteases such as FAAH amidase, LD carboxypeptidase, and cytomegalovirus protease adopt active homodimers in solution [25]. Herpesvirus proteases, which belong to a unique class of serine proteases that contain a Ser-His-His catalytic triad, exist in monomeric-dimeric equilibrium in solution, and the homodimer is its active form [30,31]. Interestingly, the active form of a mitochondrial serine protease HtrA2 is a pyramid-shaped homotrimer [32].

To characterize the oligomeric state of recombinant tPA in solution, size exclusion chromatography experiments were performed using tPA. Elution peaks were observed at volume of 72.1 ml; this corresponds to an estimated molecular weight of ~66.68 kDa, which is similar to the calculated molecular mass (~60.8 kDa). This suggests that tPA exists mostly as a monomer under native conditions. Importantly, this is the first study to report the aggregation state of full-length tPA. Interestingly, Lee et al. also observed that the P domain of tPA folded correctly as a monomeric protein in solution [10]. Collectively, these results strongly suggest that a monomer of tPA might be required for its proper physiological function.

Recombinant tPA is one of the most promising clinical treatments for improper blood clotting, which might result in a heart attack and stroke [2]. Compared with expression and purification in a eukaryotic system, the scheme described here for full-length tPA is relatively affordable, simple, and, more importantly, is suitable for the large-scale industrial production of active proteins. To our knowledge, this is the first report of the soluble and functional expression of full-length tPA in *E. coli*. The preparation of full-length tPA in its active conformation will allow its three-dimensional structure to be determined and its physiological roles to be characterized.

Conclusion

In summary, we described a novel method for the soluble expression of large amounts of full-length untagged human tPA in *E. coli*. To the best of our knowledge, this is the first report of the successful large-scale heterologous expression of correctly folded active tPA using a prokaryotic expression system. We used PSP to obtain a highly pure protein preparation with the addition of only Gly and Pro residues at the N-terminus, and a C-terminally fused RGDS to improve the targeting efficiency of tPA in thrombolysis. The typical biophysical properties of tPA, such as its protease activity, were maintained in the purified protein. tPA exists as a monomer in solution, which was verified using size exclusion chromatography, suggesting that the monomer is the structural and active unit of tPA.

Methods

Bacterial strains, plasmids, and growth conditions

For recombinant protein expression experiments, chemically competent *E. coli* strains BL21 (DE3), Origami 2 (DE3), and Rosetta™ 2 (DE3) pLysS were transformed using standard protocols (Novagen, USA). Unless otherwise stated, the strains were grown at 37°C in either Luria-Bertani (LB) broth or auto-induction medium with vigorous shaking. They were also grown on tryptone yeast agar [14]. When necessary, 50 μg ml^{-1} kanamycin was added.

Construction of the fusion expression vector

The gene *tPA* (accession number NM_000930) was amplified from human liver cDNA using the primers tPA-F (5′-AT*GGATCC*ATGCTGGAAGTT***CTGTTCCAGGGGCC***CTCTTACCAAGTGATCTGCAGAGAT-3′; the *Bam*HI site is *italics*, and the PSP site is bold-italics) and tPA-R (5′-AC*AAGCTT*TTA***GCTATCCCCTCGCGAATCCCCTCG***CGGTCGCATGTTGTCACGAATCCA-3′; the *Hin-d*III site is *italics* and the RGDS peptide is bold-italics). The PSP site enables the proteolytic removal of the His-tag, which can be separated subsequently from the receptor using a Ni^{2+}-NTA column (GE Healthcare, Sweden). *Bam*HI and *Hind*III restriction sites were introduced into the upstream and downstream oligonucleotide primers, respectively (Figure 1). PCR was performed using *Pfu* DNA polymerase (Promega, USA) with the following conditions: denaturation at 94°C for 5 min, followed by 25 cycles of 94°C for 40 s, 58°C for 40 s, and 72°C for 180 s, with a final elongation step at 72°C for 10 min. Two blanks that contained all the reaction components except the primers or cDNA, respectively, were used as controls. The PCR fragments were double-digested with *Bam*HI and *Hind*III, and then subcloned into pET28a expression vector (Novagen, USA) that had been pre-digested with the same enzymes. The vector contained an N-terminal His-tag and a C-terminal RGDS sequence. The presence of the insert in the recombinant plasmid was verified using DNA sequencing. The resulting plasmid for the expression of tPA in *E. coli* was named pET28-HisTag-tPA-RGDS.

Soluble expression and purification of tPA

The *E coli* strains BL21 (DE3), Origami 2 (DE3) and Rosetta™ 2 (DE3) pLysS were transformed with the recombinant pET28-HisTag-tPA-RGDS plasmid. The recombinant protein was then expressed using auto-induction at 37°C overnight, as described previously [14]. The transformed cells were selected on LB medium plates supplemented with 50 μg ml^{-1} kanamycin. For pre-cultures, 5 ml of LB medium was inoculated using a single colony picked directly from agar plates, and then cultivated at 37°C with shaking at 180 rpm for 12 h. For the main cultivation, five separate 200 ml cultures of auto-induction media containing kanamycin (50 μg ml^{-1}) were inoculated with 2 ml of the pre-culture in flasks shaking at 250 rpm for 15 h at 37°C. For temperature optimization, the bacterial cultures were grown at four different temperatures (25, 33, 35, and 37°C) and auto-induced for 30, 24, 20, and 15 h, respectively.

At the end of the cultivation period, cells were harvested by centrifugation at 4000 rpm for 10 min. The bacterial pellets were resuspended in 100 ml cold lysis buffer (20 mM Tris pH 8.0 and 300 mM NaCl), and disrupted by sonication (20 min at 30% amplitude, with 3 s on and 6 sec off cycles) in an ice-bath to avoid overheating the samples. All purification steps were performed at 4°C. The cell debris was removed by centrifugation at 15 000 rpm for 30 min at 4°C. The supernatants containing soluble recombinant tPA protein were applied to Ni^{2+}-NTA affinity resin (Qiagen, Germany) that had been equilibrated with lysis buffer. Non-specifically bound proteins were eluted using 200 ml wash buffer (20 mM Tris pH 8.0, 300 mM NaCl, and 30–50 mM imidazole), and the target protein was eluted in ~60 ml elution buffer (lysis buffer containing a constant concentration of 300 mM imidazole). The eluted protein was then purified further using ion exchange chromatography (DEAE, GE Healthcare) and separated using a linear gradient elution of NaCl (0–500 mM with 20 mM Tris–HCl pH 8.0). Then, fractions containing the protein were combined and incubated with PSP (made in our laboratory) overnight at 4°C to cleave the His-tag from tPA. The reaction products were centrifuged, and the supernatant was separated using size-exclusion chromatography (Superdex 200, GE Healthcare) with a mixture of 200 mM NaCl and 20 mM Tris–HCl pH 8.0. Finally, the fractions containing tPA were combined, concentrated, and buffer-exchanged into the final buffer (5 mM Tris–HCl pH 8.0 and 100 mM NaCl) using a 5-kDa cutoff Millipore Amicon concentrator. They were then stored at 193 K for subsequent studies. The concentration of the final purified recombinant protein was ~1.0 mg ml^{-1}.

SDS electrophoresis

To check the purity of the tPA preparations, SDS-PAGE was performed using a 5% stacking gel and a 10%

separating gel in Tris-glycine running buffer. Unless otherwise stated, samples were diluted to a protein concentration of 0.1–0.5 mg ml^{-1} before loading. Samples were boiled 1:2 with loading buffer (70 mM SDS, 100 mM dithiothreitol [DTT], 10% [v/v] glycerol, 0.05 M Tris pH 6.8, 0.06% [w/v] pyronin G). Gels were run in a vertical Mini-PROTEAN Tetra MP4 apparatus (Bio-Rad, USA) and stained with Coomassie Brilliant G250.

Western blotting

Proteins were separated using SDS-PAGE in Tris-glycine buffer, and transferred to polyvinylidene difluoride (PVDF) membranes overnight using a semi-dry electroblotting apparatus (Bio-Rad). The membranes were washed with TBST (10 mM Tris–HCl pH 8.0, 150 mM NaCl, and 0.05% Tween 20) and blocked with 3% (w/v) bovine serum albumin (BSA) for 1 h at room temperature. Subsequently, the membranes were incubated overnight with goat-anti-tPA antibodies (sc-5239, Santa Cruz, CA, USA) at 4°C with permanent shaking, and were then washed and incubated with horseradish peroxidase (HRP)-conjugated secondary antibodies. Finally, the bound antibodies were detected using enhanced chemiluminescence (Millipore, Billerica, MA, USA).

Oligomeric state determination

Size exclusion chromatography was performed using a Superdex 200 column (GE Healthcare) as described previously [33]. The protein of interest or molecular mass standards were applied to the Superdex 200 (GE Healthcare) column, and eluted using 50 mM Tris–HCl pH 8.0 and 150 mM NaCl. The protein molecular weight standards used were albumin bovine V (66.2 kDa), chicken egg albumin (44.2 kDa), chymotrypsinogen A (24.5 kDa), and lysozyme (14.4 kDa). The peak elution volumes (measured at 280 nm) were used to calculate the standard curve using the equation $\log MW = -0.01712\ Ve + 6.05667$, with $R^2 = 0.9726$. tPA was diluted to 1 mg ml^{-1} in buffer (50 mM Tris–HCl pH 8.0 and 150 mM NaCl), and then loaded on the column. tPA was eluted as a single peak in a volume of 72.1 ml, corresponding to an estimated molecular weight of 66.68 kDa.

Assay to assess the activity of the recombinant protein

The plasminogen activation activity of tPA was assessed using agarose-fibrin plates according to methods described previously with minor modifications [34-36]. The agarose-fibrin plates were prepared as follows: 1.0 IU thrombin, 0.75 g plasminogen, and 30 IU human fibrinogen were added to 15 ml of 1.0% agarose gel that had been dissolved in normal saline at 45–55°C. The mixture was incubated at room temperature for 30 min. The sample was then loaded onto the plate, which was incubated at 37°C for 24 h. To determine the activity of the recombinant tPA, the standard (urokinase) was serially diluted and spotted onto the fibrin plates. The activity of the recombinant tPA was then measured according to the diameter of the clear zone on the fibrin plates.

Competing interests

A patent application has been filed for the technology disclosed in this publication.

Authors' contributions

ZJ and WD conceived and designed the experiments. LX and GY was involved in all aspects of the experimental design, data collection, analysis and interpretation. LX and LM performed protein purification, partial proteolysis and serine protease assay experiments, and drafted the manuscript. ZS, ZH, BL, WS and HQ helped to design the experiment, analyzed data, reviewed and revised the manuscript. CK and HA contributed reagents, materials and analysis tools. All authors read and approved the final manuscript.

Acknowledgments

Financial support for this project was provided by research grants from the Chinese National Natural Science Foundation (grant Nos. 81301395), the National Science and Technology Major Projects (2013ZX10002002), the Natural Science Foundation Project of CSTC (grant Nos. 2011BB5124), International Cooperation Project of CSTC (grant Nos. CSTC2012GG-GJHZ0025).

Author details

[1]Department of Cardiology, The First Affiliated Hospital of Chongqing Medical University, Chongqing 400016, China. [2]Key Laboratory of Molecular Biology on Infectious Disease (Ministry of Education), The Second Affiliated Hospital, Chongqing Medical University, Chongqing 400016, China. [3]Key Laboratory of Clinical Laboratory Diagnostics (Ministry of Education), College of Laboratory Medicine, Chongqing Medical University, Chongqing 400016, PR China. [4]Department of Clinical Laboratory, Yubei District People's Hospital, Chongqing 401120, PR China.

References

1. Rouf SA, Moo-Young M, Chisti Y. Tissue-type plasminogen activator: characteristics, applications and production technology. Biotechnol Adv. 1996;14:239–66.
2. Collen D, Lijnen HR. The Tissue-Type Plasminogen Activator Story. Arterioscler Thromb Vasc Biol. 2009;29:1151–5.
3. Collen D, Stassen JM, Marafino BJ, Builder S, DeCock F, Ogez J, et al. Biological properties of human tissue-type plasminogen activator obtained by expression of recombinant DNA in mammalian cells. J Pharmacol Exp Ther. 1984;231:146–52.
4. Manosroi J, Tayapiwatana C, Götz F, Werner RG, Manosroi A. Secretion of Active Recombinant Human Tissue Plasminogen Activator Derivatives in *Escherichia coli*. Appl Environ Microbiol. 2001;67:2657–64.
5. Mattes R. The production of improved tissue-type plasminogen activator in *Escherichia coli*. Semin Thromb Hemost. 2001;27:325–36.
6. Lijnen HR, Collen D. Strategies for the improvement of thrombolytic agents. Thromb Haemost. 1991;66:88–110.
7. Vindigni A, Cera ED. Role of P225 and the C136-C201 disulfide bond in tissue plasminogen activator. Protein Sci. 1998;7:1728–37.
8. Hua ZC. Renaturation and purification of recombinant tissue-type plasminogen activator expressed in *E. coli*. Biochem Mol Biol Intl. 1997;41:815–20.
9. Wilhelm OG, Jaskunas SR, Vlahos CJ, Bang NU. Functional properties of the recombinant kringle-2 domain of tissue plasminogen activator produced in *Escherichia coli*. J Biol Chem. 1990;265:14606–11.
10. Lee H-J, Im H. Soluble Expression and Purification of Human Tissue-type Plasminogen Activator Protease Domain. Bull Korean Chem Soc. 2010;31:2607–12.
11. Studier FW, Moffatt BA. Use of bacterio phage T7 RNA polymerase to direct selective high-level expression of cloned genes. J Mol Biol. 1986;189:113–30.

12. Dubendorff JW, Studier FW. Controlling basal expression in an inducible T7 expression system by blocking the target T7 promoter with *lac* repressor. J Mol Biol. 1991;219:45–59.
13. Grossman TH, Kawasaki ES, Punreddy SR, Osburne MS. Spontaneous cAMP-dependent derepression of gene expression in stationary phase plays a role in recombinant expression instability. Gene. 1998;209:95–103.
14. Studier FW. Protein production by auto-induction in high-density shaking cultures. Protein Expr Purif. 2005;41:207–34.
15. Yamada T, Shimada Y, Kikuchi M. Integrin-specific tissue-type plasminogen activator engineered by introduction of the Arg-Gly-Asp sequence. Biochem Biophys Res Commun. 1996;228:306–11.
16. Baca AM, Hol WG. Overcoming codon bias: a method for high-level overexpression of Plasmodium and other AT-rich parasite genes in Escherichia coli. Int J Parasitol. 2000;30:113–8.
17. Bessette PH, Aslund F, Beckwith J, Georgiou G. Efficient folding of proteins with multiple disulfide bonds in the Escherichia coli cytoplasm. Proc Natl Acad Sci USA. 1999;96:13703–8.
18. Li Z, Kessler W, van den Heuvel J, Rinas U. Simple defined autoinduction medium for high-level recombinant protein production using T7-based Escherichia coli expression systems. Appl Microbiol Biotechnol. 2011;91:1203–13.
19. Ploug M, Plesner T, Rønne E, Ellis V, Høyer-Hansen G, Hansen NE, et al. The receptor for urokinase-type plasminogen activator is deficient on peripheral blood leukocytes in patients with paroxysmal nocturnal hemoglobinuria. Blood. 1992;79:1447–55.
20. Obukowicz MG, Gustafson ME, Junger KD, Leimgruber RM, Wittwer AJ, Wun TC, et al. Secretion of active kringle-2-serine protease in *Escherichia coli*. Biochemistry. 1990;29:9737–45.
21. Qiu J, Swartz JR, Georgiou G. Expression of active human tissue-type plasminogen activator in Escherichia coli. Appl Environ Microbiol. 1998;64:4891–6.
22. Geng Y, Wang S, Qi Q. Expression of active recombinant human tissue-type plasminogen activator by using in vivo polyhydroxybutyrate granule display. Appl Environ Microbiol. 2010;76:7226–30.
23. Molloy PE, Harris WJ, Strachan G, Watts C, Cunningham C. Production of soluble single-chain T-cell receptor fragments in *Escherichia coli* trxB mutants. Mol Immunol. 1998;35:73–81.
24. Waldo GS, Standish BM, Berendzen J, Terwilliger TC. Rapid protein-folding assay using green fluorescent protein. Nat Biotechnol. 1999;17:691–5.
25. Ekici OD, Paetzel M, Dalbey RE. Unconventional serine proteases: variations on the catalytic Ser/His/Asp triad configuration. Protein Sci. 2008;17:2023–37.
26. Pennica D, Holmes WE, Kohr WJ, Harkins RN, Vehar GA, Ward CA, et al. Cloning and expression of human tissue-type plasminogen activator cDNA in E. coli. Nature. 1983;30:214–21.
27. Desai BJ, Boothello RS, Mehta AY, Scarsdale JN, Wright HT, Desai UR. Interaction of thrombin with sucrose octasulfate. Biochemistry. 2011;50:6973–82.
28. Iwata A, Maruyama M, Akagi T, Hashikawa T, Kanazawa I, Tsuji S, et al. Alpha-synuclein degradation by serine protease neurosin: implication for pathogenesis of synucleinopathies. Hum Mol Genet. 2003;12:2625–35.
29. Biswas T, Small J, Vandal O, Odaira T, Deng H, Ehrt S, et al. Structural insight into serine protease Rv3671c that Protects M. tuberculosis from oxidative and acidic stress. Structure. 2010;18:1353–63.
30. Waxman L, Darke PL. The herpesvirus proteases as targets for antiviral chemotherapy. Antivir Chem Chemother. 2000;11:1–22.
31. Batra R, Khayat R, Tong L. Molecular mechanism for dimerization to regulate the catalytic activity of human cytomegalovirus protease. Nat Struct Biol. 2001;8:810–7.
32. Li W, Srinivasula SM, Chai J, Li P, Wu JW, Zhang Z, et al. Structural insights into the pro-apoptotic function of mitochondrial serine protease HtrA2/Omi. Nat Struct Biol. 2002;9:436–41.
33. Tang J, Luo M, Niu S, Zhou H, Cai X, Zhang W, et al. The Crystal Structure of Hexamer RraA from *Pseudomonas Aeruginosa* Reveals Six Conserved Protein–Protein Interaction Sites. Protein J. 2010;29:583–90.
34. Granelli-Piperno A, Reich E. A study of proteases and proteaseinhibitor complexes in biological fluids. J Exp Med. 1978;148:223–34.
35. Jiao J, Yu M, Ru B. Characterization of a recombinant chimeric plasminogen activator with enhanced fibrin binding. Biochim Biophys Acta. 2001;1546:399–405.
36. Ding FX, Yan HL, Lu YM, Xue G, Mei Q, Huang JJ, et al. Cloning, purification and biochemical characterization of a thrombus-ditargeting thrombolytic agent, comprised of annexin B1, ScuPA-32 K and fibrin-adherent peptide. J Biotechnol. 2006;1126:394–405.

Comparison of mcl-Poly(3-hydroxyalkanoates) synthesis by different *Pseudomonas putida* strains from crude glycerol: citrate accumulates at high titer under PHA-producing conditions

Ignacio Poblete-Castro[1,2*], Danielle Binger[3], Rene Oehlert[2] and Manfred Rohde[4]

Abstract

Background: Achieving a sustainable society requires, among other things, the use of renewable feedstocks to replace chemicals obtained from petroleum-derived compounds. Crude glycerol synthesized inexpensively as a byproduct of biodiesel production is currently considered a waste product, which can potentially be converted into value-added compounds by bacterial fermentation. This study aimed at evaluating several characterized *P. putida* strains to produce medium-chain-length poly(3-hydroxyalkanoates) (mcl-PHA) using raw glycerol as the only carbon/energy source.

Results: Among all tested strains, *P. putida* KT2440 most efficiently synthesized mcl-PHA under nitrogen-limiting conditions, amassing more than 34% of its cell dry weight as PHA. Disruption of the PHA depolymerase gene (*phaZ*) in *P. putida* KT2440 enhanced the biopolymer titer up to 47% PHA (%wt/wt). The low biomass and PHA titer found in the mutant strain and the wild-type strain KT2440 seems to be triggered by the high production of the side-product citrate during the fermentation process which shows a high yield of 0.6 g/g.

Conclusions: Overall, this work demonstrates the importance of choosing an appropriate microbe for the synthesis of mcl-PHA from waste materials, and a close inspection of the cell metabolism in order to identify undesired compounds that diminish the availability of precursors in the synthesis of biopolymers such as polyhydroxyalkanoates. Future metabolic engineering works should focus on reducing the production of citrate in order to modulate resource allocation in the cell's metabolism of *P. putida*, and finally increase the biopolymer production.

Keywords: *Pseudomonas putida* strains, mcl-polyhydroxyalkanoates, Raw glycerol, PHA depolymerase, Metabolic engineering, Citrate

Background

The consumption of non-renewable materials for industrial production of chemicals has given rise to environmental and energy concerns within society. Plastic products are considered essential materials to meet the needs of our current lifestyle and modern manufacturing. As an attempt to move towards a sustainable society, research has been undertaken to develop the microbial fermentation of renewable resources for the synthesis of biopolymers [1,2]. An example of the sustainable production of biopolymers is the large-scale production of polyhydroxyalkanoates (PHAs), which exhibit similar physical and mechanical properties to oil-based thermoplastics [1]. Bacteria have the capability to naturally accumulate PHAs as carbon/energy storage materials in their cytoplasmic space, usually when there is a nutrient imbalance in the environment: a high carbon concentration accompanied by the limitation of an inorganic nutrient [3]. Taking advantage of this evolutionary bacterial trait, commercialized PHAs are currently being produced by Metabolix (USA), Meridian Inc. (USA),

* Correspondence: ignacio.poblete@unab.cl
[1]Facultad de Ciencias Biológicas, Center for Bioinformatic and Integrative Biology, Biosystems Engineering Laboratory, Universidad Andrés Bello, Santiago 8340176, Chile
[2]Helmholtz Centre for Infection Biology, Microbial Drugs Group, Braunschweig D-38124, Germany
Full list of author information is available at the end of the article

Biocycle (Brazil), and Biomer (Germany), to name a few. They use sugars (cane or beet) and plant-based fatty acids as carbon feedstocks for the fermentation process in stirred-tank bioreactors, and obtain a diverse assortment of biopolymers. Despite the advances, one of the main drawbacks of PHAs for better market positioning is the high costs related to the production process [4,5]. To truly develop the synthesis of these polyesters and make them more cost-competitive against petroleum-based thermoplastics, it is crucial to find novel sustainable alternatives for large-scale production. Recently, production of PHAs using waste materials from agriculture residues and waste water treatment streams as a carbon substrate has shown to be a very promising alternative [6-8], since raw materials account for the majority of production costs in the bacterial synthesis of PHAs [4,9]. These processes have mainly focused on the synthesis of Poly(3-hydroxybutyrate) (PHB), the first and best-characterized PHA [1]. Despite the large range of applications possible with PHB and its co-polymer Poly(3-hydroxybutyrate-valerate) (PHBV), there is still an urgent demand for other types of biopolymers with unique physical and mechanical properties, with a particular emphasis on medical applications [10]. Medium chain length polyhydroxyalkanoates (mcl-PHAs) are polyesters synthesized mainly by gram-negative bacteria; most studies on their synthesis have focused on the use of Pseudomonas species as microbial cell factories [11-14]. By applying different metabolic engineering strategies [15-17] and fermentation conditions [18,19] the titer of this biopolymer was successfully improved, and it was demonstrated also that the monomer composition of the produced mcl-PHAs can be tuned [20-22]. In addition, given the high metabolic versatility shown by some *Pseudomonas putida* strains [23], waste materials from different streams have been used as carbon substrate to produce mcl-PHAs. Efficient synthesis of mcl-PHAs was obtained from animal wastes [24], PET [25] and raw glycerol [26]. Also, the conversion of biomass-derived compounds into mcl-PHA has been recently achieved [27,28]. One of the most promising raw materials for the synthesis of biobased chemicals is glycerol. The commercial price of this polyol has drastically decreased in the last decade due to the large industrial production of biodiesel from fatty acids, where glycerol is produced as a byproduct of the esterification process. For every 10 tons of biodiesel, 1 ton of glycerol is formed. Because of this, there is currently an oversupply of this commodity worldwide which has resulted in its classification as a waste product instead of a valuable good [29]. *P. putida* strains have been recently described to be capable of synthesizing mcl-PHAs from raw glycerol [25,26]. In this work, several *P. putida* strains were challenged to growth in raw glycerol as carbon and energy source and compared for their capability to produce mcl-PHA. The best PHA-producing *P. putida* strain was further engineered to enhance the synthesis of mcl-PHA in the cell. An exo-metabolome analysis was performed to identify and quantify byproduct formation during the fermentation process in bioreactors, and finally the causes for the low yield of biomass and biopolymers in *Pseudomonas putida* fed with glycerol were elucidated.

Results and discussion

Raw glycerol is produced as a byproduct of the chemical conversion of fatty acids into biodiesel, a process that results in the following basic products: glycerol (60-80%), water (10%), organic matter (8%), ash (8%), and methanol (0.5-20%). Although methanol concentration seems to be minor, it exerts a high degree of toxicity to this feed stock material, thus limiting its use in industrial biotechnology. Bacteria from the genus *Pseudomonas* — especially *Pseudomonas putida*— possess a high metabolic versatility which allow them to colonize harsh and toxic environments [30]. In an attempt to test whether this raw material is suitable for the efficient production of medium chain length polyhydroxyalkanoates (mcl-PHAs), we selected three well-known PHA-producing strains: *Pseudomonas putida* KT2440, KT2442, and F1, and the solvent-tolerant *P. putida* S12 (Table 1). All of them have been previously reported to be able to grow in pure glycerol [12,31,32]. As shown in Figure 1, each *P. putida* strain grew well in minimal medium supplemented with 3 (g/L) of raw glycerol as carbon source. It is noteworthy to mention that we have not filtered or in any way modified the industrial byproduct glycerol. As already reported, *P. putida* KT2440 showed a long lag-phase (8 hours, this work) when grown on glycerol [33]. Escapa and co-workers found that by adding a small

Table 1 Features of *Pseudomonas putida* strains used in this study

Strain	Genome size (bp)	GC content (%)	Isolated in	Solvent-tolerant	Biotechnological applications
P. putida KT2440	6.181.863	61.5	Japan	No	Bioremediation Synthesis of biomaterials and chemicals
P. putida KT2442	N. K	N. K	Spontaneous rifampicin resistant	No	Bioremediation Synthesis of biomaterials and chemicals
P. putida F1	5.959.964	61.9	USA	No	Degradation of chloroaromatic compounds
P. putida S12	6.284.656	61.5	The Netherlands	Yes	Synthesis of added-value chemicals

N.K. Not known.

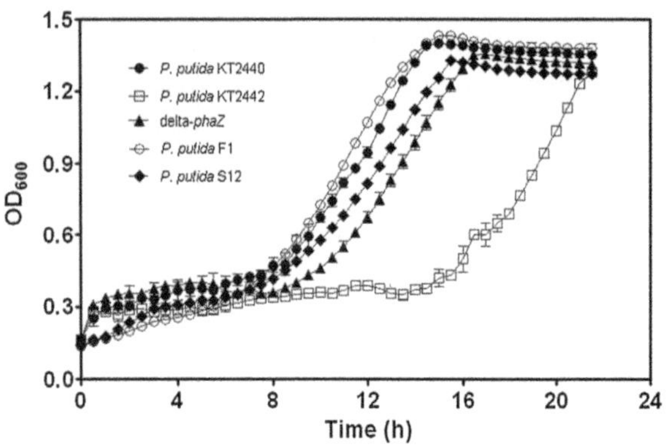

Figure 1 ***P. putida*** **strains grown on raw glycerol (3 g/L) in batch cultures.** Values are the means of three independent biological experiments.

amount of octanoate or glucose, the length of the lag-phase can be reduced. In addition, elimination of the lag-phase was also achieved by knocking out the transcriptional regulator *glpR* (PP_1074). In this work, all tested *P. putida* strains have a similar lag-phase to that shown by KT2440, except for KT2442, which took more than 15 hours to begin the exponential growth phase (Figure 1).

Evaluation of PHA synthesis on different *Pseudomonas putida* strains in shake flask experiments

We next challenged *Pseudomonas putida* KT2440, KT2442, S12, and F1 to grow in minimal medium (M9) supplemented with 30 (g/L) of crude glycerol as the sole carbon source in batch cultures. After 72 hours of cultivation, cells were harvested so as to quantify the biomass and mcl-PHA content. *P. putida* KT2442, S12, and F1 were found to have a similar biomass concentration (Table 2). This was not the case for PHA production, where *P. putida* KT2442 produced twice as much biopolymer in comparison to the other two strains (Table 2). *P. putida* KT2440 was by far the most efficient PHA-producing organism among all tested strains, amassing 35% of its CDW as PHA, and achieving a biomass yield of 0.048 (g/g). This demonstrates the intrinsic metabolic differences between *P. putida* KT2440 and KT2442, which highly influence their PHA production capacity in the presence of PHA-unrelated carbon sources (glycerol, gluconate, and glucose). In this regard, a similar response has been found when both strains were compared in batch cultures using gluconate as carbon source; the production of mcl-PHA was 10-fold higher for KT2440 relative to KT2442 [34]. This phenomenon might be attributed to an inefficient transport capacity of 2-ketogluconate in *P. putida* KT2442, which would diminish the amount of available carbon within the cell, and negatively impact PHA synthesis. As glycerol is metabolized via a different metabolic route compared to the one used for gluconate, no gluconate-related oxidized compounds, e.g. 2-ketogluconate, are formed in the upper catabolic pathways. This may explain the reduced amount of synthesized biomass and mcl-PHA in *P. putida* KT2442.

Deletion of the *phaZ* gene in *P. putida* KT2440 enhances mcl-PHA synthesis using raw glycerol as a carbon and energy source

One of the first strategies for maintaining the titer of PHA inside of cells during the fermentation process is to remove the gene responsible for the depolymerization process of PHA (PHA depolymerase). However, the role of PhaZ proteins in the synthesis of mcl-PHA in *P. putida* strains is not well understood. In a previous work, disruption of the PHA depolymerase gene (PP_5004, *phaZ*) in *P. putida* KT2442 led to an increase in PHA synthesis using fatty acids as carbon source, but not when growing on glucose or gluconate [15], where the mutant strain produced less PHA in comparison to the wild-type strain. One could argue then, since glycerol is also a non-related PHA carbon source, it should show the same PHA-accumulation pattern as glucose or gluconate, resulting ultimately in no increase of the biopolymer. Nevertheless, in the study done by [15], they generated a *phaZ* mutant strain in KT2442, and as demonstrated in the research presented in this paper, this strain produces less mcl-PHA from no-alkanoate substrates. Therefore we generated a knockout mutant strain in the best PHA-producing strain, *P. putida* KT2440. The chromosomal

Table 2 Monomer composition of medium chain length PHA produced by different *P. putida* strains

Strain	CDW (g/L)	PHA (%wt)	PHA (g/L)	Monomer composition (%)					
				C6	C8	C10	C12:1	C12	C14:0
P. putida **KT2440**	4.23 ± 0.11	34.5	1.46 ± 0.21	0.83	16.2	74.6	2.9	6.0	0.59
P. putida **KT2442**	3.45 ± 0.07	26.5	0.91 ± 0.13	t.d.	7.8	78.5	3.6	9.2	0.95
P. putida **F1**	3.50 ± 0.14	10.3	0.36 ± 0.01	t.d.	14.4	71.3	3.9	11.4	t.d
P. putida **S12**	3.20 ± 0.07	12.6	0.40 ± 0.05	t.d.	4.0	70.9	4.5	20.1	t.d.
Δ*phaZ*	4.20 ± 0.21	46.8	1.94 ± 0.17	1.1	16.3	72.3	3.4	6.9	t.d

The data were determined by GC/MS and are given as relative molar fraction (%) of C6: 3-hydroxyexanoate, C8: 3-hydroxyoctanoate, C10:3-hydroxydecanoate, C12: 3-hydroxydodecanoate, C12:1: 3-hydroxy-5-cis-dodecanoate, and C14: 3-hydroxytetradecanoate.

deletion of *phaZ* (PP_5004) in *P. putida* KT2440 was performed as described in Methods. The Δ*phaZ* mutant strain was then subjected to PHA-production conditions in flask cultures in the same manner as described above with the wild type strain KT2440. There was no difference in total biomass yield (Table 2) between Δ*phaZ* and *P. putida* KT2440 wild type, however and most importantly, the first improved the PHA titer by 34% (Table 2). PHA depolymerase is believed to be required for the efficient production of the polyester —under PHA-accumulating conditions in batch culture— in *P. putida* KT2442 and *P. putida* U using fatty acids as the carbon substrate [35-37]. However, the opposite result has been reported for the same mutant strain (Δ*phaZ*) in *P. putida* KT2442 [15], and *P. putida* U [38]. Because we measured more mcl-PHA in Δ*phaZ* than its parent strain KT2440, we conclude that PHA depolymerase does not impose a negative effect on the PHA synthesis machinery. The PHA cluster genes orchestrates a simultaneous process of PHA synthesis and hydrolysis [35], making it impossible for the cell to degrade the PHA granule and further mobilize 3-hydroxyalkanoic acids for energy generation; this may be the primary reason for the greater accumulation of mcl-PHA in the Δ*phaZ* mutant strain when grown on glycerol.

PHA morphology and monomer composition of PHA in *P. putida* strains grown on pure and raw glycerol

As raw glycerol contains methanol and other toxic compounds, it may impose a stressful environment to cells. To get insight into the structural changes of cells and the accumulated PHA in both strains while growing in raw and pure glycerol, field emission scanning and transmission electron microscopy was applied. Micrographs were taken of cells at 72 h of cultivation, where maximum accumulation of PHA was obtained. Cells of the mutant and the wild-type strain did not show changes in their morphology when grown on raw or pure glycerol (Figure 2A,B,C,D), instead cells cultivated on crude glycerol stick together, where some of the impurities contained in the growing substrate can be observed (Figure 2B,D). The deletion of *phaZ gene* in *P. putida* KT2440 had no effect on PHA granules (Figure 2G,H). A different scenario was observed for PHA granules produced in both strains on raw glycerol, where the number of inclusion bodies was higher and more irregular in comparison to granules synthesized in cells cultivated on pure glycerol (Figure 2I,J,K,L).

The dominant monomer characterized using GC-MS was 3-OH decanoic acid (C10) for all of the tested strains (Table 2). This has been shown to be true for various *P. putida* strains when they synthesize PHA from gluconate, glucose, or glycerol [12,16]. First, glycerol is converted in several steps to yield acetyl-CoA. Next, the glycerol is either used in the TCA cycle or used for the synthesis of *de novo* fatty acids (Figure 3). These *de novo* fatty acids are the precursor for PHA synthesis. Since 3-hydroxyacyl-ACP is restricted to 10 carbon substrates [39], a high content of 3-hydroxydecanoic acid is produced in the PHA polymerization cycle [16].

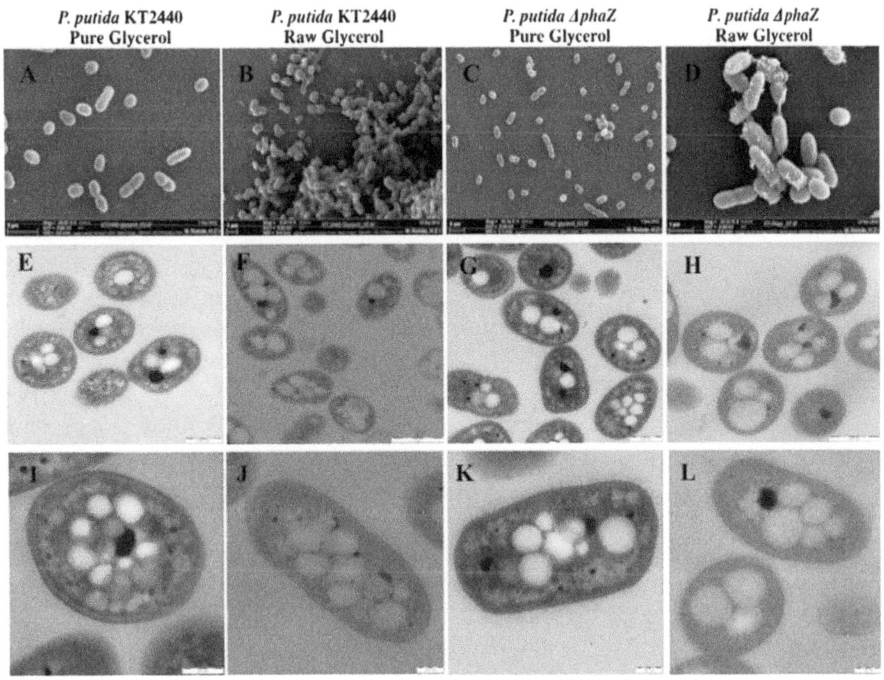

Figure 2 Transmission electron micrographs of thin section of *P. putida* KT2440 (E,I, F,J) and Δ*phaZ* mutant strain (G,H,K,L). Field emission scanning electron microscopy of *P. putida* KT2440 **(A,B)** and Δ*phaZ* mutant strain **(C,D)**.

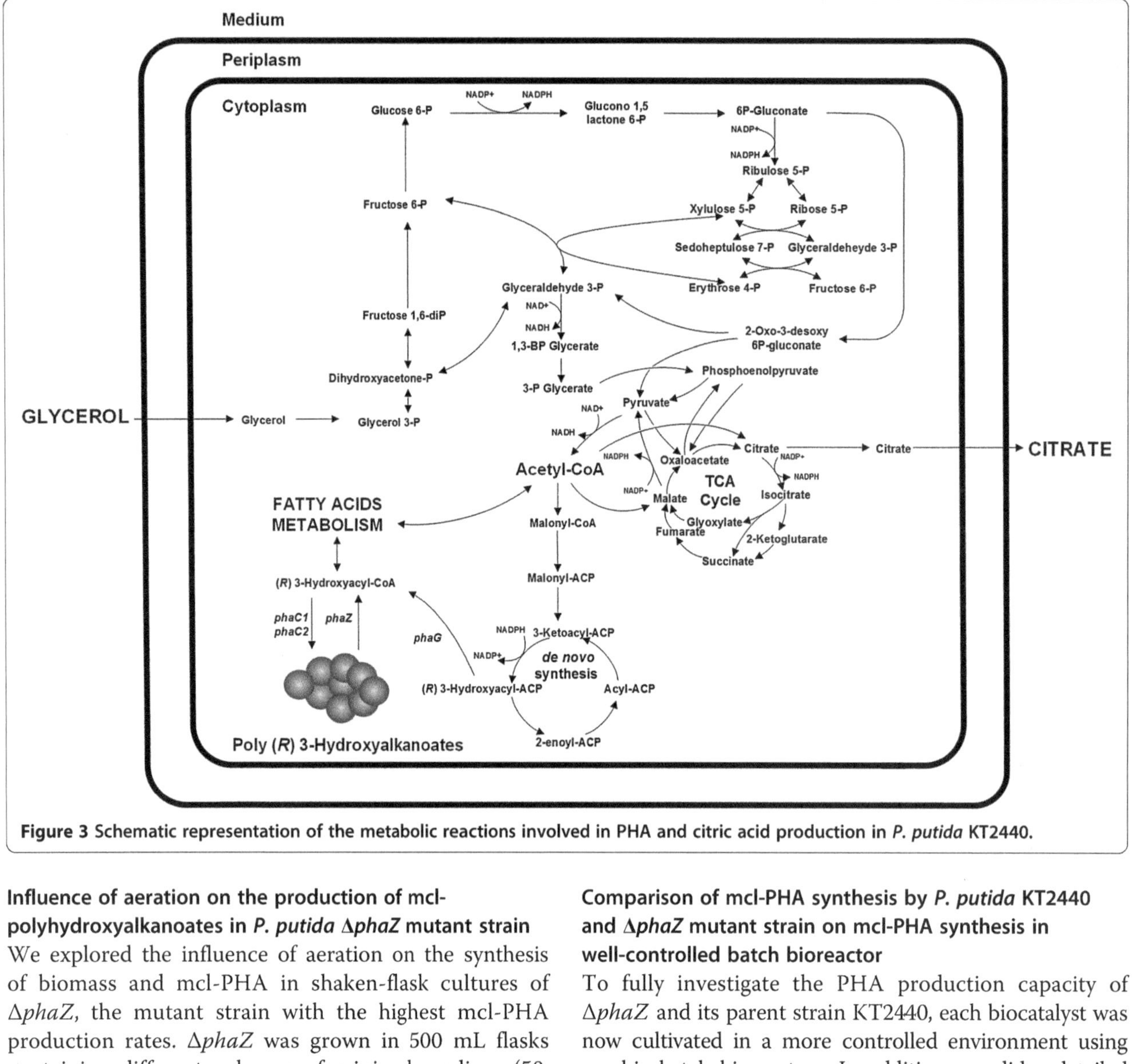

Figure 3 Schematic representation of the metabolic reactions involved in PHA and citric acid production in *P. putida* KT2440.

Influence of aeration on the production of mcl-polyhydroxyalkanoates in *P. putida* Δ*phaZ* mutant strain

We explored the influence of aeration on the synthesis of biomass and mcl-PHA in shaken-flask cultures of Δ*phaZ*, the mutant strain with the highest mcl-PHA production rates. Δ*phaZ* was grown in 500 mL flasks containing different volumes of minimal medium (50, 100, 200, 300, 400 mL, which correspond to a 0.1, 0.2, 0.4, 0.6, and 0.8 volume ratio, respectively) supplemented with 30 g/L raw glycerol. It was found that volume ratios higher than 0.2 exert a negative influence on biomass formation (Figure 4). In other words, as less oxygen is available with increasing the volume ratio, the oxidation state of the cell is disturbed, which resulted in inefficient energy generation, negatively impacting the production of biomass. PHA synthesis, in turn, also suffers from less oxic conditions, showing a 6-fold decrease of final PHA concentration (Figure 4). From this experiment we can conclude that PHA synthesis benefits from a well-aerated environment when raw glycerol is used as carbon source. This is crucial when developing an efficient PHA production process since biomass formation is critical to the production of biopolymers [19].

Comparison of mcl-PHA synthesis by *P. putida* KT2440 and Δ*phaZ* mutant strain on mcl-PHA synthesis in well-controlled batch bioreactor

To fully investigate the PHA production capacity of Δ*phaZ* and its parent strain KT2440, each biocatalyst was now cultivated in a more controlled environment using aerobic batch bioreactors. In addition, we did a detailed inspection of the exo-metabolome of the fermentation broth throughout the entire cultivation period via HPLC in order to characterize and quantify possible byproducts (citrate, formate, pyruvate, succinate, malate, acetate, see Methods section) produced by the cell. Both strains had an initial lag-phase of 7 h, after that time they started growing to a maximum specific growth rate of 0.22 h^{-1}. Ammonium was below the detection limit at 18 h and 24 h for Δ*phaZ* and *P. putida* KT2440, respectively. As a consequence of nitrogen limitation, both strains began rapidly accumulating PHA until glycerol was completely consumed (Figure 5). Although Δ*phaZ* and *P. putida* KT2440 showed a similar total biomass yield when grown on glycerol (Table 2), Δ*phaZ* synthesized more PHA (~36% higher) in comparison to KT2440. It can be clearly seen in Figure 5 that Δ*phaZ* strain accumulates

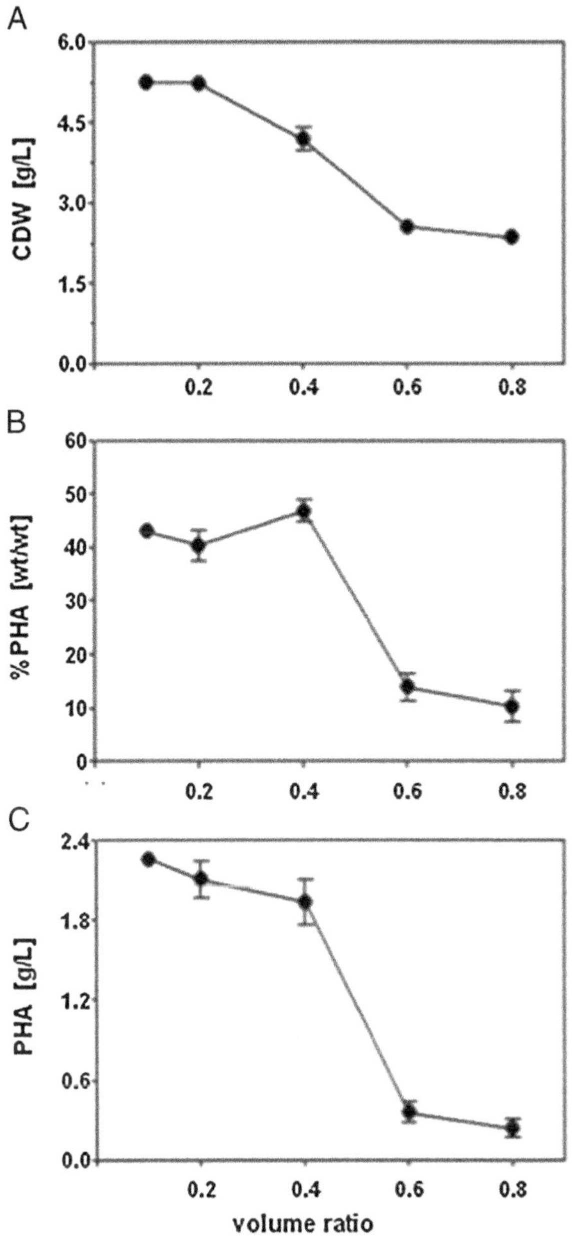

Figure 4 CDW (A), %PHA (B), and PHA (C) content of Δ*phaZ* mutant strain grown on 30 g/L raw glycerol with different volume ration in flask batch experiments. Values are the mean of three independent biological experiments.

the biopolymer while consuming the available carbon source in the culture broth. Thus, the absence of this *loci* (PP_5004, *phaZ*) is not crucial, as previously believed, for the efficient synthesis of mcl-PHA in *Pseudomonas putida*. Another important finding was the detection of citric acid via HPLC, which reached a concentration of more than 20 g/L at the end of the cultivation period (Figure 5). It is important to mention that no other organic acid was detected in the broth culture during the fermentation process. Citrate has a final yield of 0.6 [g/g] for both the mutant and the wild type *P. putida* strain when grown on glycerol. This seems to be the principal reason for the observed biomass and PHA yield for strains grown on glycerol. Once glycerol is metabolized, the pool of acetyl-CoA replenishes both the TCA cycle and the *de novo* synthesis of fatty acids (Figure 3). It seems that more carbon was delivered to the TCA cycle since a higher amount of carbon was found in the form of citric acid than mcl-PHA. As methanol concentration in the culture broth is 0.188 g/L (see Methods section), this compound cannot be the principal factor influencing the final titer of citrate obtained at the end of the fermentation period (Figures 3, 5). A possible explanation of the high accumulation of citrate under the tested conditions is that PHA-producing environments (nitrogen limitation) increase the ratio of NAPH/NADP and/or NADH/NAD into the cell [40,41], which eventually results in repression of several enzymes belonging to the TCA cycle. We recently demonstrated that these cofactors play an important role in the synthesis of mcl-PHA in *P. putida* KT2440, where NADPH is consumed to increase the biopolymer production within the cell, resulting in a high level of NADH/NAD ratios [17]. One of the most affected proteins under high NADH levels is isocitrate dehydrogenase (PP_4012 and PP_4011), which catalyzes the conversion of isocitrate into oxalosuccinate. This effect limits the proper functioning of the Krebs cycle to further oxidize citrate. We have previously reported that intracellular concentrations of citrate were higher under elevated PHA levels in continuous cultivations using decanoate as carbon source, whereas both enzymes isocitrate dehydrogenase and lyase had low expression levels at the transcriptome and proteome level [13]. It is well documented that citric acid-producing microorganisms like the yeast *Yarrowia lipolytica* and the industrial mold *Aspergillus niger* (Table 3) accumulate large amount of citric acids (>150 g/L) and lipids under nitrogen limiting conditions [42-45]. As they do not posses the PHA machinery to naturally produce the polyester, they synthesize lipids instead, mainly in the form of triacylglycerol, reaching up to 50% of its CDW as lipids [46]. In yeast, N-limitation environments have been found to trigger the drop in the concentration of adenosine monophosphate (AMP) [47], thus inhibiting the isocitrate dehydrogenase leading to the accumulation of citrate and isocitrate [42]. As explained above, it seems that a similar mechanism drives the metabolic response of *P. putida* KT2440 under the PHA-producing conditions tested in this study. In order to fully confirm this hypothesis, further studies should aim at evaluating the intracellular flux distribution as well as the concentration of AMP, NADP/NADPH, and NAD/NADH ratios, in a more appropriate system such as continuous cultures (chemostat), where there are no changes in the specific growth rate or interferences by

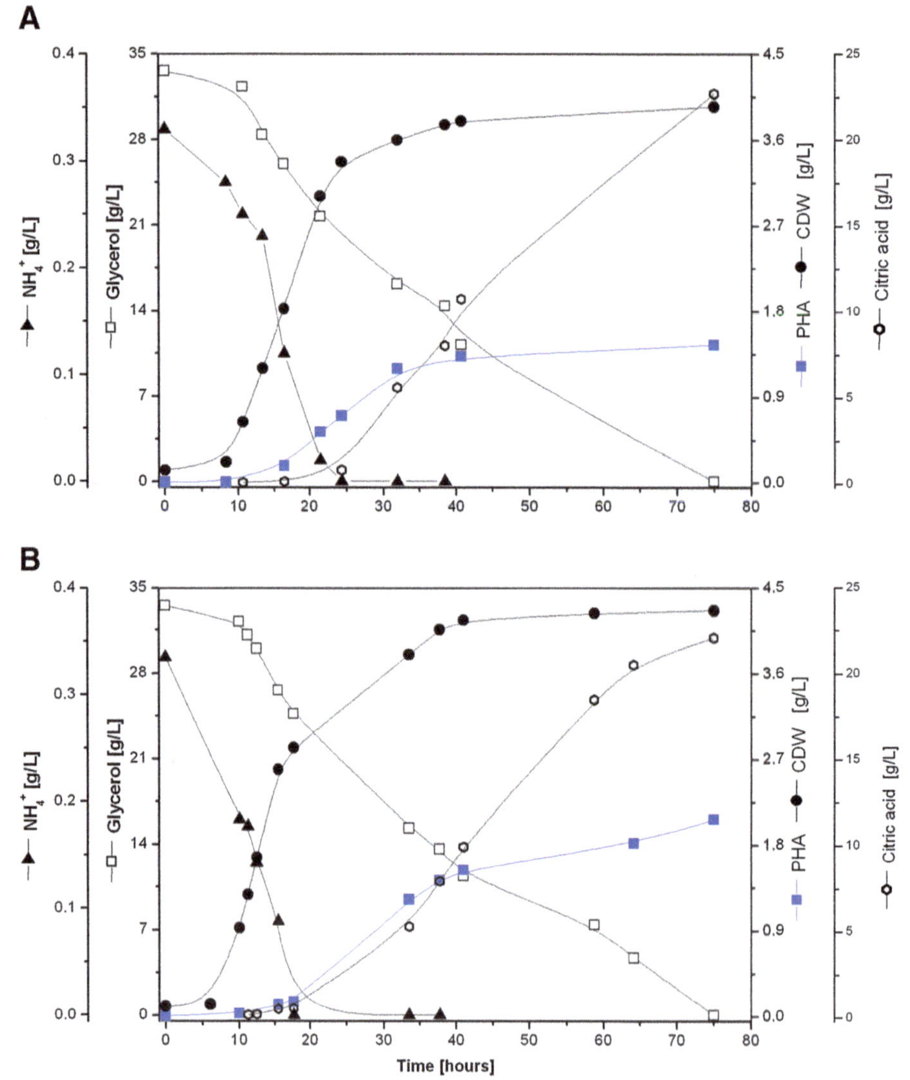

Figure 5 Fermentation profile of *P. putida* KT2440 (A) and Δ*phaZ* mutant strain (B) in batch cultures using raw glycerol as carbon source (30 g/L). Values are the means of two independent biological experiments.

carbon catabolic repression [48], which affect the levels of the redox carrier compounds in the cell [49,50].

Conclusions

This study shows that bacteria from the genus *Pseudomonas* can cope with the adverse conditions imposed by the chemical mixture of which raw glycerol from biodiesel production is composed. It also demonstrates the use of this cheap material in the efficient synthesis of mcl-PHA. Among all tested strains, *P. putida* KT2440 was the most efficient at mcl-PHA synthesis, amassing more than 34% of its CDW as PHA using raw glycerol

Table 3 Citric acid-producing microorganisms using different carbon sources

Strain	Carbon substrate	Fermentation mode	Ycitrate/S (g/g)	Citrate (g/L)	Ref.
***Y. lipolytica* mutant**	Rapseed Oil	Batch	1.5	175	[51]
***Y. lipolytica* A-101**	Raw Glycerol	Batch	0.6	112	[52]
***A. niger* mutant**	Beet, Cane Molasse	Batch	0.7	113.5	[53]
A. niger	Glucose	-	0.9	200	[54]
***P. putida* KT2440**	Raw Glycerol	Batch	0.6	20	This study
P. putida* Δ*phaZ	Raw Glycerol	Batch	0.6	20	This study

as its only energy/carbon source. Disruption of the PHA depolymerase gene (*phaZ*) in *P. putida* KT2440 enhanced the biopolymer titer up to 47% (%PHA/wt). The low biomass and PHA titer found in *P. putida* KT2440 and the Δ*phaZ* strain on glycerol seems to be caused by the production of high levels of citrate during the fermentation process. Future metabolic engineering works should focus on eliminating or reducing the production of this organic acid in order to modulate resource allocation in the cell's metabolism, and finally increase the specific volumetric productivity of PHA that is crucial for a cost-effective biopolymer production.

Methods

Bacterial strains

P. putida KT2440 (DSM 6125) and F1 (DSM 6899) were obtained from the German Collection of Microorganisms and Cell Culture (DSMZ, Braunschweig, Germany). *P. putida* KT2442 [55] and *P. putida* S12 [56] were obtained from CSIC Madrid (Spain) and RWTH Aachen (Germany), respectively.

Construction of the *phaZ* knock-out mutant strain in *P. putida* KT2440

All PCR fragments generated for vector construction were subcloned into plasmid pCR®2.1 (Invitrogen, CA, USA), transformed into *E. coli* DH5α (Invitrogen, CA, USA), and validated by sequencing. Plasmid pEMG [57] was used for the construction of pEMG_ΔPP5004. Approximately 600 bp of the upstream and downstream regions of gene PP5004 were amplified using primers PP5004_UP_fw, PP5004_UP_rev, PP5004_DOWN_fw and PP5004_DOWN_rev (Table 4), Taq DNA polymerase (Qiagen, Venlo, The Netherlands) and genomic DNA from strain *P. putida* KT2440. Both PCR fragments were fused by PCR [58] and the resulting fusion PCR product was integrated into pEMG via *Xba*I-*Kpn*I restriction sites to generate pEMG_ΔPP5004. The resulting vector pEMG_ΔPP5004 was transformed into *E. coli* CC118λpir to generate donor cells for later tri-parental mating with *P. putida* KT2440 [55] and the helper strain *E. coli* HB101 as described by [59].

To generate a single mutant of gene PP5004, genome editing was applied [57]. Therefore, vector pEMG_ΔPP5004 was co-integrated by a single crossover into the chromosome of *P. putida* KT2440 using tri-parental mating with E. coli HB101 as helper strain, as well as the donor strain E. coli CC118λpir pEMG_ΔPP5004. Successful homologous integration of the vector DNA and the successful genomic deletions was confirmed by PCR (data not shown).

Cultivation conditions

Flask experiments

P. putida strains, kept as frozen stock in 25% glycerol at −80°C, were streaked on Luria Bertani agar plates and incubated for one day at 30°C. Single colonies were then picked from the plate and inoculated it into a 50 mL shake flask containing 10 ml of the above described medium (M9) and incubated overnight under aerobic conditions at 30°C and 180 rpm (Innova, Enfield, USA) set. By taking a calculated volume of the obtained cell suspension from the pre-culture (to begin the PHA-accumulating process with an initial OD of 0.05), 500 mL baffled shake flasks containing 100 mL of culture medium were inoculated and placed in a rotary shaker under aerobic conditions at 30°C. Each culture was carried out by triplicate. *P. putida* strains were grown in a defined M9 mineral medium consisting of (per liter) 12.8 g $Na_2HPO_4{\cdot}7H_2O$, 3 g KH_2O_4, 1 g NH_4Cl, and 0.5 g NaCl. This basic solution was autoclaved and subsequently supplemented with 0.12 g of $MgSO_4{\cdot}H_2O$, trace elements (mg/L): 6.0 $FeSO_4{\cdot}7H_2O$, 2.7 $CaCO_3$, 2.0 $ZnSO_4{\cdot}H2O$, 1.16 $MnSO_4{\cdot}H_2O$, 0.37 $CoSO_47H_2O$, 0.33 $CuSO_4{\cdot}5H_2O$, 0.08 H_3BO_3 (all filter-sterilized). 30 g/L of raw glycerol was then added to the medium which was previously autoclaved. The raw glycerol used in all experiments was obtained from the company Cremer Oleo (Hamburg, Germany). It contained 80% glycerol, 0.5% methanol, 10% ash, 3% organic matter, and 6.5% water.

Bioreactor fermentations

P. putida strains were grown in M9 medium supplemented with 30 g/L raw glycerol. Bioreactor batch fermentations were carried out in a 2 L top-bench BIOSTAT B1 bioreactor (Sartorius B Systems GmbH, Melsungen, Germany) with a working volume of 1.5 L, at 30°C. The aeration rate was set to 500 mL/Lmin using a mass flow controller (PR4000, MKS Instruments, Wilmington, MA,

Table 4 Primers used in this study

Primer	Sequence (5'-3')
PP5004_UP fw	TCTAGAGACATCCTGTTCTGGAAC
PP5004_UP_rev	CCCCTGTCAGGCCGCAGCTGGCACGTGACTCTTGGGTGAAGTAAAC
PP5004_DOWN_fw	CACCCAAGAGTCACGTGCCAGCTGCGGCCTGACAGGGGAAATGGATC
PP5004_DOWN rev	GGTACCGTTGAACAGCTCCTTGAC

USA). The dissolved oxygen level was kept above 20% air saturation by control of the agitation speed up to a maximum of 700 rpm. The pH was maintained at 7.0 by automatic pH controlled addition of 0.5 M H_2SO_4 or 1 M of KOH.

Analytical procedures

Cell growth was recorded as optical density (OD) at 600_{nm} (Ultraspec 2000, Hitachi, Japan). The cell dry weight was determined gravimetrically after collection of 10 mL culture broth for 10 min at 4°C and 9,000 × *g* (Eppendorf 5810 R, Hamburg, Germany) in pre-weighed tubes, including a washing step with distilled water, and drying of the obtained pellet at 100°C until constant weight. The ammonium concentration in cell-free supernatant was measured by a photometric test (LCK 303 kit, Hach Lange, Danaher, USA).

HPLC analysis

The glycerol and organic acids concentration (citrate, isocitrate, succinate, fumarate, malate, pyruvate, and citrate) in cultivation supernatant was analyzed by HPLC Agilent 1260 (Agilent, Krefeld, Germany) equipped with an 8 mm Rezex ROA-organic acid H column (Phenomenex, USA) at 65°C with 0.013 N H_2SO_4 as the mobile phase (0.5 mL·min^{-1}) followed by detection using a RID detector (Agilent serie1260).

PHA characterization and quantification

PHA compositions of the polymer produced, as well as the cellular PHA content concentration were determined by gas chromatography (GC) and mass spectrometry (MS) of the methanolyzed polyester. First, 10 mL of the culture broth was placed in a falcon tube and centrifuged for 10 min at 4°C and 9,000 × *g* (Eppendorf 5810 R, Hamburg, Germany), following by a washing step with distilled water. The supernatant was poured away by pipetting and the cell pellet kept at −20°C for further process. Methanolysis was carried out by suspending 5–10 mg of lyophilized aliquots in 2 mL of chloroform and 2 mL of methanol containing 15% sulfuric acid and 0.5 mg/mL 3-methylbenzoic acid as internal standard, respectively, followed by incubation at 100°C for 4 h. After cooling, 1 mL of demineralized water was added and the organic phase containing the resulting methyl esters of monomers was analyzed by GC-MS. Analysis was performed in Varian GC-MS system 450GC/240MS ion trap mass spectrometer (Varian Inc., Agilent Technologies) and operated by the software MS Workstation 6.9.3 (Varian Inc., Agilent Technologies). An aliquot (1 mL) of the organic phase was injected into the gas chromatograph at a split ratio of 1:10. Separation of compounds of interest (i.e. the methyl esters of 3-hydroxyexanoate, 3-hydroxyoctanoate, 3-hydroxydecanoate, 3-hydroxydodecanoate, 3-hydroxy-5-cis-dodecanoate, 3-hydroxytetradecanoate) was achieved by a FactorFour VF-5 ms capillary column (30 m × 0.25 mm i.d. × 0.25 mm film thickness). Helium was used as carrier gas at a flow rate of 0.9 mL/min. The injector and transfer line temperature were 275°C and 300°C respectively. The oven temperature program was: initial temperature 40°C for 2 min, then from 40°C up 150°C at a rate of 5°C min^{-1} and finally up to 280°C at a rate of 10°C min^{-1}. Positive ions were obtained using electron ionization at 70 eV and mass spectra were generated by scanning ions of *m/z* 50 to *m/z* 650. The PHA content (wt%) was defined as the percentage of the cell dry weight (CDW) represented by the polyhydroxyalkanoate.

Transmission electron microscopy

Bacteria were fixed by chilling the cultures to 4°C and addition of glutaraldehyde (2%) and formaldehyde (5%). They were then washed with cacodylate buffer (0.01 mol l)1 cacodylate, 0.01 mol l)1 $CaCl_2$, 0.01 mol l)1 $MgCl_2$ 6H2O, 0.09 mol l)1 sucrose, pH 6/9) and stained with 1% aqueous osmium for 1 h at room temperature. Samples were then dehydrated with a graded series of acetone (10, 30, 50, 70, 90 and 100%) with incubation for 30 min at each concentration, except for the 70% acetone, which contained 2% uranyl acetate and was performed overnight. Samples were infiltrated with an epoxy resin, according to the Spurr formula for hard resin, for several days with pure resin. Ultrathin sections were cut with a diamond knife, counterstained with uranyl acetate and lead citrate and examined in a TEM910 transmission electron micro- scope (Carl Zeiss, Oberkochen, Germany) at an accelera- tion voltage of 80 kV. Images were taken at calibrated magnifications using a line replica. Images were recorded digitally with a Slow-Scan CCD-Camera (ProScan, 1024 × 1024, Scheuring, Germany) with ITEM-Software (Olympus Soft Imaging Solutions, Munster, Germany).

Field emission scanning electron microscopy

Samples were fixed as above, washed with cacodylate buf- fer and then washed with TE-buffer (20 mmol l)1 TRIS, 1 mmol l)1 EDTA, pH 6.9). 50 ll of washed bacteria were applied to poly-l-lysine precoated cover slips (12 mm in diameter), which were left for 5 min, washed in TE-buffer, incubated with 2% glutaraldehyde in TE-buffer for 15 min and washed again with TE-buffer. Dehydration was carried out with a graded series of ace- tone (10, 30, 50, 70, 90, 100%) on ice for 15 min for each step, followed by 100% acetone at room temperature and critical-point drying with liquid CO_2 (CPD 30; Bal-Tec, Balzers, Liechtenstein). Samples were then gold shadowed by sputter coating (SCD 500; Bal-Tec) and examined with a field emission scanning electron microscope Zeiss DSM 982 Gemini (Carl Zeiss, Oberkochen, Germany), using the Everhart Thornley SE detector and the inlens detector in a 50: 50 ratio at an acceleration voltage of 5 kV. Images were recorded onto a MO-disc. Contrast and brightness were adjusted with Adobe Photoshop CS3.

Competing interest
The authors declare that they have no competing interests.

Authors' contributions
IPC designed the study and composed the manuscript. DB, RH, MR, and IPC have made substantial contributions to the acquisition of data. IPC analyzed and interpreted the data. All authors read and approved the final version of the manuscript.

Acknowledgements
We would like to thank Dr. Nick Wierckx for providing the *P. putida* S12 strain used in this study. The authors thank Carie M. Frantz for carefully reading this manuscript. Ignacio Poblete-Castro acknowledges the financial support by the program "Convenio de Desempeño Apoyo a la Innovación en Educación Superior" (PMIUAB 1301).

Author details
[1]Facultad de Ciencias Biológicas, Center for Bioinformatic and Integrative Biology, Biosystems Engineering Laboratory, Universidad Andrés Bello, Santiago 8340176, Chile. [2]Helmholtz Centre for Infection Biology, Microbial Drugs Group, Braunschweig D-38124, Germany. [3]Helmholtz Centre for Infection Biology, Systems and Synthetic Biology, Braunschweig D-38124, Germany. [4]Helmholtz Centre for Infection Biology, Molecular Mechanism of Streptococci Group, Braunschweig D-38124, Germany.

References
1. Madison LL, Huisman GW: **Metabolic engineering of poly(3-Hydroxyalkanoates): from DNA to plastic.** *Microbiol Mol Biol Rev* 1999, **63**(1):21–53.
2. Chen G-Q: **A microbial polyhydroxyalkanoates (PHA) based bio- and materials industry.** *Chem Soc Rev* 2009, **38:**2434–2446.
3. Anderson AJ, Dawes EA: **Occurrence, metabolism, metabolic role, and industrial uses of bacterial polyhydroxyalkanoates.** *Microbiol Rev* 1990, **54**(4):450–472.
4. Choi J, Lee SY: **Factors affecting the economics of polyhydroxyalkanoate production by bacterial fermentation.** *Appl Microbiol Biotechnol* 1999, **51:**13–21.
5. Frazzetto G: **White biotechnology.** *EMBO Rep* 2003, **4:**835–837.
6. Salehizadeh H, Van Loosdrecht MCM: **Production of polyhydroxyalkanoates by mixed culture: recent trends and biotechnological importance.** *Biotechnol Adv* 2004, **22:**261–279.
7. Van-Thuoc D, Quillaguamán J, Mamo G, Mattiasson B: **Utilization of agricultural residues for poly(3-hydroxybutyrate) production by Halomonas boliviensis LC1.** *J Appl Microbiol* 2008, **104:**420–428.
8. Koller M, Bona R, Braunegg G, Hermann C, Horvat P, Kroutil M, Martinz J, Neto J, Pereira L, Varila P: **Production of polyhydroxyalkanoates from agricultural waste and surplus materials†.** *Biomacromolecules* 2005, **6:**561–565.
9. Serafim L, Lemos P, Albuquerque ME, Reis MM: **Strategies for PHA production by mixed cultures and renewable waste materials.** *Appl Microbiol Biotechnol* 2008, **81:**615–628.
10. Zinn M, Witholt B, Egli T: **Occurrence, synthesis and medical application of bacterial polyhydroxyalkanoate.** *Adv Drug Deliv Rev* 2001, **53:**5–21.
11. Zinn M, Durner R, Zinn H, Ren Q, Egli T, Witholt B: **Growth and accumulation dynamics of poly(3-hydroxyalkanoate) (PHA) in Pseudomonas putida GPo1 cultivated in continuous culture under transient feed conditions.** *Biotechnol J* 2011, **6:**1240–1252.
12. Huijberts GN, Eggink G, de Waard P, Huisman GW, Witholt B: **Pseudomonas putida KT2442 cultivated on glucose accumulates poly (3-hydroxyalkanoates) consisting of saturated and unsaturated monomers.** *Appl Environ Microbiol* 1992, **58**(2):536–544.
13. Poblete-Castro I, Escapa IF, Jäger C, Puchalka J, Chi Lam C, Schomburg D, Prieto M, Martins dos Santos VA: **The metabolic response of P. putida KT2442 producing high levels of polyhydroxyalkanoate under single- and multiple-nutrient-limited growth: highlights from a multi-level omics approach.** *Microb Cell Fact* 2012, **11:**34.
14. O'Leary ND, O'Connor KE, Ward P, Goff M, Dobson ADW: **Genetic characterization of accumulation of polyhydroxyalkanoate from styrene in pseudomonas putida CA-3.** *Appl Environ Microbiol* 2005, **71**(8):4380–4387.
15. Cai L, Yuan M-Q, Liu F, Jian J, Chen G-Q: **Enhanced production of medium-chain-length polyhydroxyalkanoates (PHA) by PHA depolymerase knockout mutant of Pseudomonas putida KT2442.** *Bioresour Technol* 2009, **100:**2265–2270.
16. Poblete-Castro I, Binger D, Rodrigues A, Becker J, Martins Dos Santos VAP, Wittmann C: **In-silico-driven metabolic engineering of Pseudomonas putida for enhanced production of poly-hydroxyalkanoates.** *Metab Eng* 2013, **15:**113–123.
17. Borrero-de Acuña JM, Bielecka A, Häussler S, Schobert M, Jahn M, Wittmann C, Jahn D, Poblete-Castro I: **Production of medium chain length polyhydroxyalkanoate in metabolic flux optimized Pseudomonas putida.** *Microb Cell Fact* 2014, **13:**88.
18. Sun Z, Ramsay J, Guay M, Ramsay B: **Carbon-limited fed-batch production of medium-chain-length polyhydroxyalkanoates from nonanoic acid by Pseudomonas putida KT2440.** *Appl Microbiol Biotechnol* 2007, **74:**69–77.
19. Poblete-Castro I, Rodriguez AL, Lam CMC, Kessler W: **Improved production of medium-chain-length polyhydroxyalkanoates in glucose-based fed-batch cultivations of metabolically engineered Pseudomonas putida strains.** *J Microbiol Biotechnol* 2014, **24:**59–69.
20. Liu W, Chen G-Q: **Production and characterization of medium-chain-length polyhydroxyalkanoate with high 3-hydroxytetradecanoate monomer content by fadB and fadA knockout mutant of Pseudomonas putida KT2442.** *Appl Microbiol Biotechnol* 2007, **76:**1153–1159.
21. Sun Z, Ramsay J, Guay M, Ramsay B: **Fed-batch production of unsaturated medium-chain-length polyhydroxyalkanoates with controlled composition by Pseudomonas putida KT2440.** *Appl Microbiol Biotechnol* 2009, **82:**657 662.
22. Escapa I, Morales V, Martino V, Pollet E, Avérous L, García J, Prieto M: **Disruption of β-oxidation pathway in Pseudomonas putida KT2442 to produce new functionalized PHAs with thioester groups.** *Appl Microbiol Biotechnol* 2011, **89:**1583–1598.
23. Poblete-Castro I, Becker J, Dohnt K, Dos Santos VM, Wittmann C: **Industrial biotechnology of *Pseudomonas putida* and related species.** *Appl Microbiol Biotechnol* 2012, **93:**2279–2290.
24. Muhr A, Rechberger EM, Salerno A, Reiterer A, Malli K, Strohmeier K, Schober S, Mittelbach M, Koller M: **Novel description of mcl-PHA biosynthesis by pseudomonas chlororaphis from animal-derived waste.** *J Biotechnol* 2013, **165:**45–51.
25. Kenny S, Runic J, Kaminsky W, Woods T, Babu R, O'Connor K: **Development of a bioprocess to convert PET derived terephthalic acid and biodiesel derived glycerol to medium chain length polyhydroxyalkanoate.** *Appl Microbiol Biotechnol* 2012, **95:**623–633.
26. Fu J, Sharma U, Sparling R, Cicek N, Levin DB: **Evaluation of medium-chain-length polyhydroxyalkanoate production by Pseudomonas putida LS46 using biodiesel by-product streams.** *Can J Microbiol* 2014, **60:**461–468.
27. Davis R, Kataria R, Cerrone F, Woods T, Kenny S, O'Donovan A, Guzik M, Shaikh H, Duane G, Gupta VK, Tuohy MG, Padamatti RB, Casey E, O'Connor KE: **Conversion of grass biomass into fermentable sugars and its utilization for medium chain length polyhydroxyalkanoate (mcl-PHA) production by Pseudomonas strains.** *Bioresour Technol* 2013, **150:**202–209.
28. Linger JG, Vardon DR, Guarnieri MT, Karp EM, Hunsinger GB, Franden MA, Johnson CW, Chupka G, Strathmann TJ, Pienkos PT, Beckham GT: **Lignin valorization through integrated biological funneling and chemical catalysis.** *Proc Natl Acad Sci* 2014, **111**(33):12013–12018.
29. Almeida JR, Favaro LC, Quirino B: **Biodiesel biorefinery: opportunities and challenges for microbial production of fuels and chemicals from glycerol waste.** *Biotechnol Biofuels* 2012, **5:**48.
30. Timmis KN: **Pseudomonas putida: a cosmopolitan opportunist par excellence.** *Environ Microbiol* 2002, **4:**779–781.

31. Wang Q, Tappel RC, Zhu C, Nomura CT: **Development of a new strategy for production of medium-chain-length polyhydroxyalkanoates by recombinant escherichia coli via inexpensive non-fatty acid feedstocks.** *Appl Environ Microbiol* 2012, **78**(2):519–527.
32. Meijnen J-P, Verhoef S, Briedjlal A, de Winde J, Ruijssenaars H: **Improved p-hydroxybenzoate production by engineered Pseudomonas putida S12 by using a mixed-substrate feeding strategy.** *Appl Microbiol Biotechnol* 2011, **90:**885–893.
33. Escapa IF, del Cerro C, García JL, Prieto MA: **The role of GlpR repressor in Pseudomonas putida KT2440 growth and PHA production from glycerol.** *Environ Microbiol* 2013, **15:**93–110.
34. Follonier S, Panke S, Zinn M: **A reduction in growth rate of Pseudomonas putida KT2442 counteracts productivity advances in medium-chain-length polyhydroxyalkanoate production from gluconate.** *Microb Cell Fact* 2011, **10:**25.
35. De Eugenio LI, Escapa IF, Morales V, Dinjaski N, Galán B, García JL, Prieto MA: **The turnover of medium-chain-length polyhydroxyalkanoates in Pseudomonas putida KT2442 and the fundamental role of PhaZ depolymerase for the metabolic balance.** *Environ Microbiol* 2010, **12:**207–221.
36. Ren Q, de Roo G, Witholt B, Zinn M, Thony-Meyer L: **Influence of growth stage on activities of polyhydroxyalkanoate (PHA) polymerase and PHA depolymerase in Pseudomonas putida U.** *BMC Microbiol* 2010, **10:**254.
37. Solaiman DKY, Ashby RD, Foglia TA: **Effect of inactivation of poly (hydroxyalkanoates) depolymerase gene on the properties of poly (hydroxyalkanoates) in Pseudomonas resinovorans.** *Appl Microbiol Biotechnol* 2003, **62:**536–543.
38. Arias S, Bassas-Galia M, Molinari G, Timmis KN: **Tight coupling of polymerization and depolymerization of polyhydroxyalkanoates ensures efficient management of carbon resources in Pseudomonas putida.** *Microb Biotechnol* 2013, **6:**551–563.
39. Fujita Y, Matsuoka H, Hirooka K: **Regulation of fatty acid metabolism in bacteria.** *Mol Microbiol* 2007, **66:**829–839.
40. Lee IY, Kim MK, Park YH, Lee SY: **Regulatory effects of cellular nicotinamide nucleotides and enzyme activities on poly(3-hydroxybutyrate) synthesis in recombinant Escherichia coli.** *Biotechnol Bioeng* 1996, **52:**707–712.
41. Sánchez AM, Andrews J, Hussein I, Bennett GN, San K-Y: **Effect of overexpression of a soluble pyridine nucleotide transhydrogenase (UdhA) on the production of poly(3-hydroxybutyrate) in Escherichia coli.** *Biotechnol Prog* 2006, **22:**420–425.
42. Ochoa-Estopier A, Guillouet SE: **D-stat culture for studying the metabolic shifts from oxidative metabolism to lipid accumulation and citric acid production in Yarrowia lipolytica.** *J Biotechnol* 2014, **170:**35–41.
43. Papagianni M: **Advances in citric acid fermentation by Aspergillus niger: biochemical aspects, membrane transport and modeling.** *Biotechnol Adv* 2007, **25:**244–263.
44. Kristiansen B, Sinclair CG: **Production of citric acid in continuous culture.** *Biotechnol Bioeng* 1979, **21:**297–315.
45. Anastassiadis S, Aivasidis A, Wandrey C: **Citric acid production by Candida strains under intracellular nitrogen limitation.** *Appl Microbiol Biotechnol* 2002, **60:**81–87.
46. Papanikolaou S, Aggelis G: **Lipid production by Yarrowia lipolytica growing on industrial glycerol in a single-stage continuous culture.** *Bioresour Technol* 2002, **82:**43–49.
47. Ratledge C, Wynn JP: **The biochemistry and molecular biology of lipid accumulation in oleaginous microorganisms.** In *Adv Appl Microbiol. Volume Volume 51*. Edited by Allen I. Laskin JWB and GMGBT-A in AM. Academic Press; 2002, **51:**1–52.
48. Daran-Lapujade P, Jansen MLA, Daran J-M, van Gulik W, de Winde JH, Pronk JT: **Role of transcriptional regulation in controlling fluxes in central carbon metabolism of Saccharomyces cerevisiae: A CHEMOSTAT CULTURE STUDY.** *J Biol Chem* 2004, **279**(10):9125–9138.
49. Canelas AB, van Gulik WM, Heijnen JJ: **Determination of the cytosolic free NAD/NADH ratio in Saccharomyces cerevisiae under steady-state and highly dynamic conditions.** *Biotechnol Bioeng* 2008, **100:**734–743.
50. Vemuri GN, Altman E, Sangurdekar DP, Khodursky AB, Eiteman MA: **Overflow metabolism in Escherichia coli during steady-state growth: transcriptional regulation and effect of the redox ratio.** *Appl Environ Microbiol* 2006, **72**(5):3653–3661.
51. Kamzolova S, Lunina J, Morgunov I: **Biochemistry of citric acid production from rapeseed oil by Yarrowia lipolytica yeast.** *J Am Oil Chem Soc* 2011, **88:**1965–1976.
52. Rymowicz W, Fatykhova A, Kamzolova S, Rywińska A, Morgunov I: **Citric acid production from glycerol-containing waste of biodiesel industry by Yarrowia lipolytica in batch, repeated batch, and cell recycle regimes.** *Appl Microbiol Biotechnol* 2010, **87:**971–979.
53. Lotfy WA, Ghanem KM, El-Helow ER: **Citric acid production by a novel Aspergillus niger isolate: I. Mutagenesis and cost reduction studies.** *Bioresour Technol* 2007, **98:**3464–3469.
54. Cullen D: **The genome of an industrial workhorse.** *Nat Biotechnol* 2007, **25:**189–190.
55. Bagdasarian M, Lurz R, Rückert B, Franklin FCH, Bagdasarian MM, Frey J, Timmis KN: **Specific-purpose plasmid cloning vectors II. Broad host range, high copy number, RSF 1010-derived vectors, and a host-vector system for gene cloning in Pseudomonas.** *Gene* 1981, **16:**237–247.
56. Hartmans S, Smits JP, van der Werf MJ, Volkering F, de Bont JAM: **Metabolism of styrene oxide and 2-phenylethanol in the styrene-degrading xanthobacter strain 124X.** *Appl Environ Microbiol* 1989, **55**(11):2850–2855.
57. Martínez-García E, de Lorenzo V: **Engineering multiple genomic deletions in Gram-negative bacteria: analysis of the multi-resistant antibiotic profile of Pseudomonas putida KT2440.** *Environ Microbiol* 2011, **13:**2702–2716.
58. Shevchuk NA, Bryksin AV, Nusinovich YA, Cabello FC, Sutherland M, Ladisch S: **Construction of long DNA molecules using long PCR-based fusion of several fragments simultaneously.** *Nucleic Acids Res* 2004, **32**(2):e19–e19.
59. De Lorenzo V, Timmis KN: **[31] Analysis and construction of stable phenotypes in gram-negative bacteria with Tn5- and Tn10-derived minitransposons.** In *Bact Pathog Part A Identif Regul Virulence Factors. Volume Volume 235*. Edited by Virginia L. Clark PMBBT-M in E. Academic Press; 1994:386–405.

24

Auto-induction expression of human consensus interferon-alpha in *Escherichia coli*

Nawal Abd EL-Baky[1], Mustafa H Linjawi[2] and Elrashdy M Redwan[1,3*]

Abstract

Background: Isopropyl-β-D-1-thiolgalactopyranoside (IPTG)-inducible expression of recombinant proteins in *E. coli* is commonly used and effective. Nevertheless, unintended induction was encountered as a problem when using these bacterial expression systems, generating cultures that give reduced or variable protein yields. Auto-induction allows for production of much higher target protein yield and cell mass than conventional procedures using induction with IPTG without monitoring cell growth then adding IPTG at the appropriate cell density. This method involves special media recipes that promote growth to high density and automatically induce expression of target protein from T7 promoter. Consensus interferon is a synthetic artificially engineered interferon having an amino acid sequence that is a rough average of the sequences of all natural human alpha interferon subtypes and has greater potency than other interferons even the pegylated versions. The purpose of this study was high-level expression of human consensus interferon-alpha (cIFN-α) in *E. coli* using an auto-induction protocol. The cIFN-α gene was cloned into pET101/D-TOPO expression vector under the T7 promoter transcriptional regulation. Expression was optimized with respect to temperature and length of incubation in shake flask cultures. The antiviral potency and anticancer activity of cIFN-α were evaluated in comparison to IFN-α2a.

Results: The expressed cIFN-α protein in auto-induction T7 system was found mostly in soluble fraction of the cell lysate (about 70% of yield in total cell lysate) after lowering incubation temperature to 25°C or 30°C. Protein expression was maximal after 24 h incubation at 25°C or 30°C. After purification via single-step chromatography using DEAE-Sepharose, the yield was 270 mg/L in shake flask *E. coli* cultures which is much higher than IPTG-inducible T7 expression system and other systems according to available data. The synthesized cIFN-α was biologically active as confirmed by its anticancer and antiviral effects and was significantly more potent than IFN-α2a.

Conclusions: The auto-induction process was reliable and convenient for production of cIFN-α protein in *E. coli*, and can be adapted for large-scale therapeutic protein production.

Keywords: Auto-induction, Human consensus interferon-alpha, IPTG-inducible expression system, T7 expression system

Background

Interferons (IFNs) are cytokines secreted by vertebrates' cells when stimulated by viruses and several other agents [1,2]. This family of glycoproteins exhibits antiviral, antiproliferative and immunomodulation functions [3]. Based on their receptor specificity, structure and functional properties, IFNs are divided into 3 major groups (type I (IFN-α and IFN-β), type II (IFN-γ) and type III (IFN-λ)) [4]. Consensus interferon is a synthetic type of IFN-α containing 166 amino acids that was engineered to comprise the most frequently observed amino acids among all natural non-allelic IFN-α subtypes in humans [5] to form a consensus molecule. *In vitro* studies have shown that consensus interferon yields more potent antiviral and antiproliferative effects than other standard interferons (IFN-α2a, IFN-α2b, and pegylated versions) [6]. This artificial interferon was used in combination with ribavirin to treat non-responder population or relapsers to a previous interferon regime [7,8].

* Correspondence: redwan1961@yahoo.com
[1]Therapeutic and Protective Proteins Laboratory, Protein Research Department, Genetic Engineering and Biotechnology Research Institute, City for Scientific Research and Technology Applications, New Borg EL-Arab, 21934 Alexandria, Egypt
[3]Biological Sciences Department, Faculty of Science, King Abdulaziz University, P.O. Box 80203, Jeddah 21589, Saudi Arabia
Full list of author information is available at the end of the article

There have been many efforts to express the artificial gene of consensus interferon in *E. coli* [9-12]. Fieschko and Ritch [10] developed a semisynthetic medium for the production of human consensus interferon-α (IFN-alfacon-1) via tightly regulated expression system and a controlled feeding schedule. They obtained yields up to 7.6×10^{12} U per liter of fermentation broth, however this result was not reproducible. Curless *et al.* [12] designed cyclic, two-stage fed-batch cultivation schemes for the expression of human consensus IFN-α under the λP_L promoter transcriptional control. They achieved a yield of 5.5 g/L interferon at a cell density of 68 g of dry cell weight (DCW)/L but maintaining two reactors may have restricted the use of this design. We have previously reported heat-inducible expression of cIFN-α protein in *E. coli* with a λP_L promoter system [13]. Most of the expressed protein aggregated into inclusion bodies, therefore needed *in vitro* refolding. After refolding and purification, the yield was 70 mg/L in shake flask cultures.

The T7 system comprising the T7 promoter and phage T7 RNA polymerase is one of the most widely used expression systems [14]. In host *E. coli* strains such as BL21 (DE3), T7 RNA polymerase is expressed under the control of the IPTG-inducible *lac*UV5 promoter. Expression of the target protein from cloned gene is controlled by the T7 promoter. This promoter is recognized specifically by T7 RNA polymerase. After sufficient amount of T7 RNA polymerase is produced, it binds to the T7 promoter and transcribes the gene of interest. Various recombinant proteins were overexpressed in *E. coli* using this system. A problem in using this expression system is that there is always some basal expression of T7 RNA polymerase from the *lac*UV5 promoter even in the absence of inducer [14]. If the target protein is toxic to the *E. coli* host, basal expression of the target gene possibly will result in plasmid instability and/or cell death. To reduce basal expression, a *lac* operator sequence (a binding site for *lac* repressor) is placed downstream of the T7 promoter, constructing a T7*lac* promoter [15]. The *lac* repressor binds to the *lac* operator and functions to further repress T7 RNA polymerase-induced basal transcription of the target gene in BL21 (DE3) cells. In addition, IPTG induction frequently leads to formation of inclusion bodies [16].

Recently, Studier [17] developed an auto-induction process for production of recombinant proteins under the control of the T7 promoter in *E. coli*. Auto-inducing media are made from a well-balanced combination of different carbon sources and other essential nutrients, allowing cultures to grow to high cell densities, and support high-level expression of the target protein. These media contain glucose, which promotes fast cellular growth in the early stages of the culture, while also preventing uptake and metabolism of inducing sugar. Lactose is also present in these media, and is consumed after glucose exhaustion, inducing recombinant protein expression. To achieve sustained growth during the induction phase, glycerol is provided in auto-inducing media together with lactose. Formulation of auto-inducing media achieved higher recombinant protein yields than those using IPTG as inducer [17]. The aim of this work was expression of cIFN-α protein in *E. coli* by auto-induction procedure in shake flasks for the first time, purification of recombinant cIFN-α and determination of its antiproliferative and antiviral activities.

Methods

Cloning of the cIFN-α gene into pET101/D-TOPO expression vector

The coding sequence of the synthetic cIFN-α gene was assembled and amplified by two-step PCR-based gene synthesis method using long (45–50 nucleotides) overlapped primers and GC-Rich PCR system (Roche Molecular Biochemicals, Mannheim, Germany) as described previously [13,18]. The blunt-ended PCR product of cIFN-α gene was cloned into the pET101/D-TOPO vector (Invitrogen, Carlsbad, USA) as recommended by the manufacturer to yield pET-cIFNα expression construct. To allow pET directional TOPO cloning, the forward PCR primer used in amplification of the cIFN-α gene was designed to contain the sequence, CACC, at the 5′ end of the primer directly precedes an initiation ATG codon. The 4 nucleotides, CACC, base pair with the overhang sequence, GTGG, in pET-TOPO vector. A stop codon (TAA) was included in the reverse primer. The resulting pET-cIFNα plasmid was transformed into BL21-CodonPlus (DE3) chemically competent *E. coli* (Stratagene, Heidelberg, Germany) for protein expression regulated by the T7 promoter. Then the cells were plated onto Luria Bertani (LB) agar plates containing ampicillin (100 μg/ml). Plasmid was purified using Miniprep kit (Qiagen, Hilden, Germany) from overnight culture of picked single colonies then *XbaI/SacI* enzymes digested. Correct clones were subjected to nucleotide sequencing.

Auto-induction for consensus interferon production in *E. coli*

The Overnight Express Autoinduction System 1 (Novagen, Darmstadt, Germany) was employed for high-level cIFN-α protein expression by auto-induction from the constructed pET-cIFNα plasmid. The Overnight Express Autoinduction System 1 medium was prepared aseptically by adding 20 ml OnEx solution 1 (induction solution), 50 ml OnEx solution 2 (buffering solution), and 1 ml OnEx solution 3 (magnesium solution) to 929 ml sterile LB broth. A single clone was cultured in the Overnight Express Autoinduction System medium supplemented with 100 μg/ml ampicillin in a shaker incubator at 30°C and 300 rpm.

After overnight or 24 h incubation, cells were harvested by centrifugation at 4000 rpm for 15 min at 4°C. Each cell pellet collected from 1 ml of the bacterial culture was resuspended in 100 µl of 5× electrophoresis sample buffer, boiled for 10 min, and then loaded onto sodium dodecyl sulfate polyacrylamide gel electrophoresis (SDS-PAGE). Auto-induced cultures were grown and induced at different incubation temperatures (25, 30, and 37°C) for 24 h, then soluble cell lysates of auto-induced cells were analyzed by SDS-PAGE.

Two immunoassays were performed to confirm the identity of the expressed protein in auto-induction T7 system; enzyme-linked immunosorbent assay (ELISA) and western blot. ELISA was performed as previously described [13,19]. Briefly, ELISA microtiter plate (Costar, Cambridge, USA) was coated with 50 µl of total cell lysate at a concentration of 5 µg/ml in carbonate buffer solution (pH 9.6) for 24 h at 4°C. After washing 5 times with 0.12 M NaCl, 0.04 M sodium phosphate, pH 7.2 buffer (PBS), the plate was blocked by adding 100 µl of blocking buffer (2% w/v gelatin in PBS) for 1 h at 37°C. Then the plate was washed 5 times with PBS and 50 µl of monoclonal mouse anti-human IFN-α antibody (PBL InterferonSource, Piscataway, USA) diluted 1:2000 was added to each well. After 1 h of incubation at 37°C, the plate was washed 5 times with PBS, and 50 µl of alkaline phosphatase-conjugated anti-mouse IgG (BIO-RAD, Alfred Nobel, Hercules, USA) diluted 1:1000 was added, followed by an incubation of 1 h at 37°C. After washing five times, *p*-Nitrophenyl phosphate (*p*-NPP) was added for color development and optical density was measured at 405 nm using an ELISA microtiter plate reader (MicroPlate Reader, BIO-RAD, USA). For western blotting, the protein bands were separated on 12% SDS-PAGE and transferred onto a nitrocellulose membrane in a buffer of pH 8.4 containing 1.93 g/L Tris base and 9 g/L glycine. Transfer was done for 8 h at 30 Volt. After transfer, the membrane was blocked using 2% w/v bovine serum albumin (BSA) in PBS for 1 h. After a brief wash with PBS, the membrane was incubated with monoclonal anti-human IFN-α antibody diluted 1:2000 for 1 h at room temperature with gentle shaking. The membrane was washed again three times with PBS for 10 min each, and then incubated with anti-mouse IgG-peroxidase conjugate (BIO-RAD, USA) diluted 1:1000 at room temperature for 1 h with gentle shaking. Washing was done again with the same procedure described above. Protein bands were developed with 3,3 Diaminobenzidine (DAB) [20].

Optimization of auto-induction expression

To grow the cells to stationary phase ($OD_{600} = 6–7$) and optimize the yield of cIFN-α production in auto-induction system, cultures were incubated at different temperatures (25, 30, and 37°C) and samples were taken at different time points (8, 10, 12, and 24 h). Yields of cIFN-α (volumetric yields (g/L)) were calculated.

Purification

The supernatant and the pellet (inclusion bodies) of the cell lysate were prepared from 1 liter of optimized auto-induced shake flask culture and then subjected to SDS-PAGE analysis as previously described [21]. The cell lysate supernatant was loaded at 1 ml/min onto a DEAE-Sepharose CL-6B anion-exchange column (Pharmacia Biotech, Uppsala, Sweden) previously equilibrated with 20 mM Tris–HCl, pH 8. The column was washed with ten bed volumes of the same buffer (flow rate 2 ml/min). The bound proteins were eluted with 100 ml linear gradient of 0–1 M NaCl generated using AKTAprime plus FPLC protein separator system (GE Healthcare Life Sciences Products, Cardiff, UK) at a flow rate of 1 ml/min. Twenty fractions of 5 ml each were collected. The eluted protein in the fractions was monitored by an online UV detector. Purified cIFN-α protein was analyzed by SDS-PAGE and ELISA. Purified cIFN-α protein content was determined by Lowry method [22]. The endotoxin content was checked [23] to avoid its mitogenic effects on the cell culture system. All cIFN-α preparations used were free of endotoxin (data not shown).

Cell culture

Hepatoma cell line HepG2 HB-8065 (ATCC, USA) was cultured in RPMI-1640 medium (Lonza, Verviers, Belgium) supplemented with 2% v/v L-glutamine, 10% v/v fetal bovine serum (FBS), and 1% v/v penicillin-streptomycin. The cultures were maintained at 37°C in a humidified atmosphere with 5% CO_2; 95% air in 25 cm^2 flasks (Greiner, Frickenhausen, Germany). For subculturing, culture medium was discarded, and then the adherent cells were detached from the surface of the flask using 1 ml of Trypsin-EDTA solution (0.25% w/v Trypsin- 0.53 mM EDTA) for 5 to 15 min at 37°C. The action of Trypsin was stopped by the addition of 3 ml RPMI-1640 medium. The cells were scraped and collected in a 15 ml conical tube, then washed twice by RPMI-1640 supplemented medium, and centrifuged at 1200 rpm for 5 min at 4°C after each wash. The pellet was resuspended in 3 ml RPMI-1640 medium, and then appropriate aliquots of the cell suspension were added to new culture vessels.

Anticancer activity assay

Recombinant cIFN-α expressed in auto-induction system was tested for anticancer activity against hepatoma cell line HepG2 and compared with commercial recombinant human IFN-α2a (Shenyang Pharmaceutical Co, Guoyaozhunzi, China) by using 3-(4, 5-dimethylthiazol-2-yl)-2, 5-diphenyltetrazolium bromide (MTT) rapid colorimetric assay as described by Mosmann [24] and El-Baky *et al.* [19]. Interferons were diluted with RPMI-1640 medium to desired concentrations (3, 3.33, 3.75, 4.28, 5, 6, 7.5, 10, 15 and 30 μg/ml). HepG2 cells were seeded in flat-bottomed 96-well cell culture plates at a density of 25×10^4 cells/ml in final volume of 200 μl RPMI-1640 supplemented medium per well, and incubated for 24 h at 37°C in 5% CO_2; 95% air. After 24 h, cells were treated with various concentrations of the tested interferons in 4 replicates. After further 24 h, the medium was discarded and cell monolayers were washed with sterile PBS three times. Twenty microliters of 5 mg/ml MTT solution (Sigma, St. Louis, USA) were added to each well and incubated at 37°C for 3 h. The formed insoluble purple formazan product was dissolved with 180 μl of dimethyl sulfoxide (DMSO). Optical density was measured at 560 nm, and then the percentage of cytotoxicity compared to the untreated cells was determined with the following equation:

$$\%\,\text{Cytotoxicity} = \frac{\text{Absorbance of cells without treatment} - \text{Absorbance of cells with treatment}}{\text{Absorbance of cells without treatment}}$$

The plot of % cytotoxicity versus sample concentration was used to calculate the concentration lethal to 50% of the cells (LD_{50}) value.

Isolation of human blood lymphocytes and selectivity index calculation

The peripheral blood mononuclear cells (PBMCs) were isolated as previously described by Lohr *et al.* [25] and El-Fakharany *et al.* [26]. In brief, peripheral blood samples collected from a single healthy volunteer were diluted with 5 volumes of a freshly prepared red blood cell (RBC) lysis buffer of pH 8 containing 38.8 mM NH_4Cl, 2.5 mM $KHCO_3$, and 1 mM EDTA, incubated at room temperature for 10 min and centrifuged at 1500 rpm for 5 min. The nucleated cells were precipitated and washed with PBS. Proliferation of human lymphocytes in response to the tested interferons was assessed by MTT assay. The selectivity index (SI) value was calculated from the LD_{50} ratio in lymphocytes over HepG2 cells. The SI value indicates the selectivity of interferon samples to the HepG2 cell line [27].

Antiviral assay

The antiviral activity of the tested interferons was determined using cytopathic effect (CPE) assay against herpes simplex virus 1 (HSV-1) as reported by Rashad *et al.* [28]. Vero cells grown to confluency in 96-well plates were infected with 100 μl of stock virus. After an adsorption period of 2 h at 37°C, virus was removed and the following concentrations of the tested interferons were added: 0.003, 0.03, 0.3, 3, and 30 μg/ml for recombinant cIFN-α and 0.1, 1, 10, 100, and 1000 IU/ml for commercial IFN-α2a, then maintenance Dulbecco's modified Eagle medium (DMEM) with 2% v/v FBS was added (100 μl/well). The cultures were further incubated at 37°C for 24 h, until complete CPE was observed in the infected and untreated virus control. The determination of the anti-HSV-1 activity of the tested interferons was based on virus-induced cytopathogenicity of HSV-1 on Vero cells, measured at day 2 post virus infection by the MTT colorimetric method. An absorbance of formazan was detected at 560 nm. The results were expressed as the 50% effective concentration (EC_{50}). The EC_{50} was defined as the interferon concentration required for protecting 50% of the virus-infected cells against viral cytopathogenicity. The concentrations of recombinant cIFN-α and commercial IFN-α2a which exhibited 50% cytotoxicity (LD_{50}) were also determined. The therapeutic index was calculated by dividing LD_{50} on EC_{50}.

Statistical analysis

Differences between the variants were tested using Student's *t*-test and McNemar's test [29]. A *P*-value of <0.05 was considered statistically significant.

Results

Plasmid construction for auto-induction expression

The blunt-ended PCR product of the synthetic gene coding for cIFN-α was directionally cloned into pET101/D-TOPO vector by adding four bases (CACC) to the forward primer. The overhang in the vector (GTGG) invades the 5′ end of the PCR product, anneals to the added bases, and stabilizes the PCR product in the correct orientation. The constructed pET-cIFNα plasmid was transformed into *E. coli* BL21 (DE3) for cIFN-α auto-induction expression under the control of the T7 promoter. One transformant harbored recombinant pET-cIFNα was analyzed by restriction digestion to confirm the presence of the insert and nucleotide sequencing. The sequencing data indicated that the transformant carried the correct construct (data not shown). As showed in Figure 1, the construct has the correct sized cIFN-α gene.

Expression of consensus interferon by auto-induction

The recombinant human cIFN-α protein was overexpressed in auto-induced cultures using the pET system and lactose as inducer. After separation by 12% SDS-PAGE, total proteins obtained from auto-induced cells contained a thick protein band of about 19 kDa size, corresponding to the calculated size of cIFN-α protein (Figure 2A, lanes 3 and 4). SDS-PAGE analysis of crude proteins from auto-induced cultures or cultures induced by IPTG revealed that the auto-induced cultures yielded more cIFN-α protein than IPTG-induced cultures (Figure 2A, lanes 2–4). Figure 2A, lane 3 shows that maximum

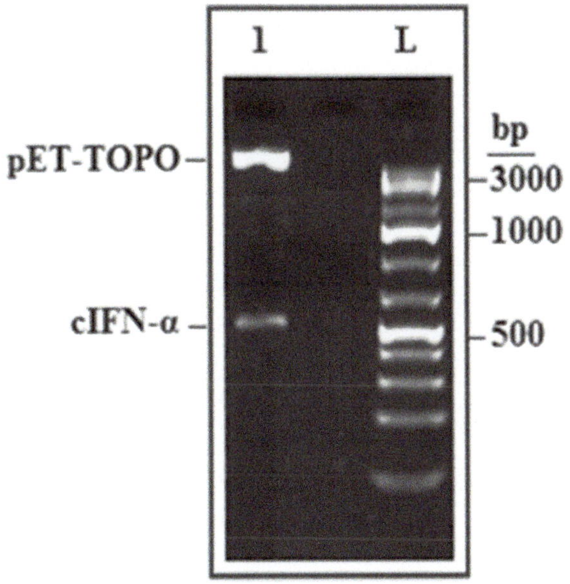

Figure 1 A 2% agarose gel showing the result of restriction digestion of recombinant pET-cIFNα. Recombinant pET-cIFNα was digested overnight at 37°C using *Xba*I and *Sac*I enzymes. Lane 1 shows the pET-TOPO vector fragment (5.73 kb) and the cIFN-α band (534 bp, with 26 bp added to the actual size of the gene from the vector cloning site and 10 bp introduced by gene-specific primers in the gene to enable pET directional TOPO cloning). L pointed GeneRuler 100 bp Plus DNA Ladder.

expression was reached by incubation of auto-induced cultures for 24 h at 30°C. Lowering incubation temperature to 25°C or 30°C improved the yield of soluble cIFN-α protein (Figure 2B). The effects of temperature and length of incubation on cIFN-α production level in auto-induction system are demonstrated in Figure 3.

The identity of the overexpressed protein was verified by ELISA using monoclonal anti-human IFN-α antibody as presented in Table 1. The cIFN-α activity of cellular extract prepared from auto-induced *E. coli* BL21 (DE3) cells was presented as the mean ± standard deviation (SD) of three replicates. As demonstrated in Table 1, the mean absorbance values for cIFN-α in auto-induced cells and IFN-α2a were respectively 0.637 ± 0.022 and 0.934 ± 0.016, compared with the value of cIFN-α in cellular extract prepared from uninduced cells as negative control of 0.046 ± 0.008. A significant difference ($P < 0.05$) in ELISA signals was found between commercial IFN-α2a or recombinant cIFN-α in auto-induced cells and the negative control. ELISA results revealed that the auto-induced cultures yielded more cIFN-α protein than IPTG-induced cultures. Western blot analysis of expressed protein revealed that recombinant cIFN-α expressed in auto-induced cells was strongly and specifically recognized by monoclonal anti-human IFN-α antibody at the expected 19 kDa molecular weight (Figure 4).

Purification of expressed cIFN-α

Both the supernatant and the pellet of the cell lysate after sonication were examined by SDS–PAGE to detect the solubility of expressed cIFN-α protein. As shown in

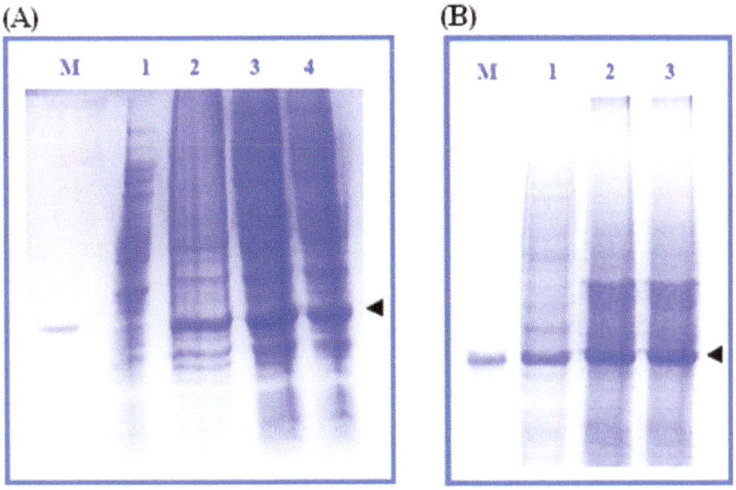

Figure 2 SDS-PAGE analysis of auto-induction expression of cIFN-α. (A) Auto-induction expression of cIFN-α at 30°C for 12 and 24 h. Lane M: commercial IFN-α2a as molecular weight marker of 19,2 kDa size. Lane 1: total proteins from uninduced cells. Lane 2: total proteins from IPTG-induced cells. Lane 3: total proteins from auto-induced cells after 24 h incubation at 30°C. Lane 4: total proteins from auto-induced cells after overnight incubation at 30°C. **(B)** Auto-induction expression of soluble cIFN-α. Lane M: commercial IFN-α2a as molecular weight marker of 19,2 kDa size. Lanes 1–3, soluble cell lysates of auto-induced cells after 24 h incubation at 37°C, 30°C, and 25°C, respectively. The arrow heads pointed the position of cIFN-α.

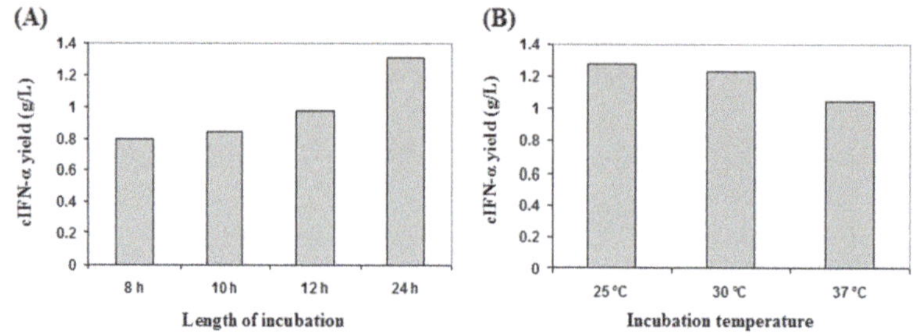

Figure 3 Time and temperature optimization of cIFN-α auto-induction expression. (A) Time optimization of cIFN-α auto-induction expression. Cultures were incubated for 8, 10, 12, and 24 h at 30°C. **(B)** Temperature optimization of cIFN-α auto-induction expression. Cultures were incubated at 25, 30, and 37°C for 24 h.

Figure 5, the main fraction of cIFN-α produced in auto-induction system was found in the soluble form after 24 h of incubation at 25°C or 30°C. Yields of cIFN-α in different cell preparations are reported in Table 2.

Our strategy for purifying expressed cIFN-α in auto-induction system involved separation of the soluble fraction of cell lysate and purification of cIFN-α by ion-exchange chromatography. The cell lysate supernatant was loaded onto a DEAE-Sepharose column previously equilibrated with 20 mM Tris–HCl, pH 8. After washing, the bound proteins were eluted with 100 ml linear gradient of 0–1 M NaCl. The elution profile of cIFN-α is shown in Figure 6A. Figure 6B shows SDS-PAGE profile of samples before and after cIFN-α purification. ELISA analysis of samples before (cell lysate supernatant containing cIFN-α) and after cIFN-α purification indicated that there is a significant difference ($P < 0.05$) in ELISA signals between the cell lysate supernatant containing cIFN-α (mean OD at 405 nm ± SD was 0.476 ± 0.023) and the purified cIFN-α (mean OD at 405 nm ± SD was 0.851 ± 0.009). The data revealed that cIFN-α was purified to near homogeneity, utilizing single-step ion-exchange chromatography on DEAE-Sepharose from cell lysate supernatant. The yield was 270 mg of purified cIFN-α from 1 liter of shake flask culture as estimated by Lowry method which is much higher than IPTG-inducible T7 expression system

Table 1 Reactivity of monoclonal anti-human IFN-α antibody against recombinant cIFN-α expressed in *E. coli* BL21 (DE3) cells by auto-induction

Sample	OD at 405 nm (mean ± SD)
Commercial IFN-α2a	0.934 ± 0.016
Recombinant cIFN-α (uninduced cells)	0.046 ± 0.008[a]
Recombinant cIFN-α (IPTG-induced cells)	0.447 ± 0.014[b]
Recombinant cIFN-α (auto-induced cells)	0.637 ± 0.022[a,b]

[a]Monoclonal anti-human IFN-α antibody reacted significantly ($P < 0.05$) with expressed cIFN-α protein in auto-induced cells.
[b]A significant difference ($P < 0.05$) in ELISA signals was found between recombinant cIFN-α in auto-induced cells and recombinant cIFN-α in IPTG-induced cells.

Figure 4 Western blot analysis of expressed cIFN-α using the specific monoclonal anti-human IFN-α antibody. Lane 1: total cell lysate of uninduced cells. Lane 2: commercial IFN-α2a as positive control and molecular weight marker of 19,2 kDa size. Lane 3: total cell lysate of auto-induced cells.

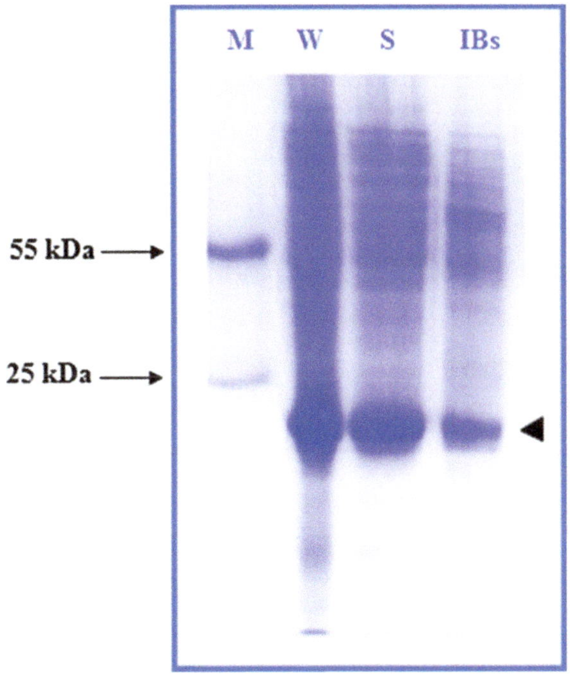

Figure 5 SDS-PAGE analysis of cIFN-α expressed in the soluble and insoluble inclusion forms. Lane M: lab-made protein molecular weight marker, W (whole cells), S (soluble fraction of cell lysate), and IBs (inclusion bodies). The arrow head pointed the position of cIFN-α.

that yielded 150 mg/L (unpublished results) and heat-inducible expression system that achieved a yield of 70 mg/L (as we have previously reported). These results suggest that auto-inducing media achieved 1.8- and 3.8-fold higher recombinant cIFN-α protein yield than IPTG-inducible T7 expression system and heat-inducible expression system, respectively.

Anticancer activity of cIFN-α expressed by auto-induction

To evaluate the anticancer (antiproliferative) activity of recombinant cIFN-α and commercial IFN-α2a, the inhibitory effect on hepatoma cell line HepG2 was studied using MTT colorimetric assay. The amount of formazan produced by HepG2 cells treated with different concentrations of the tested interferons was compared with the amount of formazan produced by untreated cells, the efficiency of tested interferons in causing death or changing metabolism of cancer cells could be deduced through the significant decrease in the absorbance of purple formazan product at 560 nm after 24 h from treatment with various concentrations of samples.

Table 2 Yields of cIFN-α expressed in optimized auto-induction system

Sample	cIFN-α yield (g/L)
Total cell protein	1.24
Total cell lysate	1.08
Soluble cell lysate	0.76[a] (70.37%)
Solubilized inclusion bodies	0.217 (20.09%)

[a]Majority of the expressed cIFN-α protein was in the cell lysate supernatant (about 70% of the yield in total cell lysate).

The results were expressed as percentage of cytotoxicity compared with the untreated cells. The relationship between percent cytotoxicity and the sample concentration is illustrated in Figure 7. This plot of % cytotoxicity versus sample concentration was used to calculate LD_{50}. Recombinant cIFN-α and commercial IFN-α2a demonstrated a potent inhibitory (antiproliferative) effect on HepG2 cell line after 24 h of treatment; ($LD_{50} = 2.79$ μg/ml and $LD_{50} = 4.16$ μg/ml, respectively). This was equivalent to 33% (1.5-fold) enhancement in activity. Figure 7 clearly demonstrated the superior anticancer activity of cIFN-α over the commercial IFN-α2a.

Selectivity index calculation

The effect of cIFN-α and IFN-α2a on the proliferation of human lymphocytes was evaluated after 24 h incubation by MTT assay (Figure 8). Both interferons showed significantly lower antiproliferative activity ($P < 0.05$) against PBMCs than that against HepG2 cells. The SI values were calculated using the ratio:

$$\begin{aligned} SI &= LD_{50} \text{ of tested samples in normal cell (lymphocyte)} \\ &\quad \div LD_{50} \text{ of tested samples in HepG2 cells.} \end{aligned}$$

The SI values for cIFN-α and IFN-α2a were 3.81 and 2.44, respectively. Recombinant cIFN-α showed a significantly higher selectivity ($P < 0.05$) to HepG2 cells than selectivity to HepG2 cells observed in case of commercial IFN-α2a.

Antiviral activity assay

The antiviral potency of recombinant cIFN-α and commercial IFN-α2a was determined via an *in vitro* assay, by using Vero cells challenged with HSV-1. The results were expressed as EC_{50}. The EC_{50} value for recombinant cIFN-α was 30.85 ng/ml. Protection obtained with a total of 30.85 ng/ml of recombinant cIFN-α was equivalent to the protection afforded by 100.35 IU/ml of commercial IFN-α2a. The EC_{50}, LD_{50}, and therapeutic index values for cIFN-α and IFN-α2a are presented in Table 3. Based on the definition of cIFN-α unit as the ability to protect 50% of cells from HSV-1-induced cytopathogenicity, it was calculated that the final purified cIFN-α has specific antiviral activity of 3.25×10^6 IU/mg protein as determined by cytopathogenicity assay; the units were determined using commercial IFN-α2a.

Discussion

Consensus interferon is a synthetic interferon that reflects most of the human type I interferons. This interferon

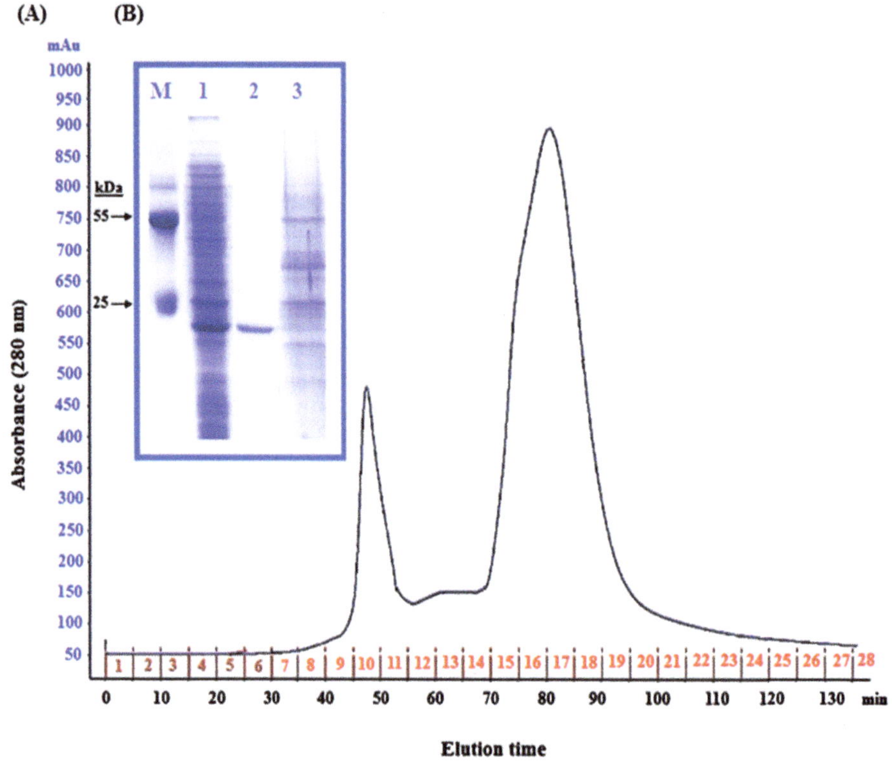

Figure 6 FPLC purification of recombinant cIFN-α expressed in auto-induction system on a DEAE-Sepharose column. (A) Protein elution profile (chromatogram). The profile clearly showed three peaks (fractions 8–11, 12–14, and 15–20). Numbers 1–28: number of collected fractions. **(B)** SDS-PAGE analysis of samples before and after cIFN-α purification. Lane M: lab-made protein molecular weight marker. Lane 1: cell lysate supernatant of auto-induced cells after 24 h incubation at 30°C. Lane 2: recovered cIFN-α protein after purification. Lane 3: cell lysate supernatant of uninduced cells.

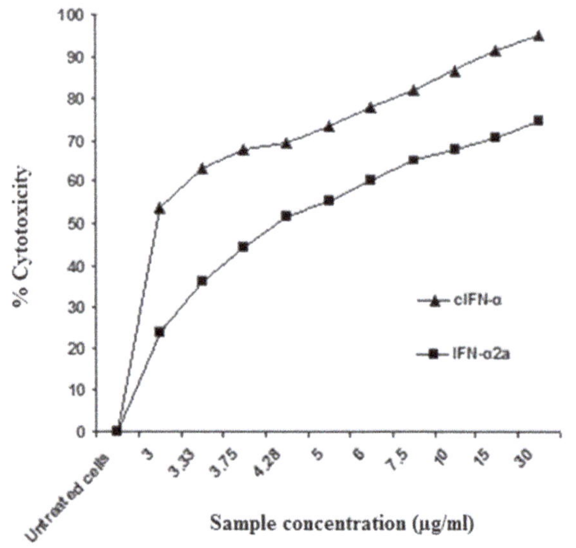

Figure 7 The inhibitory effect (antiproliferative activity) of cIFN-α synthesized in auto-induction system and commercial IFN-α2a on HepG2 cells. Both cIFN-α and IFN-α2a have an inhibitory effect on HepG2 cells after 24 h treatment; ($LD_{50} = 2.79$ μg/ml); ($LD_{50} = 4.16$ μg/ml), respectively.

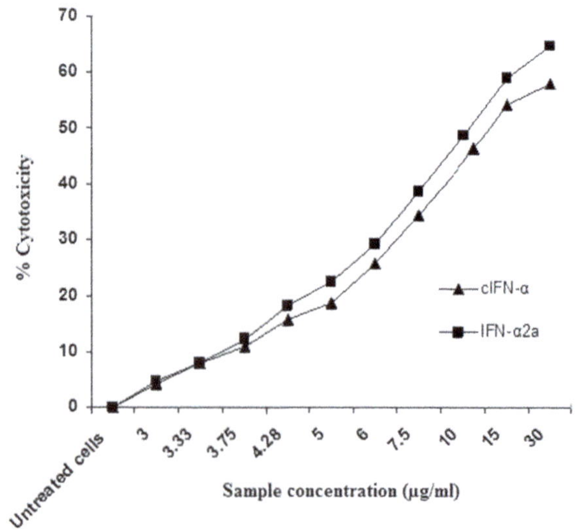

Figure 8 The inhibitory effect (antiproliferative activity) of cIFN-α synthesized in auto-induction system and commercial IFN-α2a on PBMCs. The LD_{50} values for cIFN-α and IFN-α2a were 10.63 and 10.16, respectively.

Table 3 LD_{50}, EC_{50} and therapeutic index values of recombinant cIFN-α determined via an *in vitro* assay, by using Vero cells challenged with HSV-1

Interferon	LD_{50}	EC_{50}	Therapeutic index (LD_{50}/EC_{50})
Recombinant cIFN-α	12.68 μg/ml	30.85 ng/ml	411.02
Commercial IFN-α2a	992.81 IU/ml	100.35 IU/ml	9.89

demonstrates a higher biological capacity *in vitro*. Consensus interferon has been used internationally in treatment of patients with chronic hepatitis C who fail to respond to conventional IFN-α therapy or those who relapse after treatment cessation [30].

We have previously used a heat-inducible, phage λP_L promoter-based expression system for high-level expression of recombinant human cIFN-α in shake flask *E. coli* cultures [13]. This system is highly productive, avoids the use of chemical inducers and is easily scalable. However, heat shock and high expression rates resulted in accumulation of the expressed protein into insoluble aggregates in a misfolded and biologically inactive form. Although insoluble misfolded cIFN-α was successfully refolded and purified, *in vitro* refolding yielded reduced amounts of biologically active product (70 mg/L). In another study, we reported synthesis of consensus interferon in cell-free extracts after 2–6 h [19]. Reaction conditions were altered to solve insolubility problems. The cell-free system achieved high-yield expression (400 μg/ml after purification process, which can be scaled up), and remarkably was able to efficiently support the complete maturation of cIFN-α. While cell-free protein synthesis proposes several advantages over cell-based methods, this system is not suitable for large-scale protein production. In the present study, we focused on avoiding the problems associated with the use of IPTG induction or heat induction for high-level expression of cIFN-α through the application of auto-inducing media to achieve higher biologically active product yields.

The results showed that the auto-induction system in which lactose was used as inducer produced higher cIFN-α yield than those with IPTG addition (150 mg/L) or heat induction (70 mg/L). We obtained 270 mg highly purified soluble cIFN-α protein from 1 liter of batch shake flask culture, which is the highest reported yield according to available data with the exception of the yield reported by Babu *et al.* [31] which was 300 mg of purified recombinant human IFN-α obtained from 1 liter of high cell density fed-batch *E. coli* culture. This can be explained by the fact that lactose is not only an inducer but also serves as a source of carbon and energy. In contrast, when IPTG is used as an inducer, the cells experience nutrient starvation, affecting protein expression. In addition, heat induction leads to formation of inclusion bodies that need *in vitro* refolding before purification which increases downstream product loss.

The synthesized cIFN-α in auto-induction system was biologically active as confirmed by its anticancer and antiviral effects. The anticancer effect of cIFN-α exceeds that of the monotype interferon (IFN-α2a), with higher selectivity to the tested cancer cell line, which may agree with the previous results of antiproliferative effect of consensus interferon in comparison to that of natural recombinant alpha interferons [6]. Purified cIFN-α has specific antiviral activity of 3.25×10^6 IU/mg protein.

Conclusions

In this study, higher product yield was achieved by using the auto-induction process for production of cIFN-α protein in *E. coli*. Auto-induction can be applied to increase the yield of other therapeutic recombinant proteins whose expression in *E. coli* using conventional expression systems is still challenging.

Abbreviations

BSA: Bovine serum albumin; cIFN-α: Human consensus interferon-alpha; CPE: Cytopathogenicity effect; DAB: 3,3 Diaminobenzidine; DCW: Dry cell weight; DMEM: Dulbecco's modified Eagle medium; DMSO: Dimethyl sulfoxide; EC_{50}: 50% effective concentration; ELISA: Enzyme-linked immunosorbent assay; FBS: Fetal bovine serum; HSV-1: Herpes simplex virus 1; IFNs: Interferons; IPTG: Isopropyl-β-D-1-thiolgalactopyranoside; LB: Luria Bertani; LD_{50}: Concentration lethal to 50% of the cells; MTT: 3-(4, 5-dimethylthiazol-2-yl)-2, 5-diphenyltetrazolium bromide; PBMCs: Peripheral blood mononuclear cells; *p*-NPP: *p*-Nitrophenyl phosphate; RBC: Red blood cell; SDS-PAGE: Sodium dodecyl sulfate polyacrylamide gel electrophoresis; SI: Selectivity index.

Competing interests

The authors declare that they have no competing interest.

Authors' contributions

NAE participated in experiments design, carried out the experiments, and wrote the draft. MHL collected the literature, organized it, and participated in statistical data analysis. EMR conceived the study, participated in its design and coordination, providing advice, criticism and final editing of the manuscript. All authors read and approved the final manuscript.

Acknowledgements

This study was funded by the Science and Technology Development Fund (STDF-ID269) for EMR.

Author details

[1]Therapeutic and Protective Proteins Laboratory, Protein Research Department, Genetic Engineering and Biotechnology Research Institute, City for Scientific Research and Technology Applications, New Borg EL-Arab, 21934 Alexandria, Egypt. [2]College of Applied Medical Sciences, King Abdulaziz University, P.O. Box 80203, Jeddah 21589, Saudi Arabia. [3]Biological Sciences Department, Faculty of Science, King Abdulaziz University, P.O. Box 80203, Jeddah 21589, Saudi Arabia.

References

1. Stewart IWE. The Interferon System. New York: Springer; 1979.
2. DeMaeyer E, DeMaeyer-Guignard J. Interferons and Other Regulatory Cytokines. Inc: John Wiley and Sons; 1988.
3. Pestka S, Langer JA, Zoon KC, Samuel CE. Interferons and Their Actions. Annu Rev Biochem. 1987;56:727–77.

4. Pestka S, Baron S. Definition and classification of the interferons. Methods Enzymol. 1981;78:3–14.
5. Blatt LM, Davis JM, Klein SB, Taylor MW. The biological activity and molecular characterization of a novel synthetic interferon-alpha species, consensus-interferon. J Interferon Cytokine Res. 1996;16:489–99.
6. Ozes ON, Reiter Z, Klein S, Blatt LM, Taylor MW. A comparison of interferon-Con1 with natural recombinant interferons alpha: antiviral, antiproliferative, and natural killer-inducing activities. J Interferon Res. 1992;12:55–9.
7. Cornberg M, Hadem J, Herrmann E, Schuppert F, Schmidt HH, Reiser M, et al. Treatment with consensus interferon (CIFN) plus ribavirin in non-responder patients with chronic hepatitis C: a randomized open-label pilot study. J Hepatol. 2006;44:291–301.
8. Kaiser S, Hass H, Lutze B, Gregor M. Comparison of daily consensus interferon versus peginterferon alfa-2a extended therapy of 72 weeks for peginterferon/ribavirin relapse patients with chronic hepatitis C [abstract]. J Gastroenterol. 2006;130:S1060.
9. Alton K, Stabinsky Y, Richards R, Ferguson B, Goldstein L, Altrock B, et al. Production, characterization and biological effects of recombinant DNA derived human IFN-α and IFN-γ analogs. In: De Maeyer E, Schellekens H, editors. *The Biology* of the *Interferon System*. Amsterdam: Elsevier; 1983. p. 119–28.
10. Fieschkoa J, Ritch T. Production of human alpha consensus interferon in recombinant *Escherichia coli*. Chemical Engineer Commun. 1986;45:229–40.
11. Klein ML, Bartley TD, Lai PH, Lu HS. Structural Characterization of Recombinant Consensus Interferon-α. J Chromatogr. 1988;454:205–15.
12. Curless C, Fu K, Swank R, Fieschko J, Tsai L. Design and evaluation of a two stage, cyclic, recombinant fermentation process. Biotechnol Bioeng. 1991;38:1082–90.
13. Mohammed Y, El-Baky NA, Redwan EM. Expression, purification, and characterization of recombinant human consensus interferon-alpha in *Escherichia coli* under λP_L promoter. Prep Biochem Biotechnol. 2012;42:426–47.
14. Studier FW, Moffatt BA. Use of bacteriophage T7 RNA polymerase to direct selective high-level expression of cloned genes. J Mol Biol. 1986;189:113–30.
15. Dubendorff JW, Studier FW. Controlling basal expression in an inducible T7 expression system by blocking the target T7 promoter with *lac* repressor. J Mol Biol. 1991;219:45–59.
16. Lin WJ, Huang SW, Chou CP. DegP-coexpression minimizes inclusion-body formation upon overproduction of recombinant penicillin acylase in *Escherichia coli*. Biotechnol Bioeng. 2001;73:484–92.
17. Studier FW. Protein production by auto-induction in high density shaking culture. Protein Expr Purif. 2005;41:207–34.
18. Redwan EM, Matar SM, El-Aziz GA, Serour EA. Synthesis of the human insulin gene: Protein expression, scaling up and bioactivity. Prep Biochem Biotechnol. 2008;38:24–39.
19. EL-Baky NA, Omar SH, Redwan EM. The anti-cancer activity of human consensus interferon-alpha synthesized in cell-free system. Protein Expr Purif. 2011;80:61–7.
20. Almehdar HA, Adel-Sadek MA, Redwan EM. Immunoreactivity and two-dimensional gel-electrophoresis characterization of Egyptian cobra venom proteome. Pak J Pharm Sci. 2015;28:59–64.
21. Mohammed Y, El-Baky NA, Redwan NA, Redwan EM. Expression of human interferon-α8 synthetic gene under P_{BAD} Promoter. Biochemistry (Mosc). 2012;77:1210–9.
22. Lowry OH, Rosebrough NJ, Farr AL, Randall RJ. Protein measurement with the Folin phenol reagent. J Biol Chem. 1951;193:265–75.
23. Redwan EM. Simple, sensitive and quick protocol to detect less than 1 ng of bacterial lipopolysaccharide. Prep Biochem Biotechnol. 2012;42:171–82.
24. Mosmann T. Rapid colorimetric assay for cellular growth and survival: application to proliferation and cytotoxicity assays. J Immunol Meth. 1983;65:55–63.
25. Lohr HF, Goergen B, Meyer zum Buschenfelde KH, Gerken G. HCV replication in mononuclear cells stimulates anti-HCV-secreting B cells and reflects nonresponsiveness to interferon-alpha. J Med Virol. 1995;46:314–20.
26. El-Fakharany EM, Haroun BM, Ng TB, Redwan ER. Oyster mushroom laccase inhibits hepatitis C virus entry into peripheral blood cells and hepatoma cells. Protein Pept Lett. 2010;17:1031–9.
27. El-Fakharany EM, Tabll A, El-Wahab AA, Haroun BM, Redwan EM. Potential activity of camel milk-amylase and lactoferrin against hepatitis C virus infectivity in HepG2 and lymphocytes. Hepatitis Monthly. 2008;8:101–9.
28. Rashad AE, Shamroukh AH, Abdel-Megeid RE, Mostafa A, El-Shesheny R, Kandeil A, et al. Synthesis and screening of some novel fused thiophene and thienopyrimidine derivatives for anti-avian influenza virus (H5N1) activity. Eur J Med Chem. 2010;45:5251–7.
29. Conover WJ. Practical nonparametric statistic. 2nd ed. New York: John Wiley and Sons; 1971.
30. El-Baky NA, Redwan EM. Therapeutic alpha-interferons protein: structure, production, and biosimilar. Prep Biochem Biotechnol. 2015;45:109–27.
31. Babu KR, Swaminathan S, Marten S, Khanna N, Rinas U. Production of interferon-alpha in high cell density cultures of recombinant *Escherichia coli* and its single step purification from refolded inclusion body proteins. Appl Microbiol Biotechnol. 2000;53:655–60.

Critical assessment of influenza VLP production in Sf9 and HEK293 expression systems

Christine M Thompson[1,2], Emma Petiot[1,3], Alaka Mullick[1], Marc G Aucoin[4], Olivier Henry[2] and Amine A Kamen[1,5*]

Abstract

Background: Each year, influenza is responsible for hundreds of thousand cases of illness and deaths worldwide. Due to the virus' fast mutation rate, the World Health Organization (WHO) is constantly on alert to rapidly respond to emerging pandemic strains. Although anti-viral therapies exist, the most proficient way to stop the spread of disease is through vaccination. The majority of influenza vaccines on the market are produced in embryonic hen's eggs and are composed of purified viral antigens from inactivated whole virus. This manufacturing system, however, is limited in its production capacity. Cell culture produced vaccines have been proposed for their potential to overcome the problems associated with egg-based production. Virus-like particles (VLPs) of influenza virus are promising candidate vaccines under consideration by both academic and industry researchers.

Methods: In this study, VLPs were produced in HEK293 suspension cells using the Bacmam transduction system and Sf9 cells using the baculovirus infection system. The proposed systems were assessed for their ability to produce influenza VLPs composed of Hemagglutinin (HA), Neuraminidase (NA) and Matrix Protein (M1) and compared through the lens of bioprocessing by highlighting baseline production yields and bioactivity. VLPs from both systems were characterized using available influenza quantification techniques, such as single radial immunodiffusion assay (SRID), HA assay, western blot and negative staining transmission electron microscopy (NSTEM) to quantify total particles.

Results: For the HEK293 production system, VLPs were found to be associated with the cell pellet in addition to those released in the supernatant. Sf9 cells produced 35 times more VLPs than HEK293 cells. Sf9-VLPs had higher total HA activity and were generally more homogeneous in morphology and size. However, Sf9 VLP samples contained 20 times more baculovirus than VLPs, whereas 293 VLPs were produced along with vesicles.

Conclusions: This study highlights key production hurdles that must be overcome in both expression platforms, namely the presence of contaminants and the ensuing quantification challenges, and brings up the question of what truly constitutes an influenza VLP candidate vaccine.

Keywords: Influenza vaccines, Virus like particles (VLPs), Insect cells, Mammalian cells, Process development, VLP characterization

Background

As reported by the World Health Organization, seasonal influenza is responsible for approximately 500 million cases of infection and between 250,000 to 500,000 deaths each year [1]. Currently, vaccination remains the most proficient strategy to prevent infection and to battle the persistent threat of influenza epidemics. Egg-based production has remained the standard method to produce seasonal influenza vaccines since the 1950s; however, the serious threat of an outbreak of pandemic avian flu and the influenza H1N1 pandemic of 2009 have highlighted the limitations associated with this manufacturing method [2,3]. These drawbacks include the dependence on the availability of eggs for production and their associated limitations with regards to scalability (approximately 1 vaccine dose per egg), the relatively long 6-month period from strain isolation to final dose formulation and validation [4], and the

* Correspondence: amine.kamen@mcgill.ca
[1]National Research Council Canada, Human Health Therapeutics, Montréal, Canada
[5]Department of Bioengineering, McGill University, 817 Sherbrooke St. W. Macdonald Engineering Building, Room 387, Montréal, Canada
Full list of author information is available at the end of the article

difficulties with growing isolated avian strains. Cell culture based production has been proposed to overcome such limitations; these processes are easier to scale-up and, through the standardization of production methods, allow for increased vaccine production. Moreover, cell culture processes have the potential to reduce manufacturing time by several weeks with a faster start-up time [5].

Seasonal influenza vaccines derived from cell culture are gaining increased attention. In November 2012, the first seasonal vaccine produced in cell culture (Flucelvax, Novartis) was approved by the FDA. In January 2013, the first trivalent influenza vaccine, Flublok (Protein Science Corporation) made in insect cells using a recombinant baculovirus expression system, was also approved. These recent advances reflect an important trend of adopting modern cell culture manufacturing in the influenza vaccine industry. It is highly supported by public health and regulatory agencies that promote strategies to improve responses to emerging infectious diseases.

One of the most promising approaches is the use of recombinant-based virus-like particle (VLP) vaccines [6,7]. Influenza VLPs are non-infectious and non-replicating empty viral particles lacking viral genetic material, which display intact and biochemically active antigens on their surface. They can be composed of one, or different combinations, of the viral antigens and structural influenza proteins: HA, NA, M1 and M2 [6,8-11].

Immunizations with influenza VLPs to protect against either seasonal or pandemic influenza strains have shown promising results [8,12-14]. VLPs have a stronger safety profile than native infectious viruses produced either in egg-based or cell-based systems. No infectious virus potentially harmful to humans is used as a seed inoculant or produced, thus production processes do not require the use of high level containment facilities. With more clinical trials underway for VLP vaccines [15,16], there is a critical need for the scientific and engineering communities to tackle the specific challenges associated with the bioprocessing aspects of VLP production.

Thus far, different production processes have been studied for influenza VLPs. They vary according to the viral strain produced, the type of gene delivery system used and the host-cell expression system. Influenza VLPs have been produced in mammalian, insect and plant cell platforms using a variety of vectors and gene delivery techniques [17-19]. The majority of VLP productions have been done in insect cells with the baculovirus expression vector system (BEVS) [8,18,20,21]. However, this system suffers from contamination with baculovirus (BV) particles [14,22] in final vaccine preparations. This contamination has been known for a long time and can represent up to 5% of the total proteins in final purified VLP preparations [14] and 4.5×10^6 BEVS per vaccine dose [22]. Baculovirus are non-replicative in mammalian cells, but they have been reported to trigger an innate immune response. Although there is some evidence that baculovirus is safe for immunization [22,23], in terms of process development and evaluation of the immunization efficiency from VLPs alone, such high level of contaminant presents a problem. Alternative methods of VLP production in mammalian cells, using either plasmid transfection or baculovirus gene transfer in mammalian cells (BacMam) have thus been developed. The second problem posed by the insect cell platforms is product glycosylation, as insect cells do not allow proper sialylation of the proteins [24]. Different groups are now working on the introduction of such glycosylation pathways in insect cell systems.

Most studies thus far have focused on the proof of concept of VLP production and the evaluation of influenza VLPs immunogenicity to support their candidacy as potential vaccines [8,12] and as a model to gain insight into the requirements for influenza virus budding [25]. Nevertheless, to date, limited attention has been paid to the specific challenges posed by the manufacturing of VLPs. However, studies of this nature are starting to surface [18,19,22,26].

In line with the theme of studying the bioprocessing of influenza VLPs, this study assessed two platforms for VLP production, the insect (Sf9) and mammalian (HEK293) suspension and serum-free production platforms with influenza gene transfer by baculovirus infection and transduction, respectively. VLP production and contaminant levels (VLP/mL, BV/mL, μg HA/mL, HA units/mL), as well as the challenges associated to each production systems are discussed.

Methods

Cells and medium

The HEK293 cell line used for VLP production was previously adapted to suspension and serum-free culture [27]. Working cell banks were made from a vial of the master cell bank, which was developed under Good Manufacturing Practices (GMP). HEK293 cells used for VLP production were grown in shake flasks at 37°C and 5% CO_2, in animal component and serum free SFM4Transfx 293™ (HyQ) medium (HyClone, Waltham, MA, USA). Insect *Spodoptera frugiperda* (Sf9) cells were maintained in serum free Sf900 II medium (GIBCO, Burlington, ON, Canada), in shake flasks at 27°C with an agitation rate set at 110 rpm. Cell density was monitored using the Cedex Cell Counter (Innovatis Roche Applied Science, Penzberg, Germany).

Construct design of the gene transfer system

Baculovirus BacMam for mammalian cell production system (HEK293)

The recombinant baculovirus used for HEK293 cell transduction (referred as BacMam PR8) was previously described in Tang et al. [19] and kindly donated by Dr. Ted Ross (University of Pittsburgh). One recombinant baculovirus was used to drive the expression of HA, NA and M1 genes from H1N1 A/Puerto Rico/8/1934 influenza strain under the control of individual CMV promoter. BacMam PR8 also contained a Green fluorescent protein (GFP) reporter gene under the CMV promoter and a VSV-G protein under polyhedron promoter control in order to improve cell transduction. A working stock of BacMam PR8 baculovirus was produced by generation of P0 BacMam PR8 stock in Sf9 cells with bacmid transfection and two subsequent BacMam PR8 amplifications in Sf9 cells (the VP/mL and IVP/mL of each stock can be found in the Additional file 1: Table S1).

Baculovirus construction for insect cells production system (Sf9 cells)

For production in insect cells, co-infection with three baculoviruses each carrying influenza proteins, HA, NA or M1 was chosen. The influenza proteins were under polyhedron promoter (polh) control for expression in insect cells. The construction of vectors for further generation of P0 baculovirus stocks through Sf9 cells transfection was performed as follows: the DNA sequence of H1N1 A/Puerto Rico/8/1934 HA (AB671289.1), NA (AB671290.1) and M1 (CY033578.1) were obtained from NCBI's influenza database. Influenza gene sequences were inserted between XbaI and BglII enzyme restrictions sites and further in a pUC plasmid by BioBasic (Markham, Canada). Each influenza gene, flanked by XbaI and BglII sites, was further inserted in the pVL1393 plasmid, belonging to the commercial BaculoGold™ system (BD bioscience, Franklin Lakes, USA) used for the construction of baculovirus allowing expression of recombinant proteins in Sf9 cells. Each of the three plasmids, respectively named pVL1393-HA, pVL1393-NA and pVL1393-M1, were co-transfected with baculovirus DNA to produce recombinant baculoviruses referred to as Bac-HA, Bac-NA, and Bac-M1. Similarly, to produce the BacMam PR8 virus stock, each Bac-HA, Bac-NA and Bac-M were passaged twice in Sf9 cells to produce a P2 working viral stock used for VLP production (VP/mL and IVP/mL of each stock can be found in the Additional file 1: Table S1).

VLP production

For both cellular platforms, during the production optimization phases, cultures were sampled once to twice a day and cell density, viability and average cell diameter were measured using the Cedex Cell Counter (Innovatis Roche Applied Science, Penzberg, Germany).

Mammalian HEK293 cell platform

For VLP production in HEK293 cells, transduction with BacMam PR8 was performed at a cell density of 1.5-2.0×10^6 cells/ml. An MOI of 60 was used (MOI calculations were based on infectious baculovirus titers). Production runs were carried out in shake flasks with working volumes of 60–400 ml. In production runs, butyric acid was added at a final concentration of 5 mM at the time of infection. After 48 hours post-infection, culture broth samples were clarified by slow centrifugation at 300 × g for 5–10 minutes. Culture supernatant samples were concentrated via sucrose cushion ultracentrifugation and analysed by Western blot, HA assay, Negative Stain Transmission Electron Microscopy (NSTEM) and Single Radial Immuno-diffusion assay (SRID).

Insect Sf9 cell platform

VLP productions in Sf9 cells were performed in shake flasks with working volumes of 60–400 ml. Cells were infected with P2 working stocks at a density of 2.0-2.5 × 10^6 cells/ml with a total MOI of 0.3-2.1 at 27°C (MOI calculations were based on infectious baculovirus titers). Cells were harvested after 48 hours post-infection, when the viability started to decline below 70-50%. Upon harvest, the supernatant samples were collected with slow centrifugation at 300 × g for 5–10 minutes, then concentrated via sucrose cushion ultracentrifugation and analysed by Western blot, HA assay, NSTEM and SRID.

VLPs extraction from cell pellet

The cell pellets obtained after slow centrifugation were re-suspended in a solution of PBS and 10 μg/ml of TPCK-trypsin and slowly agitated at 37°C for 30 minutes to produce cell pellet wash. Cells were pelleted via slow centrifugation and the supernatant was collected for analysis by Western blot, HA assay and NSTEM. A cell lysis extraction was also performed after this first treatment by one freeze-thaw cycle. The cell lysis supernatant was collected by centrifugation at 1000 × g for 5 min and then analysed using Western blot, HA assay and NSTEM.

Influenza virus production

Influenza H1N1 A/PR/8/1934 virus was produced in house as described in Petiot et al. [28] and used as standard for the Western blot, HA assay and NSTEM images.

VLP concentration and purification

Sucrose cushion ultracentrifugation

The VLP supernatants were sublayered with 10 ml of a 25% sucrose solution prepared in 20 mM Tris–HCl (pH 7.5). The maximal VLP volume loaded was of 200 ml of culture supernatants, centrifuged at 4°C and 37 000 × g for 3 hours (Sorvall Discovery SE 100 ultracentrifuge, A621 rotor, Thermo Scientific, Waltham, USA). The ultracentrifugation supernatants were discarded and the VLPs were collected as pellets, re-suspended in 20 mM Tris–HCl, 1% sucrose and 2 mM $MgCl_2$. Ultracentrifugation purified samples were stored overnight at 4°C, then aliquoted and stored either at −80°C or 4°C. This purification step allowed to concentrate VLP production samples up to 20-75×.

Iodixanol gradient ultracentrifugation

Gradient iodixanol purification was done from two working stock solutions of iodixanol, one with 40% w/v and one with 25.5% w/v. These solutions were prepared from Optiprep iodixanol solution (Axis-Sheild, Oslo, Norway) with 60 mM Tris–HCl (pH 7.5). VLP concentrated solutions from the sucrose cushion purification step were combined to the 25.5% iodixanol working stock and subjected to slow mixing to give a final iodixanol concentration of 20% and a VLP dilution of 4.7×. The VLP-iodixanol solution was then loaded into 13 ml ultracentrifuge tubes, sealed and centrifuged at 350 000 × g for 6 hours at 4°C (Sorvall Discovery SE 100 ultracentrifuge, 65 V13 rotor, Thermo Scientific, Waltham, USA). After ultracentrifugation, 1 ml fractions were taken from the bottom of the tubes and weighed to determine the density and iodixanol concentration. Each fraction was then analysed by Western blot, HA assay and NTSEM.

VLP quantification and detection

Western blot

Samples were loaded on a Bio-Rad mini-protean Tris-Glycine 4-15% gel with 1× running buffer (25 mM Tris, 192 mM Glycine, 3.5 mM SDS). The gel was run for 40 minutes at 200 V and then washed with Towbin transfer buffer (25 mM Tris, 192 mM Glycine, 20% Methanol pH 8.3). Semi-dry protein transfer to nitrocellulose membrane was performed using a Bio-Rad Trans-Blot 3D semi-dry transfer cell (Hercules, CA, USA) for 1 hour at 10 V. Membranes were then blocked for 1 hour in 5% milk solution (1× TBS, pH 7.5: 50 mM Tris, 150 mM NaCl, 0.1% Tween-20, 5% milk). The membrane was washed with 0.1% Tween TBS at pH 7.5 and then incubated overnight at 4°C with its respective antibody (conc. 1/1000). The secondary antibody incubation (conc. 1/1000) lasted 1 hour at 4°C. Primary antibodies used for HA, NA, M1 and GP64 were as follows: anti-HA and anti-NA, polyclonal sheep serum (ref. 03–242 and 04–230, respectively, NIBSC, London, UK); anti-HA monoclonal mouse (ref. sc-80550, Santa Cruz Biotechnology, USA); anti-M1 monoclonal mouse antibody (ref. ab22396, Abcam, Cambridge, USA); anti-GP64 monoclonal mouse antibody (ref. 14-6995-81, eBioscience, San Diego, USA). The secondary antibodies used were Donkey anti-sheep and Donkey anti-mouse (ref. 713-005-003, 715-005-150, respectively) IgG HRP from Jackson ImmunoResearch (Westgrove, USA). Images were acquired with a DS Image Station 440CF (Kodak, Rochester, USA).

Hemagglutination assay

The HA assay was completed in 96 well v-bottom plates. Wells were filled with 100 μl of 1× PBS solution in rows A-H, columns 2–12. Next, 29.3 μl of 1× PBS was loaded into wells B1, B2, D1, D2, F1, F2, H1 and H2. Each sample took up two rows (A and B, C and D, etc.) and was serially diluted with VLP samples from the supernatant, sucrose cushion and iodixanol purifications. 100 μl of VLP sample was loaded into wells 1 and 2 of rows A, C, E, and G, then 70.7 μl into wells 1 and 2 of rows B, D, F and H. The VLP sample was then serially diluted and 100 ul of 5-day old chicken red blood cells at a concentration 2×10^7 cell/ml were added to each well in the plate. The plates were left in a covered plastic container for 3 hours to overnight at room temperature and scored. The amount of HA in units HA/ml was calculated with the following correlation:

$$\frac{Log\ HA\ units}{100\ ul} = \log(dilution\ factor\ at\ last\ well\ agglutinated)$$

Single radial immunodiffusion assay

A 1% agarose gel solution was prepared and equilibrated at 50°C in a water bath. Agarose gel solution was added with anti-HA sheep serum (ref. 03–242, NIBSC, London, United Kingdom) with a final concentration of 1/1000. The agarose-antibody mixture was casted and left to cool in a Bio-Rad gel-casting module for at least 15 minutes. In the gel, 4 mm wells were punched before addition of 20 μl of sample in each well. For all the SRID experiments, a calibration curve was determined with standard samples of purified recombinant HA of A/H1N1/Puerto Rico/8/1934 strain from Protein Sciences Corporation (Meriden, USA) at concentrations ranging from 7 to 30 μg/ml. All the samples were subjected to detergent treatment for 15 minutes on a rocker platform (1% Zwittergent final concentration). Sample diffusion in the gel took place for 18–24 hours at room temperature. The gel was dried using Whatman #3 filter paper (Kent, United Kingdom), rinsed with deionized water, and stained with Coomassie Blue R-250 for 15 minutes and de-stained using a 7.5% acetic acid and 10% ethanol solution. Result analyses were performed on gel photograph where the diameter of the reference standards and samples were measured using Image-J software (Pixel-aspect ratio = 1, known distance = 4 mm) allowing further calculation of sample concentration through linear correlation.

Negative Stain Transmission Electron Microscopy (NSTEM)

NSTEM analysis were conducted at Institut Armand Frappier (Laval, Canada) adapting a method previously described [29]. For each samples, 2 micrographs were counted with samples pre-diluted to have at least 25 VLPs on the grid. The range of VLPs counted was between 25 and 300 particles. The particle count was quantified using the following equation: VLP/ml = (VLP count/latex bead count) × (latex bead concentration × virus dilution).

Results and discussion

It is important to note that, for all the analysis performed (NSTEM, Western Blot, HA assay, SRID), the samples were concentrated by a sucrose cushion ultracentrifugation step. This concentration step was necessary as the available quantification methods are not sensitive or specific enough to directly quantify total VLPs and protein composition of non-purified culture supernatants. This allowed us to i) reach detectable levels and ii) and reduce the contamination by free influenza and host cell proteins in the sample. Consequently, most of the assays were performed with at least a 40× concentrated sucrose cushion purified samples and all the values presented in the paper correspond to their relative normalized production levels (per/mL).

VLP morphology and particle quantification

Influenza virus and VLP morphology were analysed by visual inspection on NSTEM micrographs, and the concentration of VLP particles were assessed for the comparison of VLP production efficiency in both cellular platforms. Previous studies of influenza virus production in HEK293 cell culture described the virus as pleomorphic particles containing spikes with an average size of 100 +/− 20 nm [30]. Influenza virus consisted mainly of spherical and elongated particles [31]. An example of H1N1 A/Puerto Rico/8/1934 influenza viruses produced in-house on the same HEK293 cell platform [31] is shown in Figure 1. On this basis, VLPs were defined as

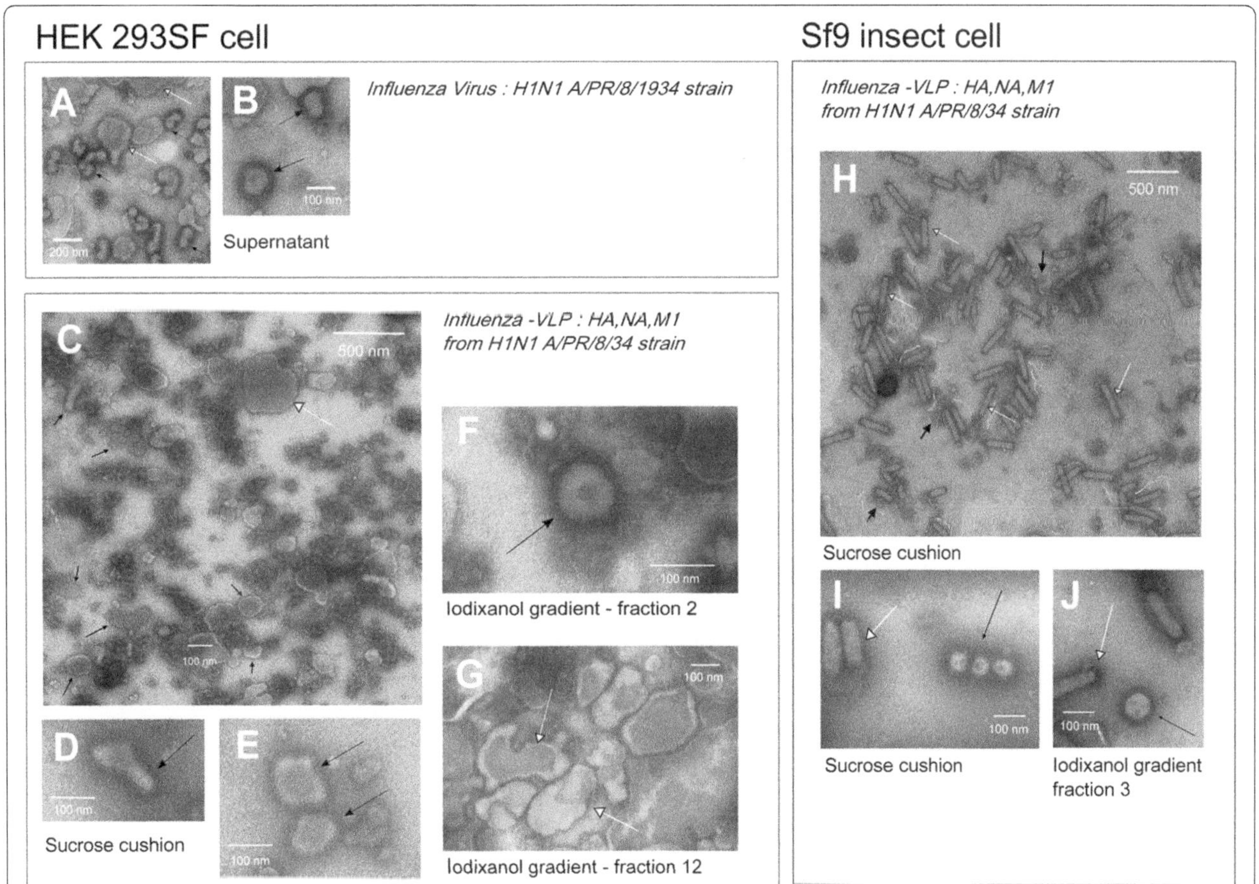

Figure 1 NSTEM images at 40,000× magnification of H1N1 A/Puerto Rico/8/34 **(A & B)**, H1N1 influenza-VLPs from HEK 293SF cell production **(C-G)** or Sf9 cell production **(H-J)**. VLPs or influenza virus are pointed by black arrows, baculovirus and cell vesicles are pointed by white arrows. **A** & **B** - Supernatant of H1N1 A/Puerto Rico/8/1934 influenza virus produced in HEK 293SF suspension cells as in Petiot et al. [28]. Cell vesicles carrying influenza glycoproteins could be identified. **C**, **D** & **E** - Sucrose cushion purified samples of HEK 293SF cell production; large view **(C)** and zoom-in **(D** & **E)** images of NSTEM grid. **F** & **G** - Iodixanol purified samples of HEK 293SF cell production. Typical shapes of VLPs and vesicles identified in high density (fraction 12: 1.03 g/ml) and low density (fraction 2:) iodixanol fractions. **H** & **I** - Sucrose cushion purified samples of Sf9 cell production. **J** - Iodixanol purified samples fraction 3, which present the highest number of VLP particles. Baculovirus were co-purified with VLPs in all the iodixanol fractions, even if they were more concentrated in the high density fractions.

particles with a fringe of spikes on their membrane, assumed to be influenza glycoproteins HA and NA, with a size range of 100 +/− 50 nm. NSTEM images of VLPs produced with both HEK293 SF and Sf9 cells are shown in Figure 1. Pictures presented are all from purified samples, either sucrose cushion (C, D, E H & I) or sucrose cushion and iodixanol purified (F, G & J).

Figure 1D, E & F present zoomed-in images of VLPs produced with HEK293 cells. The VLPs were similar to influenza virus particles, although their average size ranged from 100 to 500 nm and they displayed the typical morphology of host cell vesicles. VLPs were counted to be at a concentration of 1.5×10^8 VLPs/ml.

Figure 1I & J present zoomed-in images of VLPs produced in Sf9 cells. Their morphology seems to be uniform and close to influenza virus, consisting of spherical particles with an average size close to 100 nm. For these samples, a concentration of 5.85×10^9 VLPs/ml was quantified for Sf9 VLP production.

VLP composition

VLPs were expected to contain HA, NA and M1 influenza proteins. HA and NA are the antigen proteins located on the particle surface and M1 is the matrix protein that is the structural backbone of the viral particle. M1 is assumed to be located on the internal surface of the membrane envelope. For both production systems, VLP samples recovered in the culture supernatant and pre-purified by sucrose cushion were examined by Western blot. The presence of HA and NA in sucrose cushion was confirmed for VLPs produced in both the HEK293 and Sf9 cell platforms (Figure 2A & B). Surprisingly, M1 was present in the Sf9 VLPs but not found in HEK293 VLP samples (Figure 2C).

In order to confirm expression of M1 in HEK293 cells, anti-M1 Western blots of cell pellet were performed on both Sf9 and HEK293 cells (Figure 2C). M1 protein was present in both cell pellets, albeit in higher quantities in the Sf9 pellet than in HEK293. This indicated that M1

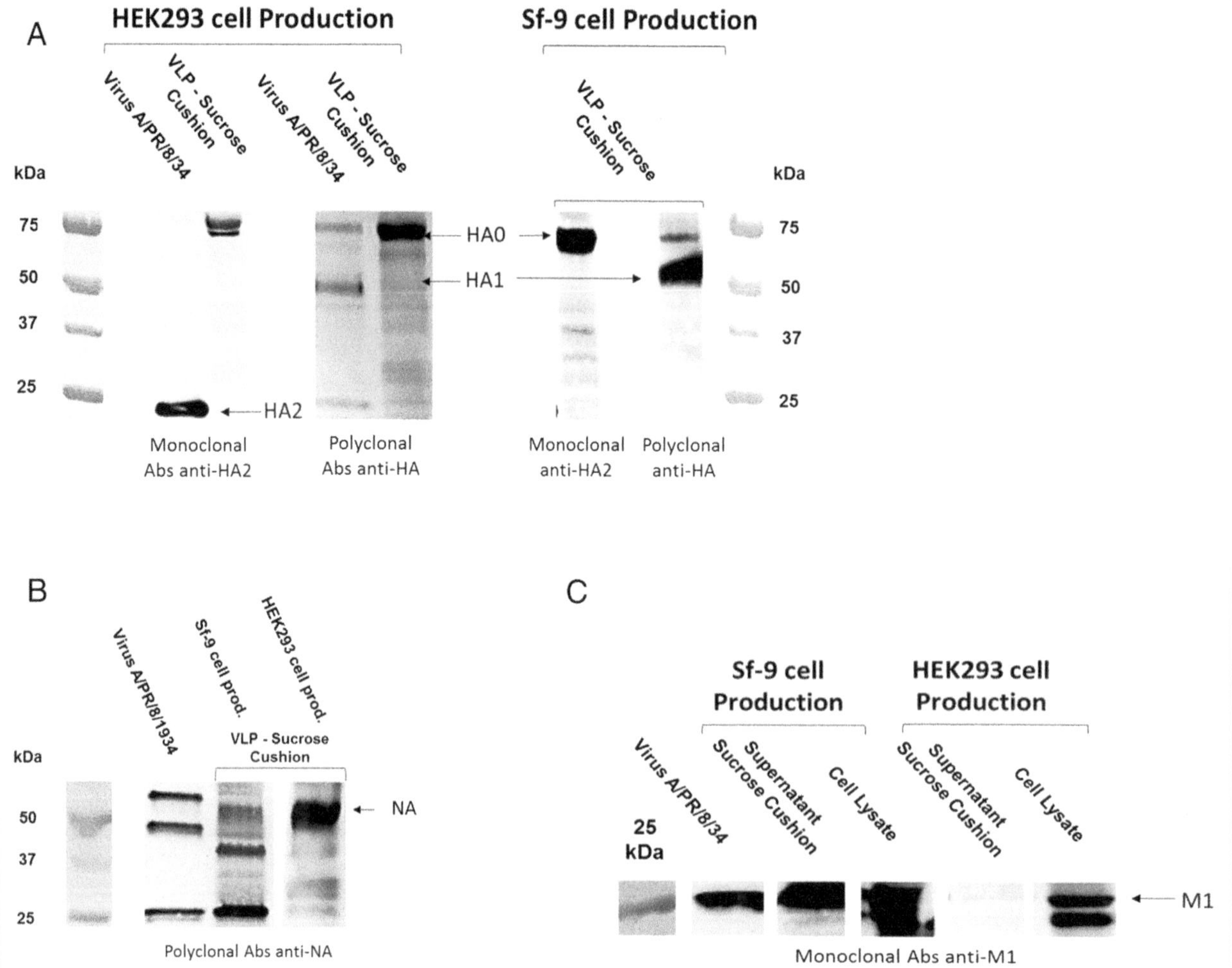

Figure 2 Western blot of Sucrose cushion VLP samples from HEK 293SF and Sf9 cells. **A** - Anti-HA western blot with two different antibodies, a monoclonal Abs against HA2 (Santa Cruz, sc-80550) and a polyclonal antibody against HA0 (anti-HA NIBSC sheep; 03–242). **B** - Anti-NA western blot using anti-NA NIBSC sheep (04–230). **C** - Anti-M1 western blot using monoclonal mouse antibody (Abcam, ab22396).

produced in HEK293 cells was not correctly incorporated into the VLPs. Integration of M1 into the Sf9 VLPs could explain for the more spherical and uniform particle shape observed with this system. M1 has been known for a long time to play an important role in determining the shape of influenza particles, especially for the formation of filamentous or spherical types of viral particles [32-34]. The M1 protein affects influenza virus morphology in a number of ways; firstly though a number of its amino acid residues [34] that interact with both HA and NA to drive the budding process [35]; and secondly, with interactions with the NP protein and the vRNP complex [36].

Studies of bud formation and viral particle morphology from the last decade proved that M1 is not the sole protein playing a role in viral particle formation and their morphology. The M2 protein, which is an ion channel protein, contains a highly conserved tail among different viral strains that was not necessarily responsible for their specific activity, but was later related to budding efficiency and particle morphology [35,37]. Besides influenza itself, host-cell heat-shock protein 70 (Hsp70) has also been described to play an important role in bud formation [38]. Actin was also reported in different studies as potentially impacting the virus produced shape [34], and a number of other cellular proteins of Vero cells have been recently detected in A/WSN/33 H1N1 influenza virions produced in this system [39]. It has also been shown for different insect cell production platforms tested (Sf9 versus Tn5 cells), that the cell type has a great impact on homogeneity of the VLPs produced [24]. Consequently, the cellular platforms used for either influenza virus or VLP production, and their membrane compositions, are expected to have a great impact on the quality of VLPs produced.

A careful review of previous influenza VLP studies indicated many contradictory results. Most of the influenza-VLP production studies consider that M1 is an important factor for correct VLP budding and is integrated into most of the constructions [19]. Chen et al. have shown that co-expression of M1 and HA proteins increases VLP production, rather than HA expressed alone [25]. But other studies have demonstrated that VLPs without M1 can be successfully produced, such as VLPs produced in HEK293T cells that only contain NA from both H5N1 and H1N1 strains [40]. Finally, M1 was also shown to form VLPs on its own in COS-1 and Sf9 cells [12,41]. This observation confirms the importance of the cellular platform on the level of incorporation of M1 in the VLPs and on particle formation, which appears to be cell line and expression system dependent.

VLP HA activity

The particles produced as VLPs by both systems were also evaluated for their activity. The HA assay was performed to demonstrate the agglutination activity of sucrose cushion purified VLPs and the HA protein content (μg/ml) was evaluated by the standard quantification method used for influenza vaccine dose release, that is the SRID assay. Results are presented in the comparative Table 1. For the HEK293 cell production system, VLPs reached HA titers of 13 HAU/ml whereas for Sf9 cell produced VLPs, the HA activity was at 335 HAU/ml, which is more than 25 times higher. As for HA protein content, results obtained for HEK293 cells were again lower than the Sf9 cell system, with a production level of 0.092 μg HA/ml for HEK293 VLP samples and 0.38 μg HA/ml for Sf9 VLP samples (4-fold increase). It is interesting to note from a product qualification point of view that the ratio of total VLPs counted from both systems is not correlated to the results obtained from the SRID and HA assay. This is due, as discussed in a previous review of influenza-VLP quantification methods [42], to the particularity of each quantification method and their intrinsic bias. It should be emphasized that the hemagglutination or SRID assays, routinely used to quantify VLPs in past studies, do not have the ability to make a distinction between real virus-like particles and impurities [19,22]. Vesicles or baculovirus carrying HA proteins will be quantified the same way by such methods. Presently, only visual direct identification will allow for a reliable assessment of the influenza-VLP particle produced from each cellular platform. That is why this work also presents elements to qualitatively assess the purity by direct visual inspection of the samples by NSTEM in the following section.

Table 1 Summary of the production performances for the two cell platform, Sf9 insect cells and HEK293SF mammalian cells

Cell line	Sucrose cushion samples				Iodixanol purified samples		
	HA protein activity (HA assay)	HA protein content (SRID assay)	VLP conc. (NSTEM)	Baculovirus contamination (NSTEM)	HA protein activity (HA assay)	VLP conc. (NSTEM)	Baculovirus contamination (NSTEM)
Insect Sf9	335 HAU/ml	0.38 μg/ml	5.85×10^9 VLPs/ml	1.12×10^{11} BV/ml	32.4 HAU/ml	2.39×10^9 VLPs/ml	1.73×10^{10} BV/ml
Mammalian HEK 293	13 HAU/ml	0.092 μg/ml	1.50×10^8 VLPs/ml	3.07×10^7 BV/ml	8.11 HAU/ml	7.71×10^7 VLPs/ml	Not quantified

Iodixanol fractions presented for both systems are the ones containing more VLP particles, i.e. fraction 3 for Sf9 cells (1.13 g/ml) and fraction 1 (1.16 g/ml) for HEK293SF cells. It should be noticed that for Sf9 and HEK293SF cells, sucrose cushions were concentrated respectively of 40 times and 50 times, and for iodixanol fractions of 110 times and 140 times. The values presented in this table represent the production levels (i.e. non concentrated samples).

VLP purity and recovery

Zoomed-out images of sucrose cushion purified VLPs for both Sf9 cells and HEK293 cells are presented in Figure 1C & H.

VLP purity

For Sf9 cell productions, influenza-VLPs are identified by black arrows in Figure 1H. However, co-produced baculoviruses are also clearly identifiable in the picture (rod shaped particles, indicated by white arrows). Baculoviruses counted in this NSTEM micrograph sample were at a concentration of 1.12×10^{11} BV/ml, which corresponds to 20 times more than influenza-VLPs.

In order to further purify influenza-VLPs from baculoviruses, iodixanol gradient ultracentrifugation was applied and a picture of the third fraction collected (density: 1.13 g/ml) is shown on Figure 1J. Baculovirus contamination is still visible and present at a concentration of 1.73×10^{10} BV/ml. As a confirmation, in both sucrose cushion and further iodixanol purified samples, the presence of envelope protein GP64 from baculovirus was observed on Western blot analysis (data not shown).

Several previous studies have reported the presence of baculovirus contamination in VLP samples produced either in Sf9 or High Five cell production platforms. The level of baculovirus contamination ranged from 10^6 to 10^8 PFU/ml [18,22] as measured by plaque assays, which is significantly lower than the one found here. In the present work, a visual assessment of baculovirus present in the sample was preferred in order to show the total contamination present and to have a better correlation with visual VLP quantification. Baculovirus contamination in the influenza-VLP product poses a significant problem, as these viruses share structural and biophysical similarities with influenza-VLPs, thereby complicating the development of efficient and appropriate purification processes [24,26].

As previously mentioned by Tang et al., the main motivation behind producing VLPs in HEK293 cells was to avoid or reduce the baculovirus contamination [19]. In our case, and as expected, baculoviruses were present in much smaller amounts in HEK293 production samples. However, another kind of contamination was visible on HEK293 VLPs images. Cell or particle debris (dark clusters) as well as large vesicles size (>200 nm) carrying the characteristic influenza fringe can be seen. The presence of cell vesicles may also complicate purification steps in downstream processing, resulting in a similar contamination problem that exists with VLP production in the baculovirus-insect cell system (similar size of particles, vesicles carrying HA). When purified by iodixanol gradient, cell vesicles are also co-purified with lower size influenza-VLPs. Nevertheless, as indicated by the picture 1-G, high-density fractions were concentrating large cell vesicles (>300 nm).

Downstream processing and purification have now been repeatedly pointed out as a critical step and a bottleneck in the development of such new antigen design for a number of different VLPs production types [24,43]. But our observations and those of others raise additional concerns: i) the correct cut-off size for particles to be qualified as VLPs and ii) the immune response generated from co-produced vesicles as it is the case of contaminating baculoviruses.

VLP recovery

As we applied a concentration step by sucrose cushion ultracentrifugation prior to NSTEM analyses, in order to detect and count enough material, it was quite difficult to evaluate the proportion of VLPs recovered from each production. Although ultracentrifugation is a method broadly used to purify VLPs, it has the potential to be damaging for fragile particles. To our knowledge, VLP stability during this process has not yet been evaluated.

Nonetheless, the most likely explanation for such a high level of debris in HEK293 cell sucrose cushion samples (Figure 1D), considering that cell viability was comparable in both production systems, is that some VLPs or vesicles were disrupted by the ultracentrifugation step. The images of VLPs produced in HEK293 cells showed more debris than those of VLPs produced in Sf9 cells. This might be due to the fact that VLPs produced by HEK293 cells were more fragile than those produced in Sf9 cells. This might also be related to the observations previously made on the incorporation/lack of M1 in VLPs produced in Sf9 and HEK293 cells, suggesting the importance of M1's presence for VLP stability. Although M1 may not be essential for VLP formation, the present study suggests its importance from a bioprocessing standpoint and the associated need for stable particles able to withstand downstream processing steps.

In order to achieve higher recovery yields, especially in the case of HEK293 cells, cell pellets from production runs were also examined. Indeed, influenza-VLPs could have been entrapped on cells due to the adhesion of HA proteins to sialic acid cellular receptors. Additionally, HEK293 cells were subjected to cell clumping, which could eventually inhibit influenza-VLPs release from the cell membrane.

First, pellets of HEK293 cells were washed with a solution containing trypsin to counteract cell clumping. Second, cells were lysed. The cell pellet wash and lysate were analysed by an HA assay and by NSTEM to reveal the presence of VLPs. A treatment of the cell pellet with exogenous neuraminidase was also tested, but no noticeable effect was observed on VLP release from a prescreening with Western blot (data not shown). A wide range of particles with a fringe were present in the sample from the cell pellet wash with trypsin on NSTEM images, including very large particles up to 500 nm (Figure 3A). A cluster of round particles approximately 100 nm in diameter could clearly be identified as influenza-VLPs. VLP concentration was counted to be 3.07×10^8 VLPs/ml. Particles with a fringe were also present in the cell lysate at a concentration of 1.43×10^8 VLPs/ml, and looked very similar to those presented in Figure 3A. HA assay confirmed that active HA protein was released from the cell pellet wash and in the cell lysate (Figure 3B).

Similar analyses of Sf9 cell pellet wash and cell lysate did not demonstrate comparable HA activity to the supernatant. This result was expected, as cell clumping did not occur during production. These results indicate that, for some cell production platforms, an extraction step prior to harvest should allow to increase the final VLP yields.

Influenza particles are known to bud from the cell membrane [44] and are expected to be found in the supernatant. However, other viruses, such as HIV, were reported to occasionally bud from internal multi-vesicular bodies [45], even though they usually bud from the plasma membrane. This explanation may hold some merit in our case considering that i) more than one influenza protein has the potential to induce budding [44], and ii) vesicles carrying HA and/or NA identified as VLPs may also have come from internal membranes and were released upon cell death.

Comparison of cell platforms for VLPs production

From a basic point of view, this comparative study suggests that the Sf9 system is a better production platform. Indeed, it is more productive than the HEK293 system with 35 times more VLP particles quantified on

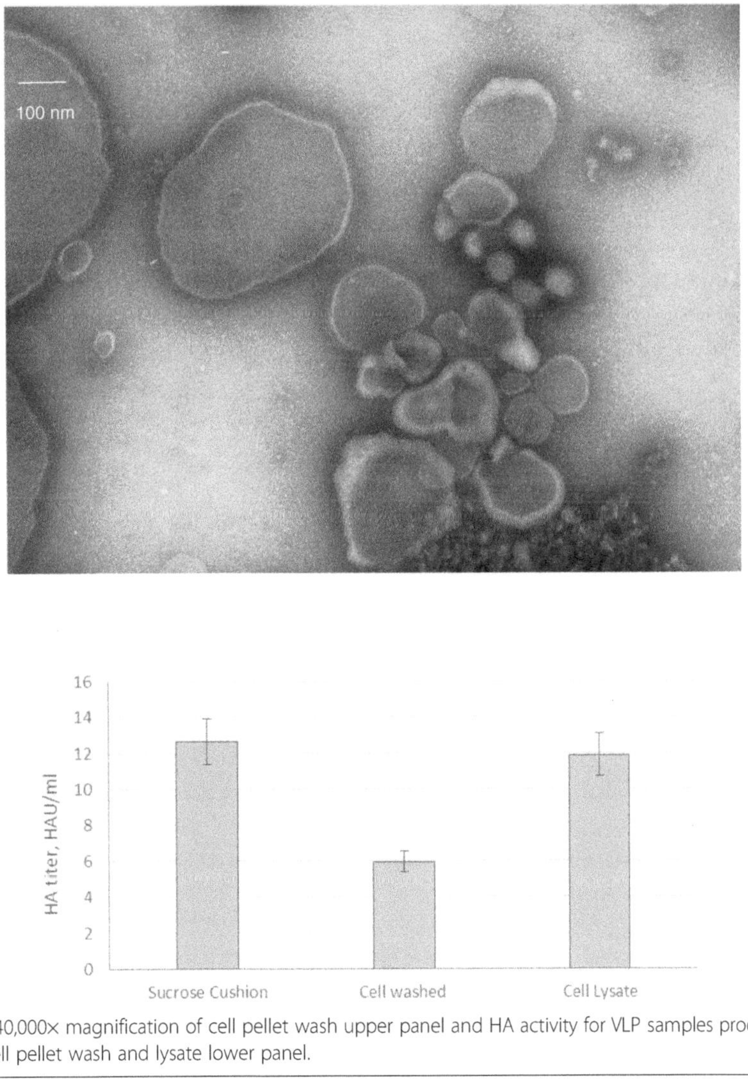

Figure 3 NSTEM image at 40,000× magnification of cell pellet wash upper panel and HA activity for VLP samples produced in HEK293SF cells with culture supernatant, cell pellet wash and lysate lower panel.

NSTEM micrographs. The particle counts allow for comparison of cell specific productivity, as both mammalian and insect cell runs were performed at 2.0-2.5 × 10^6 cells/ml over a course of 48 hours. Therefore, the Sf9 system also presents a significant advantage in terms of VLPs/cell productivity. Moreover, VLPs produced in insect cells were more homogeneous and resembled influenza particles produced in cell culture, whereas the VLPs produced in mammalian cells were heterogeneous in shape and size. These observations were also confirmed by HA protein quantification assays, which showed that the Sf9 cell samples had approximately 25 times more HA activity and presented higher amounts of HA concentration (μg/ml) than the HEK293 cell production system.

However, this conclusion should be nuanced by the M1 protein results and purity evaluation. If the HEK293 cell system could incorporate M1 into the influenza-VLPs, or if other matrix proteins (e.g. murine leukemia virus structural gag) would have been explored, as already done previously [46], the outcomes may have been different. In Haynes et al. [46], approximately 2×10^{12} VLP/ml were produced compared to 2×10^{8} PFU/ml of baculovirus. The increased amount of VLPs from this system compared to the present study (Table 1) could provide a better starting point when considering downstream purification of contaminates and VLP recovery for an industrial manufacturing process.

In addition to production yields and product quantification, potential safety concerns must also be considered for both production systems. Regarding cell vesicles released, it is known that they perform a variety of functions in the cell's secretory pathway [47]. Vesicles may contain cell waste, including proteins and nucleic acids. In vaccine products, host-cell products are strongly controlled and have to be quantified and/or inactivated as indicated in regulatory agency guidelines. To our knowledge, the content of purified influenza VLPs and/or their accompanying vesicles have never been thoroughly studied. Baculovirus have also been proved to strongly initiate an immune response in mice and present adjuvanting effects when injected even at low levels with VLPs [22,48]. These effects could either benefit and boost the vaccine efficacy or create negative synergistic effects on the target VLPs immunologic response [26]. Study of baculovirus effect on mammal's immune system is still on-going [49], and it is not presently possible to determine the safety of such contaminant in VLP products. Therefore, a lot of attention should be dedicated to the separation of VLPs from baculovirus contaminants.

Conclusions

In this work, we have characterized VLPs using a number of established influenza assays aiming to better compare the production capabilities of two expression platforms. Such work is critical to understand and identify the bioprocessing challenges that still need to be addressed to further optimize VLP production. Moreover, the present study provides a baseline for VLP production values (Table 1) using HEK 293 and Sf9 expression systems, an information critically missing in the prevailing literature. Both systems were able to produce what is referred to as VLPs in the literature. However, this work has raised questions on what may be needed to produce structurally sound influenza VLPs, contamination issues that both systems face, and what should be defined as a virus-like particle. Influenza VLP should be a particle closely resembling influenza virus particles i.e. with the same size and morphology with the exclusion of vesicles from the cellular production system. From comparing the two production systems, it was found that the Sf9 system produced VLPs at a level 35 times more than the HEK293 system, however there remains the challenge of baculovirus contamination. Additionally, two important points should be underlined: Firstly, the incorporation of M1 in the VLP may have a significant influence on the integrity and size of VLPs produced, although further work is needed to elucidate how M1 impacts VLP stability. Secondly, the HEK293 system also presents contaminating cell vesicles, visible in a wide variety of sizes in the cell supernatant.

Additional file

Additional file 1: Table S1. Total particle (VP/mL) and infectious particle (IVP/mL) baculovirus titres quantified by FACS and Easy titer method, respectively.

Abbreviations

WHO: World Health Organization; VLPs: Virus-like particles; HEK293 cells: Human Embryonic Kidney 293; Sf-9 cells: Insect *Spodoptera frugiperda* cells; Tn5 cells: Insect High-Five cells; Bacmam: Baculovirus for gene transfer in mammalian cells; HA: Hemagglutinin; NA: Neuraminidase; M1: Matrix Protein 1; M2: Matrix Protein 2; NP: Nucleoprotein (NP); vRNP: Virus ribonucleoprotein; GP64: Glycoprotein 64; BV: Baculovirus; GFP: Green fluorescent protein (GFP); HIV: Human immunodeficient virus; SRID: Single radial immunodiffusion assay; HA assay: Hemagglutination assay; NSTEM: Negative staining transmission electron microscopy; PFU: Plaque forming units; FDA: Food and Drug Administration; GMP: Good Manufacturing Practices; CMV promotor: Cytomegalovirus promotor; polh: Polyhedron promotor; MOI: Multiplicity of infection.

Competing interests

The authors declare that they have no competing interests.

Authors' contributions

CT carried out design of experiments, experimental laboratory work, data analysis and drafted the manuscript. EP participated in design of experiments, data analysis and drafting of the manuscript. AM contributed to baculovirus construct design and production, MA and OH participated in design of experiments and helped to draft the manuscript and AK conceived the study, participated in experimental design, data analysis, and helped draft the manuscript. All authors read and approved the final manuscript.

Acknowledgements

The authors would like to emphasize the expert continuing support provided by Julia Transfiguracion, Johnny Montes and Danielle Jacob from NRC, Montreal site for the completion of many aspects of the experimental work reported in this manuscript and Micheline Letarte, from INRS, Laval, who did the NSEM imaging of the viral structures.

Author details

[1]National Research Council Canada, Human Health Therapeutics, Montréal, Canada. [2]Ecole Polytechnique de Montréal, Montréal, Canada. [3]Laboratoire Virologie et pathologies Humaine (VirPath), EA4610 Lyon, France. [4]University of Waterloo, Waterloo, Canada. [5]Department of Bioengineering, McGill University, 817 Sherbrooke St. W. Macdonald Engineering Building, Room 387, Montréal, Canada.

References

1. World Health Organization (WHO). Influenza (Seasonal) fact sheet. In: WHO media cent. 2014.
2. Cohen J. Out of Mexico? Scientists ponder swine flu's origins. Science. 2009;324:700–3.
3. Michaelis M, Doerr HW, Cinatl J. Novel swine-origin influenza A virus in humans: another pandemic knocking at the door. Med Microbiol Immunol. 2009;198:175–83.
4. Gerdil C. The annual production cycle for influenza vaccine. Vaccine. 2003;21:1776–9.
5. Johnson TD. Next generation of flu vaccines comming of age: cell-based technology may replace egg-based flu vaccines. Nations Health. 2010;40:1–27.
6. Haynes JR. Influenza virus-like particle vaccines. Expert Rev Vaccines. 2009;8:435–45.
7. Kang S-M, Song J-M, Quan F-S, Compans RW. Influenza vaccines based on virus-like particles. Virus Res. 2009;143:140–6.
8. Pushko P, Tumpey TM, Bu F, Knell J, Robinson R, Smith G. Influenza virus-like particles comprised of the HA, NA, and M1 proteins of H9N2 influenza virus induce protective immune responses in BALB/c mice. Vaccine. 2005;23:5751–9.
9. Wu C-Y, Yeh Y-C, Yang Y-C, Chou C, Liu M-T, Wu H-S, et al. Mammalian expression of virus-like particles for advanced mimicry of authentic influenza virus. PLoS One. 2010;5, e9784.
10. Quan F-S, Vunnava A, Compans RW, Kang S-M. Virus-like particle vaccine protects against 2009 H1N1 pandemic influenza virus in mice. PLoS One. 2010;5, e9161.
11. Kang S-M, Yoo D-G, Lipatov AS, Song J-M, Davis CT, Quan F-S, et al. Induction of long-term protective immune responses by influenza H5N1 virus-like particles. PLoS One. 2009;4, e4667.
12. Latham T, Galarza JM. Formation of Wild-Type and Chimeric Influenza Virus-Like Particles following Simultaneous Expression of Only Four Structural Proteins. Journal of Virology. 2001;75(3):6154–65.
13. Pushko P, Pearce MB, Ahmad A, Tretyakova I, Smith G, Belser JA, et al. Influenza virus-like particle can accommodate multiple subtypes of hemagglutinin and protect from multiple influenza types and subtypes. Vaccine. 2011;29:5911–8.
14. Smith GE, Flyer DC, Raghunandan R, Liu Y, Wei Z, Wu Y, et al. Development of influenza H7N9 virus like particle (VLP) vaccine: homologous A/Anhui/1/2013 (H7N9) protection and heterologous A/chicken/Jalisco/CPA1/2012 (H7N3) cross-protection in vaccinated mice challenged with H7N9 virus. Vaccine. 2013;31:4305–13.
15. Landry N, Ward BJ, Trépanier S, Montomoli E, Dargis M, Lapini G, et al. Preclinical and clinical development of plant-made virus-like particle vaccine against avian H5N1 influenza. PLoS One. 2010;5, e15559.
16. López-Macías C, Ferat-Osorio E, Tenorio-Calvo A, Isibasi A, Talavera J, Arteaga-Ruiz O, et al. Safety and immunogenicity of a virus-like particle pandemic influenza A (H1N1) 2009 vaccine in a blinded, randomized, placebo-controlled trial of adults in Mexico. Vaccine. 2011;29:7826–34.
17. D'Aoust M-A, Couture MM-J, Charland N, Trépanier S, Landry N, Ors F, et al. The production of hemagglutinin-based virus-like particles in plants: a rapid, efficient and safe response to pandemic influenza. Plant Biotechnol J. 2010;8:607–19.
18. Krammer F, Schinko T, Palmberger D, Tauer C, Messner P, Grabherr R. Trichoplusia ni cells (High Five) are highly efficient for the production of influenza A virus-like particles: a comparison of two insect cell lines as production platforms for influenza vaccines. Mol Biotechnol. 2010;45:226–34.
19. Tang X-C, Lu H-R, Ross TM. Baculovirus-produced influenza virus-like particles in mammalian cells protect mice from lethal influenza challenge. Viral Immunol. 2011;24:311–9.
20. Bright RA, Carter DM, Daniluk S, Toapanta FR, Ahmad A, Gavrilov V, et al. Influenza virus-like particles elicit broader immune responses than whole virion inactivated influenza virus or recombinant hemagglutinin. Vaccine. 2007;25:3871–8.
21. Quan F-S, Huang C, Compans RW, Kang S-M. Virus-like particle vaccine induces protective immunity against homologous and heterologous strains of influenza virus. J Virol. 2007;81:3514–24.
22. Margine I, Martinez-Gil L, Chou Y, Krammer F. Residual baculovirus in insect cell-derived influenza virus-like particle preparations enhances immunogenicity. PLoS One. 2012;7, e51559.
23. Cox MMJ, Hollister JR. FluBlok, a next generation influenza vaccine manufactured in insect cells. Biologicals. 2009;37:182–9.
24. Liu F, Wu X, Li L, Liu Z, Wang Z. Use of baculovirus expression system for generation of virus-like particles: successes and challenges. Protein Expr Purif. 2013;90:104–16.
25. Chen BJ, Leser GP, Morita E, Lamb RA. Influenza virus hemagglutinin and neuraminidase, but not the matrix protein, are required for assembly and budding of plasmid-derived virus-like particles. J Virol. 2007;81:7111–23.
26. Vicente T, Roldão A, Peixoto C, Carrondo MJT, Alves PM. Large-scale production and purification of VLP-based vaccines. J Invertebr Pathol. 2011;107(Suppl):S42–8.
27. Côté J, Garnier A, Massie B, Kamen A, Cote J. Serum-free production of recombinant proteins and adenoviral vectors by 293SF-3 F6 cells. Biotechnol Bioeng. 1998;59:567–75.
28. Petiot E, Jacob D, Lanthier S, Lohr V, Ansorge S, Kamen AA. Metabolic and kinetic analyses of influenza production in perfusion HEK293 cell culture. BMC Biotechnol. 2011;11:1–12.
29. Alain R, Nadon F, Séguin C, Payment P, Trudel M. Rapid virus subunit visualization by direct sedimentation of samples on electron microscope grids. J Virol Methods. 1987;16:209–16.
30. Le Ru A, Jacob D, Transfiguracion J, Ansorge S, Henry O, Kamen AA. Scalable production of influenza virus in HEK-293 cells for efficient vaccine manufacturing. Vaccine. 2010;28:3661–71.
31. Moulès V, Terrier O, Yver M, Riteau B, Moriscot C, Ferraris O, et al. Importance of viral genomic composition in modulating glycoprotein content on the surface of influenza virus particles. Virology. 2011;414:51–62.
32. Roberts PC, Lamb RA, Compans RW. The M1 and M2 proteins of influenza A virus are important determinants in filamentous particle formation. Virology. 1998;240:127–37.
33. Calder LJ, Wasilewski S, Berriman JA, Rosenthal PB. Structural organization of a filamentous influenza A virus. Proc Natl Acad Sci U S A. 2010;107:10685–90.
34. Elleman CJ, Barclay WS. The M1 matrix protein controls the filamentous phenotype of influenza A virus. Virology. 2004;321:144–53.
35. McCown MF, Pekosz A. Distinct domains of the influenza a virus M2 protein cytoplasmic tail mediate binding to the M1 protein and facilitate infectious virus production. J Virol. 2006;80:8178–89.
36. Bialas KM, Bussey KA, Stone RL, Takimoto T. Specific nucleoprotein residues affect influenza virus morphology. J Virol. 2014;88:2227–34.
37. Chen BJ, Leser GP, Jackson D, Lamb RA. The influenza virus M2 protein cytoplasmic tail interacts with the M1 protein and influences virus assembly at the site of virus budding. J Virol. 2008;82:10059–70.
38. Watanabe K, Shimizu T, Noda S, Tsukahara F, Maru Y, Kobayashi N. Nuclear export of the influenza virus ribonucleoprotein complex: Interaction of Hsc70 with viral proteins M1 and NS2. FEBS Open Bio. 2014;4:683–8.
39. Shaw ML, Stone KL, Colangelo CM, Gulcicek EE, Palese P. Cellular proteins in influenza virus particles. PLoS Pathog. 2008;4, e1000085.
40. Lai JCC, Chan WWL, Kien FF, Nicholls JM, Peiris JSM, Garcia J-M. Formation of virus-like particles from human cell lines exclusively expressing influenza neuraminidase. J Gen Virol. 2010;91(Pt 9):2022–30.
41. Gómez-Puertas P, Mena I, Castillo M, Vivo A, Pérez-Pastrana E, Portela A. Efficient formation of influenza virus-like particles: dependence on the expression levels of viral proteins. J Gen Virol. 1999;80(Pt 7):1635–45.

42. Thompson C, Petiot E, Lennaertz A, Henry O, Kamen A. Analytical technologies for influenza virus-like particle candidate vaccines: challenges and emerging approaches. Virology Journal. 2013;10:141; 14 pages (open access).
43. Valle JR, Cha S, Medina F, Angel RM. Heat shock protein 90 and heat shock protein 70 are components of dengue virus receptor complex in human cells. Society. 2005;79:4557–67.
44. Rossman JS, Lamb RA. Influenza virus assembly and budding. Virology. 2011;411:229–36.
45. Welsch S, Müller B, Kräusslich H-G. More than one door - Budding of enveloped viruses through cellular membranes. FEBS Lett. 2007;581:2089–97.
46. Haynes JR, Dokken L, Wiley J, Cawthon AG, Bigger J, Harmsen AG, et al. Influenza-pseudotyped Gag virus-like particle vaccines provide broad protection against highly pathogenic avian influenza challenge. Vaccine. 2009;27:530–41.
47. Denzer K, Kleijmeer MJ, Heijnen HF, Stoorvogel W, Geuze HJ. Exosome: from internal vesicle of the multivesicular body to intercellular signaling device. J Cell Sci. 2000;113(Pt 19):3365–74.
48. Hervas-Stubbs S, Rueda P, Lopez L, Leclerc C. Insect baculoviruses strongly potentiate adaptive immune responses by inducing type I IFN. J Immunol. 2007;178:2361–9.
49. Molinari P, Crespo MI, Gravisaco MJ, Taboga O, Morón G. Baculovirus capsid display potentiates OVA cytotoxic and innate immune responses. PLoS One. 2011;6, e24108.

Permissions

All chapters in this book were first published in BMCB, by BioMed Central; hereby published with permission under the Creative Commons Attribution License or equivalent. Every chapter published in this book has been scrutinized by our experts. Their significance has been extensively debated. The topics covered herein carry significant findings which will fuel the growth of the discipline. They may even be implemented as practical applications or may be referred to as a beginning point for another development.

The contributors of this book come from diverse backgrounds, making this book a truly international effort. This book will bring forth new frontiers with its revolutionizing research information and detailed analysis of the nascent developments around the world.

We would like to thank all the contributing authors for lending their expertise to make the book truly unique. They have played a crucial role in the development of this book. Without their invaluable contributions this book wouldn't have been possible. They have made vital efforts to compile up to date information on the varied aspects of this subject to make this book a valuable addition to the collection of many professionals and students.

This book was conceptualized with the vision of imparting up-to-date information and advanced data in this field. To ensure the same, a matchless editorial board was set up. Every individual on the board went through rigorous rounds of assessment to prove their worth. After which they invested a large part of their time researching and compiling the most relevant data for our readers.

The editorial board has been involved in producing this book since its inception. They have spent rigorous hours researching and exploring the diverse topics which have resulted in the successful publishing of this book. They have passed on their knowledge of decades through this book. To expedite this challenging task, the publisher supported the team at every step. A small team of assistant editors was also appointed to further simplify the editing procedure and attain best results for the readers.

Apart from the editorial board, the designing team has also invested a significant amount of their time in understanding the subject and creating the most relevant covers. They scrutinized every image to scout for the most suitable representation of the subject and create an appropriate cover for the book.

The publishing team has been an ardent support to the editorial, designing and production team. Their endless efforts to recruit the best for this project, has resulted in the accomplishment of this book. They are a veteran in the field of academics and their pool of knowledge is as vast as their experience in printing. Their expertise and guidance has proved useful at every step. Their uncompromising quality standards have made this book an exceptional effort. Their encouragement from time to time has been an inspiration for everyone.

The publisher and the editorial board hope that this book will prove to be a valuable piece of knowledge for researchers, students, practitioners and scholars across the globe.

List of Contributors

Guillermo Marini
Bio-Manguinhos (Instituto de Tecnologia em Imunobiológicos) – Fundação Oswaldo Cruz (FIOCRUZ) - VDTEC (Vice-Diretoria de Desenvolvimento Tecnológico), Av. Brasil 4365, Pavilhão Rockfeller Sala 202 - 21040-360, Manguinhos, Rio de Janeiro, RJ, Brazil
Programa de Engenharia Química - COPPE - Universidade Federal do Rio de Janeiro (UFRJ), Av. Horácio Macedo 2030, Bloco G, Sala 115 - 21941-972, Ilha do Fundão, Rio de Janeiro, RJ, Brazil

Mateus Dalcin Luchese
Bio-Manguinhos (Instituto de Tecnologia em Imunobiológicos) – Fundação Oswaldo Cruz (FIOCRUZ) - VDTEC (Vice-Diretoria de Desenvolvimento Tecnológico), Av. Brasil 4365, Pavilhão Rockfeller Sala 202 - 21040-360, Manguinhos, Rio de Janeiro, RJ, Brazil

Ana Paula Correa Argondizzo
Bio-Manguinhos (Instituto de Tecnologia em Imunobiológicos) – Fundação Oswaldo Cruz (FIOCRUZ) - VDTEC (Vice-Diretoria de Desenvolvimento Tecnológico), Av. Brasil 4365, Pavilhão Rockfeller Sala 202 - 21040-360, Manguinhos, Rio de Janeiro, RJ, Brazil

Ana Carolina Magalhães Andrade de Góes
Bio-Manguinhos (Instituto de Tecnologia em Imunobiológicos) – Fundação Oswaldo Cruz (FIOCRUZ) - VDTEC (Vice-Diretoria de Desenvolvimento Tecnológico), Av. Brasil 4365, Pavilhão Rockfeller Sala 202 - 21040-360, Manguinhos, Rio de Janeiro, RJ, Brazil

Ricardo Galler
Bio-Manguinhos (Instituto de Tecnologia em Imunobiológicos) – Fundação Oswaldo Cruz (FIOCRUZ) - VDTEC (Vice-Diretoria de Desenvolvimento Tecnológico), Av. Brasil 4365, Pavilhão Rockfeller Sala 202 - 21040-360, Manguinhos, Rio de Janeiro, RJ, Brazil

Tito Lívio Moitinho Alves
Programa de Engenharia Química - COPPE - Universidade Federal do Rio de Janeiro (UFRJ), Av. Horácio Macedo 2030, Bloco G, Sala 115 - 21941-972, Ilha do Fundão, Rio de Janeiro, RJ, Brazil

Marco Alberto Medeiros
Bio-Manguinhos (Instituto de Tecnologia em Imunobiológicos) – Fundação Oswaldo Cruz (FIOCRUZ) - VDTEC (Vice-Diretoria de Desenvolvimento Tecnológico), Av. Brasil 4365, Pavilhão Rockfeller Sala 202 - 21040-360, Manguinhos, Rio de Janeiro, RJ, Brazil

Ariane Leites Larentis
Bio-Manguinhos (Instituto de Tecnologia em Imunobiológicos) – Fundação Oswaldo Cruz (FIOCRUZ) - VDTEC (Vice-Diretoria de Desenvolvimento Tecnológico), Av. Brasil 4365, Pavilhão Rockfeller Sala 202 - 21040-360, Manguinhos, Rio de Janeiro, RJ, Brazil
ENSP (Escola Nacional de Saúde Pública Sergio Arouca) - Fundação Oswaldo Cruz (FIOCRUZ) - Centro de Estudos da Saúde do Trabalhador e Ecologia Humana (CESTEH), Av. Leopoldo Bulhões 1480, Prédio 1º de Maio - 21041-210, Manguinhos, Rio de Janeiro, RJ, Brazil

Alastair D Sutherland
Department of Biological and Biomedical Sciences, School of Health and Life Sciences, Glasgow Caledonian University, Cowcaddens Road, Glasgow G4 0BA, Scotland, UK
Centre of Marine Sciences, University of Algarve, Campus de Gambelas, 8005-139, Faro, Portugal

Joao C Varela
Centre of Marine Sciences, University of Algarve, Campus de Gambelas, 8005-139, Faro, Portugal

Thorny Cecilie Bie Andersen
Department of Anatomy, Institute of Basal Medical Sciences, University of Oslo, Oslo, Norway

Kjersti Lindsjø
Department of Anatomy, Institute of Basal Medical Sciences, University of Oslo, Oslo, Norway

Cecilie Dahl Hem
Department of Anatomy, Institute of Basal Medical Sciences, University of Oslo, Oslo, Norway

Lise Koll
Department of Anatomy, Institute of Basal Medical Sciences, University of Oslo, Oslo, Norway

Per Eugen Kristiansen
Department of Biosciences, University of Oslo, Oslo, Norway

Lars Skjeldal
Department of Chemistry, Biochemistry and Food science, Norwegian University of Life Sciences, Ås, Norway

Amy H Andreotti
Department of Biochemistry, Biophysics and Molecular Biology, Iowa State University, Ames, Iowa, USA

Anne Spurkland
Department of Anatomy, Institute of Basal Medical Sciences, University of Oslo, Oslo, Norway

Yong Tang
Department of Chemistry and Chemical Engineering, Beijing Forestry University, Beijing, China
GuangXi Key Laboratory of Chemistry and Engineering of Forest Products, Nanning 530006, China

Fuhou Lei
GuangXi Key Laboratory of Chemistry and Engineering of Forest Products, Nanning 530006, China

Carrasco Cristhian
Department of Chemical Engineering, Lund University, Lund, Sweden

Zuguang Liu
GuangXi Key Laboratory of Chemistry and Engineering of Forest Products, Nanning 530006, China

Hailong Yu
Department of Chemistry and Chemical Engineering, Beijing Forestry University, Beijing, China

Jianxin Jiang
Department of Chemistry and Chemical Engineering, Beijing Forestry University, Beijing, China

Ana Maria Vélez
Chemical Engineering Department, Federal University of São Carlos (UFSCar), Rodovia Washington Luís, km 235, C.P. 676, CEP 13565-905 São Carlos, SP, Brazil

Adilson José da Silva
Chemical Engineering Department, Federal University of São Carlos (UFSCar), Rodovia Washington Luís, km 235, C.P. 676, CEP 13565-905 São Carlos, SP, Brazil

Antonio Carlos Luperni Horta
Institute of Chemistry, Federal University of Goiás, Campus Samambaia, C.P. 131, CEP 74001-970 Goiânia, GO, Brazil

Cintia Regina Sargo
Chemical Engineering Department, Federal University of São Carlos (UFSCar), Rodovia Washington Luís, km 235, C.P. 676, CEP 13565-905 São Carlos, SP, Brazil

Gilson Campani
Chemical Engineering Department, Federal University of São Carlos (UFSCar), Rodovia Washington Luís, km 235, C.P. 676, CEP 13565-905 São Carlos, SP, Brazil

Gabriel Gonçalves Silva
Chemical Engineering Department, Federal University of São Carlos (UFSCar), Rodovia Washington Luís, km 235, C.P. 676, CEP 13565-905 São Carlos, SP, Brazil

Raquel de Lima Camargo Giordano
Chemical Engineering Department, Federal University of São Carlos (UFSCar), Rodovia Washington Luís, km 235, C.P. 676, CEP 13565-905 São Carlos, SP, Brazil

Teresa Cristina Zangirolami
Chemical Engineering Department, Federal University of São Carlos (UFSCar), Rodovia Washington Luís, km 235, C.P. 676, CEP 13565-905 São Carlos, SP, Brazil

Sun-Young Kang
Chemical Biology Research Center, Korea Research Institute of Bioscience and Biotechnology(KRIBB), 30 Yeongudanji-ro, Ochang-eup, Chungbuk 363-883, Republic of Korea
Department of Pharmacy Graduate School, Chungbuk National University, Cheongju 361-763, Republic of Korea

Jae Kyoung Lee
Chemical Biology Research Center, Korea Research Institute of Bioscience and Biotechnology(KRIBB), 30 Yeongudanji-ro, Ochang-eup, Chungbuk 363-883, Republic of Korea
Department of Pharmacy Graduate School, Chungbuk National University, Cheongju 361-763, Republic of Korea

Oksik Choi
Chemical Biology Research Center, Korea Research Institute of Bioscience and Biotechnology(KRIBB), 30 Yeongudanji-ro, Ochang-eup, Chungbuk 363-883, Republic of Korea

Cha Young Kim
Eco-friendly Bio-Material Research Center, KRIBB, Jeongeup 580-1853, Republic of Korea

Jae-Hyuk Jang
Chemical Biology Research Center, Korea Research Institute of Bioscience and Biotechnology(KRIBB), 30 Yeongudanji-ro, Ochang-eup, Chungbuk 363-883, Republic of Korea

Bang Yeon Hwang
Department of Pharmacy Graduate School, Chungbuk National University, Cheongju 361-763, Republic of Korea

Young-Soo Hong
Chemical Biology Research Center, Korea Research Institute of Bioscience and Biotechnology(KRIBB), 30 Yeongudanji-ro, Ochang-eup, Chungbuk 363-883, Republic of Korea

Haibo Zhang
CAS Key Laboratory of Biobased Materials, Qingdao Institute of Bioenergy and Bioprocess Technology, Chinese Academy of Sciences, No.189 Songling Road, Laoshan District, Qingdao, PR 266101, China

Mingle Cao
University of Marburg, Marburg, Germany

Xinglin Jiang
CAS Key Laboratory of Biobased Materials, Qingdao Institute of Bioenergy and Bioprocess Technology, Chinese Academy of Sciences, No.189 Songling Road, Laoshan District, Qingdao, PR 266101, China

Huibin Zou
CAS Key Laboratory of Biobased Materials, Qingdao Institute of Bioenergy and Bioprocess Technology, Chinese Academy of Sciences, No.189 Songling Road, Laoshan District, Qingdao, PR 266101, China

Cong Wang
CAS Key Laboratory of Biobased Materials, Qingdao Institute of Bioenergy and Bioprocess Technology, Chinese Academy of Sciences, No.189 Songling Road, Laoshan District, Qingdao, PR 266101, China

Xin Xu
CAS Key Laboratory of Biobased Materials, Qingdao Institute of Bioenergy and Bioprocess Technology, Chinese Academy of Sciences, No.189 Songling Road, Laoshan District, Qingdao, PR 266101, China

Mo Xian
CAS Key Laboratory of Biobased Materials, Qingdao Institute of Bioenergy and Bioprocess Technology, Chinese Academy of Sciences, No.189 Songling Road, Laoshan District, Qingdao, PR 266101, China

Ying Wang
State Key Laboratory of Microbial Technology, School of Life Science, Shandong University, Jinan 250100, P. R. China

Yue-Zhong Li
State Key Laboratory of Microbial Technology, School of Life Science, Shandong University, Jinan 250100, P. R. China

Magdalena Zuk
Faculty of Biotechnology, Wrocław University, Przybyszewskiego 63/77, 51-148 Wroclaw, Poland
Linum Fundation, Stabłowicka 147/149, 54-066 Wrocław, Poland

Agata Dorotkiewicz-Jach
Institute of Genetics and Microbiology, Wroclaw University, Przybyszewskiego 63/77, 51-148 Wroclaw, Poland

Zuzanna Drulis-Kawa
Institute of Genetics and Microbiology, Wroclaw University, Przybyszewskiego 63/77, 51-148 Wroclaw, Poland

Malgorzata Arendt
Faculty of Biotechnology, Wrocław University, Przybyszewskiego 63/77, 51-148 Wroclaw, Poland

Anna Kulma
Faculty of Biotechnology, Wrocław University, Przybyszewskiego 63/77, 51-148 Wroclaw, Poland

Jan Szopa
Faculty of Biotechnology, Wrocław University, Przybyszewskiego 63/77, 51-148 Wroclaw, Poland
Linum Fundation, Stabłowicka 147/149, 54-066 Wrocław, Poland

Yukihiro Hiramatsu
Faculty of Pharmaceutical Sciences, Fukuoka University, 8-19-1, Nanakuma, Jonan-ku, Fukuoka, Japan

Masatatsu Yamamoto
Department of Molecular Oncology, Graduate School of Medical and Dental Sciences, Kagoshima University, 8-35-1, Sakuragaoka, Kagoshima, Japan

Tomomitsu Satho
Faculty of Pharmaceutical Sciences, Fukuoka University, 8-19-1, Nanakuma, Jonan-ku, Fukuoka, Japan

Keiichi Irie
Faculty of Pharmaceutical Sciences, Fukuoka University, 8-19-1, Nanakuma, Jonan-ku, Fukuoka, Japan

Akiko Kai
Faculty of Pharmaceutical Sciences, Fukuoka University, 8-19-1, Nanakuma, Jonan-ku, Fukuoka, Japan

Saori Uyeda
Faculty of Pharmaceutical Sciences, Fukuoka University, 8-19-1, Nanakuma, Jonan-ku, Fukuoka, Japan

Yuki Fukumitsu
Faculty of Pharmaceutical Sciences, Fukuoka University, 8-19-1, Nanakuma, Jonan-ku, Fukuoka, Japan

Akihisa Toda
Daiichi College of Pharmaceutical Sciences, 22-1, Tamagawa-cho, Minami-ku, Fukuoka, Japan

Takeshi Miyata
Division of Molecular Functions of Food, Department of Biochemistry and Biotechnology, Kagoshima University, 1-21-24, Korimoto, Kagoshima, Japan

Fumio Miake
Faculty of Pharmaceutical Sciences, Fukuoka University, 8-19-1, Nanakuma, Jonan-ku, Fukuoka, Japan

Takeshi Arakawa
Molecular Microbiology Group, Department of Tropical Infectious Diseases, COMB, Tropical Biosphere Research Center, University of the Ryukyus, 1 Senbaru, Nishihara, Okinawa, Japan Division of Host Defense and Vaccinology, Department of Microbiology, Graduate School of Medicine, University of the Ryukyus, 207 Uehara, Nishihara, Okinawa, Japan

Nobuhiro Kashige
Faculty of Pharmaceutical Sciences, Fukuoka University, 8-19-1, Nanakuma, Jonan-ku, Fukuoka, Japan

Yulan Yi
State Key Laboratory of Bioreactor Engineering, East China University of Science and Technology, 130 Meilong Road, Shanghai 200237, China

Huakai Sheng
State Key Laboratory of Bioreactor Engineering, East China University of Science and Technology, 130 Meilong Road, Shanghai 200237, China

Zhimin Li
State Key Laboratory of Bioreactor Engineering, East China University of Science and Technology, 130 Meilong Road, Shanghai 200237, China
Shanghai Collaborative Innovation Center for Biomanufacturing Technology, Shanghai, China

Qin Ye
State Key Laboratory of Bioreactor Engineering, East China University of Science and Technology, 130 Meilong Road, Shanghai 200237, China

Justin Smith
Department of Process Engineering, Stellenbosch University, Private Bag X1, Matieland 7602, South Africa

Eugéne van Rensburg
Department of Process Engineering, Stellenbosch University, Private Bag X1, Matieland 7602, South Africa

Johann F Görgens
Department of Process Engineering, Stellenbosch University, Private Bag X1, Matieland 7602, South Africa

Yuanyuan Bu
Key Laboratory of Saline-Alkali Vegetation Ecology Restoration in Oil Field (SAVER), Ministry of Education, Alkali Soil Natural Environmental Science Center (ASNESC), Northeast Forestry University, Hexing Road No. 26, Xiangfang District, Harbin City, Heilongjiang Province 150040, China

Mengqing Zhao
Key Laboratory of Saline-Alkali Vegetation Ecology Restoration in Oil Field (SAVER), Ministry of Education, Alkali Soil Natural Environmental Science Center (ASNESC), Northeast Forestry University, Hexing Road No. 26, Xiangfang District, Harbin City, Heilongjiang Province 150040, China

Bo Sun
Key Laboratory of Saline-Alkali Vegetation Ecology Restoration in Oil Field (SAVER), Ministry of Education, Alkali Soil Natural Environmental Science Center (ASNESC), Northeast Forestry University, Hexing Road No. 26, Xiangfang District, Harbin City, Heilongjiang Province 150040, China

Xinxin Zhang
Key Laboratory of Saline-Alkali Vegetation Ecology Restoration in Oil Field (SAVER), Ministry of Education, Alkali Soil Natural Environmental Science Center (ASNESC), Northeast Forestry University, Hexing Road No. 26, Xiangfang District, Harbin City, Heilongjiang Province 150040, China

Tetsuo Takano
Asian Natural Environmental Science Center (ASNESC), The University of Tokyo, Nishitokyo, Tokyo 188-0002, Japan

Shenkui Liu
Key Laboratory of Saline-Alkali Vegetation Ecology Restoration in Oil Field (SAVER), Ministry of Education, Alkali Soil Natural Environmental Science Center (ASNESC), Northeast Forestry University, Hexing Road No. 26, Xiangfang District, Harbin City, Heilongjiang Province 150040, China

Silvia Sommaruga
Dipartimento di Biotecnologie e Bioscienze, Università di Milano-Bicocca, Piazza della Scienza 2, Milano 20126, Italy

Elisabetta Galbiati
Dipartimento di Biotecnologie e Bioscienze, Università di Milano-Bicocca, Piazza della Scienza 2, Milano 20126, Italy

Jesus Peñaranda-Avila
Dipartimento di Biotecnologie e Bioscienze, Università di Milano-Bicocca, Piazza della Scienza 2, Milano 20126, Italy

Chiara Brambilla
Dipartimento di Biotecnologie e Bioscienze, Università di Milano-Bicocca, Piazza della Scienza 2, Milano 20126, Italy

Paolo Tortora
Dipartimento di Biotecnologie e Bioscienze, Università di Milano-Bicocca, Piazza della Scienza 2, Milano 20126, Italy

Miriam Colombo
Dipartimento di Biotecnologie e Bioscienze, Università di Milano-Bicocca, Piazza della Scienza 2, Milano 20126, Italy

Davide Prosperi
Dipartimento di Biotecnologie e Bioscienze, Università di Milano-Bicocca, Piazza della Scienza 2, Milano 20126, Italy
Istituto di Scienze e Tecnologie Molecolari, CNR, via Fantoli 16/15, Milano 20138, Italy

Bahgat Fayed
Department of Biology, University of York, York YO10 5DD, UK

Ellen Younger
School of Medical Sciences, Institute of Medical Sciences, University of Aberdeen, Aberdeen AB25 2ZD, UK

Gabrielle Taylor
Department of Biology, University of York, York YO10 5DD, UK

Margaret C M Smith
Department of Biology, University of York, York YO10 5DD, UK

Jaakko M Itkonen
Centre for Drug Research, Division of Pharmaceutical Biosciences, Faculty of Pharmacy, University of Helsinki, P.O. Box 56, Viikinkaari 5E, Helsinki 00014, Finland

Arto Urtti
Centre for Drug Research, Division of Pharmaceutical Biosciences, Faculty of Pharmacy, University of Helsinki, P.O. Box 56, Viikinkaari 5E, Helsinki 00014, Finland
School of Pharmacy, University of Eastern Finland, Kuopio, Finland

Louise E Bird
Division of Structural Biology, Henry Wellcome Building for Genomic Medicine, University of Oxford, Roosevelt Drive, Oxford OX3 7BN, UK

OPPF-UK, Research Complex at Harwell, R92 Rutherford Appleton Laboratory, Harwell Oxford, Didcot, Oxford OX11 OFA, UK

Sanjay Sarkhel
Centre for Drug Research, Division of Pharmaceutical Biosciences, Faculty of Pharmacy, University of Helsinki, P.O. Box 56, Viikinkaari 5E, Helsinki 00014, Finland

Juhan Kim
Department of Molecular, Cellular, and Developmental Biology, University of Colorado Boulder, Boulder, CO 80309, USA
Cooperative Institute for Research in Environmental Sciences, University of Colorado Boulder, Boulder, CO 80309, USA

Anthony M Webb
Department of Molecular, Cellular, and Developmental Biology, University of Colorado Boulder, Boulder, CO 80309, USA

Jamie P Kershner
Department of Molecular, Cellular, and Developmental Biology, University of Colorado Boulder, Boulder, CO 80309, USA
Cooperative Institute for Research in Environmental Sciences, University of Colorado Boulder, Boulder, CO 80309, USA

Stephen Blaskowski
Department of Molecular, Cellular, and Developmental Biology, University of Colorado Boulder, Boulder, CO 80309, USA

Shelley D Copley
Department of Molecular, Cellular, and Developmental Biology, University of Colorado Boulder, Boulder, CO 80309, USA
Cooperative Institute for Research in Environmental Sciences, University of Colorado Boulder, Boulder, CO 80309, USA

Fei Xiang Zhan
Hubei Collaborative Innovation Center for Green Transformation of Bio-Resources, College of Life Science, Hubei University, Wuhan 430062, People's Republic of China

Qin Hong Wang
Tianjin Institutes of Industrial Biotechnology, Chinese Academy of Science, Tianjin 300308, China

Si Jing Jiang
Hubei Collaborative Innovation Center for Green Transformation of Bio-Resources, College of Life Science, Hubei University, Wuhan 430062, People's Republic of China

Yu Ling Zhou
Hubei Collaborative Innovation Center for Green Transformation of Bio-Resources, College of Life Science, Hubei University, Wuhan 430062, People's Republic of China

Min Zhang
Hubei Collaborative Innovation Center for Green Transformation of Bio-Resources, College of Life Science, Hubei University, Wuhan 430062, People's Republic of China

Yan He Ma
Tianjin Institutes of Industrial Biotechnology, Chinese Academy of Science, Tianjin 300308, China

Juozas Nainys
Institute of Biotechnology, Vilnius University, Graiciuno 8, LT-02241 Vilnius, Lithuania

Rita Lasickiene
Institute of Biotechnology, Vilnius University, Graiciuno 8, LT-02241 Vilnius, Lithuania

Rasa Petraityte-Burneikiene
Institute of Biotechnology, Vilnius University, Graiciuno 8, LT-02241 Vilnius, Lithuania

Jonas Dabrisius
Institute of Biotechnology, Vilnius University, Graiciuno 8, LT-02241 Vilnius, Lithuania

Raimundas Lelesius
Institute of Microbiology and Virology, Veterinary Faculty of Veterinary Academy, Lithuanian University of Health Sciences, Tilzes 18, LT-47181 Kaunas, Lithuania

Vilimas Sereika
Institute of Microbiology and Virology, Veterinary Faculty of Veterinary Academy, Lithuanian University of Health Sciences, Tilzes 18, LT-47181 Kaunas, Lithuania

Aurelija Zvirbliene
Institute of Biotechnology, Vilnius University, Graiciuno 8, LT-02241 Vilnius, Lithuania

Kestutis Sasnauskas
Institute of Biotechnology, Vilnius University, Graiciuno 8, LT-02241 Vilnius, Lithuania

Alma Gedvilait
Institute of Biotechnology, Vilnius University, Graiciuno 8, LT-02241 Vilnius, Lithuania

Nooshin Rahnama
Department of Bioprocess Technology, Faculty of Biotechnology and Biomolecular Sciences, Universiti Putra Malaysia, 43400 UPM Serdang, Selangor, Malaysia

Hooi Ling Foo
Department of Bioprocess Technology, Faculty of Biotechnology and Biomolecular Sciences, Universiti Putra Malaysia, 43400 UPM Serdang, Selangor, Malaysia
Institute of Bioscience, Universiti Putra Malaysia, 43400 UPM Serdang, Selangor, Malaysia

Nor Aini Abdul Rahman
Department of Bioprocess Technology, Faculty of Biotechnology and Biomolecular Sciences, Universiti Putra Malaysia, 43400 UPM Serdang, Selangor, Malaysia
Bioprocessing and Biomanufacturing Research Center, Faculty of Biotechnology and Biomolecular Sciences, Universiti Putra Malaysia, 43400 UPM, Serdang, Selangor, Malaysia

Arbakariya Ariff
Department of Bioprocess Technology, Faculty of Biotechnology and Biomolecular Sciences, Universiti Putra Malaysia, 43400 UPM Serdang, Selangor, Malaysia
Bioprocessing and Biomanufacturing Research Center, Faculty of Biotechnology and Biomolecular Sciences, Universiti Putra Malaysia, 43400 UPM, Serdang, Selangor, Malaysia

Umi Kalsom Md Shah
Department of Bioprocess Technology, Faculty of Biotechnology and Biomolecular Sciences, Universiti Putra Malaysia, 43400 UPM Serdang, Selangor, Malaysia
Institute of Tropical Forestry and Forest Products, Universiti Putra Malaysia, 43400 UPM Serdang, Selangor, Malaysia
Bioprocessing and Biomanufacturing Research Center, Faculty of Biotechnology and Biomolecular Sciences, Universiti Putra Malaysia, 43400 UPM, Serdang, Selangor, Malaysia

Elizabeth Minogue
Nucleic Acid Diagnostics Research Laboratory (NADRL), Microbiology, School of Natural Sciences, National University of Ireland Galway, Galway, Ireland

Nina L Tuite
Nucleic Acid Diagnostics Research Laboratory (NADRL), Microbiology, School of Natural Sciences, National University of Ireland Galway, Galway, Ireland

Cindy J Smith
Marine Microbial Ecology Laboratory, Microbiology, School of Natural Sciences, National University of Ireland Galway, Galway, Ireland

Kate Reddington
Nucleic Acid Diagnostics Research Laboratory (NADRL), Microbiology, School of Natural Sciences, National University of Ireland Galway, Galway, Ireland

Thomas Barry
Nucleic Acid Diagnostics Research Laboratory (NADRL), Microbiology, School of Natural Sciences, National University of Ireland Galway, Galway, Ireland

Xiaobin Long
Department of Cardiology, The First Affiliated Hospital of Chongqing Medical University, Chongqing 400016, China
Key Laboratory of Molecular Biology on Infectious Disease (Ministry of Education), The Second Affiliated Hospital, Chongqing Medical University, Chongqing 400016, China

Yeran Gou
Department of Cardiology, The First Affiliated Hospital of Chongqing Medical University, Chongqing 400016, China
Key Laboratory of Molecular Biology on Infectious Disease (Ministry of Education), The Second Affiliated Hospital, Chongqing Medical University, Chongqing 400016, China

Miao Luo
Key Laboratory of Molecular Biology on Infectious Disease (Ministry of Education), The Second Affiliated Hospital, Chongqing Medical University, Chongqing 400016, China
Key Laboratory of Clinical Laboratory Diagnostics (Ministry of Education), College of Laboratory Medicine, Chongqing Medical University, Chongqing 400016, PR China
Department of Clinical Laboratory, Yubei District People's Hospital, Chongqing 401120, PR China

Shaocheng Zhang
Key Laboratory of Molecular Biology on Infectious Disease (Ministry of Education), The Second Affiliated Hospital, Chongqing Medical University, Chongqing 400016, China
Key Laboratory of Clinical Laboratory Diagnostics (Ministry of Education), College of Laboratory Medicine, Chongqing Medical University, Chongqing 400016, PR China

Hongpeng Zhang
Key Laboratory of Molecular Biology on Infectious Disease (Ministry of Education), The Second Affiliated Hospital, Chongqing Medical University, Chongqing 400016, China
Key Laboratory of Clinical Laboratory Diagnostics (Ministry of Education), College of Laboratory Medicine, Chongqing Medical University, Chongqing 400016, PR China

Lei Bai
Department of Cardiology, The First Affiliated Hospital of Chongqing Medical University, Chongqing 400016, China
Key Laboratory of Molecular Biology on Infectious Disease (Ministry of Education), The Second Affiliated Hospital, Chongqing Medical University, Chongqing 400016, China

Shuang Wu
Key Laboratory of Molecular Biology on Infectious Disease (Ministry of Education), The Second Affiliated Hospital, Chongqing Medical University, Chongqing 400016, China
Key Laboratory of Clinical Laboratory Diagnostics (Ministry of Education), College of Laboratory Medicine, Chongqing Medical University, Chongqing 400016, PR China

Quan He
Department of Cardiology, The First Affiliated Hospital of Chongqing Medical University, Chongqing 400016, China

Ke Chen
Key Laboratory of Molecular Biology on Infectious Disease (Ministry of Education), The Second Affiliated Hospital, Chongqing Medical University, Chongqing 400016, China

Ailong Huang
Key Laboratory of Molecular Biology on Infectious Disease (Ministry of Education), The Second Affiliated Hospital, Chongqing Medical University, Chongqing 400016, China

Jianzhong Zhou
Department of Cardiology, The First Affiliated Hospital of Chongqing Medical University, Chongqing 400016, China

Deqiang Wang
Key Laboratory of Molecular Biology on Infectious Disease (Ministry of Education), The Second Affiliated Hospital, Chongqing Medical University, Chongqing 400016, China
Key Laboratory of Clinical Laboratory Diagnostics (Ministry of Education), College of Laboratory Medicine, Chongqing Medical University, Chongqing 400016, PR China

Ignacio Poblete-Castro
Facultad de Ciencias Biológicas, Center for Bioinformatic and Integrative Biology, Biosystems Engineering Laboratory, Universidad Andrés Bello, Santiago 8340176, Chile
Helmholtz Centre for Infection Biology, Microbial Drugs Group, Braunschweig D-38124, Germany

Danielle Binger
Helmholtz Centre for Infection Biology, Systems and Synthetic Biology, Braunschweig D-38124, Germany

Rene Oehlert
Helmholtz Centre for Infection Biology, Microbial Drugs Group, Braunschweig D-38124, Germany

Manfred Rohde
Helmholtz Centre for Infection Biology, Molecular Mechanism of Streptococci Group, Braunschweig D-38124, Germany

Nawal Abd EL-Baky
Therapeutic and Protective Proteins Laboratory, Protein Research Department, Genetic Engineering and Biotechnology Research Institute, City for Scientific Research and Technology Applications, New Borg EL-Arab, 21934 Alexandria, Egypt

Mustafa H Linjawi
College of Applied Medical Sciences, King Abdulaziz University, P.O. Box 80203, Jeddah 21589, Saudi Arabia

Elrashdy M Redwan
Therapeutic and Protective Proteins Laboratory, Protein Research Department, Genetic Engineering and Biotechnology Research Institute, City for Scientific Research and Technology Applications, New Borg EL-Arab, 21934 Alexandria, Egypt
Biological Sciences Department, Faculty of Science, King Abdulaziz University, P.O. Box 80203, Jeddah 21589, Saudi Arabia

Christine M Thompson
National Research Council Canada, Human Health Therapeutics, Montréal, Canada
Ecole Polytechnique de Montréal, Montréal, Canada

Emma Petiot
National Research Council Canada, Human Health Therapeutics, Montréal, Canada
Laboratoire Virologie et pathologies Humaine (VirPath), EA4610 Lyon, France

Alaka Mullick
National Research Council Canada, Human Health Therapeutics, Montréal, Canada

Marc G Aucoin
University of Waterloo, Waterloo, Canada

Olivier Henry
Ecole Polytechnique de Montréal, Montréal, Canada

Amine A Kamen
National Research Council Canada, Human Health Therapeutics, Montréal, Canada
Department of Bioengineering, McGill University, 817 Sherbrooke St. W. Macdonald Engineering Building, Room 387, Montréal, Canada